AF342083

# Intraspecific Genetic Diversity

Yu. P. Altukhov

Yuri P. Altukhov

# Intraspecific Genetic Diversity

## Monitoring, Conservation, and Management

With 136 Figures and 71 Tables

 Springer

Professor Dr. Yuri P. Altukhov
N.I. Vavilov Institute of General Genetics
Russian Academy of Sciences
Gubkin Street 3
119991 Moscow, GSP-1
Russia

*The original Russian edition (ISBN 5-94628-083-X) was published
by PTC Academkniga, Moscow, Russia, 2003.*

Library of Congress Control Number: 2005929633

ISBN-10 3-540-25490-0 Springer Berlin Heidelberg New York
ISBN-13 978-3-540-25490-4 Springer Berlin Heidelberg New York



**Springer is a part of Springer Science + Business Media**
springeronline.com

© Springer-Verlag Berlin Heidelberg 2006
Printed in Germany



Editor: Dr. Dieter Czeschlik, Heidelberg, Germany
Desk editor: Dr. Andrea Schlitzberger, Heidelberg, Germany
Cover design: *design&production*, Heidelberg, Germany
Typesetting and production: LE-TeX Jelonek, Schmidt & Vöckler GbR, Leipzig, Germany
31/3152 - 5 4 3 2 1 0 - Printed on acid-free paper

*To the 250th anniversary
of Lomonosov Moscow State University*

# Foreword

Population and evolutionary genetics have been quickly developing fields of biological research over the past decades. This book compiles our current understanding of genetic processes in natural populations. In addition, the book provides the author's original ideas and concepts based on the data obtained by himself and his close coworkers. The author introduces his pioneering concept of *population genetic stability*, and much of the book is concerned with *the factors and conditions of such stability*.

Why does genetic stability matter so much? Altukhov argues that the sustainable use of natural resources, including genetic resources of populations, critically depends on the maintenance of their stability. The preservation of well-adapted genetic characteristics from one generation to the next is essential for this stability. Traditionally, population genetics has been focused on evolution and the role of evolutionary factors in shaping genetic structures of populations. While the idea of a population as a dynamic unit of evolution has been widely accepted, the significance of genetic stability and its implications for the long-term survival of populations and species have not been fully appreciated.

Altukhov is well aware of the importance of studying the role of evolutionary factors in population dynamics, but his main focus is on the components of the genetic stability of populations that are essential for the survival of species. This aspect deserves particular attention since in many cases today natural populations are no longer units of natural evolution because of significant human influences. Global change due to anthropogenic impacts on ecosystems threatens the existence of populations and ultimately species. The author emphasizes that a sound understanding of the principles ensuring the survival of populations is mandatory for conserving and managing biological diversity at all levels, from molecules to ecosystems.

An important aspect of genetic stability is the *subdivision of populations into partially isolated units of limited reproduction-effective size*. Subpopulations are transient and highly unstable, both in time and space. Changes in environmental conditions affect their adaptation and reproduction as well as migration patterns. Stability of the overall system and, hence, survival of populations as systems is facilitated by dynamic processes driven by counteracting evolutionary factors and resulting in complex internal structures.

We regard the innovative ideas presented in this book and the unconventional concepts proposed by the author as major assets. Altukhov's views met strong and numerous objections when he first published them during the late 1960s and 1970s of the last century. However, his ideas also stimulated many interesting studies proving the validity of his concepts. This book reviews most of the relevant experimental data in detail. Studied organisms include various plants and animals from fishes to forest trees, as well as humans and model species such as *Drosophila*. Experimental approaches are complemented by computer simulations.

The author concludes that *the main result of adaptive evolution is an optimum of genetic diversity in natural populations as a principal condition for their stable reproduction over generations and as a measure of their maximum adaptation to the actual environment in which they have been shaped and now inhabit*. Both the reduction and the increase in genetic diversity as a consequence of evolutionary processes within natural populations can be detrimental for their long-term survival. Strategies for monitoring, conserving, and wisely utilizing natural populations are proposed, and principles for stabilizing gene pools of agricultural crops are outlined. Finally, the author discusses genetic processes in modern human populations.

This book is a valuable resource for students and researchers interested in a concise introduction to population genetics, evolution, and the importance of genetic processes for species survival and ecosystem functioning. Both theoretical and experimental approaches are adequately covered, allowing appreciation of the complex dynamics of genetic structures in time and space. Geneticists and evolutionary biologists will find interesting and challenging ideas and concepts proposed by the author. The focus on the sustainable management of genetic resources in agriculture, fisheries, and forestry will be of great value to environmental scientists and conservation biologists. The publication of the book in English will have profound effects on the international research community working in the relevant basic and applied fields of biology.

*Konstantin V. Krutovsky, Associate Professor*
*Texas A&M University, College Station, Texas, USA*

*Reiner Finkeldey, Professor*
*Georg-August-University, Göttingen, Germany*

# Preface

This book is not a textbook, nor is it a compilation of knowledge in the scientific field which it covers. The author has set himself a different task, that of communicating the results of his own work, and of his closest researchers, within the framework of *an approach to revealing the factors and conditions of genetic stability*.

Why is this matter so important? Because from its beginning, the main interests of population genetics have been connected with the theme of evolution. This has led to the formation of a carefully developed probabilistic concept of the evolutionary process, but at the same time has diverted the attention of scientists from the fact that a natural population is not only a unit of evolution, but frequently also an object of practical activity – one need only mention fishery, hunting, or forestry. The important factor is that the successful practical utilization of natural populations is bound up not so much with their evolution as with their stability, with preservation of their essential and historically formed genetic characteristics from generation to generation.

This conclusion acquires special significance in our times when anthropogenic influences upon nature have reached a scale unprecedented in mankind's history. It must be clearly realized that the life of a population, as that of any other biological level of organization, from molecular to biospheric, conforms to its own laws or principles. These cannot be abolished but, once known, they may be used for the good of mankind.

One of the most important principles for the functioning of many populations is their *systemic organization, their subdivision into partially isolated subpopulations of limited reproductive size*. On experiencing the effects of well-known evolutionary factors such as natural selection, migration, and random genetic drift, any subpopulation formation can become highly unstable in both time and space. Taken as a whole, however, the system of such subpopulations displays dynamic stability resulting from the reciprocal balance of the pressures of microevolutionary factors on their internal structure. Usually, a population organization of this kind is regarded as having the greatest potential for evolutionary transformation. However, our comparative investigations have shown that this view has no basis. An alternative hypothesis has been advanced – that of evolution as a process in which periods of long stability give way to phases of swift species transformation through qualitative genome changes marked by monomorphic, functionally important traits. On the basis of this model, protein and DNA polymorphisms in populations cannot be regarded as a transitory phase of molecular evolution, but must be seen as an adaptive strategy of nature, enabling populations and species overall to remain themselves in a normally fluctuating environment.

These views, postulated at the end of the 1960s and the beginning of the 1970s, met with immediate objections. Even today some authors do not share them, although there is a tendency for some viewpoints to converge.

This book examines all these questions in detail, drawing on the findings of field and experimental research as well as the results of computer simulation. The key conclusion to be drawn from the work carried out is that *the main result of adaptive evolution is an optimum genetic diversity of natural populations as the principal condition for their stable reproduction over generations and as a measure of their maximum adaptation to the actual environment in which they have been shaped and which they now inhabit.* The reduction or, conversely, the increase in genetic diversity resulting from different types of external influences is equally disadvantageous to the normal functioning of biological populations. The book analyzes a strategy for monitoring, conservation and rationally utilizing natural populations in various forms of economic activity; it discusses principles of stabilizing the gene pools of agricultural populations; and it devotes attention to the specific nature of the genetic processes occurring in modern human populations.

The main part of this book repeats the structure and logical presentation of the material contained in the author's monograph *Genetic Processes in Population*, published by Nauka in 1983 and 1989. Because new theoretical and experimental publications have appeared in the intervening years, the version now translated into English has had to be revised and updated. However, the author has not attempted to take into account all the latest literature, the more so as English-speaking readers are probably familiar with recent publications in the West. We have considered it essential to focus attention chiefly on Russian authors whose research covers the approach and problems dealt with in this book, and which may not be so well known to an English-speaking audience. In general, these are works by my closest colleagues and students: E.A. Salmenkova, O.L. Kurbatova, E.Yu. Pobedonostseva, A.G. Bernashevskaya, L.A. Zhivotovsky, I.I. Suskov, T.V. Malinina, Yu.E. Dubrova, K.V. Krutovsky, K.I. Afanas'ev, G.D. Ryabova, B.A. Kalabushkin, D.V. Politov, and others. Without their continued intensive efforts and close collaboration this book could not have appeared. They are all co-authors with me, but I alone accept responsibility for any omissions or mistakes.

The author is deeply grateful to his colleagues and to all those who read the book in manuscript and offered useful advice, namely, L.A. Zhivotovsky, E.A. Salmenkova, Yu.G. Rychkov, L.I. Korochkin, Yu.E. Dubrova, B.A. Kalabushkin, and O.V. Kalnina, and V.V. Kalnin. I have made several corrections in the light of criticisms contained in D.A. Aleksandrov's very stimulating review in 1985 of the first Russian edition. I express my particular gratitude to Springer-Verlag, the publishers who arranged for my book to be translated into English, and for their comprehensive scientific editing of the translation. In its very essence and by virtue of its underlying historical tradition, science is international, and publications of this kind are important for extending contacts and exchanging ideas and facts between scientists in different countries.

I hope that this translation will enable the English reader to form an impression of a trend in population genetics that is developing in modern Russia.

*Yu.P. Altukhov*

# Contents

# Introduction

One of the basic characteristics of life is its natural differentiation at subordinate levels, structurally reflected to a large extent by the biological sciences. Thus, biochemistry and molecular biology study macromolecules, cytology endeavors to understand the cell, while anatomy and physiology are occupied with the entire body. The achievements in these and many other scientific trends, both in the elaboration of a wide range of theoretical problems and in making practical use of the results obtained, are well known. A comprehensive investigation of populations is equally important.

The populational level of organization has particular significance for the existence and development of life on earth as it implements the genetic continuity of generations and regulates important biological factors such as fertility, numbers, resistance to disease, and so on. All of these factors are directly determined by the special features of gene pools – the accumulated, inherited information that is persistently transmitted from parents to offspring and preserved in time in a normally fluctuating environment. When, however, environmental conditions change drastically, genetic population structure is rearranged and considerable alterations take place.

At the same time, since nature and society represent a single dynamic system, biospheric changes cannot fail to affect the biological nature of the human species. Unfavorable results following from disturbances to ecological balances are already in evidence today and may become even more intensified in successive generations for several reasons: reduced genetic diversity of the biosphere, increased genetic loads of populations and species, changes in historically formed demographic structure, and so forth.

How should one interpret the specific features of these phenomena and processes, elucidate the relationships of cause and effect on which they are based, and formulate the principles of long-term forecasts and methods of preventing undesirable consequences? The superficial answer is simple: the effects of the environment on biological systems have increased so much that they exceed the adaptive capacity of such systems; consequently, without our entire strategy of reciprocity with nature undergoing a radical revision and a decisive change, no partial measures can be effective.

One must, of course, agree with this view, but not entirely, since it does not indicate ways and means of overcoming the negative tendencies. In order to find a solution, one must realize for a start that although the biosphere experiences anthropogenic effects as a whole, it is biological populations – elementary self-reproducing structural units, ensuring the succession and development of living matter – which act as the point of application for the corresponding external influences. It is precisely in studies at this "average" level of organized life that one may avoid both the "aberration of closeness", concomitant with an analysis at a molecular level, and the "aberration of distance", inescapable when examining the biosphere as a whole – that is to say, distortions that prevent one from understanding the specific aspects of the processes occurring in biological systems as a consequence of our activities that transform nature.

It is also obvious that a scientist can only solve the problem of forecasting and management by basing the necessary calculations on a clearly described concept of the *"norm"* – a *normal state*, or a *normal process* ("zero reference level"). Is this possible when one is dealing with a system that is constantly changing and has been inadequately studied? The answer to this question can only be in the negative, so that elucidating the factors and conditions of the genetic stability of populations and species assumes particular importance. This is the main purpose of the present book, representing many years of research by the author and his colleagues in population genetics. On the basis of this experience, we shall try to indicate how an appropriate approach can be made to solving these problems and formulating a more general strategy for human interaction with the biosphere.

One of the theses propounded in this book is that the concept of evolution, now universally accepted, requires a serious critical analysis in the light of new data from comparative population genetics. Existing methods of utilizing economically valuable populations, whether they involve industrial exploitation of nature's biological resources or very extensive selection projects in agriculture, should also be examined from an appropriate viewpoint. All these approaches have a positive effect in the short term, but frequently underestimate population-stability factors associated with the need to preserve and support these populations' inner genetic diversity. These questions are examined in the book's final chapter (Chap. 7). Because this kind of conclusion has been made possible primarily through studying genetic polymorphism, a considerable part of this book is devoted to analyzing the phenomenon and the mechanisms that support it in natural and experimental populations, with most attention being paid to protein and DNA polymorphism (Chaps. 2 and 5). The very high levels of this heritable variation within populations – discovered thanks to the development of molecular methods – is today attracting a lot of attention from geneticists and students of evolution, and is discussed in its most varied aspects. The fact remains, however, that besides polymorphic proteins and DNA fragments we have genetically invariant, monomorphic gene loci. This monomorphism is a real phenomenon in nature, and the analysis of it is fruitful both for molecular genetics and speciation theories (Chap. 6), as well as for the problem of genetic loads of populations (Chap. 7). Chapter 1 is introduced by a review of the book's principal contents, with a concise exposition of the main theoretical principles of population genetics, while Chapter 2 discusses the objectives and methods of biochemical (molecular) population genetics as one of the leading modern trends in this division of science. These chapters may be regarded as complementary.

Note also that major generalizations and conclusions are made in terms of biochemical populations genetics, i.e., inferred from protein polymorphisms. Data on DNA polymorphism were used to a lesser extent. This latter area, which has been explosively developing in the last decade, still does not allow in-depth understanding of population-genetic processes, while biochemical population genetics, as a methodologically mature field, provides such possibilities. However, a simultaneous study of both protein and DNA polymorphism even now permits one to discover phenomena and answer questions that until recently were considered unresolvable.

The new expression "comparative population genetics" is employed in this book. As will become clear, we use it for one purpose only: to emphasize that although genetic population principles are universal and can be applied to any bisexual species, it may not be feasible to compare research findings if the historically evolved systemic organization of natural populations is ignored.

When interpreting population genetic data and attempting to understand the specific features of the genetic processes in natural populations, one should be clear about the *level of the organizational structure of a population system* at which the research is conducted. Only in this way, by studying the populations of different species within the framework of a single comparative approach, can we be sure that our conclusions are sufficiently broadly based in a biological sense. The development of population genetics has increased the potential of this method in recent times, making possible the study of homologous genes in as wide a circle of organisms as one likes. However, before turning to the relevant facts one must first examine the theoretical bases of population genetics, albeit in a compressed way.

# 1 The Theoretical Principles of Population Genetics

Before turning to an analysis of genetic processes in populations, we will review in general terms the main theoretical principles of population genetics. These are of importance for understanding and evaluating the contents of this and following chapters. Population genetics, perhaps the most theoretically advanced field of research, occupies a special place in modern genetics and biology as a whole.

In population genetics, quantitative theory has existed for a long time and continues to improve. It is a theory involving natural factors, under pressure of which a population either remains constant or changes in successive generations with a concomitant change in biologically important traits. In other words, population genetics has mathematical models. The models may or may not correspond closely to nature, but they are important since they permit one to plan research in a certain way. Furthermore, if we observe the conformity with the natural situation of a model, it becomes possible to evaluate numerically the changes in populations and to predict the possible consequences. Because a large number of diversified papers and books have been published on this theme in recent years, we shall examine only the main population genetic terms, models, and approaches. We base our further discussion on works by Sewall Wright (1931, 1951, etc.), Neel and Schull (1958), Ehrlich and Holm (1963), Dobzhansky (1970), Kimura (1983), Kimura and Ohta (1971), Cavalli–Sforza and Bodmer (1971), Nei (1975, 1987), Li (1976), and several others to be mentioned in the text.

## 1.1 Estimation of Gene Frequencies

To a first approximation, a population may be defined (as did Dobzhansky) as *an aggregate of freely interbreeding individuals that share a common gene pool.* Because the number of segregating loci in the genome is large, one can understand the difficulties that confront the researcher in attempting to give an adequate description of this pool of inherited information. But at the same time it is evident that, however great these difficulties may be, there is only one way of obtaining such a description: by defining the frequencies of allelic genes at each single locus. Knowledge of the spatial and temporal distribution of gene frequencies enable a quantitative assessment of how genetic processes in populations are influenced by given external and internal factors.

There are several carefully formulated methods for evaluating gene frequencies in populations. We shall examine two of them, the first applied to a situation without dominance and the second to inheritance with dominance. The situation relates to a pair of alleles at a single autosomal locus.

**Absence of Dominance.**   Let us assume that of $N$ diploid individuals $N_1$, carrying only the allele $A$, that is, are $AA$; $N_2$ carrying the heterozygotes $AB$; and $N_3$ carrying the homozygotes $BB$, so that $N_1 + N_2 + N_3 = N$, and the total number of genes is $2N$. Each $AA$ homozygote has two $A$ genes, and each $AB$ heterozygote has only one gene of this kind. Consequently, the total number of $A$ genes in the group under study equals $2N_1 + N_2$ and the fraction (frequency) of this gene is

$$pA = \frac{2N_1 + N_2}{2N} = \frac{N_1 + \left(\frac{1}{2}\right) N_2}{N} .$$

The frequency of gene $B$ is defined in exactly the same way:

$$qB = \frac{2N_3 + N_2}{2N} = \frac{N_3 + \left(\frac{1}{2}\right) N_2}{N} ,$$

so that $p + q = 1$. We can also apply the same method to loci with more than two alleles.

Of course, a complete examination of natural populations is generally impossible, which in practical terms means that sampling is necessary. Hence, the reliability of estimates of gene frequency depends very much on the numbers sampled. Such estimates should be characterized by the least possible error or dispersion factor, that is, they should satisfy the so-called *criterion of effectiveness*. Thus, the requisite sample size depends on the genetic population structure, which can be established by preliminary research. The most dependable results come from the method of "directly calculating" genes, which was formulated by Fisher and which we have used in the above example. However, if there is **dominance** of one allele ($A$) over the other ($a$), then only two distinguishable phenotypes are present in a population, and of them only one phenotype – the homozygote ($aa$) for the recessive allele – corresponds to only one genotype.

The method of directly determining the allelic frequency is inapplicable to a genetic situation of this kind, and one must allow the hypothesis that the Hardy–Weinberg equation (see the next section) holds in the population; namely, that the distribution of genotypes in random mating conforms to the coefficients of the binomial expansion $p^2 + 2pq + p^2 = 1$.

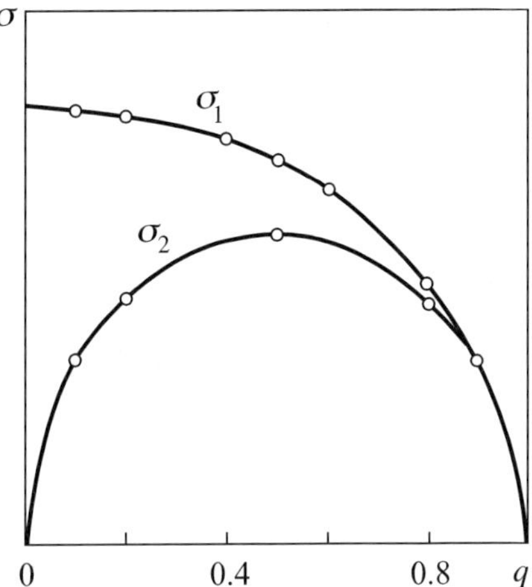


**Fig. 1.1.** Distributions of the standard errors of the two values of a gene frequency for a single pair of alleles without dominance (based on Neel and Schull 1958). $\sigma_1 = \sqrt{(1 - q^2)/4N}$ (the gene frequency found by extracting the square root of the fraction of homozygous genotype); $\sigma_2 = \sqrt{q(1 - q)/2N}$ (the same allele frequency determined by the direct calculation method)

It follows that in order to obtain an effective estimate of the frequency ($q$) of a recessive gene ($a$) we should extract the square root of the fraction of $aa$ individuals in our sampling: $q = \sqrt{aa}$. Accordingly, $pA = 1 - q$. In principle, this method could also be used to estimate allelic frequency in the absence of dominance, but it only gives least biased estimates at high frequency values (Fig. 1.1).

The reader can find more general examples of estimating gene frequencies (for example, for multiple alleles or linked loci) in Li's book (1978) and other publication on theoretical population genetics (e.g., Zhivotovsky 1991; Weir 1995; Hedrick 1999).

A number of software packages used in population genetics for estimating gene frequencies, intra- and intergroup components of gene diversity, genetic distances, for constructing phylograms, etc., can be found on the Internet at the following addresses:

| Software package | Site address |
| --- | --- |
| Analysis of Molecular Variance – Amova | http://anthropologie.unique.ch/ftp/com/win/amova/ |
| Arlequin | http://anthropologie.unige.ch/arlequin |
| Genetic Data Analysis (GDA) | http://lewis.eeb.uconn.edu/lewishome/software.html |
| GENEPOP | ftp://ftp.cefe.cnrs-mop.fr/PC/MSDOS/GENEPOP/ |
| Molecular Evolutionary Genetics Analysis (MEGA) | http://www/megasoftware.net/ |

| Software package | Site address |
| --- | --- |
| Phylogeny Inference Package (PHYLIP) | http://evolution.genetics.washington.edu/phylip.html |
| POPGENE | http://www.ualberta.ca/$\sim$fyeh/ |
| FSTAT | http://www.unil.ch/izea/softwares/fstat.html |
| Genetic Analysis in Excel (GenAlEx V5) | http://www.anu.edu.au/BoZo/GenAlEx/ |
| Multilocus Mating System Program (MLTR) | http://genetics.forestry.ubc.ca/ritland/programs.html |
| Structure | http://pritch.bsd.uchicago.edu/ |
| PCAGEN | http://www2.unil.ch/popgen/softwares/pcagen.htm |

## 1.2
## The Hardy–Weinberg Rule

The field of population genetics examines the principles regulating the maintenance and dynamics of population genotypic structure in time and space. The Hardy–Weinberg principle provides a theoretical basis for this view. It reflects the invariability of the genetic composition of a randomly mating (panmictic) population, unlimited in number and existing extra-environmentally, (i.e., in the absence of environmental pressure). In this structureless community, the genotype frequencies and, hence, also the gene frequencies at an autosomal locus having a pair of alleles $A$ and $a$, reach equilibrium in the generation that follows random mating.

Inasmuch as random matings signify merely a random association of gametes, it can easily be verified that this combination of $p(A)$ and $q(a)$ male gametes and $p(A)$ and $q(A)$ female gametes, when $p + q = 1$, produces an invariable distribution of $p^2AA + 2pqAa + q^2aa = 1$.

|  |  | Male gametes | |
| --- | --- | --- | --- |
|  |  | $p(A)$ | $q(a)$ |
| Female gametes | $p(A)$ | $p^2$ | $pq$ |
|  | $q(a)$ | $pq$ | $q^2$ |

The algebraic calculations supporting the Hardy–Weinberg equilibrium are given in Table 1.1. As we are concerned here with autosomal genes, the reciprocal crosses (that is of the type *male Aa×female AA* or *male AA×female Aa*, etc.) may be combined, and consequently the nine possible crosses reduce to six, given that $p + q = 1$.

It is clear that this equilibrium ratio of genotypes is provided by the symmetry of the distribution of allelic genes into male and female gametes and by the free combination of these into the zygotes formed in the process

**Table 1.1.** Types of matings and proportions of genotypes of population progeny at genetic equilibrium

| Type of mating | Frequency of mating | Proportions of genotypes among progeny | | |
|---|---|---|---|---|
| | | $AA$ | $Aa$ | $aa$ |
| $AA{\times}AA$ $(p^2 \times p^2)$ | $p^4$ | $p^4$ | $0$ | $0$ |
| $AA{\times}Aa$ $2(p^2 \times 2pq)$ | $4p^3q$ | $2p^3q$ | $2p^3q$ | $0$ |
| $Aa{\times}Aa$ $(2pp \times 2pq)$ | $4p^2q^2$ | $2p^2q^2$ | $2p^2q^2$ | $p^2q^2$ |
| $AA{\times}aa$ $2(p^2 \times q^2)$ | $2p^2q^2$ | $0$ | $2p^2q^2$ | $0$ |
| $Aa{\times}aa$ $2(2pq \times q^2)$ | $4pq^3$ | $0$ | $2pq^3$ | $2pq^3$ |
| $aa{\times}aa$ $(q^2 \times q^2)$ | $q^4$ | $0$ | $0$ | $q^4$ |
| Totals for population | $1,00$ | $p^2$ | $2pq$ | $q^2$ |

of reproduction. From this it follows that when there are no disturbances affecting a population of unlimited numbers, the frequencies of the genotypes and genes that characterize it remain unchanged in an infinitely long series of generations – the so-called absolute zero of genetic dynamics.

In Table 1.2 an elementary example is presented of testing an empirical distribution of genotypes at a diallele locus of the MN blood type (codominant expression) for goodness-of-fit to Hardy–Weinberg proportions in a British population. The gene frequencies are $pM = 0.542$ and $qN = 458$. These data show virtually ideal goodness-of-fit of the observed genotype distribution to that expected: the total chi-square value at one degree of freedom (the number of genotypes minus the number of allelic genes) is as low as 0.22, which is much lower than the threshold level of 3.84.

But ideal populations of this kind are virtually never encountered in nature; there are always natural factors that shift them from the point of equilibrium, disturbing their stability – random genetic drift, mutations, migration, and natural selection. These are the "factors of evolution" or microevolutionary forces that we shall now examine.

**Table 1.2.** Empirical distribution of genotypes at a diallele locus of the MN blood type (codominant expression) for goodness-of-fit to Hardy–Weinberg proportions in a British population

| Genotype | Observed number (obs) | Expected number (exp) | $\chi_i^2 = \frac{(\text{obs}-\text{exp})^2}{\text{exp}}$ |
|---|---|---|---|
| $MM$ | 298 | $p^2 n = 294.3$ | 0.05 |
| $MN$ | 489 | $2pqn = 496.4$ | 0.11 |
| $NN$ | 213 | $q^2 n = 209.3$ | 0.06 |
| Total $n$ | 1,000 | 1,000 | $\sum \chi_i^2 = 0.22$ |

The gene frequencies are $pM = 0.542$ and $qN = 458$

## 1.3
## Random Genetic Drift

Random genetic drift is a mathematical fact that emerges from the phenomenon of the finiteness of the numbers of any real population. Of special importance is the difference that exists between total population number and that part of it transmitting the gene pool to the next generation: a reproductive population size ($N_r$), and even more so its genetically effective population size ($N_e$) – and of the species as a whole – are virtually always less, and most often considerably less, than its total or census-population size ($N_t$). Concepts of the substantial differences between genetically effective numbers and total population number were developed both theoretically and experimentally during studies of the stochastic processes of gene frequency dynamics (Serebovsky 1927, 1930; Fisher 1930; Wright 1931; Dubinin 1931; Romashov 1931; Dubinin and Romashov 1932; Kolmogorov 1935).

Extreme age groups are excluded from the reproduction, and the $N_e$ value is influenced by population parameters such as the balance in numbers of the sexes during the reproductive period, individual fertility differences, periodic fluctuations of numbers.

### 1.3.1
### Sex Ratio and Fluctuation Size

If $N_m$ represents the male and $N_f$ the female reproductive part of the population, then the effective population size is

$$N_e = 4N_mN_f / \left(N_m + N_f\right) . \tag{1.1}$$

When the male and female fractions deviate sharply from equilibrium, the $N_e$ value depends more on the less numerous sex. The same effect occurs whether the mean value of $N_e$ is valid for associated populations scattered in space and differing in numbers, or for one and the same population during fluctuations in time (generations) of the number of mating individuals (Wright 1938). In the case of cyclical variations at an interval of $n$ generations $N_e = \widetilde{N}$, where

$$\widetilde{N} = \frac{n}{\sum\limits_{1}^{n} 1/N_i} \tag{1.2}$$

that is, $N_e$ is the harmonic mean. For instance, if the effective number in each of five generations of one population is 10, $10^2$, $10^3$, $10^4$ and $10^5$, the average harmonic $N_e$ value represents only 45 individuals. The longer time interval of observations or the more number of distributed in space populations are studied, the more stable the $N_e$ estimate.

## 1.3.2
## Variability of Individual Fertility

When population numbers are stable (the average number of progeny reaching reproductive age per one pair of parents, $\bar{k} = 2$) and there is individual variation in the number of gametes ($k$) produced by the parent population ($N$), then:

$$N_e = \left(4N - 2\right) / \left(V_k + 2\right),\tag{1.3}$$

where $V_k$ is the variance of $k$.

When the number of progeny has a Poisson distribution ($\bar{k} = V_k = 2$), the genetically effective population size is approximately equal to its reproductive number, $N_e = N_r$. However, in the majority of natural populations $V_k > \bar{k}$; hence, $N_e$ is always less than $N_r$. According to the estimates of Crow and Morton (1955), the $N_e/N_r$ ratio is approximately 0.75 for many species. There are grounds, however, for considering this estimate to be a maximum, since in those cases where this kind of definition has been applied to actual populations, the difference has been much greater (Cavalli–Sforza et al. 1964; Kerster 1964; Tinkle 1965; Rychkov 1968; Kerster and Levin 1968; Frankham 1995) and the $N_e/N_t$ ratio could even be as low as 0.30 (Rychkov and Sheremet'yeva 1979), or down to about 0.10 for different species, or still less (for details see: Frankham 1995; Altukhov et al. 2000a; Jehle et al. 2001).

Crow (see Crow 1954; Crow and Morton 1955; Kimura and Crow 1963) introduced the concept of *inbreeding effective number*, $N_e$ ($f$) and *variance effective number*, $N_e$ ($v$). The *inbreeding effective number* is defined as

$$N_e\left(f\right) = \left(N_{t-2}\bar{k} - 2\right) \Big/ \left(\bar{k} - 1 + V_k/\bar{k}\right),\tag{1.4}$$

in which $N_t - 2$ is the number of individuals two generations back, and $k$ is the number of gametes introduced by them.

When the population number is constant, $N_t - 2 = N$, $k = 2$, and consequently

$$N_e(f) = \left(4N - 4\right) / \left(V_k + 2\right).\tag{1.5}$$

It is clear that when $N$ is sufficiently large, this equation differs little from Eq. (1.3).

The *variance effective number* is defined as

$$N_e(v) = p(1-p) \,/\, 2V_{\delta p}\,, \tag{1.6}$$

where $p$ and $1-p$ are the frequencies of any pair of alleles in a population and $V_{\delta p}$ is the variance of allele frequency due to random sampling of gametes in a given generation:

$$V_{\delta p} = p(1-p) \,/\, 2N\,. \tag{1.7}$$

It is obvious that if the $N_e$ values designated by Eqs. (1.1), (1.2), (1.3), and (1.4) are derived from the results of field observations of numbers, the sex ratio, and the other biological parameters of a real population, then formula (1.6) represents a method of defining an $N_e$ value by analyzing the genetic parameters on the hypothesis of the selective neutrality of the allelic genes being studied.

All the preceding equations for $N_e$ were derived for non-overlapping generations. According to Nei and Imaizumi (1966a,b), when generations overlap in time,

$$N_e = \tau N_a\,,$$

where $N_a$ is the number of individuals who reach mean reproductive age, and $\tau$ is the generation time or mean reproductive age. For example, the $N_e$ value assigned in this way to cover the general Japanese human population represented approximately 40% of the total number.

If a population is stable, then $N_a = Nbp$, where $N$ is the total population size, $b$ the annual birthrate, and $p$ the probability of a newborn individual reaching reproductive age.

In principle, the *variance effective number* under conditions of stability is nothing other than the correlation between the random variance of gene frequencies and the effective population size – a theory developed by Wright in the 1930s. The relationship of these parameters was used as a basis for estimating effective population size $N_e$ from temporal changes in allele frequencies, i.e., from their standardized variance (Krimbas and Tsakas 1971; Nei and Tajima 1981; Pollak 1983; Waples 1990a,b). The approaches were elaborated for applying this method to populations with discrete (Nei and Tajima 1981) and overlapping generations (Jorde and Ryman 1995).

Change in gene frequency caused by sampling error during the formation of gametes constituting the following generation has a random, stochastic character. For this reason only a probability approach is feasible when assessing random genetic drift based on the genetic structure of a population, and this only makes it possible to determine the range of variance in gene frequencies. According to Wright, these non-directional fluctuations depend exclusively on the $N_e$ value and are described by Eq. (1.7).

                                                    

The distribution of probabilities of all possible $q$ values in a series of generations of a population limited in numbers approaches normal, and an allele's ultimate fate is fixation ($q = 1$) or loss ($q = 0$). This process is essentially irreversible.

### 1.3.3
### Modeling of Random Genetic Drift

Dubinin and Romashov (1932) were the first to model random genetic drift by providing "drawings" of varicolored balls that imitated different alleles in a population of randomly mating organisms with no selection. They used a statistical population model to demonstrate that the process of loss (or fixation) during genetic drift depended entirely on effective population size (Fig. 1.2).

One can easily see how, by the 40th generation, in a 50-strong population with 100% allelic diversity at the start, only six alleles at varying frequencies remain. Numbers 64 (0.44) and 92 (0.01) have maximum frequency contrast. By the 110th generation allele number 92 has vanished from the population, while allele number 64 reached a frequency of 100% in the 123rd generation (fixation).

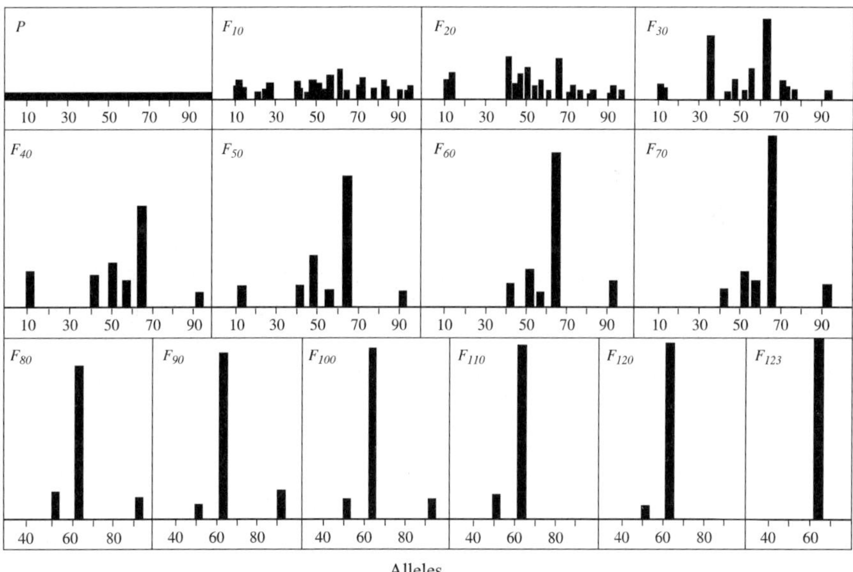


**Fig. 1.2.** Fluctuations of gene frequency in successive generations of an evenly reproducing population (from Dubinin and Romashov 1932). The effective number in the model is 50 individuals

In the absence of selection, this process of loss of genetic diversity (that is, reduction in heterozygosity) may be described by an equation enabling one to reconstruct the evolutionary time:

$$H_T = H_0\, e^{-T/2N_e} \, , \tag{1.8}$$

where $H_0$ and $H_T$ are the frequency of heterozygotes at the zero point in the process and at the instant of time $T$, and $N_e$ is the effective populations size.

The fluctuations of gene frequencies during drift are completely random, but this process acquires "direction" (or vector) after a certain number of generations: the frequencies of rarer alleles in each successive generation diminish with greater probability (the loss process), whereas frequencies of other alleles increase (the fixation process; Fig. 1.2).

The range of possible changes in the random variance of gene frequencies (Eq. 1.7) during the generations of the life of one population may be regarded as exactly the same as that for a community of many populations roughly equal in numbers and having identical allele $q_0$ frequencies in the $t_0$ generation. In the following generation the gene frequencies in each population change randomly to become $q_i$. In the process the average $\bar{q}$ frequency for the community remains equal to $q_0$, but the interpopulation variance will increase and in the $t$th generation will be:

$$V_q = p_0 q_0 \left[ 1 - \left( 1 - \frac{1}{2N} \right)^t \right] . \tag{1.9}$$

In the final stage of the process all populations become homozygous, and the intergroup variance reaches a maximum of $V_q = p_0 q_0$; that is, in some populations allele $A$ reaches a state of fixation ($p = 1$) and in others it is lost ($p = 0$). At the same time, the average allelic frequency remains unchanged as before. Thus, during random genetic drift the process of changing the probability distribution of the frequencies depends on only two factors – the population's effective size and the duration of the process, measured by the number of generations. When the effective population size is small, the distribution rapidly becomes U shaped.

Buri (1956) modeled random genetic drift in experimental populations of *Drosophila melanogaster*. He studied the distribution frequencies of *bw* and $bw^{75}$ alleles at the "brown" locus in 19 consecutive generations of 107 lines with 16 individuals in each. The picture he obtained was very similar to what might have been expected (Fig. 1.3).

The increase in the homozygosity of a population through random genetic drift indicates a direct link with *inbreeding*, since non-random association of gametes increases with time in a population with small numbers.

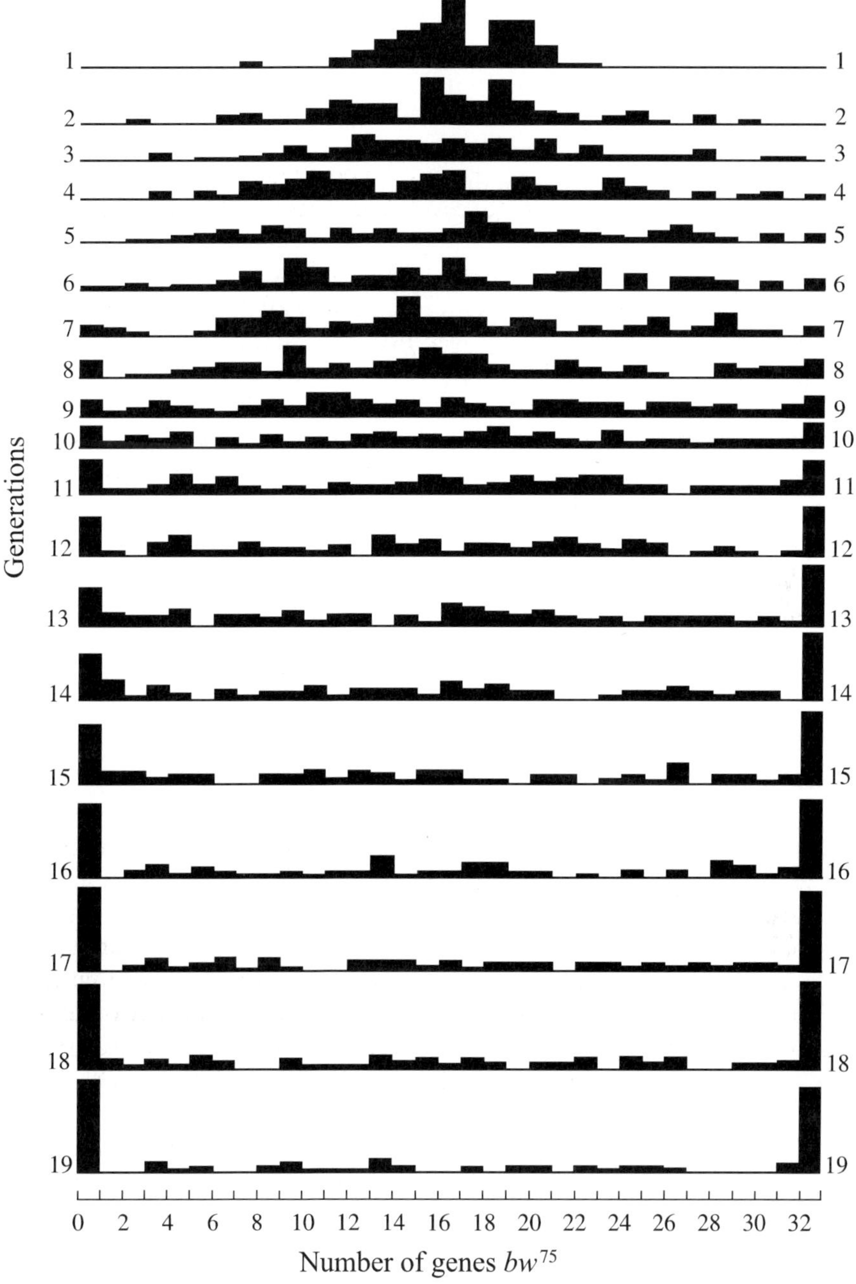



**Fig. 1.3.** The distribution of gene bw$^{75}$ frequencies in 19 generations of 107 *Drosophila melanogaster* experimental populations (according to Table 13 from Buri 1956)

By virtue of this fact, the deviation from panmixia becomes increasingly obvious; this also implies an increase in the degree of the genetic relationship among population members.

For convenience, a population with the same degree of inbreeding may be regarded as consisting of two parts, one of which is completely inbred while the other is completely panmictic. From this viewpoint, the probability of combining two $A$ gametes in random mating is $p^2$, whereas with inbreeding it should be more by a certain positive value of $\varepsilon$, namely, $p^2 + \varepsilon u$. Exactly the same probability holds for combining the $a$ gametes $q^2 + \varepsilon$, and the probability of combining $A$ with $a$ is $2pq - 2\varepsilon$.

At genetic equilibrium the inbreeding coefficient (= the correlation coefficient between uniting gametes) is $F = \varepsilon/pq$, that is, $\varepsilon = Fpq$. Thus the zygote proportions equal

$$\begin{array}{ccc} AA & Aa & aa \\ Q_1 = p^2 + Fpq & Q_2 = 2pq(1 - F) & Q_3 = q^2 + Fpq \end{array} \qquad (1.10)$$

when $Q_1 + Q_2 + Q_3 = 1$.

## 1.4
## Mutation and Migration of Genes

### 1.4.1
### Mutations

Mutations are heritable, non-directional, random changes of genetic material occurring spontaneously or under the influence of specific physical, chemical, or biological factors. The process of spontaneous mutation is the ultimate source of new alleles, resulting in increased genetic diversity.

Since the overwhelming majority of de novo (or "fresh") mutations are deleterious (Muller 1950), one naturally asks why a considerable number of alleles at a wide range of loci are to be found in populations? Such diversity is obviously maintained at the expense of part of the selectively neutral alleles. On the other hand, alleles that have recently arisen and that display no harmful effect in the heterozygotes may gradually be incorporated, step by step, in a species' gene pool which, like a sponge, absorbs them throughout its lengthy evolution (Chetverikov 1926). In the latter case of mutation, it is the raw material for natural selection that adjusts to coexist with ancient wild-type allelic genes already ground and polished by evolution, though effects on fitness depend on environment. These questions will form the subject of the discussion in Chaps. 5–7 of this book. Here, however, it is only the dynamics of selectively neutral mutations, as a process capable of eliciting deviations in genetic population structure from the Hardy–Weinberg equilibrium that are discussed.

The fate of a single mutation was first analyzed by Fisher (1930) who showed that when a population remains constant in size (the average number of progeny per family, $m = k = 2$), the probability of total loss of a mutant gene in the first generation is approximately 0.37. If this gene is not lost in the first generation, it again risks being lost in the second generation, and so on; thus the loss probability limit ($l_n$) is equal to 1.

Because the loss of de novo mutations is an essentially irreversible process, the bulk of them have no chance of becoming fixed in a population. Table 1.3 represents the overall dynamics of this process.

It should be remembered, however, that mutations constantly recur in every generation, and it is quite likely that so-called "new" mutations have occurred repeatedly. When an allele has a low frequency in a population, it is practically impossible to identify its mutation into another allele. It is only when a certain frequency has been reached that one can take the effects of the mutation process on this allele into account. Thus, if $\mu$ the mutation rate from allele $A$ to allele $a$, and $v$ is the reverse mutation rate, $a \rightarrow A$, then the value of the change in gene frequency per generation is

$$\Delta q = \mu p \ (\text{increase}) - vq \ (\text{decrease}) . \tag{1.11}$$

Thus, the increase or decrease in a gene's frequency is determined by the relative values of such increase or decrease. For instance, if the increase is more than the decrease, then at any given moment $q$ will grow. But as, due to increase of $q$, $vq$ also increases, it can only continue to the point

**Table 1.3.** Probability of loss of a neutral mutant gene (from Fisher 1930)

| Generation | Probability of loss, $l_n$ | Probability of persistence, $1-l_n$ |
|---|---|---|
| 1 | 0.3679 | 0.6321 |
| 2 | 0.5315 | 0.4685 |
| 3 | 0.6259 | 0.3741 |
| 4 | 0.6879 | 0.3121 |
| 5 | 0.7319 | 0.2691 |
| 6 | 0.7649 | 0.2351 |
| 7 | 0.7905 | 0.2095 |
| ... | ... | ... |
| 31 | 0.9411 | 0.0589 |
| 63 | 0.9698 | 0.0302 |
| 127 | 0.9847 | 0.0153 |
| ... | ... | ... |
| $\infty$ | 1.0000 | 0.0000 |

of equilibrium, after which no more changes of gene frequencies will be observed in the generations, that is, at the equilibrium point $(\hat{p}, \hat{q})$ $\Delta q = 0$.

Consequently,

$$\Delta q = 0 = \mu \left(1 - \hat{q}\right) - v\hat{q}, \quad \text{and} \quad \hat{q} = \frac{\mu}{\mu + v}. \tag{1.12}$$

Solving this equation in exactly the same way for $p$, we obtain

$$\hat{p} = \frac{v}{\mu + v}.$$

This equilibrium is stable; moreover, the allele equilibrium frequency $\hat{q}$ is independent of its initial value, being determined solely by the relation between forward and backward mutation rates: if $q > \hat{q}$ in a certain generation, the gene frequency will be reduced in the following generations; whereas if $q > \hat{q}$ it will, on the contrary, increase.

When the $q$ value deviated from $q < \hat{q}$ then the $\Delta q$ per generation assigned by Eq. (1.11) may be expressed in terms of the deviation $(q - \hat{q})$.

If in accordance with Eq. (1.12) we write $\mu = (\mu + v)\hat{q}$, then

$$\Delta q = \mu(1 - q) - vq = -(\mu + v)(q - \hat{q}). \tag{1.13}$$

Thus, the rate at which equilibrium is reached is proportional to the deviation of the actual value of $q$ from its value $\hat{q}$ at equilibrium.

Two important facts should be taken into account when discussing the effect of the mutation process on genetic equilibrium in a population. Firstly, the rate of forward gene mutation is of a higher order of magnitude than the rate of back mutation. Secondly, although the mutation rate may vary substantially for different loci, the value is extremely low in eukaryotes, on the average of the order of $1 \times 10^{-5} - 10^{-6}$ per locus per generation (for protein-coding genes). This signifies that in describing the genetic population structure of any species according to any of the classical Mendelian genes well known at the present time, the effect of newly occurring mutations can be disregarded when compared with factors of population dynamics such as natural selection, random drift, and migration.

## 1.4.2
## Migration

Migration is an important factor of population dynamics, since each population in nature interacts with other groupings of the same kind through exchanging genes (except for cases of extreme isolation). It stands to reason that if immigrants differ genetically from the population that receives them, they cause a corresponding change in gene frequency in each generation. Thus,

$$\Delta q = -m\left(q - q_m\right) = -mq + mq_m \,, \tag{1.14}$$

where $m$ is the number of immigrants divided by the size of the population receiving them, $q$ is the gene frequency of the population, and $q_m$ is the gene frequency of the immigrants.

The formula obtained is identical to Eq. (1.13), reflecting the similarity with the results of direct and reverse mutation. To be convinced of this, one need only substitute the constant $\mu$ for $mq$, $v$ for $mp = (1 - q)$, and $\mu + v$ for $m$. An equilibrium is established in precisely the same way during equal intensity of gene flow to and from the population.

We shall return to the effects of gene migration later when we examine the effects of natural selection and the combined action of well-known evolutionary factors on the genetic structure of populations.

# 1.5
# Natural Selection

In the theory of population genetics, natural selection is regarded as an extremely important factor of evolution that causes adaptive changes in genetic structure. These changes result differences in relative contributions of genotypes from reproductive individuals in a population through differential reproduction or survival.

## 1.5.1
## Basic Equations and Types of Selection

If populations inhabited space of unlimited area and resources, and if at any instant of time $t$ the birth rate $a$ exceeded the death rate $b$ by a constant value $r$, then the population number would grow continuously in an exponential manner

$$N_t = N_0\, e^{rt} \,, \tag{1.15}$$

where the parameter $r = a - b$, denoting the coefficient of the population growth, is called the Malthusian parameter; Fisher (1930) introduced it in *The Genetical Theory of Natural Selection*. But because in nature space and resources are always limited and the coefficient $r$ does not remain constant, the exponential dependence may only be observed over limited sections of time (and space), ultimately giving way to an S-shaped logistic curve

$$\Delta N_t = \frac{k}{1 + C_0\, e^{-rt}}\,, \tag{1.16}$$

where $k$ is the maximum number of individuals capable of living in a given specific environment, and the constant $C_0 = (k - N_0)/N_0$ is a correction factor – the "environmental resistance" to population growth (Fig. 1.4).

The curves of the dynamics of numbers plotted on the diagram relate to models that are *continuous in time*, that is, across generations, and their most frequent application is in ecology. Models with *discrete time* (non-overlapping generations) are chiefly used in population genetics for quantitative descriptions of natural selection, and usually operate with a similar quantity – fitness coefficient $W$. Thus, if $N_t$ is the number of adult individuals in a generation $t$, $k$ and $v$ are their fertility and viability, and $W = kv$, then the growth of numbers in a population is

$$\Delta N_t = N_{t+1} - N_t = (W - 1)N_t \text{ and } N_t = W^t N_0\,. \tag{1.17}$$

Consequently, $W^t = N_t/N_0$, that is, the population's fitness at a certain instant of time is equal to the ratio of its numbers in the subsequent and previous generations (when the coefficient of fitness and the state of the environment are stable).

Thus, the parameter $W$ is also important in a biological sense: when $W > 1$ the population size grows, when $W < 1$ it falls, and when $W = 1$ it remains the same.

In the theory of natural selection, developed by Fisher, Wright, and Haldane, the fitness of genotypes is taken most frequently as constant

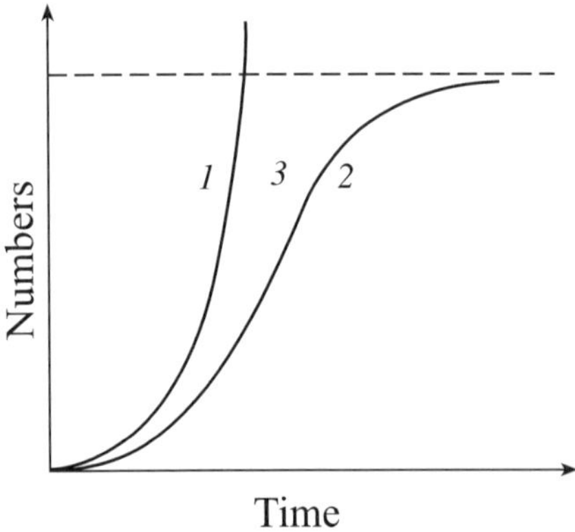


**Fig. 1.4.** Theoretical curves of population growth. *1* Exponential; *2* logistic, *3* environmental pressure "resistance"

throughout the selection cycle, and its relative value is more important than its absolute value. Here we are only concerned with the changed ratio of genotypes, which lends itself very conveniently to the study of the dynamics of gene frequencies. In this sense the symbol $W_i$ signifies the $i$-th genotype's relative fitness; in other words, a quantity that reflects its reproductive contribution to the genetic pool of the following generation through differential fertility or survival, compared with other genotypes when population numbers are stable.

If the consequences of selecting out genotypes in a population with two alleles at one locus are considered from these viewpoints, then the gene frequencies after one selection cycle are changed in the following way (see Table 1.4).

Clearly, under selection conditions only certain genotypes can have relative fitness equal to unity, and hence the average fitness of a segregating population is always less than unity (that is, less then the "optimal" genotype's fitness):

$$\overline{W} = f\, W_i = p^2 W_1 + 2pq W_2 + q^2 W_3 \,. \tag{1.18}$$

A population's fate subsequently is as follows: its evolution towards the loss ($p = 0, q = 1$) or fixation ($p = 1, q = 0$) of allele $A$, or else its transition to a condition in which both alleles remain, depends on the relative $W_i$ value of the genotypes. If $W_1 > W_2 \geq W_3$ or, on the other hand, $W_3 > W_2 \geq W_1$, then the population will inevitably reach a trivial equilibrium (a stationary state when either $p = 1, q = 0$ or $q = 1, p = 0$, corresponding to what is known as *directional selection*. A non-trivial equilibrium point ($0 < p < 1$) is reached if $W_1 < W_2 > W_3$ or $W_1 > W_2 < W_3$; that is, when the heterozygote fitness is more than or less than that of both homozygotes. In both cases, despite the continuing effects of selection, no genetic changes occur in the population. However, only in the first instance ("overdominance") will there be a stable ratio of genotype frequencies at

**Table 1.4.** Change in gene frequencies in a randomly mating population after one generation of selection

| Geno- types | Frequency before selection | Relative fitness | Frequency after selection | New gene frequencies | Frequency of genotypes before selection |
|---|---|---|---|---|---|
| $A_1A_1$ | $p^2$ | $W_1$ | $p^2 W_1$ | $p' = \dfrac{p^2 W_1 + pq W_2}{\overline{W}}$ | $p'^2$ |
| $A_1A_2$ | $2pq$ | $W_2$ | $2pq W_2$ | | $2p'q'$ |
| $A_2A_2$ | $q^2$ | $W_3$ | $q^2 W_3$ | $q' = \dfrac{pq W_2 + q^2 W_3}{\overline{W}}$ | $q'^2$ |
| Sum | 1 | | $\overline{W}$ | | 1 |

equilibrium:

$$\hat{p} = \frac{W_3 - W_2}{\left(W_1 - W_2\right) + \left(W_3 - W_2\right)} . \tag{1.19}$$

This type of selection, one of the forms of balancing selection, is called *stabilizing*.

When, however, the heterozygote is less adaptive than both homozygotes the state of equilibrium in the population is unstable and a gene with a higher frequency is evolved in the direction of fixation, while a gene with lower frequency tends toward loss. This type of selection is called *disruptive* or *diversifying*. Figure 1.5 shows how a population's average fitness changes in these cases (Li 1976).

The rate at which a population approaches the equilibrium point depends on the intensity of selection ($s = 1 - W$), and is determined by the magnitude of the change in gene frequency per generation ($\Delta p$), based on the expressions for $p'$ and $q'$ (see Table 1.4):

$$\Delta p' = p' - p = \left(pq/\overline{W}\right)\left[p\left(W_1 - W_2\right) + q\left(W_2 - W_3\right)\right]. \tag{1.20}$$

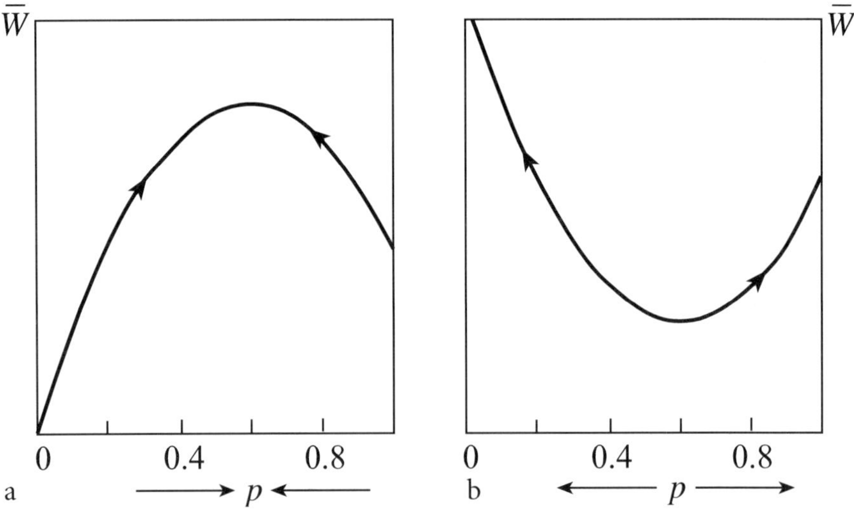


Fig. 1.5. The average fitness of a population as a function of a gene frequencies (Li 1967). **a** $w_1 = 2$, $w_2 = 4$, $w_3 = 1$, $\overline{w} = 2p^2 + 8pq + q^2$. Point 0.6 on the parabola corresponds to maximum fitness – stable equilibrium; the trivial equilibrium points ($p = 1, 0$) are unstable; **b** $w_1 = 3$, $w_2 = 1$, $w_3 = 4$, $\overline{w} = 3p^2 + 2pq + 4q^2$. Point 0.6 corresponds to minimum fitness. As unstable equilibrium is affected by disruptive selection, a population moves toward gene frequencies of $p = 0.1$; these trivial equilibrium states are stable. (Reproduced with permission of International Biometric Society, Washington, D.C. 20005–6542)

Here is a numerical example: population $0.25AA{:}0.50Aa{:}0.25aa$, moving toward the equilibrium frequencies of $\hat{p} = 0.60$, $\hat{q} = 0.40$ and influenced by the effects of selection of the relative values of $W$ for the three genotypes $AA$, $Aa$, and $aa$ in the four following situations:

$$(1, 3, 0)\ (3, 7, 1)\ (8, 10, 7)\ (101, 103, 100)\,.$$

It is clear that the types of selection in these four cases differ greatly from each other. Whereas in the first case $(1, 3, 0)$ genotype $aa$ is lethal and the reproductive contribution of the heterozygote $Aa$ is far in excess of that of the $AA$ homozygote, the fourth type of selection $(101, 103, 100)$ displays only very insignificant differences in the fitness of genotypes. The second $(3, 7, 1)$ and third $(8, 10, 7)$ models occupy an intermediate position between the two extremes. The character of the gene frequency changes in these four situations is shown in Table 1.5.

It can be seen that in the first case the rate at which the population reaches equilibrium is very great, while it is negligibly small in the last case. When $p = 1$ or $p = 0$, $\Delta p = p' - p = 0$ [see Eq. (1.20)] and the population reaches trivial equilibrium state. The factor $pq$ in Eq. (1.20) also shows that the rate at which equilibrium is approached is fairly high at intermediate allele frequencies and becomes smaller the smaller the value of $p$ and $q$.

Let us turn again to Fig. 1.5 to convince ourselves that the mean fitness of the population $\overline{W}$ under the pressure of selection always increases to its maximum value in the equilibrium state. This rule, known as the "fundamental theorem of natural selection", was formulated by Fisher in 1930 and was originally defined as: "The rate increase in fitness of any organism at any time is equal to its genetic variance in fitness at that time." In the new edition of 1941, it was rephrased: "The rate of increase in the average fitness of a population is equal to the genetic variance [that is, its additive component – the author] of fitness of that population."

Fisher based this premise on models having continuous time and logarithmic fitness. Li (1967) extended it to include models of populations with non-overlapping generations. This principle, which confirms the mono-

Table 1.5. Gene frequency in a population after one generation of selection as a function of fitness coefficients of genotypes

| Selections type | Fitness coefficient values | | | Initial gene frequency | Gene frequency after selection, | Change value |
|---|---|---|---|---|---|---|
| | $W_1$ | $W_2$ | $W_3$ | $p$ | $p'$ | $\Delta p = p' - p$ |
| I | 1 | 3 | 0 | 0.5000 | 0.57143 | 0.07143 |
| II | 3 | 7 | 1 | 0.5000 | 0.55556 | 0.05556 |
| III | 8 | 10 | 7 | 0.5000 | 0.51429 | 0.01429 |
| IV | 101 | 103 | 100 | 0.5000 | 0.50123 | 0.00123 |

tonic increase of a population fitness at any instant of time under selection pressure, plays a major role in Wright's "adaptive topography" – a component of his shifting balance theory of evolution (see Fig. 1.12).

Fisher's fundamental theorem of natural selection has several limitations (Levins 1968; Zhivotovsky 1981b). A population's fitness may be reduced as a result of inbreeding, frequency-dependent selection, mutations, and gene recombinations; hence, selection is capable of making a population extinct. [Frequency-dependent selection is a form of balancing selection (see Li 1978), in which a genotype's selective advantage changes as its frequency in a population changes. Of particular interest to us is the situation in which fitness and frequency are inversely correlated.]

Nevertheless, Fisher's theorem is justified for constant values of genotypic fitness (Zhivotovsky 1984). Obviously, a natural population existing today reached maximum adaptation in previous stages of evolution that are unknown to us, and now maintains a dynamic balance with its surroundings. The totality of these adaptations to the specific environment that populations encountered previously is recorded in patterns of contemporary genotypic structure. It represents their reserve of genetic stability despite the changing conditions of the environment.

The evolution described above of the genetic structure of a single-locus diallelic population is certainly a simplification. Each natural population of a bisexual species evolves simultaneously at a large number of loci, among which there may also be multiallelic genetic systems. The corresponding genotypes can interact in varied ways; genes may be linked or their combinations may be non-random under the effects of selection (and where there is very close linkage, the situation may be reduced effectively to a single locus and the complex of several or many separate genes regarded as a supergene in inheritance). All this creates many additional difficulties, and it follows that no mathematical description is adequate to provide a complete quantitative picture of genetic processes influenced by selection. Nonetheless, existing models are proving exceptionally useful even today: they enable one to plan research along definite lines and to quantify the results obtained. For all this information we refer the reader to several sources that discuss these problems, and the tasks ahead, comprehensively (see, for instance, Wright 1969, 1970 and others; Lewontin 1978a; Li 1978; Zhivotovsky 1984; Weir 1990; Hedrick 1999; and also Chap. 5).

## 1.5.2
## Genetic Load of Populations

The constant pressure of mutations and the migration of genes, as well as the segregation of biologically less adapted genotypes at balanced polymorphic loci, give rise to the problem of so-called "genetic loads", which is of paramount scientific and practical importance.

The concept of genetic load was introduced by Muller (1950), but the first researches – revealing a saturation of natural populations with non-adaptive mutant phenotypes that are produced in each generation from apparently normal individuals – were carried out in the twenties and thirties (Chetverikov 1926; Timofeeff–Ressovsky and Timofeeff–Ressovsky 1927; Romashov et al. 1931; and see also Dobzhansky 1970).

In his paper "Our load of mutations", Muller (1950) showed that weakly deleterious mutant genes can inflict greater damage on a population than individual mutant genes having a strong negative effect. Moreover, he suggested that each of us is on average the bearer of at least eight harmful genes concealed in the heterozygous state.

Morton, Crow, and Muller (Morton et al. 1966) determined the amount of genetic load in humans by investigating inbreeding depression. They proposed the special term "lethal equivalent" and indicated an approach to evaluating the relative contribution to the total amount of genetic burden of the mutation processes (mutational load) and segregation (segregational load) of genes; the latter relates to loci whose polymorphism is maintained at the expense of the selective advantage of heterozygotes.

Although many unresolved points remain in the problem of genetic load (Dobzhansky 1970; see also Cavalli–Sforza and Bodmer 1971), nevertheless, the concept of genetic load is of fundamental interest in quantifying selection intensity and as a parameter linked with a population fitness. From this point of view, the following definition by Crow (1958) deserves attention. According to Crow (1958) the genetic load is the proportion by which the fitness of the average genotype in a population is reduced in comparison to the best (optimal) genotype.

In this instance the genetic load for a model population with discrete time is

$$L = \frac{W_{\max} - \overline{W}}{W_{\max}}, \tag{1.21}$$

where $W_{\max}$ is the fitness of the best (optimal) genotype, and $\overline{W}$ is the average fitness of a population.

It is obvious that this approach can easily be implemented by taking into account the segregational load arising on the basis of the heterozygotes' increased fitness compared with both homozygotes ("overdominance").

Let us assume two alleles at one locus with constant and positive selection coefficients:

| Genotype | $A^1A^1$ | $A^1A^2$ | $A^2A^2$ |
|---|---|---|---|
| Fitness | $W_1(1 - s_1)$ | $W_2$ | $W_3(1 - s_2)$ |
| Frequency | $p^2$ | $2pq$ | $q^2$ |

In random mating, equilibrium is reached when $s_1 p = s_2 q$ and hence the equilibrium frequencies are:

$$\hat{p} = \frac{s_2}{s_1 + s_2}, \quad \hat{q} = \frac{s_1}{s_1 + s_2}.$$

The total amount of load equals a decrease of population fitness resulting from the selective elimination of both types of homozygotes, that is, $s_1\hat{p}^2 = s_2\hat{q}^2$, and consequently:

$$L_{OD} = s_1 \frac{s_2^2}{\left(s_1 + s_2\right)^2} + s_2 \frac{s_1^2}{\left(s_1 + s_2\right)^2} = \frac{s_1\left(s_2\right)^2 + s_2\left(s_1\right)^2}{\left(s_1 + s_2\right)^2} = \frac{s_1 s_2}{s_1 + s_2}. \tag{1.22}$$

The population's total load, segregating at $n$ loci, is

$$L_T = 1 - e^{-\sum l_i}$$

where $l_i$ is the load at the $i$ locus.

The concept of genetic load in connection with the problem of inbreeding is discussed in the book of Lynch and Walsh (1998), which I recommend for interested readers. It is necessary only to emphasize in our context that in evaluating segregational (balanced) load, great importance is attached to the question of the number of loci whose polymorphism may be maintained in a population by overdominance. It is quite obvious that a population's maximum genetic load will come to bear when both homozygotes are lethal; in that event 50% of the descendants will perish in each generation.

If the homozygotes are not lethal, the load is reduced; but even in this case the number of overdominant loci is limited by selection coefficients and population numbers.

For example, if $s_1 = s_2 = 0.02$, the equilibrium frequency $\hat{p} = 0.5$ and the number of independent overdominant loci $n = 1,000$, the probability of a specific individual reaching reproductive age is $4.3 \times 10^{-6}$. In a selection type of this nature, the view of a maximally heterozygous individual as the best adapted, optimal genotype has no meaning since such an individual is simply not encountered in a population of finite numbers. If, for instance, the number of polymorphic loci $n = 40$, then the expected frequency of a genotype, heterozygous at all loci, is less than $2^{-40}$, that is, one in a trillion.

Taking these calculations into account, Kimura and Ohta (1971) estimated, on the basis of the population size $N$ and the number $n$ of overdominant loci with multiplicative fitness effects, by how many times ($e^{\tilde{L}}$) the fertility of the most adapted heterozygous genotype would be greater than the population mean:

$$\tilde{L} = \sqrt{\sum_1^n \left(\frac{s_1 s_2}{s_1 + s_2}\right)^2 \times 2\log_e\left(0.4N\right)}. \tag{1.23}$$

If $s_1 = s_2 = 0.01$, $n = 1,000$ and $N = 25,000$, then $\tilde{L} = \sqrt{0.46} = 0.68$ and $e^{\tilde{L}} = 1.97$. This estimate is completely realistic and corresponds to the many natural situations (for example, many populations of mammals). However, when $s_1 = s_2 = 0.1$, other conditions being equal, the segregational load is already exceptionally large $\left(\tilde{L} = \sqrt{46} = 6.8\right)$ because a maximally adapted individual should produce $e^{6.8} \approx 898$ times more progeny than the population average. Clearly, this scale of fertility does not exist in populations of most species. We shall return to the problem in Chap. 5 when we examine the new data on the role of selection in maintaining biochemical polymorphism.

A population's segregational load is a constant "cost", in the form of the less adapted genotypes appearing in each generation, that a population is forced to pay for its stable existence at the maximum point of fitness.

Under directional selection, usually regarded as the most important adaptive evolutionary factor linked with the replacement of "less adapted" alleles, the total volume of the genetic load increases still more at the expense of the so-called substitutional load (Kimura's term). This problem was first posed and then mathematically investigated by Haldane (1957, 1960). His logic in solving this problem is closely allied to what forms the basis for calculating the number of overdominant loci of a population in equilibrium. By his calculations, the number of deaths resulting from a changed vector of selection are not connected with its intensity, but are determined exclusively by the initial frequency $p$ of an unfavorable allele. Nevertheless, the number of generations to which elimination distributions apply depends on the intensity of selection.

If the effects of genes are additive, the substitutional load in one generation for a single locus is $L_i = -2 \log p_0$.

For $p = 0.5$, the approximate equation for the connection between selection intensity and the number of generations necessary for an allele's substitution is $s = 30/n$ where 30 is the "cost" of a single gene replacement. This quantity shows by how many times in the selection process the total number of deaths in all generations exceeds the number of individuals in a given generation. Assuming that $s = 0.1$, corresponding to 10% selective mortality per generation, then $n = 300$; in other words, about 300 generations are required to replace a single allelic gene in the adaptive evolutionary process. Thus, according to Haldane, the rate of evolution must be very slow and the number of simultaneously evolving genes fairly small for there not to be a sharp reduction in fitness threatening the life of a population under the conditions described above. Haldane suggested that allelic substitutions at 1,000 loci are sufficient for the emergence of a new species, necessitating not less than 300,000 generations, the number of concomitantly evolving genes not exceeding 12 (see also Kimura 1960a,b).

The speciation rate along the lines of this model generally accords with paleontological findings about evolutionary tempos, at least insofar as they apply to mammals. However, numerous examples are known of exceedingly rapid speciation, which contradicts Haldane's calculations (see, for instance, Mayr 1968, 1974; Bush 1975; Dobzhansky et al. 1977; Coyne 1992; Orr 1995; Lynch and Force 2000; Danley and Kocher 2001; van Alphen and Seehausen 2001; Kocher 2004). These questions will be under consideration in Chap. 6.

The most important category of population hereditary burden – the *mutational load* – is introduced by means of recurrent harmful mutations. However, the frequency of these mutations in a population is low, since by exerting a negative influence over fitness they are eliminated by natural selection in the first generation or in those immediately following. The reduced fitness (the total load) subjected to recessive mutation is equivalent to the mutation rate, $L = \mu$; in semidominant mutations $L = 2\mu$ (Haldane 1937).

If the mutant recessive allele of a normal gene $A$ is represented by $a$ appearing in each generation with a frequency $\mu$, and the relative fitness of $A$ and $a$ is denoted by 1 and $1-s$ respectively, then the mutant frequency $q$ $aa$ is reduced by selection in each generation as follows (Kimura and Ohta 1971):

$$\Delta q = \frac{(1-s)q}{\overline{W}} - q = -\frac{sq(1-q)}{\overline{W}} \, ,$$

where $\overline{W} = 1 - sq$ is the population's average fitness.

When the mutation process is in equilibrium with selection:

$$\Delta q = \frac{(1-q)\,sq}{1-sq} + \mu\left(1-q\right) = 0 \tag{1.24}$$

Since $q$ is small, the effect of reverse mutations may be disregarded. In that case, the mutant gene's equilibrium frequency is

$$\hat{q} = \frac{\mu}{(1+\mu)\,s} \, . \tag{1.25}$$

Because $W_{max} = 1$ and $\overline{W} = 1 - sq$, the mutational load at the point of equilibrium is:

$$L_{mut} = \frac{1-\left(1-s\hat{q}\right)}{1} = s\hat{q} = \frac{\mu}{1+\mu} = \mu \, , \tag{1.26}$$

since $\mu$ is much less than unity. Hence, as has already been pointed out, as far as a recessive allele is concerned, the mutational load equals the mutation rate.

If the genotype $aa$ is lethal or almost lethal, so that $s \approx 1$, the fraction of $aa$ individuals in the population is $q^2 \approx sq^2 = \mu$, and the gene's equilibrium frequency is $\hat{q} = \sqrt{\mu}$.

Crow (1958) and Kimura (1960b) calculated the mutational load for varying degrees of dominance in a randomly mating population. If the fitness of genotypes $AA, AA^1$, and $A^1A^1$ is given as $1, 1 - hs$, and $1 - s$, respectively, the amount of mutational load corresponds approximately to

$$L_{\mathrm{mut}} = \mu \left\{ 1 - \theta \pm \sqrt{\theta(2 + \theta)} \right\} , \qquad (1.27)$$

where

$$\theta = \frac{sh^2}{2\mu(1 - 2h)} .$$

The plus and minus signs correspond to different values of $h$ – the coefficient of dominance, depending on whether it is more or less than 0.5. The load dynamics of lethal mutations ($s = 1$), where $\mu = 10^{-5}$ depending on the degree of dominance, is shown in the graph in Fig. 1.6. It can be seen that at the same time as the load increases from $\mu$ to $2\mu$, the value of $h$ changes from 0 (complete recessivity) to 1 (complete dominance). We also perceive that the mutational load reaches a maximum of $L_{\mathrm{mut}} = 2\mu$, when $h$ becomes greater than 0.5.

The total mutational load for a community of independent loci in a large panmictic population equals the simple sum of the loads of the individual loci. The total load in small populations may be more significant if, as

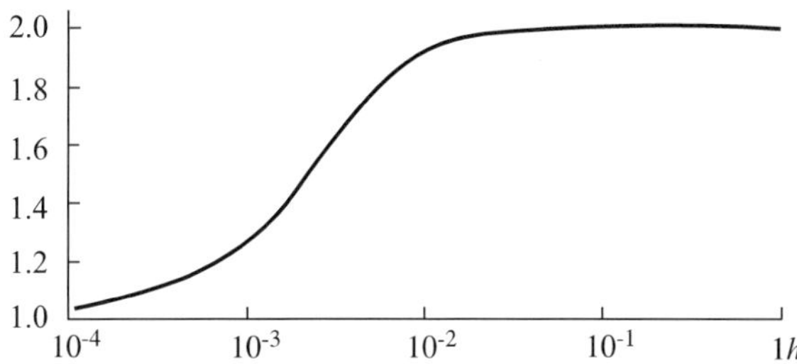

**Fig. 1.6.** Load of lethal mutations as a function of the degree of dominance ($h$; Kimura 1961). $x$-Axis: coefficient of dominance; $y$-axis: mutational load in units of mutation rate ($\mu = 10^{-5}$). (Reproduced with permission from Prinston University Press)

a result of random drift, the gene frequencies in them deviate significantly from the equilibrium values (Kimura et al. 1963).

At the same time, according to Wright, when a species is subdivided into small colonies, each of these will contain only some of the lethals characteristic of the species as a whole; moreover, the average equilibrium frequency of the lethals for a subdivided species as a whole is less than that for a species representing a large panmictic population. In the case of lethals having a mutation frequency $\mu = 10^5$ per gamete per generation, the equilibrium frequencies and distribution of number of colonies free from lethals are as follows (Dobzhansky 1970):

| Population size, $N$ | Equilibrium frequency, $\hat{q}$ | Number of subpopulations free from lethals (%) |
| --- | --- | --- |
| $10^6$ or more | 0.0032 | 0 |
| $10^5$ | 0.0030 | 0 |
| $10^4$ | 0.0020 | 15 |
| $10^3$ | 0.0008 | 87 |
| $10^2$ | 0.00026 | 99 |
| 10 | 0.00008 | 99.9 |
| Self-fertilization | 0.00002 | 99.996 |

A population's subdivision into subpopulations of a limited size also has other important consequences in influencing the particular features of the genetic process in a community of this kind.

In the preceding sections, we have concentrated on the micro-evolutionary factors that influence a population's genetic structure but are essentially external to it. Such a view, however, is incomplete without taking into consideration a population's organizational structure. In many instances natural populations are not panmictic groups but represent historically evolved communities of semi-isolated subpopulations. Subpopulations are constantly exchanging genetic material with each other, being subjected as much to genetic random drift as to different forms of selection.

Several original models have been suggested to describe these more realistic situations in population genetics; Sewall Wright (1931, 1938, 1943a,b, 1951; and others), Masatoshi Nei (1975, 1987) and others authors (Kimura 1953; Kimura and Weiss 1964; Malécot 1955, 1967; Bodmer and Cavalli–Sforza 1968; Maruyama 1970b, 1971a,b, 1972b; Nagilaki 1989; Rannala and Hartigan 1995) have distinguished themselves most in this field.

A population's subdivision into subpopulations of a limited size has important consequences on genetic structure. Let us examine this factor.

# 1.6
# The Influence of Subdivision of a Population on Its Genetic Structure

## 1.6.1
## Subdivision and Inbreeding. The Wahlund Effect

Wahlund (1928) was the first to show that if a large population is subdivided into $K$ panmictic groups, the effect is observed in this community similar to the consequences of inbreeding in a randomly mating population: the fraction of homozygotes increases to the value of the interpopulation variance of gene frequencies through the reduction in the heterozygote fraction.

Indeed, if we denote by $q_i$ the frequency of the gene in the $i$th group ($p_i + q_i = 1$) and the frequency of this same gene in the subdivided population as a whole by $q$, then the average gene frequency characteristic of it and its variance will be:

$$q = \frac{\sum q_i}{K}, \quad V_q = \frac{\sum \left(q_i - q\right)^2}{K} = \frac{\sum q_i^2}{K} - q^2 . \tag{1.28}$$

The corresponding zygote (or the genotype) frequencies equal

$$(AA) \quad \frac{\sum p_i^2}{K} = p^2 + V_q ,$$

$$(Aa) \quad \frac{2\sum p_i q_i}{K} = 2pq - 2V_q , \tag{1.29}$$

$$(aa) \quad \frac{\sum q_i^2}{K} = q^2 + V_q .$$

By comparing the frequencies of the genotypes in Eq. (1.25) with their frequencies in a population, using an inbreeding coefficient $F$ in Eq. (1.10), we obtain:

$$V_q = Fq\left(1 - q\right) \text{ or } F = \frac{V_q}{\overline{q}\left(1 - \overline{q}\right)} . \tag{1.30}$$

Since the $F$ value characterizes the subdivided population as a whole, the corresponding frequencies of the genotypes in the population are equal to those frequencies that would characterize a *separate* inbreeding colony. In other words, "a population's subdivision into separate crossing groups is equivalent in formal terms to inbreeding of the total population" (Li 1978, p. 467).

The degree of this differentiation is directly connected with the scale of the interpopulation differences of the gene frequencies – the greater the genetic differences among subpopulations, the higher the variance of $q$.

Sewall Wright (1943a,b, 1951) played a leading part in describing the local differentiation of the gene frequencies of a subdivided population in terms of $F$ statistics, by establishing several $F$ coefficients as indicators for measuring genetic differentiation:

1. $F_{IT}$ – inbreeding coefficient of an individual relative to the total ($T$) population;

2. $F_{IS}$ – inbreeding coefficient of an individual relative to the subpopulation ($S$) to which it belongs;

3. $F_{ST}$ – inbreeding coefficient of a subpopulation ($S$) relative to the total ($T$) subdivided population.

The relation among these values is given by the equation

$$F_{IT} = F_{ST} + \left(1 - F_{ST}\right) F_{IS} . \tag{1.31}$$

$F_{ST}$ as a measure of genetic differentiation of subpopulations always assumes a positive value, whereas $F_{IT}$ and $F_{IS}$ are positive at a deficit of heterozygotes and negative at their excess (Fig. 1.7).

Nei's (1975, 1987) $G_{ST}$ statistics are equivalent to $F_{ST}$ statistics, relating the total and intrapopulation gene diversity ($H_T$ and $H_S$) by the following formula:

$$G_{ST} = (H_T - \overline{H}_S)/H_T . \tag{1.32}$$

Here,

$$H_T = 1 - \sum \overline{p}_i^2 ,$$
$$\overline{H}_S = 1/n \sum H_S ,$$
$$H_S = 1 - \sum p_{is}^2 ,$$

where $p_{is}$ is the frequency of the $i$th allele in subpopulation $S$ and $\overline{p}_i$ is the mean allele frequency in the total subdivided population consisting of $n$ subpopulations. Hence, $\overline{H}_S$ is the mean heterozygosity of the subpopulation, and $H_T$ is the heterozygosity of the total subdivided population if it were converted to a single randomly mating unit.

Wright's $F_{ST}$ statistics, as a measure of genetic subdivision of population and simultaneously a measure of individual inbreeding in a subpopulation, have an important biological sense; in steady state, they reflect the balance of integration and differentiation processes in population gene pools.

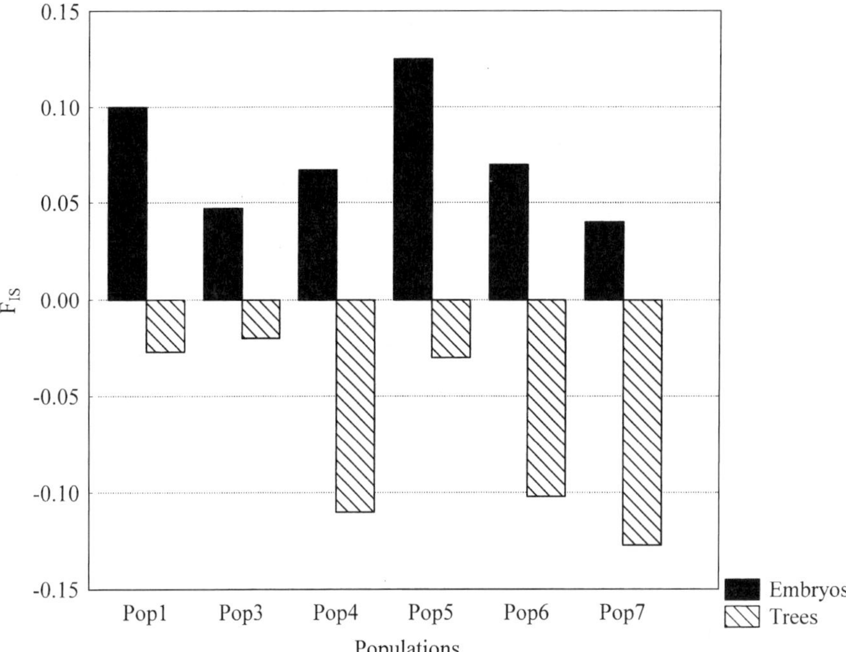


**Fig. 1.7.** $F_{IS}$ values are usually positive (heterozygote deficiency) in embryo samples of Siberian stone pine (*Pinus sibirica*) and negative (heterozygote excess) among mature trees from the same populations (modified after Politov et al. 1992)

Empirical heterozygosity for the set of loci is estimated by the equation

$$\overline{H} = \frac{1}{L} \sum_{i}^{L} \frac{n_i}{N_i},$$

where $L$ is the number of gene loci studied, $n_i$ is the number of individuals heterozygous at the $i$th locus, and $N_i$ is the total individual sampling.

The $F_{ST}$ coefficient described in Eq. (1.26) was suggested by Wright as long ago as 1943, since when it has been repeatedly used to analyze the distributions of gene frequencies in natural subdivided populations (Cavalli–Sforza et al. 1964; Nei and Imazumi 1966a,b; Rychkov 1969; Altukhov 1974; Altukhov et al. 1975a,b). This coefficient is of great interest as it enables one to analyze certain important effects of population subdivision on genetic structure.

Wright proposed two original population models for this purpose: an "island model" and "isolation by distance".

## 1.6.2
## The Island Model of Population Structure

There are two known versions of the population island model: (1) A species's subdivision into a large (infinite) number of randomly mating subpopulations and with the same effective size $N$, each of them having equal probability and identical intensity ($m$) of exchanging genes with a common gene pool; (2) a large panmictic population ("mainland") surrounded by a host of isolated, genetically differentiated small colonies ("islands") each of which receives genes from the mainland with an intensity of $m$ per generation (Fig. 1.8). The effects of back migration may be disregarded.

The intergroup variance of gene frequencies serves as a measure of the random differentiation of subpopulations in the system

$$V_q = \frac{\bar{q}\left(1 - \bar{q}\right)}{4N_e m + 1}, \tag{1.33}$$

and the state of equilibrium between genetic drift and migration in terms of $F$ statistics (see Eq. 1.26) may therefore be written as

$$F_{ST} = \frac{1}{4N_e m + 1}. \tag{1.34}$$

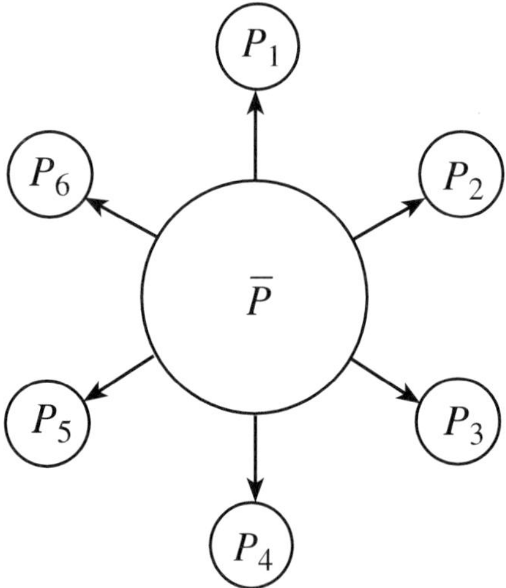


**Fig. 1.8.** A version of the island population model, in which $P$ is the gene frequency in the system averaged out for the peripheral subpopulations ("islands", *1–6*) and equal to the value of $P$ in the population "core" ("mainland")

Wright's (1943a) more exact treatment gives

$$V_q = \frac{\overline{q}(1 - \overline{q})}{2N_e - (2N_e - 1)(1 - m)^2} \,, \tag{1.35}$$

at small values of $m$ ($m \ll 1$) the difference between (1.33) and (1.35) being negligible.

Thus, the local differentiation of gene frequencies is conditioned by the parameter $Nm$. In other words, the deciding factor is not the migration coefficient or effective size of the population itself, but their product, equaling the number of individuals entering a population in a generation.

As a result of interacting drift and migration we have a probability distribution of gene frequencies. At any instant of time $T$ it is a function of $\Delta q = -m(q - qT)$ as a measure of the effect of systematic migration, and $V_{\delta q} = q(1 - q)/2N_e$ as the sampling variance of a gene frequency in one generation through isolation, that is, random drift (Wright 1938, 1939):

$$\Phi(q) = (C/V_{\delta q}) \exp\left[ 2 \int (\Delta q/V_{\delta q})\, dq \right].$$

The general formula for the stationary distribution of gene frequencies in the island model represents the $\beta$-function of the probability density as follows:

$$\Phi(q) = \frac{C}{q(1 - q)} \exp\left[ 4N \int \frac{\Delta q}{q(1 - q)}\, dq \right], \tag{1.36}$$

where $C$ is a normalizing constant chosen so that $\int_0^1 \Phi(q)dq = 1$:

$$C = \frac{\Gamma(4Nm)}{\Gamma(4Nmq)\,\Gamma[4Nm(1 - q)]}.$$

Distribution (1.36) takes the following forms, depending on what combination of random and systematic factors is assigned:

- For equilibrium of random drift by gene migration

$$\Phi(q) = Cq^{4Nm\overline{q}-1}(1 - q)^{4Nm(1-\overline{q})-1}, \tag{1.37}$$

where $p$ and $(1 - q)$ are the allelic frequencies in the subpopulation, $\overline{q}$ and $(1 - \overline{q})$ are the mean allelic frequencies for a subdivided population as a whole, $N$ is the effective population size, and $m$ is the migration coefficient.

- With the combined effects of isolation, migration and selection:

$$\phi\left(q\right) = C\overline{W}^{2N} q^{4Nm\overline{q}-1}\left(1-q\right)^{4Nm(1-\overline{q})-1}, \tag{1.38}$$

where the notation is the same as for the previous formula and $\overline{W}$ is the population's average intralocus fitness determined by summation of the genotypes' fitness according to their frequencies (see Eq. 1.18).

The relationship between the effects of random and systematic evolutionary factors upon population genetic structure can vary widely by virtue of the fact that the corresponding stationary distributions of gene frequencies may take different forms (Fig. 1.9).

Stationary distributions can describe: (1) the distributions of allelic frequencies of many loci in the same population in the case of neutrality or when subjected to nearly identical pressure of selection; (2) the distributions of gene frequencies at any locus in successive generations of the same stationary population; (3) the distribution of allelic frequencies of one or several loci in the community-isolated populations. All three types of distribution are mathematically equivalent.

In the island model, the gene migration coefficient is independent of the degree of remoteness of the populations. Wright (1943a) and Malécot (1955, 1967) employed mathematical methods to investigate a population in which the intensity of exchange among the subpopulations depends on distance.


## 1.6.3
## Isolation by Distance

This model postulates a population spread evenly over a large territory substantially exceeding the radius of an individual's reproductive activity. The special features of local differentiation in such a system depend upon the effective size or the "neighborhood" $N_N$ from which parents arise at random, as well as upon the dimensions of the area. In particular, local differentiation arises sooner or later in a one-dimensional area, if it is sufficiently large in extent. However, on a plane (two-dimensional area) the possibility of this differentiation is much reduced.

According to Wright (1951), the size of a neighborhood corresponds approximately to the number of genetically effective individuals within a circle whose radius equals twice the standard deviation ($\sigma$) of the extent of migration (i.e., the distance between the birthplaces of parents and progeny) in one direction in a given generation.

Differentiation is very great when $N_N \approx 20$, somewhat less, but still quite pronounced when $N_N \approx 200$, and verges on panmixia when $N_N \approx 2,000$.

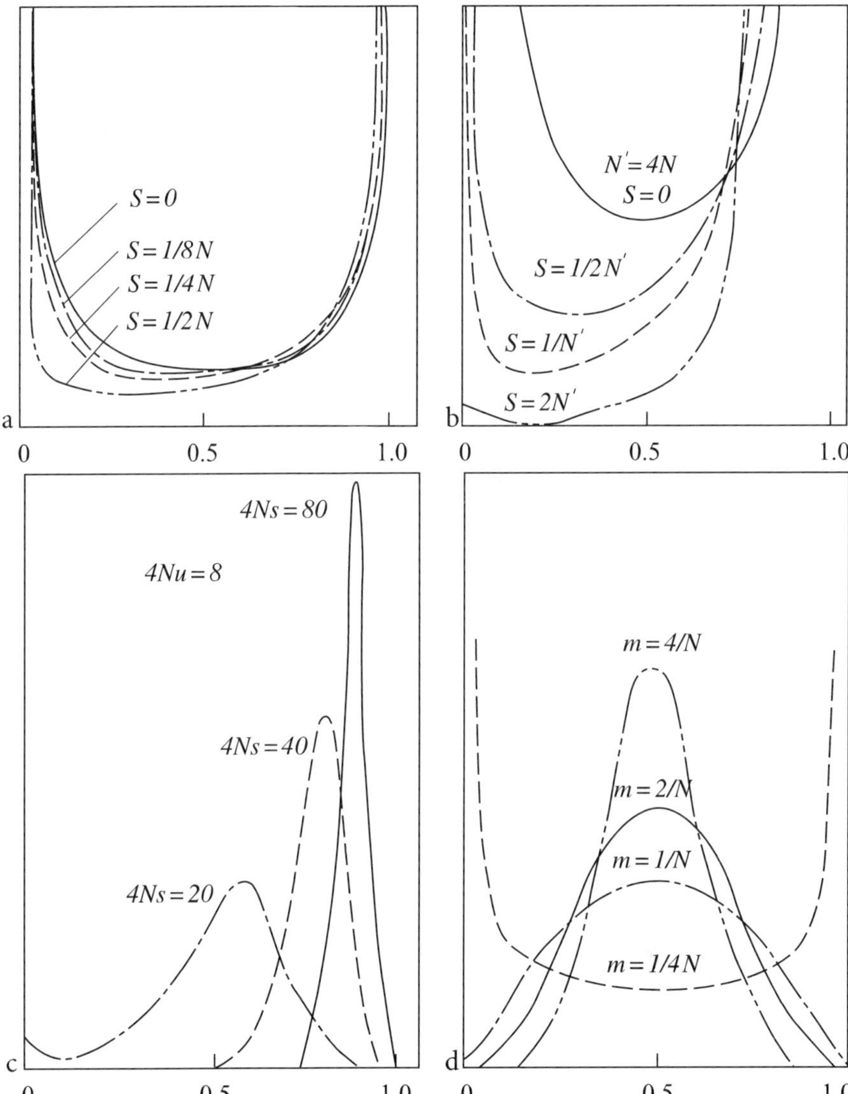


**Fig. 1.9.** Equilibrium (stationary) distributions of gene frequencies for different relationships between effective population size and the pressures of migration, mutation, and selection. Compiled from Wright (1931). The $x$-axis represents the gene frequency, the $y$-axis represents the number of subpopulations; **a** is the distribution for small and completely isolated populations affected by selection; **b** is the distribution for the same selection $s$ coefficients as in **a**, but the subpulation size is fourfold ($N' = 4N$); **c** is the equilibrium frequency distribution of mutant alleles when $u = 2/N$; $s$ as in case **a**; $N' = 40N$; **d** is the interaction between genetic drift and migration at a mean frequency of 0.5; a U-shaped distribution, reflecting allelic fixation in many populations, is observed only when $m = 1/4N$, that is, when a population receives no more than one migrant in every four generations

Kimura (1953; see also Kimura and Weiss 1964) has proposed another model, called the "stepping-stone model", representing a situation intermediate between Wright's island model and that devised by Wright and Malecot for evenly distributed populations.

## 1.6.4
## The Stepping-Stone Structure of Gene Migration

In this, as in the island model, colonies are considered together, but exchange of individuals proceeds only among neighboring colonies, as shown in Fig. 1.10; thus it is directly dependent on the colonies' distance from each other.

At equilibrium the interpopulation variance of gene frequencies is

$$V_q = \frac{\bar{p}\left(1-\bar{p}\right)}{2N_e - \left(2N_e - 1\right)\left\{1 - 2R_1 R_2 / \left(R_1 + R_2\right)\right\}} \, , \tag{1.39}$$

where $R_1 = [(1 + \alpha)^2 - (2\beta)^2]^{1/2}; R_2 = [(1 - \alpha)^2 - (2\beta)^2]^{1/2}$, in which $\alpha = (1 - m)(1 - \overline{m}_\infty)$ and $\beta = m_1(1 - \overline{m}_\infty)/2; m_1$ in these equations stands for the intensity of migrations among adjacent colonies (*short-range migration*) and $\overline{m}_\infty$ for the external effects of gene migration among all the colonies together (*long-range migration* corresponding to the coefficient $m$ in Wright's island model). If $m = 0$, then $\alpha = 1 - \overline{m}_\infty, \beta = 0$, and Eq. (1.39) reduces to Wright's Eq. (1.30). Thus the island model represents a special case of stepping-stone model in which there is no gene exchange among neighboring colonies.

In cases where $m_1$ is much higher than $\overline{m}_\infty$, Eq. (1.39) reduces approximately to:

$$V_p = \frac{\bar{p}\left(1-\bar{p}\right)}{1 + 4N_e\sqrt{2m_1\overline{m}_\infty}} \, , \tag{1.40}$$

assuming that $\overline{m}_\infty \ll m_1 \ll 1$.

The standardized genetic variance in this case is:

$$F_{ST} = \frac{V_p}{\bar{p}(1-\bar{p})} = \frac{1}{1 + 4N_e\sigma\sqrt{2\overline{m}_\infty}} \, , \tag{1.41}$$

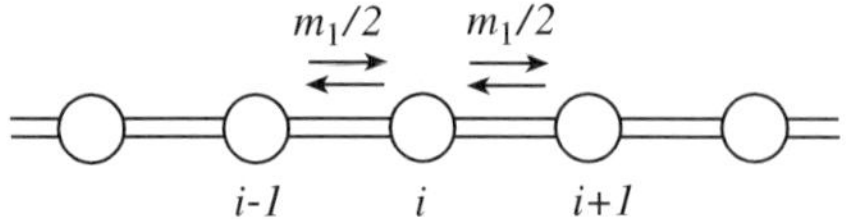


**Fig. 1.10.** One-dimensional stepping-stone model of gene migration

where $\sigma = \sqrt{m_1}$, since $m_1 = V_m$ is the migration dispersion for a distance of more than one step apart.

According to the formula obtained by Kimura and Weiss (1964), under stationary conditions where there is no selection, the variance of gene frequencies may also be found from the expression:

$$V_p = \frac{\bar{p}\left(1 - \bar{p}\right)}{1 + 4N_e m \left[1 - r(1)\right]} ,$$  (1.42)

where $r(1)$ is the correlation coefficient of gene frequencies among adjacent subpopulations.

The tendency in the stepping-stone model and in isolation by distance toward local differentiation depends greatly on the dimension of the area. Following Kimura and Weiss (1964), the correlation coefficient of gene frequencies among colonies in a one-dimensional model diminishes with distance from the exponent and is described by the formula:

$$r(d) \approx e^{-\left(\sqrt{\frac{m_\infty}{m_1}}\right)d} ,$$

where $d$ is the distance between subpopulations (in "steps") and $r(D)$ is the correlation coefficient between them.

The mathematical apparatus becomes more complicated when it involves analysis of two- or three-dimensional models, each colony exchanging genes with four or six adjacent ones (Kimura and Ohta 1971). For our purposes it is enough to stress that local differentiation of gene frequencies, all other conditions being equal, is a maximum in one dimension and quickly diminishes as the number of dimensions increases. This dependence manifests itself strikingly in research on the genetic correlation $r(d)$ among colonies based on the distance between them. When the distance increases the correlation quickly declines, a feature which is particularly characteristic of the three-dimensional model (Fig. 1.11). In accordance with Kimura and Maruyama's (1971) calculations, local differentiation is especially marked in a two-dimensional stepping-stone model when $Nm < 1$, while when $Nm > 4$ a population acts as a single panmictic unit.

For the one-dimensional case, the condition for local differentiation is $Nm > k/\pi^2 4$, in which $k$ is the number of subpopulations (Maruyama 1970a).

Thus, apart from the dimensions of an area, local genetic differentiation in populations having a stepping-stone structure depends both on the intensity of short range ($m$) and long range ($m_\infty$) migration (the parameter $m_\infty$ can combine all possible stabilizing factors: the mutation process, migration from a constant external gene pool, and selection), and on the size of the colonies and their numbers.

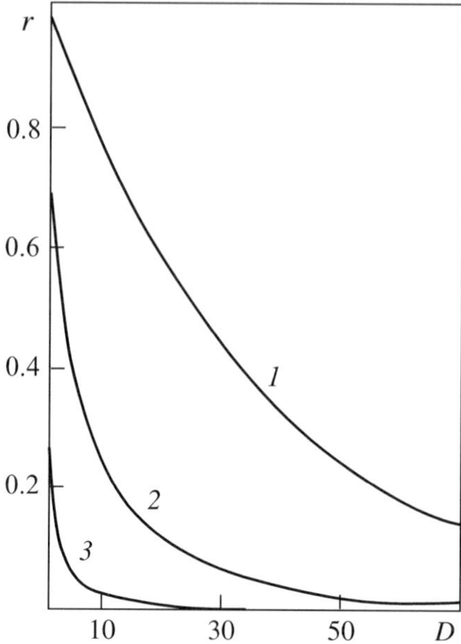

**Fig. 1.11.** Decrease in correlation $r(D)$ among gene frequencies in colonies with increased distance $(D)$ between them (Kimura and Weiss 1964). $m = 0, 1$; $m_\infty = 4 \times 10^{-5}$; $1, 2, 3$ – one, two, and three dimensions, respectively

## 1.7
## Conclusion

Almost all population genetic models and methods considered above will be employed in our following analysis. Along with this, main estimates and calculations will be connected with the subdivided population models that are mostly adequate to species structure, and, at least in part of the work, with the theory of stationary distribution of gene frequencies.

In essence, the theory of stationary distributions emphasizes yet again the important fact that, although individual evolutionary factors are also capable of causing directional genetic changes, the interaction of these factors (for example, forward and reverse mutations, genetic drift and migration, etc.) leads finally to reciprocal balance, engendering a stationary type of gene frequency dynamics. This stability may be particularly great when all the known factors of evolution act simultaneously on a population.

Another important feature of subdivision, which has also been examined theoretically, is the ability of subdivided populations to maintain significantly greater genetic diversity compared with panmictic populations of similar total size (Wright 1951; Kimura 1968b). It is believed that precisely this diversity allows a subdivided population to react more effectively to

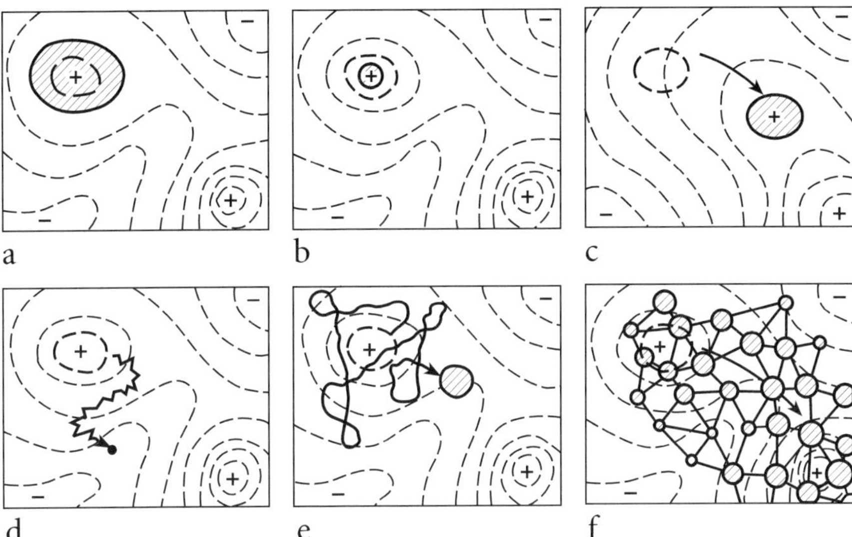

**Fig. 1.12.** Wright's two-dimensional contour maps with adaptive peaks and non-adaptive valleys in the field of gene combinations occupied by different types of populations (Wright 1932). The isolines relate to different levels of adaptation: the *thick broken line* is the population's "starting" position; *arrows* indicate evolutionary trends. **a** Reduced selection (or increased mutation rate) produces increased genetic variability and reduced average fitness. The population's (= species) evolutionary plasticity is fairly large, and with increased numbers it can occupy the lower slopes of another, more highly adaptive peak and then also conquer it entirely ($4Nu$ and $4Ns$ are very large). **b** The consequences of increased selection (or reduced mutation rate). The amount of the population's genetic variability is narrowed by selection, and the population's average fitness grows. Correspondingly, its evolutionary plasticity is reduced and its chances of conquering a neighboring peak are diminished ($4Ns$ is very large). **c** Sharp changes in the adaptive landscape: peaks are transformed into valleys, and vice versa. The result of the genetic process is determined exclusively by selection intensity and environmental rates of change. A population having a small adaptive peak prior to these shifts, and lacking an adequate reserve of genetic plasticity (variability), may remain in a valley and become extinct ($4Nu$ and $4Ns$ are large). **d** The consequences of drastic curtailment of numbers and close inbreeding. A population falls from the adaptive peak, and as a result of drift its genetic pool undergoes random fluctuations leading, however, eventually to inbreeding and, as a rule, to degeneration ($4Nu$ and $4Ns$ are very small). **e** An average-sized population subject to moderate mutations, leading to increased genetic diversity. The population shifts from its adaptive peak but cannot go far away from it. This means that the occupation of any new adaptive peak would be an exceptionally slow process ($4Nu$ and $4Ns$ have values intermediate between extreme situations). **f** The population of an extensive area subdivided into a multiplicity of interacting subpopulations, evolving rapidly by migration and mostly nonadaptive. When one population or another comes into the environs of any adaptive peak it may colonize it. After increasing in numbers, it begins to "pour" corresponding genes into other subpopulations, and entire species may become "stretched out" in the zone of this new peak. Such a picture is considered to be evolutionarily optimal in the sense that genetic variability and the capacity for further change are preserved ($4Nm$ has an intermediate value)

changes in the environment through changes in genetic structure. This thesis, which plays a decisive role in Wright's concept of evolution, is known as the shifting balance theory in which the "surface" $W$ is represented on a topographical map with peaks and valleys for one landscape of gene combinations (Fig. 1.12). The most important conclusion from this model is that the "...evolutionary process depends on a continually shifting but never obliterated state of balance between factors of persistence and change, and that the most favourable condition for this occurs where there is a finely subdivided structure in which isolation and cross-communication keep in proper balance" (Wright 1951).

Our further task is to compare this concept with what is observed in nature. However, before making this comparison it is necessary to consider classical and new approaches to the phenomenology and biological sense of the genetic diversity of populations.

# 2 Heritable Variation in Populations

From the preceding chapter it should be obvious that if we wish to use the apparatus of theoretical population genetics to analyze real (natural or experimental) populations, and to characterize the process that occurs in them in terms of genotype and gene frequencies, we must have recourse to markers for discrete gene expression, that is, make use of genetic polymorphism.

## 2.1
## Population Genetic Polymorphism and the Adaptive Norm Concept

Polymorphism is a manifestation of the individual, discontinuous variability of living organisms. The term was originally used quite broadly to denote any discontinuous variability within a species (for example, the castes of social insects, seasonal morphs, age coloration differences, sexual dimorphism, etc.). At the present time, however, it has been suggested that these differences should be given the name "polyphenism" (Mayr 1974), whereas polymorphism is to be treated only in a strictly genetic sense. The term polymorphism should also be distinguished from polytypical, which is employed to signify complex taxonomic categories (for instance, a polytypical species – a species represented by two or more subspecies – a polytypical genus, etc.). Ford (1940), the creator of the concept of genetic polymorphism, defined it as *"the occurrence together in the same locality of two or more discontinuous forms of a species in such proportions that the rarest of them cannot be maintained merely by recurrent mutation"* (Fig. 2.1).

This kind of hereditary variation, as has already been emphasized, is controlled by allelic genes (or blocks of closely linked genes – so-called supergenes); hence a somewhat different definition can be given of polymorphism, namely, "the occurrence in the same population of two or more alleles at one locus, each with appreciable frequency" (Cavalli-Sforza and Bodmer 1971, p. 118). In practice, a population is usually considered to be polymorphic if it has a heterozygote frequency at a specific locus

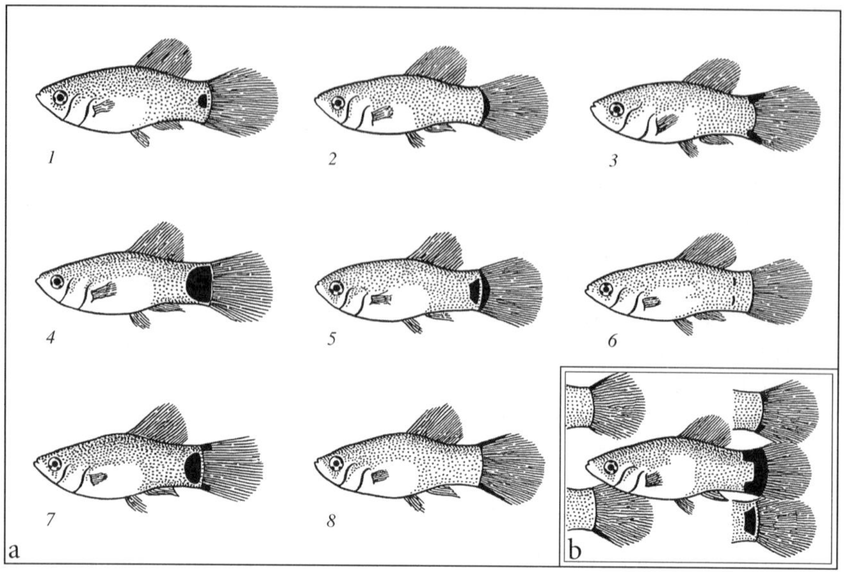

**Fig. 2.1.** Hereditary polymorphism of a pattern on the tail of Central American poeciliid fish (males *Xiphophorus maculatus* – wild relatives of our aquarium forms (from Gordon 1947, with changes). A series of seven sexlinked alleles responsible for synthesizing the melanin pigment in special cells – melanophores – leading to the existence of eight widely disseminated and five rare phenotypes (*inset*). On the basis of this data it has been established that the poeciliid fish, introduced into Europe in 1909–1911 and to become a favorite item of aquarium keepers, originated from Honduras populations. (Reproduced with permission from Elsevier)

of > 1–5 %. Analysis of polymorphic traits is the key to studying genetic processes in populations. In this connection, early research was carried out on species with individuals of well-defined polymorphic appearance (shell striation of land mollusks; color patterns of wing-case insects, fish, and mammals). Later, a wide-ranging series of works were devoted to chromosome polymorphism.

Research of this kind laid the foundations for the modern so-called "synthetic" theory of evolution; decisive contributions to its development have been made by S.S. Chetverikov, A.S. Serebrovsky, N.P. Dubinin, D.D. Romashov, R. Fisher, S. Wright, J. Haldane, Th. Dobzhansky, E. Ford, S.M. Gershenson, N.V. Timofeev-Resovsky, and other well-known scientists. A large body of research material has been accumulated on the phenomenology and genetic control of, and the mechanisms maintaining polymorphism in different species. It has been shown that this kind of morphological variability:

1. is a relatively rare phenomenon which, thanks to "a caprice of nature," has stamped the Mendelian gene series in an extremely small group

of "selected" species against a background of an enormous number of externally uniform ones;

2. has conspicuous adaptive significance; moreover, in the processes of adaptive evolution the rearrangement of population genetic structure not only involves separate loci, but is also associated with the integration of complex and highly stable polygenic systems;

3. is, in many cases, maintained in a balanced form as a result of the adaptive-fill-superiority of heterozygotes having such considerable selection coefficients that they leave virtually no room for the effects of random genetic drift;

4. should be regarded in several instances (because of the limited nature of the stability phase) as testimony of the divergence of populations to the status of new species that has occurred and continues to take place before our eyes, the clearest example of this being the well-known phenomenon of industrial melanism;

5. involves heterozygosity, which serves as a standard of the genetic diversity of populations reflecting their reserves of ecological plasticity resulting from the segregation and combination of different genotypes, whose relative fitness can change under differing environmental conditions.

The results of works of Dubinin, G.G. Tiniakov and Dobzhansky illustrate best these positions on the theory of population polymorphism.

Dubinin and Tiniakov (1946; and others) investigated the polymorphism of chromosome inversions in populations of *Drosophila funebris* over a large area and showed the pronounced cyclic character of variation in the frequency of gene arrangements. They discovered that inversion polymorphism is closely correlated with the habitat's changing temperature, and that under special conditions individuals with certain gene arrangements are most numerous, while under other conditions different gene arrangements are more common. Having shown clearly defined frequency differences among various karyotypes of populations from urban ("city race") and rural ("country race") localities, the authors carried out an experiment in introducing individuals from one ecological niche into another. This research revealed the exceptionally high inimitability and conservatism of local genotypic adaptations that prevent outside genes from penetrating into a coadaptive genetic system that has been formed by natural selection.

One hundred thousand flies homozygous for the inverted gene arrangement (II-1; in the center of Moscow 50% inversion frequency) were released in the habitat of a rural race (the Kropotovo settlement, 105 km from Moscow; 1% frequency) at the beginning of June 1945. In July (in a virtually optimal neutral environment), the frequency of this inversion in the Kropotovo population had reached 49.5%, and the numbers of the different genotypes corresponded exactly to a Hardy–Weinberg distribution, indicating free mating of migrants with local flies. However, a cytological analysis conducted in August, September and October exhibited rapid inversion displacement of the rural population as a result of selection, the respective frequencies of it being 23.8, 17.2 and 9.7%. Considerable deviations from the expected genotypic equilibrium were also found, caused by strong selection pressures against II homozygotes (Table 2.1).

By spring of the following year, after hibernation, the Kropotovo population had completely re-established its typical frequency of II (1.9%); in other words, elimination of the introduced genotypes, incapable of adapting to the new conditions, took place in the course of only a few generations. In the summer of 1946 the experiment was repeated and its results reproduced completely the dynamics just described.

A very similar picture was obtained when research was carried out on Californian populations of *Drosophila pseudoobscura*, polymorphic for the Standard (ST) and Chiricahua (CH) gene-arrangements. Since different karyotypes differ in no way morphologically from each other, it was suggested at first that this polymorphism was selectively neutral. However, subsequent experiments established the seasonal dynamics of the frequencies of a strictly cyclical character, obviously due to the "changeable" adaptation of genotypes to temperatures at different seasons of the year (Fig. 2.2). This hypothesis was confirmed by research of experimental populations: at a temperature of 16 °C the three genotypes ST/ST, ST/CH, and CH/CH did not differ in fitness, but at 25 °C the adaptive superiority of the heterozygotes was evident (Wright and Dobzhansky 1946; Dobzhansky and Pavlovsky 1953):

| Genotype | Relative fitness | Selection coefficient |
| --- | --- | --- |
| ST/CH | 1.00 | 0.00 |
| ST/ST | 0.89 | 0.11 |
| CH/CH | 0.41 | 0.59 |

Dobzhansky and his colleagues (Beardmore et al. 1960) also compared the productivity of polymorphic and monomorphic (homozygous) *D. pseudoobscura* populations containing the Arrowhead (AR) and Chiricahua (CH) gene arrangements. It turned out that while the homozygous AR and CH lines produced on average 230 and 202 individuals, respectively, per unit of resource, the corresponding figure for the polymorphic population with AR+CH was 327 flies.

Later (Dobzhansky et al. 1964b), homozygous and heterozygous gene arrangements (the same as noted above plus PP – "Pikes Peak") in experimental populations of the same species were compared for their ability to show exponential growth by evaluating the Malthusian parameter $r_m$ (see Chap. 1). The following results were obtained at 25 °C:

**Table 2.1.** Distribution of karyotypes in the phase of decline of the inversion II-1 frequency in the Kropotovo population settlement. (Dubinin 1966)

| Karyotype | Distribution of karyotypes | | Deviations from expected | |
| --- | --- | --- | --- | --- |
| | Actual observed | Expected from the Hardy–Weinberg equation | Absolute | Relative % |
| Homozygotes for the standard order of genes (SS) | 924 | 934.6 | −10.6 | 1.13 |
| Inversion heterozygotes (SI) | 359 | 337.9 | +21.1 | 6.24 |
| Inversion homozygotes (II) | 20 | 30.5 | −10.5 | 34.43 |

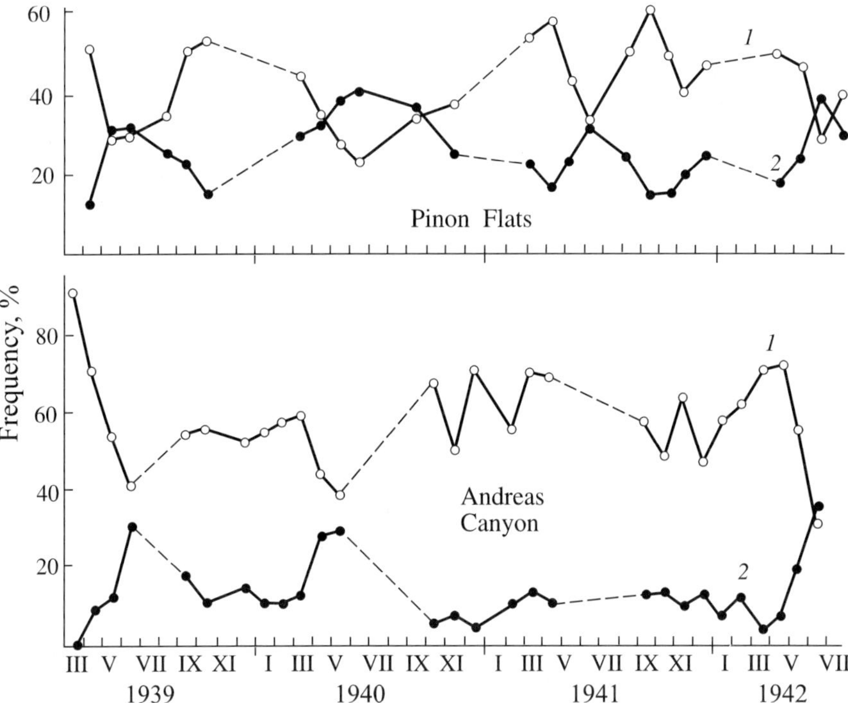


**Fig. 2.2.** Seasonal changes in the frequencies of two third chromosome gene arrangements – Standard (*light circles*) and Chiricahua (*dark circles*) in a population of *Drosophila pseudoobscura* from two localities in California (Dobzhansky 1943). *x-axis* Time; *y-axis* frequency, as a percentage

| Population | State of population | $r_m$ values |
|---|---|---|
| AR+CH | Polymorphic | 0.220 |
| AR+PP | Polymorphic | 0.214 |
| AR | Monomorphic | 0.207 |
| AR | Monomorphic | 0.205 |
| CH | Monomorphic | 0.192 |
| PP | Monomorphic | 0.170 |

The experiment was conducted in overcrowded population cages; under these conditions the adaptive advantage of heterozygosity was evident. With environmental factors close to optimal, the polymorphic and monomorphic populations do not differ reliably, although the values of the parameter $r_m$ as a whole are higher than those for a more extreme environment (Ohba 1967).

The adaptive significance of heterozygosity has been convincingly shown in Carson's (1958) experiments with *Drosophila melanogaster*.

A third chromosome of the Oregon-R line was introduced into a large cage population homozygous for five recessive genes in the third chromosome. The result was a greatly increased productivity of the experimental populations (Table 2.2).

One could easily add to these examples (see Ford 1964; Mayr 1968; Sheppard 1970; Dobzhansky et al. 1977; Sperlich and Pfriem 1986). They all clearly demonstrate the selective advantages of heterozygotes over a broad range of unfavorable or fluctuating conditions of the environment. It has also been shown that, along with obvious signs of polymorphism, there are enormous reserves of "hidden" hereditary variability in almost all species. This was first pointed out by Chetverikov (1926) in his paper "aspects of the evolutionary process from the viewpoint of modern genetics", which played an important part in changing views about the genetic content of populations and species, and facilitated the emergence of a trend subsequently defined by Dobzhansky (1955a) as the "balanced" school in population genetics.

**Table 2.2.** Numbers and productivity of homozygous and heterozygous *Drosophila* populations. (Carson 1958)

| Populations | Equilibrium duration (weeks) | Population size experiments | Productivity (raw weight, mg/week) |
|---|---|---|---|
| Control | 32 | 161.6± 6.4 | 90.3±3.0 |
| Homozygous | 20 | 154.4± 4.4 | 88.7±2.0 |
| Experimental | 20 | 454.4±13.7 | 292.5±9.1 |
| Heterozygous | 20 | 502.6±14.9 | 318.6±8.9 |

Indeed, many authors have envisaged a species as a community of individuals that are identical in the overwhelming majority of the genes represented only by a wild-type allele (that is, monomorphic), and distinguished only by a small number of mutant, harmful alleles concealed in the heterozygous state (the so-called classic school; Fig. 2.3). This typological view in fact was rejected by the balanced concept (Fig. 2.4) of a population which represents the essence of the modern theory of species and speciation and forms the basis in principle of important views about a population's adaptive norm and the phenomenon of heterosis.

The concept of the "adaptive norm" of a population rejects the typological view of a species' genetic structure, postulating that large numbers of diversified genotypes underlie the externally "normal" optimally adapted "average" phenotypes. However, their selective value can alter with chang-

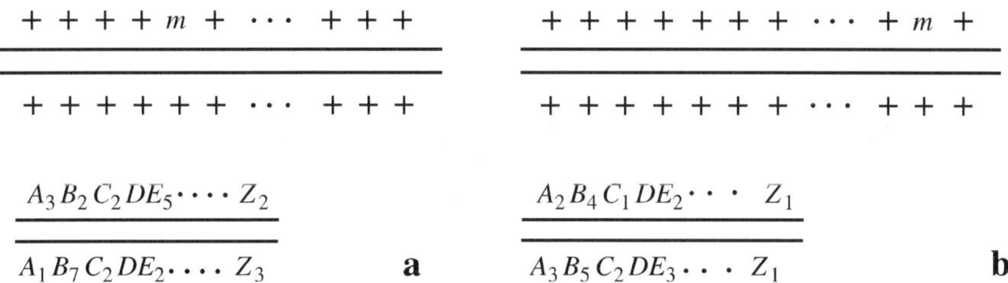

Fig. 2.3. "Classic" (*a*) and "balanced" (*b*) models of a species' genetic population structure (based on Lewontin 1978a). Explanations in the text

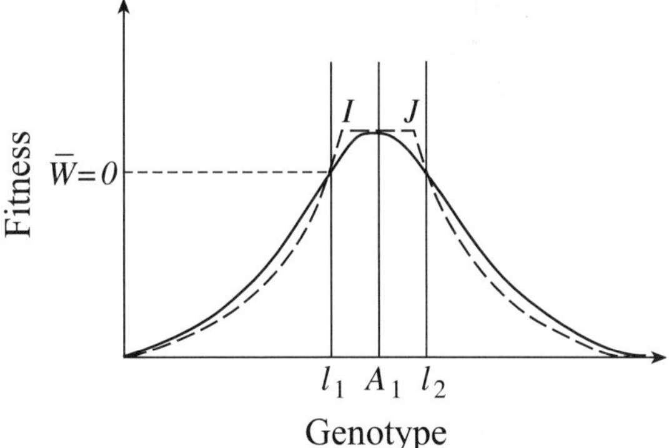

Fig. 2.4. Diagram of typological and population concepts of a species' genetic structure (Mettler and Gregg 1972). *Continuous line* Typological concept according to which the best genotype (phenotype) corresponds to point $A_1$; *vertical lines* $l_1 + l_2$ represent the survival limits of stabilizing selection. The *broken line* illustrates the concept of the adaptive norm when a large number of similar phenotypes, equally well adapted within the limits *IJ*, correspond to a large number of different genotypes

ing conditions of the environment, and some genotypes that are less well adapted at a given and specific moment of time may prove to be better adapted under other conditions. All this ensures the broad reaction norm of populations as integrated systems, and their successful adaptation to diversified fluctuations of the environment, as Fig. 2.4 illustrates.

As noted above, it has been found that in many, though not all, cases heterozygous genotypes are at an indisputable advantage, at least in environments with harsher, more fluctuating conditions, whereas homozygotes are better adapted to more restricted, specialized conditions; in a neutral environment genotypic differences of fitness subside.

In any event, it should be pointed out that many questions remain unanswered, and to this day it is not clear why a heterozygous genotype is superior to a homozygous genotype. Most authors associate heterosis with the favorable effects of dominance or overdominance. What happens in the former case is that in the hybrids the number of dominant alleles increases at loci that in the parental stocks were homozygous for unfavorable recessive alleles. Probably many situations can be explained within this framework, but it does not accommodate all cases of monogenic (one locus) heterosis.

In the latter case, the explanations rest on Haldane's (1955) biochemical hypothesis which postulates that the effect of heterosis is based on the interaction of protein products in heterozygotes having a different form of activity and resulting in the biochemical "enrichment" or greater biochemical flexibility of a hybrid cell (Kirpichnikov 1967). This multiplicity of gene products and their combinations enables a heterozygous organism to maintain the constancy of its functions over a wider range of environmental changes than is possible for homozygous genotypes ("canalization", the "buffered state" of ontogenesis) (Waddington 1942, 1970; Dubinin 1948; Lerner 1954).

Heterozygosity also enables a population to re-establish its genetic structure after the action of various forces have disequilibrated it – so-called genetic homeostasis (Lerner 1954).

While investigating *Bombyx mori*, the Chinese silkworm, Strunnikov (1974, 1986) devised an original heterosis model linking this phenomenon with the emergence of a *gene compensation complex* (GCC), which eliminates the harmful effects of semilethals in the selection process, increasing viability. When this kind of selected strain with GCC is crossed with unselected strains, it transmits a dose of GCC, which has adaptive effects, and a semilethal's depressant influence is eliminated by its conversion into a heterozygous state. As a result, the excess number of favorable genes, no longer balanced by semilethals, causes heterosis. This model is of great theoretical interest and is very important for solving several selection problems (Strunnikov 1983; Strunnikov and Strunnikova 2000). However, whatever final shape the formulation of the heterosis concept may take, the theory of population genetics makes it clear that adaptation based on the selective advantages of the heterozygotes is always accompanied by unfavorable biological effects by the segregation of less well-adapted homozygotes, and in the norm the price paid for adaptation cannot exceed a population's reproductive capacity.

The question of the number of polymorphic loci in the average genome and the mechanisms maintaining this variability are central problems of population genetics – and the main source of contradictions between the adherents of the typological and population concepts of a species' genetic structure. This problem has attracted exceptional attention in recent years because of the widespread protein polymorphism discovered in many species of plants and animals, including humans. But before analyzing the phenomenon, the general principles for revealing and genetically interpreting this kind of hereditary variability will need to be recalled.

## 2.2
## Hereditary Protein Polymorphism

The earliest works on biochemical population genetics were concerned with descriptions of hemoglobin's anomalous electrophoretic behavior in sickle-cell anemic individuals (Pauling et al. 1949) and with showing how this form of pathology is inherited as a simple Mendelian trait (Beet 1949; Neel 1949). Using the peptide map method, Ingram (1957, 1960, 1961, 1963) examined hemoglobin's tryptic hydrolysates and showed that the difference between normal and abnormal electrophoretic types is determined by the mutational replacement of glutamic acid for valine in the 6th position of the globin $\beta$ chain (Fig. 2.5).

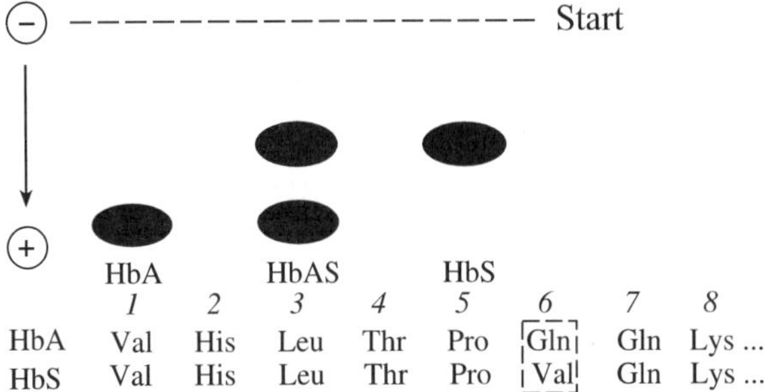


**Fig. 2.5.** Hemoglobin polymorphism in man. *Above* Schematic diagram of electrophoregrams of hemoglobins of a normal person (*A*), the heterozygous carrier of a sickle-cell gene (*AS*), an individual with the anomalous hemoglobin gene (*S*). *Below* Sections of the $\beta$-globin chain of two hemoglobin types differing in substitution of glutamic acid (the norm) for valine (pathology)

This research provided direct evidence of the effects of gene mutation on primary protein structure and laid the foundations for utilizing electrophoretic methods to reveal hereditary biochemical variability. The establishment of high-resolution analytical methods of protein electrophoresis in starch (Smithies 1955) and polyacrylamide gels (Raymond and Weintraub 1959; cited by Maurer 1971) was of fundamental importance.

In recent decades achievements in elucidating genetic theory and the mechanisms of action and interaction among genes have disclosed the main principles upon which the discovery and interpretation of hereditary protein polymorphism are based.

## 2.2.1
## The Mechanism of the Action of Genes

The synthesis of all the protein structures of which a typical organism is composed is directed by hereditary information incorporated, in a coded form, in the DNA of structural genes. All other genes controlling the rate of synthesis of protein products, their numbers, synthesis switching mechanisms, etc., may be regarded as genes with regulatory functions.

During the development of an individual, a sequence of nitrogenous bases in the DNA is transcribed by RNA polymerase in the cell nucleus into the complementary nucleotide sequence of messenger RNA (mRNA). In the transcription process, mRNA determines the linear sequence of amino acids in the polypeptide chains that form the primary structure of protein molecules. This process takes place in the cytoplasm into which mRNA, which has been synthesized in the nucleus, emerges. Here mRNA is linked with ribosomes in which several enzymes bring about assembly of proteins from the amino acids delivered by transport RNA (tRNA) to them. Each type of amino acid has highly specific tRNA molecules.

The nucleotide "text" of mRNA is "calculated" in sequence, three letters at a time beginning with a 5′-carbon end and ending with a 3′-carbon chain. Each triplet, or codon, is "translated" in sequence into one or another amino acid according to nature's laws that form the basis of the genetic code (Table 2.3). Figure 2.6 shows the general arrangement of protein synthesis.

The polypeptide chains synthesized at the ribosomes form the structure of protein molecules. They may either be monomers, consisting of single chains (subunits), or multimers, formed by two or more subunits, and their synthesis is controlled by one, two, or more allelic or non-allelic genes.

Proteins have a hierarchical structure represented by several levels of molecular organization (Fig. 2.7): the primary structure is a sequence of amino acids in a polypeptide chain, the secondary structure is a protein molecule's helical configuration; the tertiary structure is a three-

**Table 2.3.** Correlations among the 64 possible codons in mRNA and the 20 amino acids from which proteins are built (genetic code). (Ayala 1976 with supplements. Reproduced with permission from the author)

| First letter | Second letter | | | | | | | | Third letter |
|---|---|---|---|---|---|---|---|---|---|
| | Uracil | | Cytosine | | Adenine | | Guanine | | |
| Uracil | UUU | phenylalanine | UCU | serine | UAU | tyrosine | UGU | cysteine | Uracil |
| | UUC | | UCC | | UAC | | UGC | | Cytosine |
| | UUA | leucine | UCA | | UAA | terminating codons | UGA terminating codon | | Adenine |
| | UUG | | UCG | | UAG | | UGG tryptophan | | Guanine |
| Cytosine | CUU | leucine | CCU | proline | CAU | histidine | CGU | arginine | Uracil |
| | CUC | | CCC | | CAC | | CGC | | Cytosine |
| | CUA | | CCA | | CAA | glutamine | CGA | | Adenine |
| | CUG | | CCG | | CAG | | CGG | | Guanine |
| Adenine | AUU | isoleucine | ACU | threonine | AAU | asparagine | AGU | serine | Uracil |
| | AUC | | ACC | | AAC | | AGC | | Cytosine |
| | AUA | | ACA | | AAA | lysine | AGA | arginine | Adenine |
| | AUG methionine | | ACG | | AAG | | AGG | | Guanine |
| Guanine | GUU | valine | GCU | alanine | GAU | aspartic acid | GGU | glycine | Uracil |
| | GUC | | GCC | | GAC | | GGC | | Cytosine |
| | GUA | | GCA | | GAA | glutamic acid | GGA | | Adenine |
| | GUG | | GCG | | GAG | | GGG | | Guanine |

Codons in mRNA do not overlap. Amino acids are ordered in a polypeptide chain of colinear sequence of nitrogen bases in DNA of structural genes. Synthesis of a polypeptide chain ends where ribosomes encounter the terminating codons UAA, UGA, and UAG. They have no affinity with transport RNA and therefore do not participate in synthesis; they are also called "nonsense" codons. The code is "degenerate" as nearly all amino acids may be controlled by more than one codon. This redundancy is almost always linked with the third base of the encoding triplet. It has long been assumed that the genetic code is universal for all living organisms. However, in the late 1970s and early 1980s exceptions from this rule were found. For instance the UGA codon, which typically is responsible for termination of synthesis, encodes tryptophan in humans and yeast as well as in mycoplasm. More examples of this could be listed (see, e.g., Weaver and Hedrick 1997, p. 277; Klug and Cummings 2000, p. 359). Recently, principally new data have been obtained indicating the existence in eukaryotes of a specific protein factor, eRF1, associated with specific (noncanonical) code alterations in a wide range of organisms (see Lehman 2001 for details). In view of these recent findings, many authors tend to regard the genetic code as "quasi-universal"

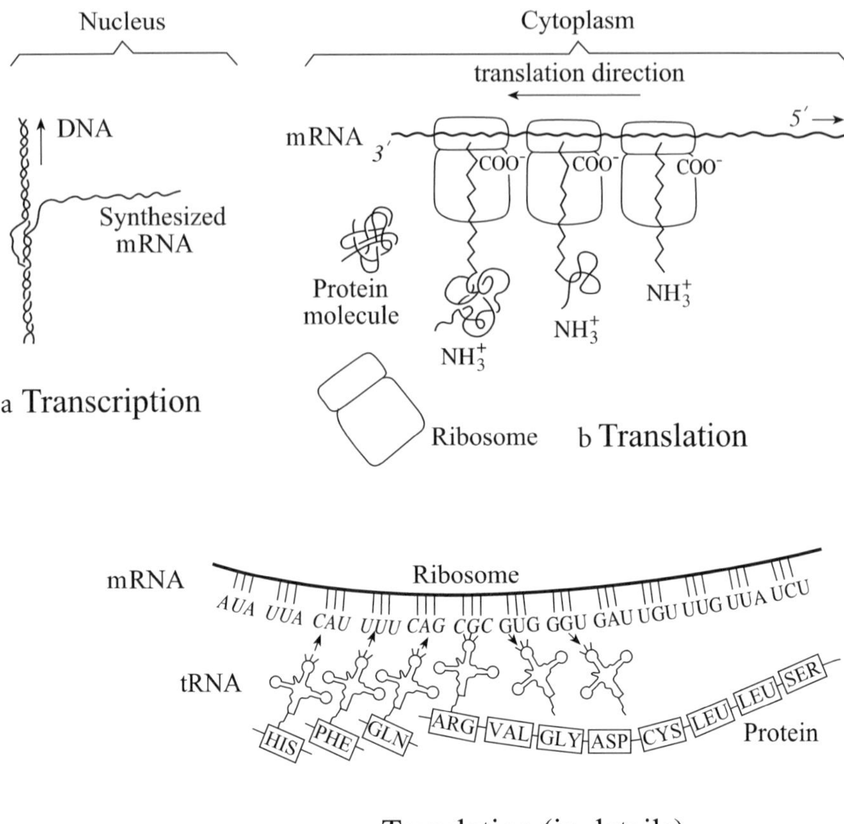



**Fig. 2.6.** Scheme of protein synthesis (Ayala 1976). **a** Transcription. One of the DNA strands represents a matrix for synthesizing a complementary copy of mRNA. **b** The translation of mRNA, synthesized in the nucleus, takes place in the cytoplasm attached to which are several ribosomes responsible for the synthesis of the polypeptides. Each codon in mRNA is distinguished by a complementary anticodon of tRNA molecules carrying particular amino acids; **c** details of the translation process. (Reprinted with permission from the author)

dimensional molecular configuration, and the quaternary structure is a multidimensional molecular organization.

The best-studied human hemoglobin (Hb A), consisting of two $\alpha$- and two $\beta$-polypeptide chains controlled by unlinked genes, is usually given as an example of a protein with a quaternary structure. The $\alpha$-chain is encoded by a genetic locus active throughout an individual's life, from the earliest ontogenetic stages, whereas the $\beta$-locus functions only in postnatal ontogenesis.

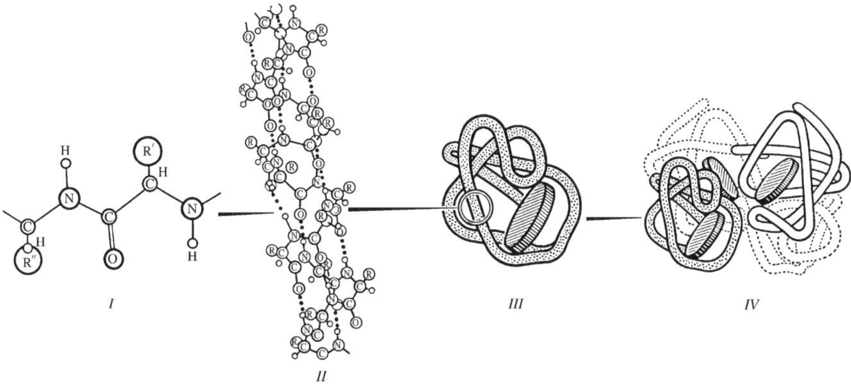

**Fig. 2.7.** Four levels of protein structure (*I–IV*) (Chapeville and Haenni 1977). Explanations in the text. (Reprinted with permission from the authors and Editions Hermann)

A further, $\gamma$-chain, controlled by another gene, is synthesized during a fetus's intrauterine life, and comprises approximately 70–80% of the total hemoglobin content in the blood of a newborn child. "Fetal" hemoglobin, $\alpha_2\gamma_2$-Hb F, is formed by association between the $\gamma$-chain and the $\alpha$-chain. Protein characteristics of adult blood are already found in fetal erythrocytes during the 13th week of development, and by the end of a child's first year of life Hb A has completely replaced Hb F.

Finally, a minor component, Hb $A_2$, exists in an adult human's blood; in this component the $\alpha$-chain combines with the $\delta$-chain, encoded by a locus that is active during postnatal ontogenesis. The $\beta$-$t$ and $\delta$-loci are closely linked.

Below we give a generalized arrangement of the genetic control of synthesis of normal human hemoglobins:

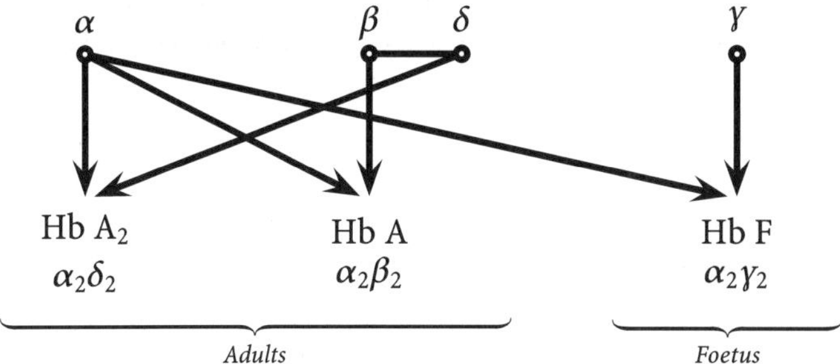


The scheme described was obtained through the development of several molecular–biological methods of analyzing protein structure. In this respect, the use of electrophoretic equipment has been especially valuable.

This arranges protein molecules in a neutral fine-pore gel according to differences in total electrostatic charge, form, or molecular weight. Since these characteristics of protein molecules are determined by their primary structure to a decisive extent, these differences discovered by electrophoresis are also confirmed by analyzing the amino acid order in polypeptide chains. This correlation for normal and abnormal hemoglobins was demonstrated in Fig. 2.5.

The study of protein genetics and biochemistry has shown that protein electrophoresis makes it possible to analyze a gene's fine structure; in this instance a protein acts as a marker for the corresponding gene locus.

One then asks, however, if differences in electrophoretic mobility of protein molecules go "hand in hand" with differences in their primary structure, or in other words, whether electrophoresis reveals all amino acid replacements in proteins. Until recently it was thought that only about 30% of the total number of amino acid replacements would be discovered electrophoretically since only 5 out of 20 amino acids influence a protein molecule's total charge, depending on the ratio between the number charged negatively (aspartic and glutamic acid), and charged positively (lysine, arginine, and histidine). In the last few years this estimate has increased to at least 45% (see Ramshaw et al. 1979; Neel et al. 1986).

The literature is full of examples of the successful detection of individual electrophoretic protein variants inherited in strict conformity with Mendelian principles. These differences among individuals are the result of point mutations. However, electrophoretic methods also enable one to record other, more considerable genome reorganizations. We will examine several examples of the kind later when we discuss the genetic bases of speciation.

## 2.2.2
## Types of Mutation and Their Effect on Protein Structure and Functions

We can distinguish the following types of genetic changes:

1. gene mutations, linked with replacement of single amino acids and altered electrophoretic protein mobility, and also determine the rate and intensity of its synthesis;

2. chromosome mutations – inversions, duplications, deletions;

3. genomic mutations – different instances of polyploidy, parthenogenesis, and changes in chromosome number.

**Gene Mutations.**    Gene (or point) mutations are divided into several types.

*Nonsense* mutations lead to a base replacement of a kind that makes a triplet meaningless (UGA, for example). Transport RNA cannot recognize such a codon, hence protein synthesis is terminated at this point. Since this may happen at any point in a protein chain, the catastrophic functional consequences that these mutational disturbances of protein structure are capable of inducing are in evidence. The frequency of these generally lethal mutations is believed to be small.

*Frame-shift mutations*, which are connected with the insertion or deletion of nucleotide pairs, also exert powerful changes on protein structure and are usually lethal. This is because there are no "commas" among codons and, hence, the mutated shift of the starting point of molecular synthesis can completely distort the meaning of the relevant information, leading to the appearance of a functionally defective polypeptide.

*Missense* mutations alter a codon's "meaning". These mutations form the basis of the genetic polymorphism of proteins (see, for instance, Hb S); their functional value, however, will vary according to whether conservative replacements occur that affect amino acids with analogous characteristics (hydrophobic or hydrophilic, for example), or whether contrasting replacements take place such as hydrophilic amino acid with hydrophobic, or vice versa. Obviously, the replacement of phenylalanine with tyrosine will not have such strong negative functional effects as does substitution of glutamic acid with valine. Even more serious disturbances may be caused by cysteine substitutions; this amino acid has an SH group, essential for the normal functioning of several enzymes, for maintaining tertiary protein structure, and for providing a full protein complement drawn from the formation of disulphide bridges (for further details see Ohno 1970a, Dubrova 1980 and Kimura 1983).

*Samesense* mutations do not lead to changes in the amino acid composition of a polypeptide chain, since they involve redundant (synonymous) bases for codons. For instance, alanine, valine, glycine, threonine, and other amino acids (see Table 2.3) have more than one RNA codon, and in all cases only the first two bases remain unchanged, while adenine, guanine, thymine, or cytosine can serve as a third base. Clearly not every mutation at a third base elicits an amino acid replacement. Since out of 20 amino acids, only tryptophan and methionine are specified by a single triplet codon and all the rest by two or more, it is obvious that there is a considerable redundancy of the code. It follows of necessity that very many mutation replacements of nucleotides have no effect at a phenotypic (protein) level and, therefore, from the point of view of classical genetics, such events should not be defined as mutations.

Nevertheless, molecular genetics still assumes that samesense mutations can play a definite evolutionary role, on the one hand as an intermediate step for missense mutations and, on the other, by changing the synthesis of polypeptide chains (Ohno 1970a).

In discussing types of gene mutations, one should also dwell on the category of so-called mutation-like events – cases of intracistron recombination in heterozygotes. When Ohno et al. (1969) was investigating polymorphism at the 6-PGD locus in the Japanese quail, he discovered several instances of intragenic recombination whose frequency was exceptionally high, having an order of magnitude of $10^{-2}$–$10^{-3}$ (Fig. 2.8). Moreover, it has been conclusively shown that at least in human, one of the *PGM1* alleles was generated by intragenic recombination (March et al. 1993).

These mutation-like events can only occur when the corresponding allele forms of a gene differ by at least two base-pairs. If the exchange of gene sections occurs in simple heterozygotes that have alleles differing in only one base-pair, the recombination has no genetic effect (Ohno 1970a):

| | |
|---|---|
| Wild type: | -Pro-Arg- |
| Mutant 1: | -His-Arg- |
| Mutant 2: | -Pro-Met- |
| Recombination between 1–2: | -His-Met- |

The frequency of intragenic recombination has proved equally great in other research situations. Thus, Wright and Atherton (1968) showed that recombination between the B′ and B″ alleles at a lactate dehydrogenase

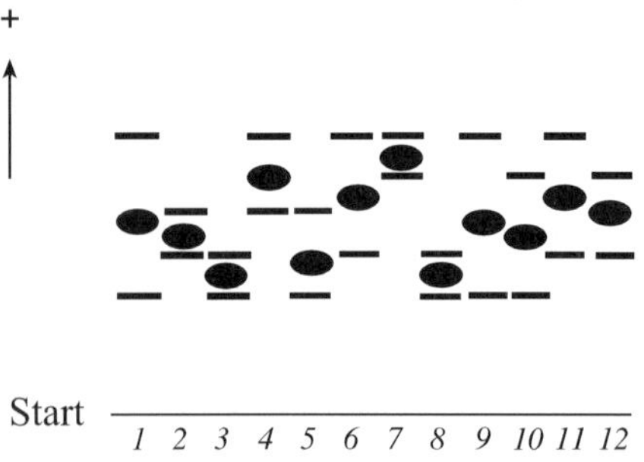

Fig. 2.8. Japanese quail 6-PGD genotypes identified by the electrophoresis of protein in starch gel for different crossing variants (Ohno 1973). This enzyme polymorphism is determined by four alleles (*A, B, C, D*) at an autosomal locus. As 6-PGD has a dimer structure, three types of molecules are formed in a heterozygote – two homodimers and one heterodimer (hybrid) occupying a middle position on the electrophoregram. In *1–7* are the genotypes of the mother (*BC, 2*), father (*AD, 1*) and their offspring (*3–7*). Four phenotypes (*AB, 3*; *CD, 4*; *AC, 5*; and *BD, 6*) correspond to what had been anticipated in conformity with the free segregation of alleles, whereas phenotype *7* is a recombinant. After receiving subunit *D* from its father, it inherited recombinant *D′* from its mother, instead of the expected *B* or *C*. In *8–12* are the results of crossing male *DD′*, heterozygous at a recombinant subunit, with female *AB* (*8*). In *9–12* are genotypes *AD* (*9*), *AD′* (*10*), *BD* (*11*) and *BD′* (*12*)

locus in trout with reversion to the wild-type B occurs at a frequency of 0.02 if female BB are crossed with male B′ B″. Values of the same order were also obtained in estimating the spontaneous recombination rate of antigens belonging to the B-system blood group in cattle (Stormont 1965), and in reciprocal skin transplants of heterozygous mouse strains (Bailey 1966).

In recent years, as a result of the excessive anthropogenic pressure and the destruction of historically formed ranges of species, significant intergenic recombination can be found in so-called hybrid populations. For instance, rare allozyme alleles occur in hybrid populations of spruce species *Picea abies* and *P. obovata* two to four times more frequently than in the original species, while for unique alleles this difference reaches ten times. In arboreal species (conifers) this trend, which is related to the post-glacial events, is well-known as "the rare allele phenomenon" (Krutovsky and Bergmann 1995), and the unusual allozyme forms were termed hybrizymes (Woodruff 1989). Hybrizymes were also described for some animal species (Barton and Hewitt 1985).

In view of the existence of ancient hybrid zones in some taxa, an old idea has been recently revived that hybridization can in some cases generate extensive variability important for adaptation processes as well as for speciation (Hewitt 2001; Rogues et al. 2001). As shown in Chaps. 6 and 7, these findings are directly related to the problems of genetics and evolution of populations (Watt 1972, Leslie and Watt 1986; Eanes and Koehn 1977).

Let us now consider in more detail phenomenology of biochemical hereditary variation determined by single amino acid substitutions.

We have already mentioned the first example of gene mutation found in a $\beta$-globin chain. Today in the human about 600 variants of so-called anomalous hemoglobins have been described that are distinguished from normal replacement only by single amino acid residues in certain sections of $\alpha$-chains as well as $\beta$-chains (Vogel and Motulsky 1997). Subsequently similar situations were also described for other proteins in many different species.

On the basis of these and other data, several conclusions can be drawn about protein synthesis gene control mechanisms:

1. One structural gene (cistron) encodes the synthesis of one polypeptide chain.

2. Polypeptide chains each encoded by a different locus produce, on combination, a protein that is indivisible in structural and functional respects.

3. Single allelic replacements in proteins, reflected by electrophoretic mobility, are easily revealed experimentally.

Such individual variants are inherited strictly in accordance with Mendelian principles, and in many instances inheritance is codominant – both alleles are detectable in the heterozygote. This variability frequently is seen to form part of genetic polymorphisms.

Much evidence has accumulated to show that a whole series of structural genes (multiple cistrons) of higher organisms are responsible for synthesizing proteins with the same enzymatic activity.

Sequencing most of the human genome has demonstrated that the number of proteins greatly exceeds the number of genes that encode them. Owing to alternative splicing, a single gene (rather than multiple gene loci) can generate hundreds and even thousands of protein varieties (Black 2000) referred to as isoforms.

By contrast, constitutive splicing results in including all (e.g., five) exons of a mosaic gene into the mature mRNA transcript. With alternative splicing, one of the five exons may be absent in the mature transcript, which leads to the formation of a truncated polypeptide chain. For instance, if the original peptide consists of 100 amino acid residues (each exon encodes 20 amino acids), after alternative splicing it will include only 80 amino acid residues because of losing an exon. Analysis of the primary structure of these two proteins shows that the 60 first and 20 last amino acid residues in them are identical (King and Stansfield 1997 p. 16). As to the majority of protein inserts produced by alternative splicing, their primary structure proved to be conserved to the same extent as that of the adjacent sequences. Hence, alternative splicing can "create" different protein isoforms only by insertion of new exons, which may be derived from noncoding intron sequences (Kondrashov and Koonin 2003).

The phenomenon of alternative splicing, its role in regulation of gene expression (Klug and Cummings 2000) and production of multiple protein isoforms by a single locus are currently at the focus of an ever growing number of studies and discussions. In particular, it has been suggested that this molecular mechanism, in the past decade found in many eukaryotes, plays an even more important role in generating multiple proteins than the common gene duplication (Black 2000; Ewing and Green 2000; Brett et al. 2002). For example, at least 50% of human genes are subject to alternative splicing. In humans, defective alternative splicing has been shown to be associated with a number of diseases, which testifies to the adaptive nature of this phenomenon.

Long before the discovery of alternative splicing, which may be in the future regarded as one of (or even the primary) mechanisms of adaptive protein synthesis in eukaryotes similar to adaptive (inducible) protein synthesis in bacteria, the phenomenon of multiple proteins or multigene families was well known in genetics, owing to the discovery of isozymes.

Genetically determined multiple molecular forms of one and the same enzyme, distinguishable by their primary structure, have been called isoenzymes or isozymes (Hunter and Markert 1957; Markert and Moller 1959). It has been suggested that the isoenzymes encoded by the same gene and manifesting intraspecific polymorphism should be called allozymes (Prakash et al. 1969).

Isoenzyme genes display differentiated activity in different tissues. When two or more loci of similar type function in the same tissue at the same stage of its development, and the protein that they encode has a subunitary (quaternary) structure, these systems frequently form "hybrid" molecules which are easily discovered electrophoretically.

For example, in many species of animals and plants malate dehydrogenase (dimer) and lactate dehydrogenase (tetramer) enzymes are encoded by a minimum of one pair of independent loci each, and electrophoretic distribution reveals correspondingly three and five zones of enzymatic activity on the basis of the free association of two subunits forming, in one instance, dimer, and in the other, tetramer molecules. Mutation at any of the loci, which influences the total molecular charge increases the multiplicity of the isoenzymes, but the genetic interpretation of this kind of codominant type of inheritance poses no problems (Fig. 2.9). The main complication is connected with the presence of so-called "null" alleles, when either synthesis does not occur or a product that lacks enzymatic activity is formed. At the same time it is clear that the effect of such a gene in homozygous form can be reliably determined if there is no zone of enzymatic activity on the electrophoregram. Such an allele may or may not be detectable in a heterozygous state depending on the degree of "dominance" of the other allele (the "dose effect") and whether the products of different alleles can unite as subunits.

We shall not immerse ourselves further in enzyme electrophoresis genetics because the results of a large number of original papers and surveys on this subject have been published (see, for instance, Korochkin et al. 1977; Whitt 1981; Ward et al. 1994). Our task is merely to examine the data that are important for understanding the contents of the following chapters. We have seen earlier how point mutations may affect protein structure. Now we must discuss the effects of large reorganizations of genetic material.

**Chromosomes and Genomic Mutations.**    Unlike gene mutations, which change a single base in a DNA molecule, chromosome mutations can lead to duplication (or deletion) of larger DNA parts and have large effects on protein structure. Unequal crossing-over is the main mechanism at the basis of this reorganization. Usually crossing-over presupposes a very precise correlation of conjugating pairs of homologous chromosomes with each other throughout their length. However, if during meiosis unequal crossing-over occurs in one or another area of DNA, one would expect "mistakes" in conjugation, leading to a changed sequence of nucleotides; duplication of genetic material in one part of a genome countered by deletion in another. Unequal

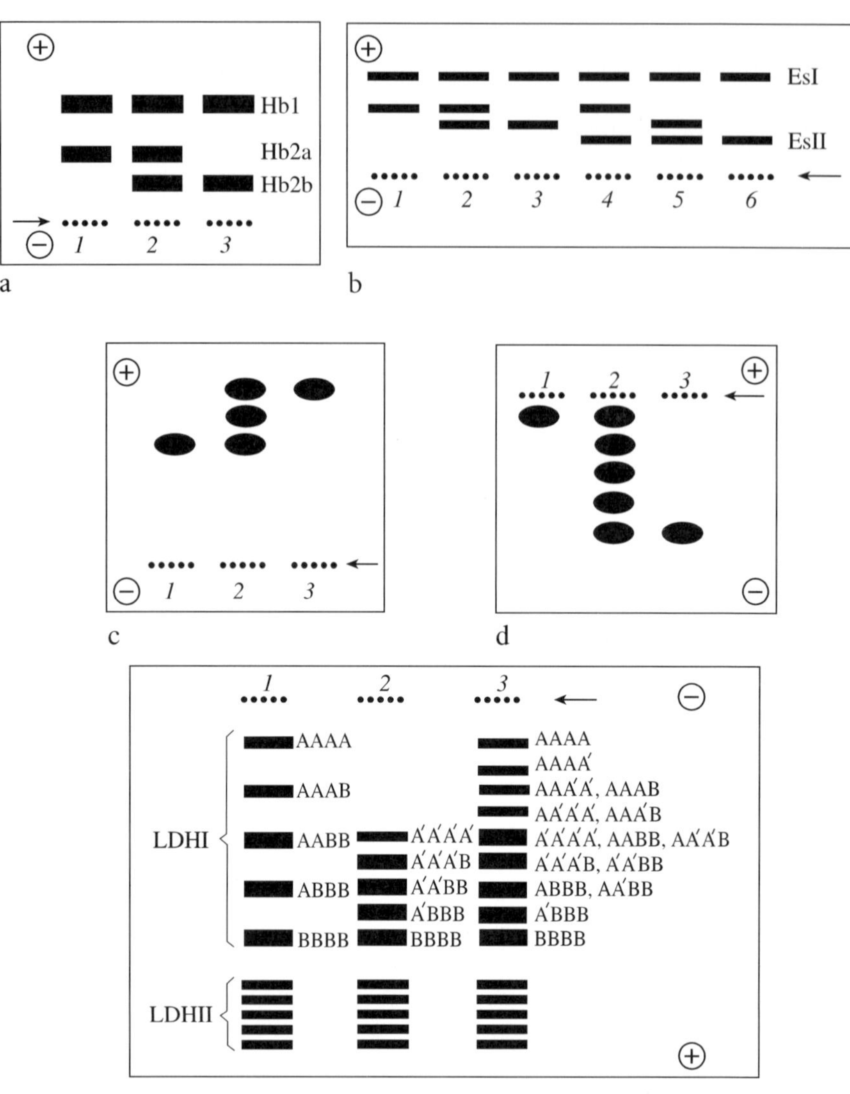



crossing-over, as is now well known, can involve sections within a gene, genes, and also blocks of them. Intragenic recombinations, which in their effect may be equated with point mutations, were examined earlier (Fig. 2.8).

The duplication of entire genes forms the basis of multiplicity of structural cistrons that encode the synthesis of the various proteins, including many isoenzyme systems. The most fully studied examples are human haptoglobins, hemoglobins and immunoglobulins, pancreatic serine proteinases (trypsin and chymotrypsin, and the isoenzyme of phosphoglucoisomerase in teleost fish), and others. In these cases the work on the

**Fig. 2.9.** Genetically controlled electrophoretic protein variants (codominant inheritance). **a** Hemoglobin of the mollusk *Anadara trapezia*. *1, 3* Homozygous genotypes; *2* heterozygote. Locus *Hb* 1 is monomorphic. Electrophoresis (e/ph) in cellulose acetate membranes (from O'Gower and Nicol 1968). **b** Esterase of *Mysis relicta* (Crustacea). *1, 3, 6* Homozygotes; *2, 4, 5* heterozygotes. Locus *Es* 1 is monomorphic. E/ph in a starch gel (Fürst and Nyman 1969). **c** 6-Phosphogluconate dehydrogenase 6-PGD of the quail *Coturnix coturnix*. *1, 3* Homozygotes; *2*: heterozygotes. The enzyme has a dimer structure, hence three bands of activity are detected in the heterozygote: that is, the "hybrid" molecules being formed, represented by the middle band, with e/ph in a starch gel (from Manwell and Backer 1970). **d** Malic enzyme of the house mouse *Mus musculus*. This enzyme, a tetramer, is composed of two subunits represented by one ("fast" or "slow") at the homozygotes (*1* and *3*). Both genes are active in heterozygotes (*2*), and free combination of four subunits from the two different subunits synthesized yields five isozymes; e/ph in a starch gel (from Shows and Ruddle 1968); **e** Lactate dehydrogenase isozymes in muscle tissue of *Oncorhynchus keta* salmon. Two isozyme groups are revealed, denoted by the curly brackets. In each group the tetrameric molecule is encoded by a pair of independent genetic loci, one of which is invariant (*LDH II*), while variation is found at locus *A* in the second group. Because with this method the mobility of several isozomes overlaps (see the lettering), nine instead of the 15 bands of activity theoretically predicted (Shaw and Barto 1963) are found in the heterozygote (*3*). E/ph in a polyacrylamide gel (from Altukhov et al. 1970, with supplements)

primary structure molecules studied has enabled us to understand the molecular bases of duplications quite clearly (Ohno 1970a; Harris 1977; Kimura 1983; Whitt 1987; Nei 1987).

By way of illustration, Fig. 2.10 reproduces Ingram's (1961) well-known diagram of the successive evolutionary emergence of four hemoglobin loci, based on a comparison of the levels of homology of polypeptide chains; the dots in the diagram are individual gene duplications. It is suggested that the more strongly a given subunit differs from others, the further the corresponding gene duplication is removed in time. This time factor may be determined knowing the average rate of the mutation process and the degree of difference in the primary structure of the polypeptide chain being compared. The following results were obtained by this calculation of the evolution of globins – hemoglobin and myoglobin ( a heme-bearing protein that ensures oxygen transport in the muscles) – originating from a common ancestor:

1. Myoglobin evidently appeared as a result of duplication that took place roughly 650 million years ago. Its molecule, a monomer, is practically indistinguishable from hemoglobin subunits both in size ($\sim$ 150 amino acids) and weight ($\sim$ 17,000). There is also a homology between the amino acid sequences of hemoglobin and myoglobin subunits.

2. Some 380 million years ago yet another duplication occurred which gave rise to the globin $\alpha$-chain. This chain differs considerably from all others

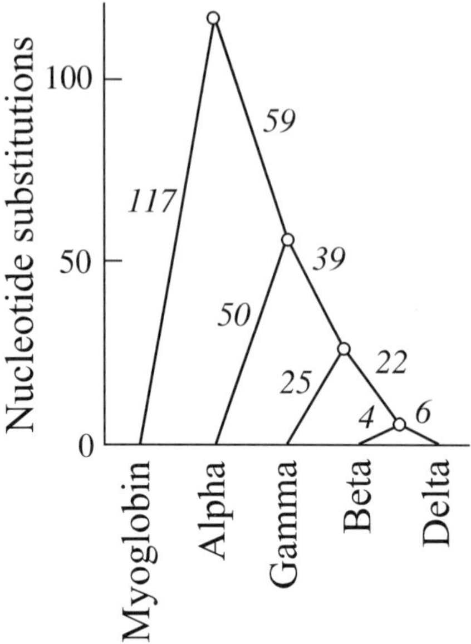


**Fig. 2.10.** The evolution of globin genes (from Fitch and Margoliash 1970). The *dots* indicate hypothetical duplications leading to new genes. The *numbers* denote the number of nucleotide replacements that differentiate five genes which now exist from each other and from a common ancestral gene

in its degree of homology and in the number of amino acids forming its structure – 141 of them; the latter difference may be connected with a small deletion.

3. Another variant of gamma hemoglobin found in human embryos (the so-called fetal hemoglobin $\alpha_2\gamma_2$) appeared later (about 200 million years ago).

4. The $\beta$- and $\delta$-chains, encoded by closely linked genes, arose relatively recently, as they only differ in 10 out of the 146 amino acids composing them.

Tandem (local) duplications, involving only an insignificant section of a genome and leading to linkage of duplicate loci, should be contrasted with polyploidy mutations affecting a genome as a whole. Simple genome duplications take place during autotetraploidy, resulting in tetrasomic inheritance of alleles at polymorphic loci. An example of this is diallelic polymorphism at an autosomic locus of serum albumin in the autotetraploid South American frog, *Odontophrinus americanus* (Beçak et al. 1968). When free assortment in meiosis of chromosomes with A and B alleles occurs, two types of gametes are formed, as in a diploid, with the following possible

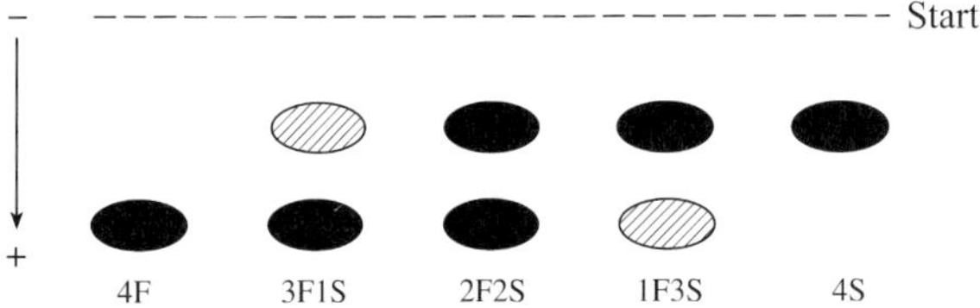


**Fig. 2.11.** Albumin polymorphism in the tetraploid frog, *Odontophrinus americanus*. Explanations in the text

combinations in zygotes: AAAA, AAAB, AABB, BBBA, and BBBB. With the additive action of genes in this instance one would expect the effects of the dose and the distribution of genotypes in a population to conform to the binomial coefficients 1:4:6:4:1, a modification of the Hardy–Weinberg distribution $[(p + q)^2]^2$. Precisely this picture has been found (Fig. 2.11).

However, although the autotetraploidy process does not necessarily lead to qualitative changes in protein structure, it sharply increases population heterozygosity through the appearance of two classes of "asymmetrical" genotypes – 3AIB and 3BIA. If, in a diploid population having equal frequencies of two alleles and segregating in accordance with the binomial coefficients 1:2:1, the homozygote fraction is 50%, then in an autotetraploid population it is only 12.5%. It is also clear that the more alleles there are at a locus, the less frequently are homozygotes encountered, while the population becomes increasingly heterozygous, and simultaneously the multiplicity of homologous proteins increases.

In recent years, polyploidy has been discovered in a large number of animal species – fish, amphibians, and reptiles. Today one can regard the important role that genome polyploidy plays in the evolution of the vertebrates as proven (Ohno 1970a,b; Bogart 1980; Fisher et al. 1980; Schultz 1980). A detailed study has been made of the biochemical genetics of the tetraploid salmon family, which has exceptional multiplicity in a whole series of proteins, both enzymic and non-enzymic (Ohno et al. 1968; Ohno 1970a,b; Altukhov et al. 1970, 1972; Altukhov 1974). The diagram (Fig. 2.12) shows how the genomic polyploidization process in these fish is reflected upon the gene control and the expression of lactate dehydrogenase isoenzymes. Four pairs of loci specific to different tissues have been found in salmon; subunits combine freely with each other to give a group of proteins that differ in electrophoretic mobility and comprise five isoenzymes.

Some authors (Allendorf and Thorgaard 1984) believe that during their differentiation from the common tetraploid ancestor, salmonids have lost up to 50% of duplicated genes, i.e., the products of these genes are impossible to discover. Other duplicated genes have diverged to an extent that their protein products, retaining their specific functions, acquired structural features allowing their separation by electrophoretic methods. Some du-

The most important features of the eukaryotic genome, distinguishing it from a prokaryotic genome, may be reduced to the following:

1. mechanisms for persistently switching genes on and off during the process of cellular differentiation;

2. organizational complexity of the genetic apparatus and the appearance of new regulation types by combining nuclear DNA with a far greater number of proteins than applies to prokaryotes, and especially with histones;

3. spatial separation of the transcription (nucleus) and translation (cytoplasm) processes;

4. frequent localization of functionally connected genes in different parts of a genome, so that possibly a regulatory gene is not linked at all with a structural gene controlled by it, as happens in bacteria, such a gene group perhaps being regulated by a single gene regulator;

5. an exceptionally high level of total or regional genome duplication.

A detailed analysis of modern views of regulating gene expression in eukaryotes may be found in the books of L.I. Korochkin (1999) and B. Lewin (1998).

New discoveries have been made in molecular biology about the "mosaic" or the "split" structure of genes in eukaryotes and about the phenomenon of alternative splicing. It appears that in the majority of eukaryote structural genes, coding regions (exons) alternate with non-coding regions (introns). Intron-derived RNA is spliced out by special enzymes after the formation of primary transcription molecules.

The origin of the intron–exon structure is under a serious debate. To date, two alternative theories exist: the exon gene theory and insertional intron theory. The main debated issue is the absence of introns in primordial genes. Population studies of the evolution of the intron–exon structure using relevant databases on nucleotide DNA sequences yielded novel interesting information. An example of this approach is provided by statistical analysis of the intron phase, i.e., its position within or among codons.

This analysis has shown considerable exon *shuffling* in DNA containing both old and new exon sequences. However, a stable association of intron position within a codon is regarded as a strong argument favoring of the exon gene theory. It is also assumed that exon reshuffling played an important role in the origin of both ancient and modern genes, and that introns were present in the protogene (protogenote; Long et al. 1995).

Several hypotheses on the intron functional significance have been advanced. One of these hypotheses relates this gene organization to the mechanism protecting eukaryotic cells against mutations (Altukhov 1982a). The facts that intron mutations arrest splicing or even alter protein structure (van Ommen et al. 1980; Spitz and Forget 1983; Kan 1986; and others) do not contradict this idea, since regions crucial for intron functioning have been also found within introns. The discovery of alternative splicing revealed another functionally significant role of introns as a source of multiple protein isoforms (Sect. 2.2.2). Based on this viewpoint, all other conditions being equal, genes with a higher number of introns are expected to produce a more stable (monomorphic) phenotype. However, this is only one of the possible approaches to analysis of mosaic genes. In what follows, according to the main line of this book, we examine the phenomenon of protein polymorphism in populations.

## 2.3
## Levels of Biochemical Polymorphism and Heterozygosity of Natural Populations

We mentioned above that traditional concepts of species uniformity with regard to prevalence of single wild-type alleles with only rare recessive mutations of some genes were considerably shaken in the 1950s. However, even in the 1960s, there was no unity among researchers in assessing the true extent of concealed genetic variability. This state of affairs was due to there being no appropriate systematic method that, in the opinion of Hubby and Lewontin (1966), would be needed to meet several criteria. Such a method should:

1. reveal the presence or absence in separate individuals of discrete phenotypic differences caused by allelic substitutions at corresponding gene loci;
2. enable one to easily distinguish allelic substitutions in a given gene as well as at different loci;
3. make it possible to treat the totality of the loci studied as an unbiased sample of genes of the genome with regard to the degree of their variation and the effect on fitness.

Having shown that not one of the then existing methods of genetic analysis satisfied all three criteria simultaneously, Hubby and Lewontin (1966) used the technique of protein electrophoretic separation as being most adequate to the task. Employing electrophoresis in a polyacrylamide gel, they examined eight enzymes and ten proteins of larval hemolymph in 43 strains isolated from five natural populations of *Drosophila pseudoobscura*, and they were the first to obtain estimates of the polymorphism and the

**Table 2.4.** Levels of polymorphism and heterozygosity in five populations of *Drosophila pseudoobscura*. (Lewontin and Hubby 1966)

| Population | Proportion of polymorphic loci | Heterozygosity per individual |
|---|---|---|
| Strawberry Canyon (California) | 0.33 | 0.148 |
| Cimarron (Colorado) | 0.28 | 0.099 |
| Wild Rose (California) | 0.28 | 0.105 |
| Mather (California) | 0.33 | 0.143 |
| Flagstaff (Arizona) | 0.28 | 0.081 |
| Average | 0.30 | 0.115 |

The locus is regarded as polymorphic if the heterozygotes' frequency is not less than 5%. Heterozygosity per individual is estimated by the formula $H = \sum h_i/n$, where $n$ is the number of loci studied, including monomorphic, and $h_i$ is heterozygosity at the $i$th locus: $h_i = 1-\left(p_1^2 + p_2^2 + ... + p_n^2\right)$, where $p_1^2...p_n^2$ are the expected frequencies of the homozygotes

heterozygosity of a reasonable sample of the totality of loci in the genome of a species (Table 2.4).

From the data in Table 2.4 it can be seen that on average one third of all the loci studied were polymorphic, while the average level of heterozygosity in an individual was roughly 12%.

Simultaneously, the paper by Harris (1966) appeared. He had studied 10 randomly chosen human blood enzymes, also electrophoretically; three of them were polymorphic, and the heterozygosity per individual averaged 0.099.

The importance of these investigations lies not so much in their similarity as in the very high degree of hereditary variability they disclose. Let us accept that there are of the order of 10,000 to 50,000 functional loci in *Drosophila* and in humans, and that the proteins studied represent a reasonable sample of a genome more or less adequately. Then at perhaps 3,000 to 15,000 loci there should be two or more allelic variants and the "average" individual is likely to be heterozygous at several thousand loci. The amount of variability increases still more if one recalls that some single amino acid replacements are not revealed electrophoretically.

Of course, it has been clear from the very beginning that sampling only some 10 – 20 protein markers is inadequate to characterize the genome as a whole. That is why, shortly after these preliminary publications, a whole series of similar research findings appeared with authors endeavoring to extend both gene sampling and the number of populations to embrace a wider spread of species and proteins. Lewontin et al. (1978) wrote a fairly complete, though not always critical review of how matters stood prior to 1974. His main conclusion was that his earlier estimates of polymorphism and heterozygosity levels remained virtually unchanged. Lewontin was at that time (1978) inclined to account for the fact that our results contradicted this conclusion – we had adduced data suggesting significantly lower levels of biochemical variability of populations (Altukhov et al. 1972) – by suggesting that it resulted from including several non-enzymatic proteins in the analysis. We shall return to this question later. It must be pointed out immediately, however, that a vast amount of information has accumulated in past years about the biochemical hereditary variability of natural populations. The most complete summaries have been compiled in Ayala's (1976) collective work, Molecular Evolution, and by Nevo (1978; Nevo et al. 1984). The latter review includes information about the levels of protein polymorphism and heterozygosity for 1,111 animal and plant species in which population samples were studied at more than 14 loci. Later, new and similar publications have appeared, and their numbers continue to grow (see Altukhov 1983; Ward et al. 1992).

Taking these data into account, one reaches the incontrovertible conclusion that it is wrong to ascribe low levels of protein polymorphism to non-representativeness of sampling or inadequacies of procedure. Of the total number of species described in the latest report by Nevo and his coworkers, in at least 150 the proportion of polymorphic loci and heterozygosity is legitimately lower than the estimates made in the first publications. Nonetheless, P and H values, averaged over all the data in the relevant table, are high, 0.28 and 0.07 respectively, with a correlation coefficient of $r = 0.79$ ($P < 0.001$).

The level of polymorphism increases still more when correction is made for electrophoretically "silent" alleles and for the allozymes revealed as having differential heat stability, kinetic characteristics, or optimum pH differences (see McIntyre and Wright 1966; Singh et al. 1974; Basset et al. 1978; Lewontin 1978b; Singh 1979; Satoh and Mohrenweiser 1979; Loukas et al. 1981; Ayala 1983; Graur 1986). It is not surprising, therefore, that the new type of analysis has revived old discussions about which of the population dynamics – random genetic drift or selection – is chiefly responsible for maintaining genetic diversity.

At present, two polar concepts stand out clearly: one of them tends to regard protein polymorphism as selectively neutral while the other regards it as adaptive variability, maintained by different forms of selection. Lewontin, whose pioneering works of 1963–66 greatly promoted the development of this research trend, holds an intermediate position, emphasizing the several difficulties in adopting one or other system of views. This attitude is most fully reflected in his well-known book, *The Genetic Basis of Evolutionary Change*. Perhaps nothing mirrors the difficulties confronting this trend so well as the epigraph to his book, expressed in words taken from Dante:

*Nel mezzo del cammin di nostra vita mi ritrovai per una selva oscura, che la diritta via era smarrita*[1].

It is no exaggeration to say that the question of the meaning and significance of biochemical polymorphism has today become central to the genetics of natural populations: more and more research workers are devoting their studies to it (see Kimura 1983; Nevo et al. 1984; Livshits and Kobylansky 1985; Nei 1987; Ward et al. 1992; Altukhov 1991, 1995; Altukhov and Salmenkova 1991; Altukhov et al. 1996, 2000a; Raybould et al. 1996; Hedrick 1999; Hey 1999; Ohta 2000; Schlötterer 2004).

What is the nature of the main difficulties? Let us examine them in order.

**1. Limitations to the Number of Overdominant Loci.**   If one considers that protein polymorphism is stable and is supported by the selective superiority of heterozygotes (overdominance), then (through the validity of the multiplicative fitness model) natural populations should carry an enormous genetic (segregational) load (see Chap. 1). The estimates discussed earlier present an important case in favor of Kimura's views, which recall Muller's view that the majority of mutations that have arisen recently are deleterious. Consequently, if such a high level of protein polymorphism exists, it should be selectively neutral (the "neoclassical school", Lewontin's term) and different alleles are disseminated within populations entirely as a result of random genetic drift.

Supporting his view, Kimura (1968a) adduces data on the mutation substitution rate in polypeptide chains, which remains very constant for analogous proteins in the most diversified phyla. From this perspective, the biochemical variability of a population represents only transitional polymorphism – *a temporary phase of molecular evolution*.

However, one cannot escape the obvious fact that, although the rates of amino acid replacements are relatively constant in various taxons within an isofunctional family of proteins, they may nevertheless vary considerably for different families. For example, whereas this rate for fibrinopeptide A is on average $4.29 \times 10^{-9}$ annual replacements per amino acid site, for histon H4 it is $10^{-11}$; that is, the rate is two orders of magnitude lower. The rate of accumulation of amino acid replacements in the molecule of cytochrome C is 20 times less than it is for fibrinopeptide A (King and Jukes 1969). The number of such examples could easily be increased (see review Fitch and Margoliash 1970).

However, the rate of amino acid replacements also varies for different sections of one and the same protein molecule. Thus amino acid replacements occur eight times less frequently in a group of amino acid bases

---

[1]Traversing this earthly life to its mid-point, I found myself in a murky wood, having lost the right path in the valley's darkness.

forming a heme "pocket" than in sections of a polypeptide chain that are less critical to function (King and Jukes 1969).

All these and many other similar facts are sometimes believed to contradict the theory of neutrality which, of course, is not the case. Kimura does not deny that a considerable proportion of mutations that have recently arisen are deleterious; latterly the proposition has arisen that 10–13% of all mutations occurring in a given generation (Kimura 1983) will be neutral.

The most important premise in the theory of neutrality asserts that the frequency of origin of new mutations, which give at least some selective advantage and hence, are supported by positive Darwinian selection, is so small that it cannot realistically be used to explain the levels of polymorphism and heterozygosity observed in natural populations.

**2. Excess Rare Alleles.**   While comparing two mathematical models postulating neutrality of polymorphisms (Kimura and Crow 1964; Ohta 1975) with empirical distributions of the allelic frequencies at protein loci in several species of Drosophila and in humans, Ohta (1975) discovered that in both species the theoretical distributions agree well with those observed over a wide frequency range, except for the 0–5% interval where a considerable excess of rare alleles is observed in natural populations (Fig. 2.13). This conclusion has been confirmed by Nei's group (Chakraborty et al. 1978) and by the work of Latter (1975).

Three possible explanations are offered for these differences. Firstly, the excess may be connected with a certain proportion of "slightly deleterious" alleles not accounted for by a neutral model; their frequency resulting from negative selection should not be great (Ohta 1975, 1976; Chakraborty et al. 1978).

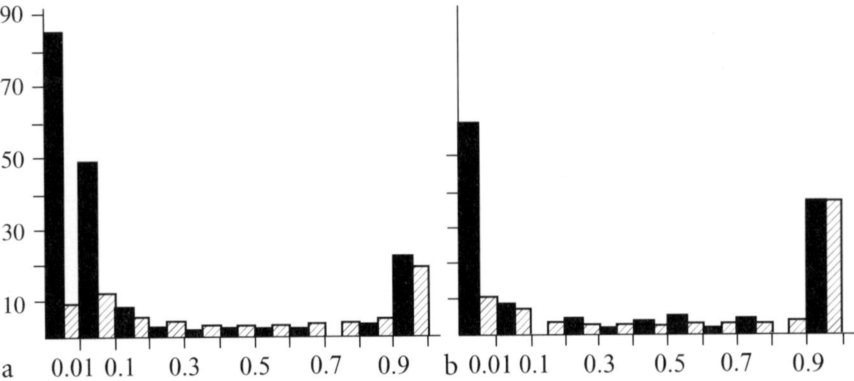


**Fig. 2.13.** Actual (*dark columns*) and expected (*light columns*) distributions of allelic frequencies at protein loci in *Drosophila willistoni* (**a**) and man (**b**; from Ohta 1976). *Abscissa* Classes of allelic frequencies; *ordinate* number of alleles

Secondly, it is possible that the species investigated have relatively recently experienced a sharp fall in numbers with a subsequent rise (the bottleneck effect). This could lead to a marked increase in population homozygosity followed by increased genetic variability in the growth phase of numbers (Chakraborty 1977); the majority of alleles newly arising in this case would be rare even given their selective neutrality (Ohta 1975; Nei 1975).

Thirdly, one must not exclude the possibility of intragenic recombination (Koehn and Eanes 1976; Strobeck and Morgan 1978), which may also be capable of causing an observed excess of rare alleles without allowing for the effects of selection at the loci investigated. These questions still remain open.

**3. Features of the Geographical Distribution of Gene Frequencies.**    The use of allozymes as genetic markers in research into natural populations has revealed three main types of spatial variability of allelic frequencies: clinal (Fig. 2.14), mosaic ("genetic kaleidoscope", Fig. 2.15), and showing considerable uniformity of patterns over areas that are sometimes huge and disconnected (Table 2.5). The latter finding indicates that it has not been possible to discover noticeable differences in the frequencies of the genes investigated even between extremely remote localities; moreover, for several loci this picture has been seen to be characteristic of the most varied species, including humans (Altukhov 1974; Korochkin et al. 1977; Lewontin 1978a; Harris and Hopkinson 1978; Altukhov et al. 1987a, 2000a).

The features revealed in the frequency distributions of allozyme genes by areas have been examined from the viewpoint of both "selectionists"

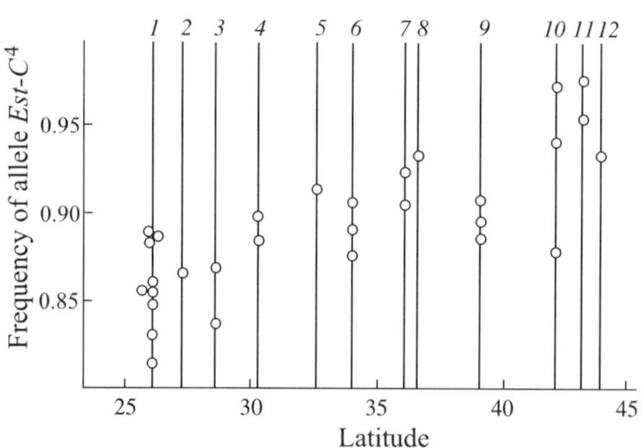

**Fig. 2.14.** Latitudinal clinal variability of allelic frequencies at the $Est\text{-}C^4$ locus in *D. melanogaster* populations of eastern USA (from Johnson and Schaffer 1973). *1* Lake Miami;. *2* Lake Placid; *3* Orlando; *4* Jacksonville; *5* Obern; *6* Columbia River; *7* Raleigh; *8* Knoxville; *9* Winchester; *10* Lake Erie; *11* Niagara Falls; *12* Portland

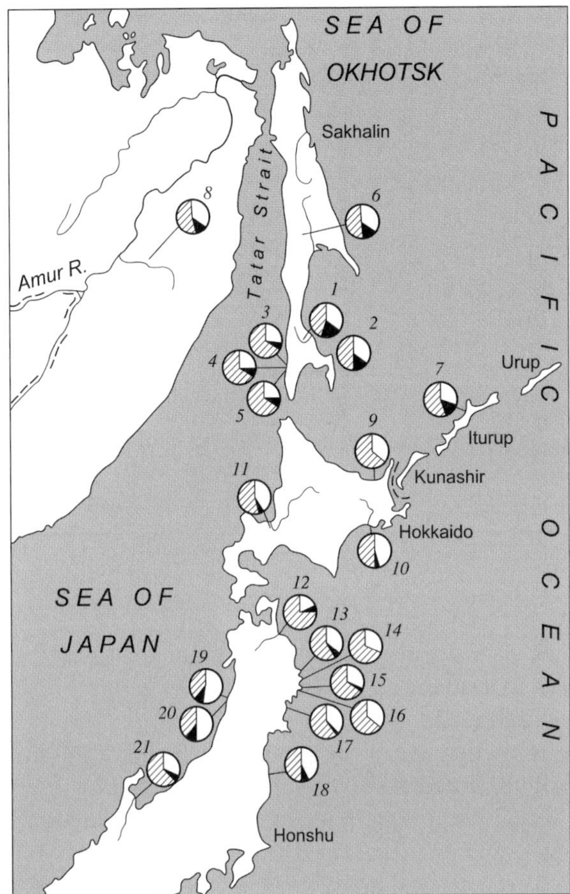



**Fig. 2.15.** Mosaic variability of allelic frequencies at a cytoplasmic isocitrate dehydrogenase locus in spawning populations (different rivers) of the chum salmon, *Oncorhynchus keta*. The *shaded and light sectors* of the *circles* are the frequencies of the alleles $Idh\text{-}A2^F$ and $Idh\text{-}A2^S$; the *dark sector* is the total frequency of other, rarer alleles. Rivers, total sampling volumes, and years of research: *1* Nayba, 992 (1976–1981); *2* Udarnitsa, 256 (1977–1979; 1981); *3* Kalininka, 1783 (1977, 1978, 1980, 1981); *4* Zavetinka, 385 (1980); *5* Yasnomorka, 294 (1980); *6* Buyuklinka (Poronay), 86 (1976); *7* Kurilka, 311 (1977–1980); *8* Anyuy 196 (1977, 1978); *9* Shari 191 (1976); *10* Kushiro, 89 (1976); *11* Chitoze, 96 (1976); *12* Oirase, 50 (1977); *13* Tsugaruishi, 90 (1977); *14* Origasa, 100 (1977); *15* Otsuchi, 81 (1977); *16* Katagisi, 91 (1977); *17* Okawa, 53 (1977); *18* Ukedo, 299 (1976, 1977); *19* Usivatari, 100 (1976 1977); *20* Takibuchi, 50 (1977); *21* Sho, 85 (1977). (Data of the Population Genetics Laboratory of the Institute of General Genetics of the Russian Academy of Sciences; and Kijima and Fujio 1979)

**Table 2.5.** Similarity of the allelic frequencies at different allozyme loci in three geographically separated populations of *Drosophila pseudoobscura*. (Prakash et al. 1969)

| Allele[a] | Strawberry Canyon (California) | Mesa Verde (Colorado) | Austin (Texas) |
|---|---|---|---|
| Xanthine dehydrogenase | | | |
| 0.90 | 0.053 | 0.016 | 0.018 |
| 0.92 | 0.074 | 0.073 | 0.036 |
| 0.99 | 0.263 | 0.300 | 0.232 |
| 1.00 | 0.600 | 0.580 | 0.661 |
| 1.02 | 0.010 | 0.032 | 0.053 |
| Malate dehydrogenase | | | |
| 1.00 | 0.97 | 0.95 | 0.97 |
| 1.20 | 0.03 | 0.05 | 0.03 |
| Leucinaminopeptidase | | | |
| 0.90 | 0.008 | 0.025 | 0.043 |
| 0.95 | 0.050 | 0.008 | 0.022 |
| 1.00 | 0.892 | 0.840 | 0.870 |
| 1.10 | 0.050 | 0.025 | 0.054 |

[a] The alleles are designated by their relative electrophoretic mobility

and "neutralists". Below we sum up in a generalized form the principal theses of the disputing sides in explaining the three types of geographical variability described above (Table 2.6).

The concept of the selective neutrality of protein polymorphism was appraised by tests in a specially planned research series conducted by Ayala and his colleagues (Ayala et al. 1971, 1974a,b; Ayala 1972, 1975, 1978; Ayala and Powell 1972; Ayala and Gilpin 1974; Ayala and Tracey 1974; and others). Attention was chiefly paid to the critical analysis of several consequences inevitably emerging from Kimura's theory, as based on strict mathematical calculations. This concept postulates: (1) the defined dependence among mutation rates, population size, and the effective number of selectively neutral alleles that are simultaneously present in a balanced population; (2) the permanency of mutation frequency for different gene loci; (3) differences of sets of alleles, both among populations known to be isolated and among different species as independent evolutionary units.

The belief in migration as a cause of the spatial similarity of allozyme frequencies in a whole series of cases is not corroborated by study of populations known to be discrete within the range of a species. Against a background of the uniformity of the allelic frequencies at allozyme loci, the same populations simultaneously and clearly differ in the frequencies of gene-arrangements within their gene-pools (Ayala et al. 1971, 1972; Sperlich and Pfriem 1986), or they reveal differences in the frequencies

**Table 2.6.** Rival explanations of geographical variability of allozyme frequencies

| Characteristics of a population's spatial genetic differentiation | Interpretation in the context of the neutral theory | Interpretation in the context of the balance hypothesis |
| --- | --- | --- |
| Clinal variability of allozyme frequencies | 1. The result of drift, reflecting the history of population structure formation in the process of the successive differentiation of the gene pool of an ancestral population (the formation of a U-shaped frequency distribution) 2. The result of intergradation of populations differing in allelic frequencies 3. The result of interaction of random genetic drift and migration when the latter's intensity decreases with distance | The result of selection along gradients of environmental factors in space and time |
| Mosaic pattern in the distribution of allozyme frequencies | Random genetic drift, interaction of drift and very weak migration. Contoured population structure of small genetically effective size | Selection in a heterogeneous environment in the area of a single panmictic population |
| Uniformity of allozyme frequencies | The migration of individuals ensuring a continuous gene flow and corresponding to panmixia in huge areas | Different forms of balancing selection whose direction and intensity are approximately identical in all populations |

of other genes that encode proteins (see, for instance, Altukhov 1974). A very similar picture is also seen in comparative electrophoretic protein analysis of species known to be different (Altukhov and Rychkov 1972; Altukhov 1974), including those whose evolutionary divergence is fairly large. In Fig. 2.16 we can see how the indices of interspecies similarity are distributed for a set of gene protein markers in a comparison among *Drosophila nebulosa* and three other *Drosophila* species of the same *willistoni* group – *D. tropicalis*, *D. willistoni* and *D. equinoxialis*. With regard to each other, these three species are siblings, but morphologically they are very different from *D. nebulosa*. In line with the neutral theory, the distribution of similarity indices should correspond to the normal with a mode in the frequency range of about 0.5; but in fact this distribution is U-shaped.

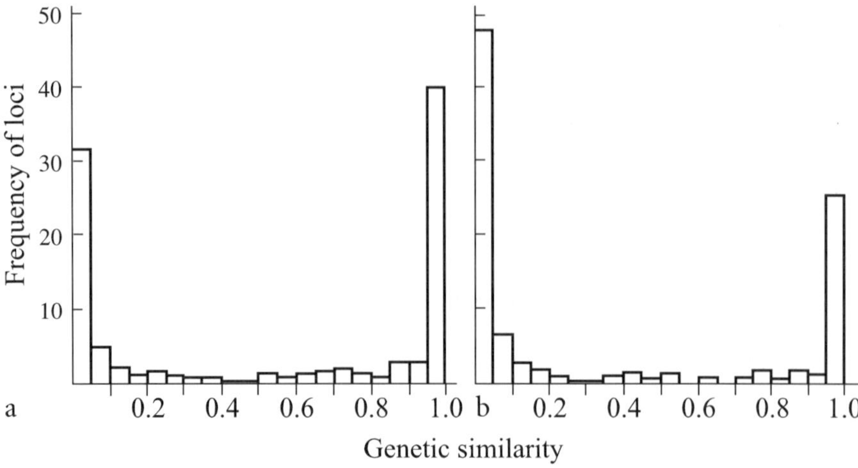

**Fig. 2.16.** Distribution of protein loci according to indices of genetic similarity in a paired comparison of four species of the *willistoni Drosophila* group (from Ayala 1974). **a** Sibling species, $I = 0.517 \pm 0.024$; **b** *D. nebulosa*, sibling species, $I = 0.352 \pm 0.023$. (Reproduced with permission from F. Ayala)

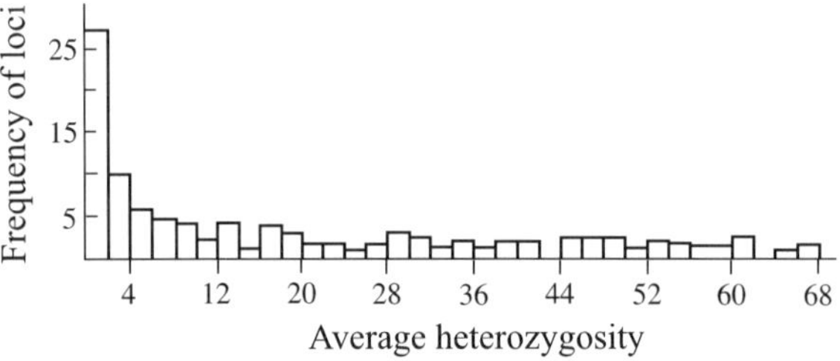

**Fig. 2.17.** Distribution of average heterozygosity at a locus in populations of five *Drosophila* species of the *willistoni* group (Ayala et al. 1974a). Average heterozygosity determined for the entire set of data, $H = 0.177 \pm 0.004$. (Reproduced with permission from F. Ayala)

The second thesis, postulating the permanency of mutation rates for the genes that encode proteins, has already been discussed in part. To this one must add that if this really were the case, the distribution of different loci according to their heterozygosity would also be close to normal, with a mean value equal to the average individual heterozygosity. In fact the picture is different: we see an asymmetrical distribution having a mode in the zero class and a long "tail" of frequencies close to each other at an interval of heterozygosities ranging from 0.02 to 0.68 (Fig. 2.17). The only possible way of explaining this distribution is if one accepts that either the mutation rates differ sharply for different genes (over two orders of magnitude), or that different loci are subject to selection in different ways.

Cavalli-Sforza (1966) first drew attention to the second possibility during analysis of the interlocus differences of the standardized variance of gene frequencies in human populations, and later the same method was used in other investigations (Rychkov 1969; Lewontin and Krakauer 1973; Rychkov and Sheremet'yeva 1976; see also Lewontin 1978a). This point will be examined again in more detail in Chap. 5, but we will now turn to evaluations of the effective number of alleles predicted by a neutral model for a population in equilibrium.

**4. The Number of Alleles Simultaneously Present in a Population.** According to Kimura and Crow (1964), the effective number of selectively neutral alleles (ne) at a locus may be found from the expression:

$$n_e = 4N_e\mu + 1\,,$$

where $N_e$ is the effective population size, and $\mu$ is the mutation rate.

For electrophoretically different alleles, whose number is always less than the total pool of alleles in a population, the preceding formula has the following form (Ohta and Kimura 1973):

$$n_e = \sqrt{8N_e\mu + 1}\,.$$

Since in research on *Drosophila* populations their effective size is unknown, in contrast to the precise values known for $\mu$ (mutation rate), verification of the applicability of this model to the natural situation is almost impossible. In this connection, the effective number of alleles expected in accordance with the theory of neutrality was estimated using the formula $n_e = 1/(1 - \overline{H})$; that is, in accordance with the fact that $n_e$ in a panmictic population corresponds to the reciprocal value of the proportion of homozygous individuals within it.

As $\overline{H} = 0.173$ (data for 36 loci) in the four species of the *Drosophila willistoni* group, $n_e = 1.209$.

For values of $N_e$ of $10^9 - 10^{10}$ (Ayala et al. 1972) and $\mu = 10^7$ (Kimura and Ohta 1971), $n_e = \sqrt{801} = 28.3$, which is in sharp contrast with the value of $n_e = 1.209$ actually found. However, for $N_e = 10^4 - 10^5$ and $\mu = 10^{-5} - 10^{-6} n_e$ lies within the interval of 1–1.34; this agrees well with the expected value of 1.209.

Clearly, an adequately reliable assessment of actual characteristics, such as genetically effective population size and the mutation rates at allozyme loci, is of decisive importance in approximations of this kind.

**5. Genetic Variability of Proteins and Their Functional Significance.**     Data on the connection between levels of genetic variability of protein loci and their functional significance also favors the idea of the adaptive value of biochemical polymorphism and speaks against neutrality (Gillespie and Kojima 1968; Altukhov 1969a,b, 1974; Altukhov and Rychkov 1972; Johnson 1973, 1974).

Gillespie and Kojima (1968; see also Gillespie and Langley 1974) discovered that the enzymes directly associated with metabolism energetics (group I) are not as variable as other water-soluble enzymes – they are less specialized and functionally less loaded (group II). As a rule, whereas enzymes of group I utilize substrates that are specific body metabolites, group II enzymes utilize extremely varied substrates emanating from the external environment (Kojima et al. 1970; Omenn et al. 1971; Cohen et al. 1973; Johnson 1974).

Lewontin (1978a) has summarized the relevant findings, and the picture obtained looks very convincing despite the relatively small samples of genes examined (Table 2.7). Later, Johnson (1973, 1974, 1976) developed similar concepts, hypothesizing that the least polymorphic enzymes are those that occupy key positions along metabolic paths and, hence, are more sensitive to mutational disturbance. He pooled the results available from several species, dividing all the enzymes studied into the same groups and subdividing the first of them into two further subgroups – those having a control function and those without. The result was the same (Table 2.8).

**Table 2.7.** Polymorphism and heterozygosity at loci coding for enzymes of energy metabolism (group I) and for other enzymes (group II) of *Drosophila*, mice, and man. (Lewontin 1978a)

| Species | Number of loci | Proportion of polymorphic loci | Hetero-zygosity | Reference |
|---|---|---|---|---|
| *Drosophila melanogaster* | | | | |
| Group I | 11 | 0.36 | 0.094 | Kojima et al. (1970) |
| Group II | 8 | 0.50 | 0.156 | |
| *D. simulans* | | | | |
| Group I | 11 | 0.36 | 0.030 | Kojima et al. (1970) |
| Group II | 7 | 1.00 | 0.364 | |
| *D. willistoni* | | | | |
| Group I | 10 | – | 0.112 | Ayala (1972) |
| Group II | 18 | – | 0.223 | |
| *Mus musculus* | | | | |
| Group I | 17 | 0.24 | 0.089 | Selander and Yang (1969) |
| Group II | 11 | 0.45 | 0.106 | |
| *Homo sapiens* | | | | |
| Group I | 24 | 0.21 | 0.048 | Harris and Hopkinson (1972) |
| Group II | 47 | 0.32 | 0.077 | |

**Table 2.8.** The connection between the metabolic function of enzymes and the levels of genetic polymorphism at the loci that code for them. (Johnson 1973, 1974)

| Class of biochemical reactions | Average heterozygosity | | |
| --- | --- | --- | --- |
| | *Drosophila* (14 species) | Small vertebrates (22 species) | Man |
| Varying substrates | 0.24 | 0.22 | 0.18 |
| Specific substrates | | | |
|   Regulator enzymes | 0.19 | 0.14 | 0.18 |
|   Non-regulator enzymes | 0.06 | 0.06 | 0.005 |
| All enzymes | 0.16 | 0.12 | 0.07 |

From our point of view, the differentiation of loci into groups depending on their functional significance is extremely important, as was stressed previously when we discussed the question of the biological significance of protein polymorphism in populations (Altukhov 1970, 1974; Altukhov and Rychkov 1972; Altukhov and Dubrova 1981).

However, when one takes into consideration the work discussed above and a series of other investigations closely allied to them, it must be admitted that the problem is still far from being resolved (see Kimura 1983; Nevo et al. 1984). Here is how Lewontin (1978a, p. 267) assesses the situation:

"How can such a rich theoretical structure as population genetics fail so completely to cope with the body of fact? Are we simply missing some critical revolutionary insight that in a flash will make it all come right, as the Principle of Relativity did for the contradictory evidence on the propagation of light? Or is the problem more pervading, more deeply built into the structure of our science? I believe it is the latter."

Although there are undoubtedly certain contradictions in the structure of theoretical population genetics, as in any other science, it is fair to postulate that the crux of the matter is not merely (and not so much) what Lewontin concludes, the inadequacy of the models utilized, but lies rather with their separate parameters as they attempt to represent the fundamental parameters of natural populations and those of the typical eukaryotic genome as a whole.

How justified are the positions held by the adherents of two such contradictory viewpoints about the population structure of a species and, above all, about its most important parameter, effective population size?

Have correct assessments been made of the rate of the mutation process for the genes that encode protein structure, as have been of the genetic effectiveness of migration links among populations?

Can the genetic content of a species be reduced to only one mode of variability – polymorphism – or do stable, monomorphic genes possessing unique species characteristics and maximal functional significance exist together with highly variable genes?

Do population geneticists work with natural populations that are historically well described, or are their evaluations predominantly based upon individual samples which, in very many cases, are in fact influenced by anthropogenic factors or which, because of their random nature, are only ephemeral stages in the evolutionary process?

For answers to these questions let us turn to the results of several projects that have been conducted over the last 30 years in the context of an approach to analyzing genetic processes at the level of populations and species. We believe that this approach combines the view of the "balanced" and "neoclassical" schools and leads to the conclusion that *the genetic differentiation of a species is not exhausted by the contribution of any one micro-evolutionary factor, whether it be selection, migration, or random genetic drift, but is conditioned by the complex binding together of their interactions.* The arguments for this kind of view have been presented in several previous publications (Altukhov et al. 1975b; Altukhov 1977, 1983, 1989a,b; Altukhov and Dubrova 1981). Here, we will substantiate the above conclusions in a more detailed way, and it will, therefore, be necessary to concentrate on the genetic processes in natural subdivided population systems and their experimental analogues.

## 2.4
## DNA Polymorphisms

In the mid-1980s, after the discovery of DNA polymorphism, a new class of genetic markers appeared due to the advent of techniques of gene isolation, cloning, and restriction. The discovery of polymerase chain reaction played the key role in the formation and development of the "new genetics". Since the majority of the genome does not take part in the known and important functions, the corresponding parts of noncoding DNA exhibit polymorphism levels that are far higher than known in any protein polymorphism.

Advances in the new field are so great that some authors propose to forsake the methodology of biochemical genetics for the sake of DNA polymorphism. The fallacy of such statements is evident even a priori, since protein and DNA polymorphisms naturally complement each other in relation to the issue examined (see, e.g., Beaumout and Nichols 1996; Hedrick 1999; Allendorf and Seeb 2000; De-Xing and Hewitt 2003; Schlötterer 2004). We discuss this problem in the following Chapter of this volume, setting forth how the insights from biochemical genetics may help in revealing natural selection that limits the range of some DNA polymorphisms. However, first we consider phenomenology of the main types of DNA polymorphism, briefly outline their application in different fields of biology, and pay attention to some techniques.

## 2.4.1
## Restriction Enzymes (Restriction Endonucleases)

Restriction enzymes were discovered by Verner Arber in 1979 (see Arber 1979) upon infecting various *Escherichia coli* strains by the $\lambda$ phage. Arber has shown that in the bacterium the phage DNA is cleaved (digested) and loses its infectiousness. Later, it was found that not only phage but any foreign DNA is neutralized in that manner upon entering a bacterial cell.

The digestion (restriction) is effected by specific enzymes, restriction endonucleases. Since these enzymes are indifferent to their own DNA, this phenomenon is regarded as a mechanism of cell protection. The question was why do restriction enzymes digest the foreign rather than their own DNA? It was found that restriction endonucleases react only with recognition sites, i.e., specific, demethylated DNA sequences consisting of four to six base pairs, which in homologous bacterial DNA are protected by methyl groups.

DNA digestion with restriction endonucleases results in the appearance of DNA fragments with the so-called "sticky" ends due to the fact that after cleavage of two-stranded DNA, one of the strands turns out to be longer than the other by several nucleotides. The sticky ends of one fragment can readily join with such ends of any other fragment, which underlies the technology of obtaining recombinant DNA molecules of different origin.

At present over several hundred restriction enzymes differing in recognition and digestion sites have been identified (Vogel and Motulsky 1997). Restriction endonucleases are used for various experimental purposes. In particular, they are employed in specific DNA amplification that requires determination of the primary structure (nucleotide sequence) of the molecule or in studying gene expression after their cloning, i.e., introduction into a bacterial cell and amplification. In this context, the use of restriction enzymes for detecting DNA polymorphism is of particular importance. It is known that DNA polymorphism is has several causes – point mutations (one-nucleotide substitutions), DNA replication errors (insertions and deletions of one to several hundred or thousand nucleotides in length), and other types of DNA reorganization among restriction sites. All changes in the primary DNA structure result in changes in the length of the fragments produced by restriction endonucleases. The corresponding method of analysis was termed *restriction-fragment-length polymorphism* (RFLP). DNA fragments here become the simplest of genetic markers.

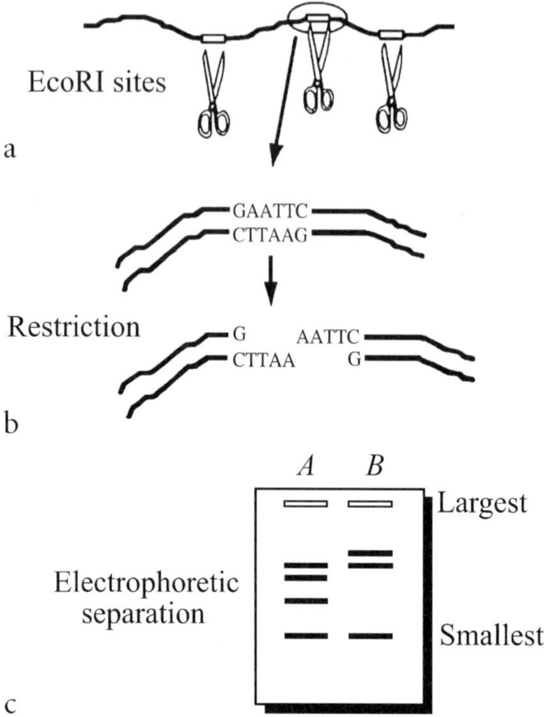



**Fig. 2.18.** Schemes of DNA digestion by a restriction enzyme and electrophoretic separation of resultant fragments. The depicted DNA molecule has three sites recognized by restriction endonuclease *Eco*R1 (**a**). Each DNA region having sequence GAATTC (shown by *open rectangles*) is cleaved by this endonuclease producing sticky ends (**b**). The resultant four fragments are separated in agarose gel ("profile" *A* in **c**). If the population contains an alternative allele lacking this restriction site (*encircled* in **a**), the two middle fragments of the *A* "profile" are absent; instead of them, a single larger fragment is present in the *B* "profile". (Park and Moran 1994) (Reprinted with permission from authors and Cluwer Academic Publishers)

A mixture of DNA fragments is separated by electrophoresis in agarose or polyacrylamide gel and visualized via radionuclide end-labeling with subsequent autoradiography or staining with ethidium bromide. In Fig. 2.18, the simplest scheme of DNA digestion and electrophoretic separation of resulting fragments is presented.

This method is ineffective for separating a mixture of numerous fragments. For selective identification of particular fragments in a gel, hybridization with a specific *probe*, usually according to Southern (*Southern blotting*), is used (Southern 1975), This probe is a cloned DNA sequence complementary to the specific sequence of the fragment and radioactively or fluorescently labeled. Due to the label, the fragment can be visualized in the subsequent procedures.

At first hybridization with restriction DNA fragments involved universal *multilocus probes*, which contained nucleotide sequences that very often occurred in the genome as repeat families (Jeffreys et al. 1988). To date, specific probes have been obtained for many organisms (e.g., introns of human myoglobin gene, phage M13 DNA, etc.). These probes simultaneously show many similar loci on the gel that are typically heterozygous due to their numerous alleles. These complicated patterns of DNA fragments, which are virtually completely specific for an individual, were termed *DNA fingerprints*. The patterns exhibit extensive variation both within and among species (Jeffreys et al. 1988; Wright 1993). DNA fingerprints are very convenient for determining relatedness or origin, but less helpful in studies of interpopulation differences because it is nearly impossible to assign the numerous allelic variants to a given locus and determine allele frequencies. Nevertheless, statistical procedures for estimating heterozygosities, genetic distances, and diversity parameters have been developed (Lynch 1990; Jin and Chakraborty 1994).

At present many specific *single locus probes* (SLPs) have been created (Taggart and Ferguson 1990; Bentzen et al. 1991; Heath et al. 1994; Prodohl et al. 1994). These considerably simplify analysis since variation detected in hybridization with SLP ("DNA profile") concerns a particular locus, corresponds to Mendelian variation, and permits estimation of allele frequencies, heterozygosity, and other population parameters (Goldstein et al. 1995; Slatkin 1995; Zhivotovsky and Feldman 1995).

Apart from restriction endonucleases, other enzymes used include polymerases (to synthesize new strands on the one-strand DNA template) and ligases (to link sticky ends of two DNA strands, e.g., in the cloning procedure).

## 2.4.2
## Polymerase Chain Reaction

Another very important approach widely used in studies of DNA polymorphism is the polymerase chain reaction (PCR) method (Mullis et al. 1986; Saiki et al. 1988). Using this technique, the required nucleotide sequences can be relatively easily and rapidly amplified from traces of plant and animal DNA, including that from fossil specimens.

In PCR, the DNA region to be amplified (more precisely, each of its two strands) is used as a template for in vitro synthesizing the complementary sequence. The reaction is catalyzed by thermostable *Taq* polymerase in the presence of two primers, i.e., synthetic oligonucleotides complementary to the sequences flanking the amplified fragment (one primer for each DNA strand), and involves several steps at different temperatures. The procedure is fully automated, and the reaction cycle is repeated many times (Fig. 2.19). The number of DNA molecules is doubled in each cycle, so that after 30 replication cycles the initial DNA amount is amplified more than a million times.

DNA amplification via PCR is more effective than *cloning*, but the latter method, unlike PCR, permits operating with larger (thousands of base pairs) DNA fragments (Datta et al. 1988). In cloning, ligase is used to insert a DNA fragment into a self-reproducing construction, or a vector. Vectors are DNA molecules of bacterial plasmids or phages capable of penetrating a microbial or yeast cell and replicating many times when it is in its reproduction cycle (Fig. 2.20).

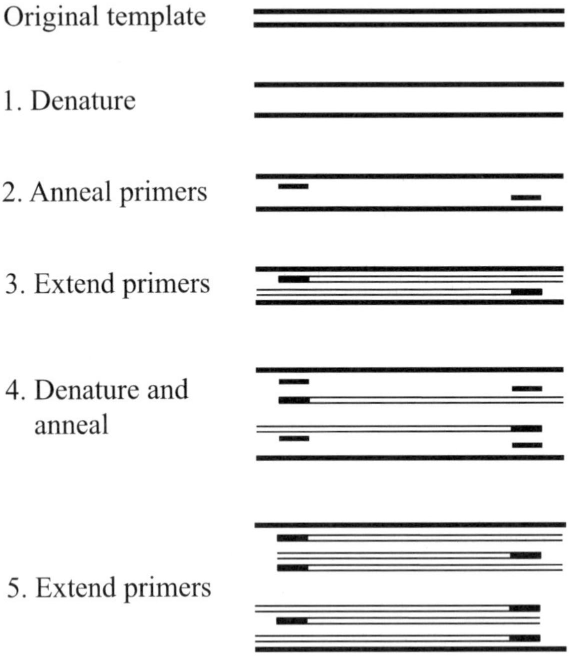

Original template

1. Denature

2. Anneal primers

3. Extend primers

4. Denature and
   anneal

5. Extend primers

**Fig. 2.19.** Scheme of polymerase chain reaction (PCR): *1* heat denaturation and cooling of the DNA molecule; *2* short synthetic primers anneal to each complementary chain; *3* primers are extended by thermostable polymerase, which results in duplication of the template; *4* the heating-cooling cycle is repeated 20–40 times; *5* in each cycle, the newly synthesized DNA strands act as templates for the next replication, thus producing more than a million-fold amplification of the original target sequence. (Park and Moran 1994) (Reprinted with permission from authors and Cluwer Academic Publishers)

## 2.4.3
## DNA Polymorphism Markers

DNA polymorphism was first described in 1978 (Kan and Dozy 1978) in relation to analysis of a DNA sequence tightly linked to the human $\beta$-globin gene, which permitted prenatal diagnostics of sickle-cell anemia. Later, it was shown that several hundred of such polymorphisms are spread throughout the genome, allowing researchers to localize genes on chromosomes if a sufficiently complete pedigree is available (see Vogel and Motulsky 1997 for details).

At present, polymorphism has been found in mitochondrial (mt) and nuclear DNA – more precisely, in the coding and noncoding parts of the latter, in its unique and repetitive sequences. The coding DNA part, which constitutes only 1% of the genome in mammals, is more conserved whereas the noncoding part is more variable as it is less constrained by selection (Nei 1987).

**Mini- and Microsatellites.**   A considerable part of *repeated (satellite) nuclear DNA* consists of *tandemly* repeated copies of the so-called core units of two to several thousand base pairs in length. Insertion/deletion mutations (indels), generated by slipping and mispairing of DNA strands at replication and by unequal crossing-over, alter the repeat number, i.e., the total length of the multicopy sequence (Levinson and Gutman 1987). This variation observed in different chromosomes and individuals was termed VNTR (*variable number of tandem repeats*).In such tandem repeat families, the attention of researchers is focused on *minisatellites* consisting of repeated copies (motifs) of 9 or 10 to 100 base pairs each (Jeffreys et al. 1985) and *microsatellites* whose copies are typically one to four, sometimes six nucleotides in length (Tautz 1989; Wright and Bentzen 1994); the latter are sometimes designated SSR (simple sequence repeat) or STR (short tandem repeat). A minisatellite locus can contain two to several hundred such repeats; microsatellite locus, ten to a hundred repeats. A hypothesis on the evolutionary origin of minisatellites from microsatellites has been advanced (Wright 1994). Individual alleles of these loci differ from one another by the number of tandemly repeated copies.

Minisatellite loci are examined by restriction fragment hybridization with a *multi-* or *single-locus probe*, which is a nucleotide sequence complementary to the repeated "motif" sequence (Armour et al. 1999).

Individual microsatellite loci are analyzed by PCR using primers complementary to the unique sequences (domains) flanking each microsatellite locus [38]. Then the "size" of its alleles is estimated in polyacrylamide gel electrophoresis by comparing it with a set of reference DNA fragments of known length (Fig. 2.21).

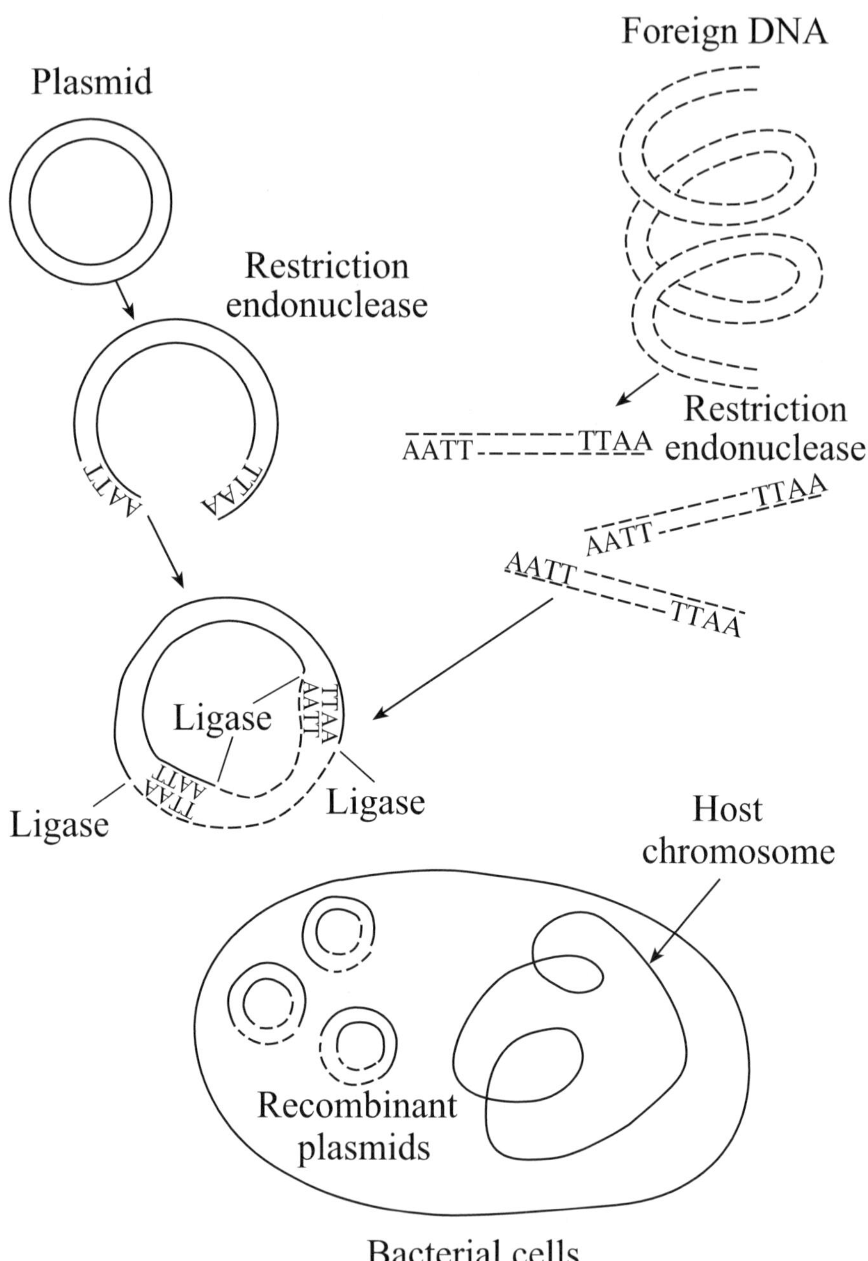

Plasmid
Foreign DNA
Restriction endonuclease
Restriction endonuclease
AATT
TTAA
AATT
TTAA
AATT
TTAA
Ligase
Ligase
Ligase
TTAA
AATT
AATT
TTAA
Host chromosome
Recombinant plasmids
Bacterial cells

**Fig. 2.20.** Scheme of cloning of a DNA fragment. A restriction fragment of a DNA molecule is joined by ligase to a preliminary cleaved plasmid. The plasmid is a nonchromosomal small circular DNA molecule of bacterial DNA or phage lambda DNA. The cleavage of the plasmid and foreign DNA is carried out by the same restriction endonuclease that provides complementation of the end sequences of the plasmid and DNA fragments and joining of their "sticky" ends. The recombinant construction enters the bacterial host where it replicates. (Vogel and Motulsky 1997) (Reprinted with permission from Springer Verlag)

The following properties make these loci very suitable genetic markers having a great potential (Wright 1993):

1. Both types of loci are very numerous and dispersed throughout the genome. For instance, rough estimates of the numbers of microsatellites, consisting of dinucleotide repeats $(GT)_n$ and $(CT)_n$ in the genome of brown trout *Salmo trutta* are 109,000 and 33,000, respectively (Estoup et al. 1993).

2. These loci are mainly located in noncoding genome regions and, consequently, must be selectively neutral. This general rule probably has exceptions when the loci in question are closely linked to adaptive genes. Moreover, although precise functions of mini- and microsatellites are unclear, some evidence testifies to the fact that they act as coding or regulatory elements (Kashi and Soller 1999); sometimes they have been found inside exons and associated with diseases (Hancock 1999).

3. These loci are characterized by rapid evolution. Spontaneous mutation rates of mini- and microsatellite loci are about $10^{-2}$ to $10^{-4}$ per locus per generation (Weber and Wong 1993), which is far higher than in allozyme genes (about $10^{-5}$–$10^{-6}$; Neel et al. 1986; Nei 1987). Hence, if the divergence of (selectively neutral) allozyme genes is caused only by genetic drift, that of mini- and microsatellite loci must be caused by both drift and mutation (Ward and Grewe 1994). Heterozygosity of minisatellite loci can be more than an order of magnitude higher than that of allozyme loci, reaching almost 100% (Jeffreys et al. 1988; Huang et al. 1992), whereas microsatellites exhibit different polymorphism levels generally higher than the allozyme ones.

4. Microsatellites and single-locus minisatellites display codominant Mendelian inheritance.

5. Microsatellites are identical in related species, which permits the use of the same primers and similar protocols. Note, however, that for creating primers in analysis of a new species, microsatellites must be isolated de novo. A review of these methods and their more rapid and simplified combination are presented in a study by Zane et al. (2002).

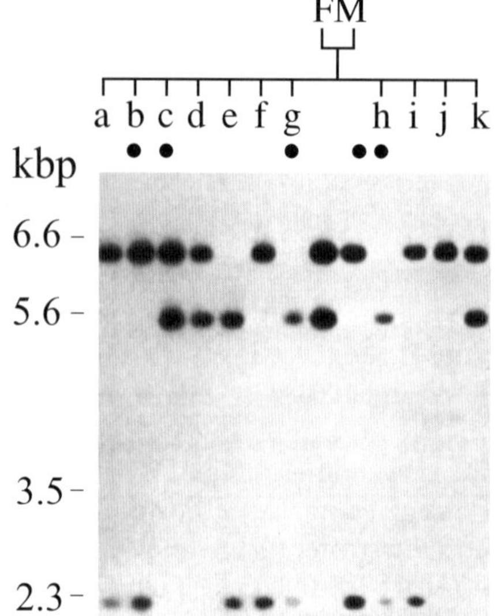

FM
a b c d e f g h i j k
kbp
6.6 -
5.6 -
3.5 -
2.3 -
a


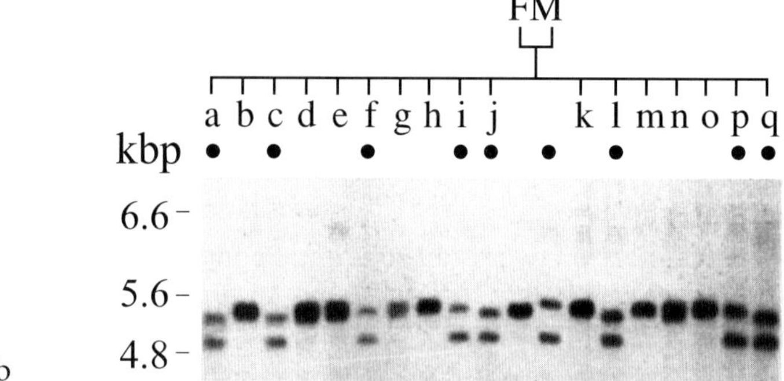

FM
a b c d e f g h i j k l m n o p q
kbp
6.6 -
5.6 -
4.8 -
b


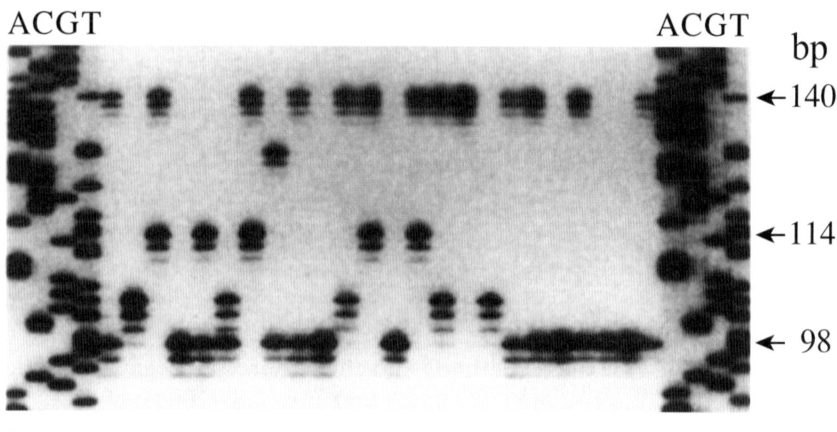

ACGT
ACGT
bp
← 140
← 114
← 98
c

Fig. 2.21. Inheritance of minisatellite DNA loci and genetic variation of microsatellite DNA loci in salmonid fishes (electrophoretic analysis). **a,b** Examples of minisatellite DNA profiles in brown trout detected with single-locus probes *pStr-A3* (**a**) and *pStr-A9* (**b**). In both cases, a brown trout family represented by a parental pair (M male; F female) and 11 (a–k) or 17 (a–q) progeny is shown. Male progeny are denoted by *black points*. DNA fragment size scale is given in kb (from Prodohl et al. 1994; reproduced with permission from Blackwell Science Ltd., Oxford). **c** PCR-amplified alleles of microsatellite locus *Omy77* in several rainbow trout individuals. The size of each allele was estimated by reference to comigrating M13 sequence fragments (A, C, G, T). (Morris et al. 1996; reproduced with permission from the National Research Council of Canada)

6. Microsatellite analysis requires only a minute amount of blood or tissue. Consequently, samples can be taken from a live animal (in fishes, for example, dry scales or otoliths can be used).

7. Automated microsatellite analysis is possible.

Microsatellite loci are currently extensively used as genetic markers (see Goldstein and Schlotterer 1999 for a review). The level of their polymorphism can be seen from the summarized evidence for fishes and other animals given in Table 2.9.

Compare these data with allozyme variation. According to Ward et al. (1994), mean population heterozygosity for allozyme genes in 49 freshwater fish species is $0.046 \pm 0.005$; in 7 species of anadromous fishes, $0.052 \pm 0.008$; in 57 species of marine fishes, $0.059 \pm 0.004$; i.e., about an order of magnitude lower than that for microsatellite loci. Although allozyme-based estimates usually also include monomorphic loci, this can hardly account for such a great difference. In a number of species the considerable difference in variation of allozyme and microsatellite loci may be associated with the fact that the restoration of genetic variability after a population or a species had passed a bottleneck occurs much faster in the case of microsatellite

**Table 2.9.** Genetic variation at minisatellite loci in fishes and other animals (from DeWoody and Avise 2000)

| Group of organisms | No. of loci | No. of species | No. of individuals | Mean number of alleles per locus and per species | Mean heterozygosity of species for loci |
|---|---|---|---|---|---|
| Freshwater fishes | 75 | 13 | 7,755 | 9.1 (6.1) | 0.54 (0.25) |
| Anadromous fishes | 43 | 7 | 5,393 | 10.8 (7.2) | 0.68 (0.12) |
| Marine fishes | 66 | 12 | 6,005 | 19.9 (6.6) | 0.77 (0.19) |
| Other animals | 340 | 46 | 20,567 | 7.7 (4.6) | 0.60 (0.16) |

*Note:* Standard errors are given in parentheses

loci because of their high mutation rate. Conversely, allozyme loci retain traces of such demographic events far longer on the historical scale. Because of this, some species are known to have very limited allozyme but high microsatellite variation (Hughes and Queller 1993). Note that the ratios among the variation estimates in the above fish species groups are similar for both marker types. In view of de Woody and Avise (2000), these ratios are likely to reflect a similar response of both types of markers to greater effective population sizes and larger gene exchange in the evolution of sea fishes which, as compared to freshwater fishes, inhabit more historically stable environment. This also may be caused by environmental conditions since sea water medium is selectively more neutral (homogeneous) than fresh water medium; in the latter, selection pressure must be stronger.

Indeed, although the localization of micro- and minisatellite loci in the noncoding genome regions suggests their selective neutrality, several examples demonstrate that some of these loci do not always act as neutral, which may be explained by possible tight linkage with adaptive genes. In particular, such situation has been described for *Semibalanus balanoides*, in which the variation at two out of six microsatellite loci examined correlated with the latitudinal variation of two allozyme loci, and the selective character of the latter was revealed in special experiments (Dufresne et al. 2002).

Although the mutation mechanism in microsatellite loci is not yet completely clear, it is supposed to correspond to the model of single-step mutation (also referred to as stepwise mutation; Levinson and Gutman 1987), which, as noted above, results in changing the tandem repeat number in the locus. Based on this model, new estimates of genetic distances and population-genetic structure have been established from microsatellite data (Goldstein et al. 1995; Shriver et al. 1995; Slatkin 1995; for comparison of different measures, see Ruzzante 1998). For instance, to estimate interpopulation divergence for microsatellites, Slatkin (1995) introduced parameter $R_{st}$, which is analogous to the $\Theta$ parameter of Weir and Cocherham (1984). $R_{st}$ differs from $\Theta$ in that the former parameter accounts for the differences in *the size* of alleles rather than identity or nonidentity of allelic states as in the infinite allele model.

**The Use of Mini- and Microsatellite Loci.**   The high mutation rate, great allele diversity, and high heterozygosity (in some cases for minisatellites attaining nearly 100%) opens incomparable prospects for individual classification, particularly in forensic medicine (DNA fingerprinting; Jeffreys et al. 1985, 1988; Huang et al. 1992), for studies of induced mutagenesis (Dubrova et al. 1996, 1997, 1998), and for various studies in the fields of demography, ecology, and conservation biology (Ryskov 1999).

Microsatellite markers often reveal genetic differentiation in the cases when allozyme markers fail to detect it, e.g., in organisms having low variation at allozyme loci (Fontaine et al. 1997), in mobile marine fishes (Shaw et al. 1999; Wirth and Bernatchez 2001), at the micro-geographic scale, in closely related populations or ecological groups within a species (Bernatchez et al. 1998; Brunner et al. 1998; Primmer et al. 1999; Banks et al. 2000). Using these markers, the origin of introduced populations can be successfully determined (Burger et al. 2000; Ruzzante et al. 2001). Statistical techniques and approaches have been developed from microsatellite variation for estimating the effects of passing a population bottleneck, and for characterizing migration (i.e., asymmetric gene flow, sex-based dispersal) and rates thereof, as well as for identifying parentage and kinship among individuals (see Estoup and Angers 1998; Luikart and England 1999; Ryskov 1999 for review) and assigning individuals to particular populations (Paetkau et al. 1995; Banks and Eichert 2000; Hansen et al. 2001).

Based on high allelic diversity of microsatellite loci, the progeny of concrete parents can be identified not only in the first but also in the subsequent generations. This opens new possibilities for examining reproductive success and fitness in individuals differing in biological and ecological characteristics. A particularly important point is that such studies are feasible not only in experimental but also in natural populations (Ferguson 1995). Using microsatellite markers to determine the degree of kinship among experimental progeny, Mousseau et al. (1998) proposed a novel method for estimating heritability of quantitative traits. This technique allows circumvention of prolonged controlled artificial reproduction in such long-lived species as, for example, salmon, by obtaining and analyzing the offspring of spawners taken directly from the wild. Heritabilities estimated by these authors for several traits in Chinook salmon from a natural population were in good agreement with the heritability estimates obtained for salmon using classical methods of quantitative genetics.

At the same time, we should like to note that microsatellite loci cannot be used in evolutionary reconstructions (phylogenies) at the interspecific and higher levels because the observed similarity in allele sizes may reflect the so-called homoplasy rather than common descent. Homoplasy is caused by a high mutation rate, due to which microsatellite alleles of the same size are generated by convergence from different number of direct or reverse mutational events. Another problem of microsatellite analysis consists in the presence of null alleles that appear by virtue of mutations at the primer binding site. This precludes accurate genotype identification; in addition, null alleles can differ in origin (Karp and Edwards 1977).

Hedrick (1999) has conclusively shown that extremely high variation of some microsatellite and similar DNA loci warrants caution in interpretation of the relevant results. This concerns the differentiation (which is underestimated) and genetic distances (which are overestimated) among groups since these estimates are greatly affected by heterozygosities of the highly variable loci. Furthermore, statistically significant intergroup differences at such loci, sometimes detected by powerful statistical tests, may be biologically meaningless (Hedrick 1999).

**Random Amplified Polymorphic DNA (RAPD) and Amplified Fragment Length Polymorphism (AFLP).**   Unlike microsatellites or single-locus minisatellites used for examining individual genome loci, markers designated *random amplified polymorphic DNA* (RAPD) and *amplified fragment length polymorphism* (AFLP), similar to multilocus minisatellites, permit investigating the total genome by obtaining the appropriate fingerprints.

For RAPD analysis, short (usually 10 to 20 nucleotides) primers with random DNA sequences are used. Anonymous DNA sequences are amplified in PCR and subsequently analyzed by electrophoresis. The number and size of the amplified fragments depend on the length and sequence of the arbitrary primer. The primer binding sites are randomly distributed throughout the genome, and polymorphism in such sites is expressed as the presence or the absence of the corresponding fragments in the gel (Williams et al. 1993).

AFLP analysis is based on selective amplification of fragments obtained by restriction of the genomic DNA. This method includes three stages: (1) DNA digestion (typically by two restriction enzymes) and binding sticky fragment ends with oligonucleotides adapters by a ligase; (2) selective PCR amplification of the set of restriction fragments; and (3) electrophoretic analysis of the amplified fragments. The nucleotide sequence of the adapter and an adjoining restriction site serves as a target for the annealing of the primer; the primer sequence complementary to the target is elongated at the 3' end by several arbitrary nucleotides. This permits selectively amplifying only the fragments whose restriction sites are flanked by the corresponding complementary nucleotides. Using AFLP or RAPD, particular sets of fragments (fingerprints) can be produced by PCR without previously knowing their nucleotide sequence. The fingerprint polymorphism is determined by the polymorphism of restriction sites and flanking nucleotides, and is manifested as the presence or absence of particular bands in the gel. The method has high resolution and, in contrast to RAPD, good repeatability (Mueller and Wolfenbarger 1999).

Both RAPD and AFLP markers exhibit Mendelian inheritance generally of a dominant type. However, if the progeny data are available, some codominant markers can also be revealed. These markers constitute 2–3% of all RAPD markers in plants (Krutovsky et al. 1998) and 4–15% of all AFLP markers in various organisms (Mueller and Wolfenbarger 1999). In their new modifications, RAPD and AFLP are combined with PCR of microsatellites; the corresponding modifications are referred to as *randomly amplified microsatellite polymorphism*, or RAMP (Schierwater et al. 1997) and microsatellite AFLP, or *selective amplification of microsatellite polymorphic loci*, SAMPL (Karp and Edwards 1997; Paglia and Morgante 1998). These combinations permit work with a great number of codominant microsatellite markers.

RAPD and AFLP markers are supposed to be located mostly in the noncoding DNA regions since these regions constitute an overwhelming part of the eukaryotic genome. Mutation rates in noncoding DNA are about two times higher than in its coding part (Nei 1987). In addition, RAPD (and possibly AFLP) markers are sometimes amplified from the repetitive DNA regions (Williams et al. 1993; Aagaard et al. 1998) and can thus reflect high rates of their mutation.

**Application of RAPD and AFLP Markers.**   Since by RAPD and AFLP methods one can obtain numerous (up to several hundred) markers dispersed throughout the genome, these markers are particularly suitable for constructing genetic maps (Krutovsky et al. 1998) or linkage maps with quantitative trait loci (QTL), for instance, with loci for commercially important traits (see review in Mueller and Wolfenbarger 1999).

This review also considers the use of AFLP markers for studies on systematics and biodiversity, population and conservation genetics, as well as for individual identification and kinship analysis. (The above discussion to some extent also refers to RAPD markers although these are not always reproducible and tend to produce artifacts). AFLP markers proved to be very helpful for detecting hidden variation in lines and closely related species that could not be discriminated on the basis of morphology or using other molecular methods. Using these markers to analyze phylogenetic relationships for higher taxa is thought to be problematic because of the very high variation of these markers.

Nevertheless, a study of RAPD fingerprints of several conifer species has revealed, along with RAPD markers exhibiting intraspecific variation, invariant markers that lacked individual and geographic variability but differentiated species within the genus (Altukhov and Abramova 2000). It was suggested to distinguish this DNA (termed random amplified monomorphic DNA, or RAMD) from polymorphic DNA and regard the former as the manifestation of genetic monomorphism, which had been discovered earlier for protein markers (Altukhov 1969b), at the DNA level. These species-specific RAMD markers may be common among other organisms and helpful in solving taxonomic issues.

**Mitochondrial DNA.**   Mitochondrial DNA (mtDNA) of vertebrates can be formally classified as repeated DNA, since a cell can contain hundreds of mitochondria, and each mitochondrium can have from two to ten copies of DNA molecules. DNA of human and animal mitochondria is a closed circular molecule (Fig. 2.22) of the size typically not exceeding 20,000 bp. In plants, cells of one plant sometimes have mtDNAs of different sizes (from several hundred to several thousand base pairs), which are represented by both circular and linear molecules; in very large mitochondrial genomes, the greatest part of redundant DNA consists of noncoding sequences (Singer and Berg 1991). In what follows, we restrict our discussion to animal mtDNA.

Complete sequencing of mtDNA in vertebrates revealed 37 genes (2 ribosomal genes, 22 genes for transport RNAs, and 13 protein-coding genes) and a noncoding control region that participates in replication, and is referred to as the D-loop (Ferris and Berg 1987). The control region consists of the central conserved sequence, usually flanked by polymorphic domains

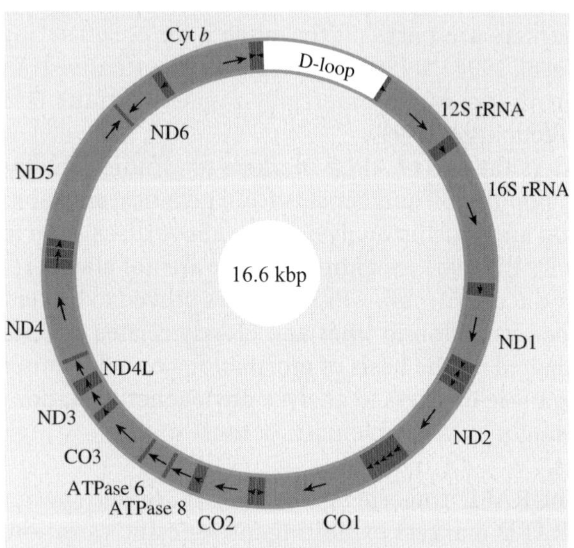


**Fig. 2.22.** Scheme of a circular molecule of human mitochondrial DNA. Genes shown in *dark grey* code for the following proteins: three subunits of cytochrome c oxidase (*CO1*, *CO2*, *CO3*), cytochrome b (*Cyt b*), subunits 6 and 8 of the ATPase complex, and subunits of the NADH-dehydrogenase complex (*ND1–ND6*). Genes coding for different ribosomal RNAs (*rRNAs*) are shown in *light grey*; *narrow segments* indicate transport RNA (*tRNA*) genes; they flank regions of protein-coding genes. Two RNAs are transcribed clockwise from the heavy (H) chain as a template; the transcripts are shown by *wavy lines* within the *circle*. The short transcript produces rRNA, and the long one, mRNA and most of tRNAs. Only one full-size transcript is synthesized on the light (L) chain. The regions shown with *dots* – ORI (H) and ORI (L) – are origins of DNA replication. The 5′ to 3′ direction on the sense chain of different genes is indicated by *arrows*. (Lewin 1998) (Reproduced with permission from Oxford University Press)

containing tandem repeats of four to several hundred base pairs. Tandem repeats of mtDNA are frequently highly polymorphic, varying in number not only among individuals but, in some organisms, even within an individual causing *heteroplasmy* (Bentzen et al. 1998), i.e., the presence of different mtDNAs in one individual. MtDNA is transmitted to the progeny with the cytoplasm, i.e., it is haploid and strictly maternally inherited. This DNA mostly lacks recombination (rare cases of heteroplasmy in higher organisms are analogous to ordinary somatic mosaicism, although currently the possibility of recombination is under discussion; see Awadalla et al. (1999) and displays clonal inheritance. Hence, although mtDNA contains more than 30 different genes, in terms of formal genetics it is regarded as one locus; mtDNA haplotypes are considered alleles of this locus or separate clones. Correspondingly, the effective population size for mtDNA is equal to 1/4 of the analogous estimate for nuclear genes (Nei and Tajima 1981).

Because of this, mtDNA variation is more subjected to random gene drift and the bottleneck effect upon a drastic decrease in population size. Moreover, the rate of nucleotide substitutions in the mitochondrial genome is at least 5 to 10 times greater than in the nuclear genome (Moritz et al. 1987). Consequently, the divergence of mtDNA, which is caused by more intense mutation and drift, must be higher than in nuclear allozyme genes. The mutation rate is maximum in the D-loop, mostly in its hypervariable segment 1 (HVI; see van Hooft et al. 2002). In addition, the high polymorphism of the tandem repeat number in the control mtDNA region observed in some species implies the mutation rate of $10^{-2}$ (Wilkinson and Chapman 1991), although this is not recorded in all species (Park et al. 1993; Churikov et al. 2001). The mutation rate in RNA genes is minimal. The average nucleotide substitution rate in mtDNA of different organisms is 1–2% per million years (Brown et al. 1979; Ferris et al. 1983).

**Application of Mitochondrial DNA.**  Due to the features of inheritance and variation described above, mtDNA is widely and successfully used in various studies of evolution and phylogeny (Cann et al. 1987; Kocher et al. 1989; Harrison 1989; Horai et al. 1995; Ingman et al. 2000), in analysis of population structure and historical biogeography (phylogeography) of the species (Avise 1994, 2000; Bernatchez and Wilson 1998), and in analyses of hybridization, of introgression of mitochondrial genome, of consequences of introduction, and of acclimatization (Billington and Hebert 1991; Fontaine et al. 1997).

Note also that the traces of former isolation of populations are preserved in mtDNA longer than in nuclear DNA. Furthermore, if males of a given species during reproduction migrate more often than females, the population divergence may not manifest in allozyme variation, but will be expressed in maternally inherited mtDNA. With equal migration of both sexes, a fourfold greater migration exchange is required in mtDNA as compared to nuclear genes in order to prevent divergence (Billington and Hebert 1991).

An important property of mtDNA, in contrast to allozyme and VNTR alleles, lies in the ability of its haplotypes to be linked together by a series of consecutive evolutionary transformations. This infers a minimum number of stepwise mutational changes required for transformation of haplotypes via the appearance/loss of restriction sites or any other alterations of the primary sequence (Fig. 2.23). Having the data on all used restriction endonucleases or on the nucleotide sequence, a phylogeny (genealogy) of the complex haplotypes can be constructed by specially designed computer programs (McElroy et al. 1991; Felsenstein 1993). The molecular distances among mtDNA haplotypes can be included in the estimates of genetic differentiation (Excoffier et al. 1992). Based on the haplotype's phylogeny, additional information on genetic relatedness

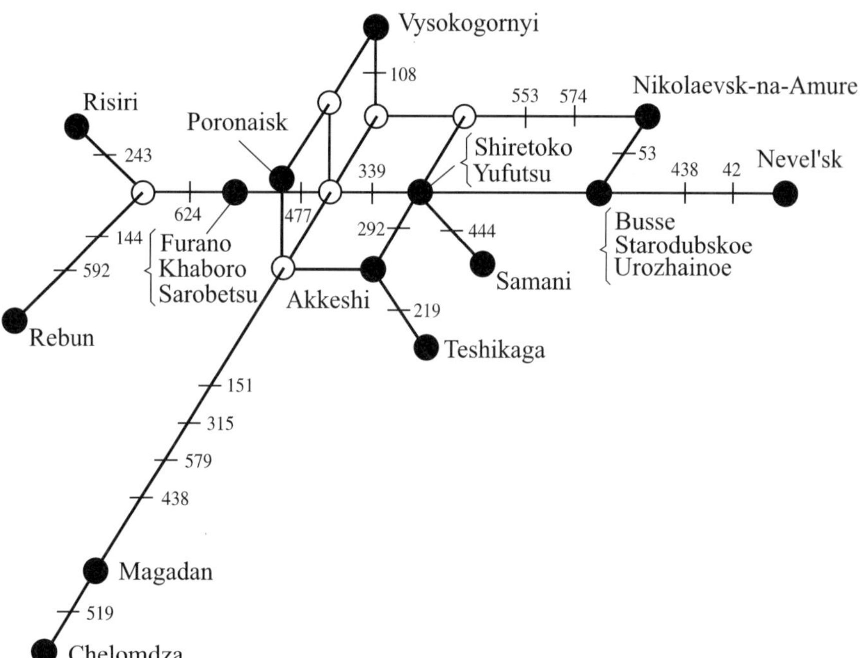


**Fig. 2.23.** Phylogenetic relationships among 14 haplotypes identified by nucleotide sequences (630 nucleotides) of the mitochondrial cytochrome b gene in 19 individuals of *Sorex gracillimus*. *Open circles* indicate undetected putative haplotypes; *shaded circles* indicate haplotypes found in the marked localities. The *numerals* by the *short lines* on the branches show nucleotide substitution positions. (Ohdachi et al. 2001; reproduced with permission from Blackwell Science Ltd., Oxford)

and historical relationships among species and populations can be obtained.

**Y-Chromosomal DNA.**  The use of mtDNA polymorphism in evolutionary studies is well supplemented by analysis of Y-chromosomal DNA polymorphism, since it provides information both on the maternal and paternal contributions to the evolutionary process (Cavalli-Sforza 1998). Due to haploidy, the Y chromosome mostly lacks meiotic recombination, being transferred from father to son as an integral entity. Hence, each particular set of loci making up the nonrecombining part of the Y chromosome (NRY) is regarded as one haplotype (Santos and Tyler-Smith 1996). If the numbers of males and females are equal, the total number of Y chromosomes in a population is equal to one-fourth of the copy number of any autosome, and, consequently, the effective population size inferred from the Y chromosome must be equal to one-fourth of the analogous value for the autosomes. In addition, the effective population size in this case is reduced due to high variability of the progeny number in males (Santos and Tyler-Smith 1996; Jobling and Tyler-Smith 2000). The relatively lower level of Y-chromosomal DNA polymorphism is related to the lower effective size and positive selection on NRY (Underhill et al. 2000). This polymorphism is represented by (1) unique point mutations ($10^{-9}$ per base pair per year) or, for example, an insertion of the Alu sequence (one of the types of dispersed repeats in the genome; Hammer 1994), and (2) polymorphism of micro- and minisatellite sequences of Y-chromosomal DNA (with the mutation rate of $10^{-2}$–$10^{-3}$ per locus per generation (see Santos and Tyler-Smith 1996).

Due to the absence of the "shattering" effect of recombination in the main segment of the Y chromosome and to the fourfold smaller effective size, the historic changes in the corresponding DNA molecule are easier to trace than in autosomal DNA (Jobling and Tyler-Smith 2000). For this reason, Y-chromosomal DNA is a unique, paternally inherited genetic system, which is particularly successfully used in such fields as the origin and evolution of humans, migration, and genetic contacts of populations (Underhill et al. 1996; Cavalli-Sforza 1998). For instance, in the native human populations, males generally more often stay at the place of their birth than females (so-called patrilocality), which is expressed in more pronounced geographical and probably social clustering of Y-chromosomal variants (see Jobling and Tyler-Smith 2000). Patrilocality explains a type of differentiation frequent in human populations: high diversity in mtDNA and low diversity in Y-chromosomal DNA within groups, but great differences in Y-chromosomal DNA and small differences in mtDNA among groups (Seielstad et al. 1998). Interestingly, exactly these and directly opposite pictures of genetic differentiation were recorded in two different groups of tribes (villages) in northern Thailand differing in that in one group, men, and in the other, women, were traditionally more settled (Oota et al. 2001).

The Y chromosome determines male sex via the effect of one gene, SRY (sex-determining region Y; Sinclair et al. 1990). In addition, numerous genes participating in the key cell processes have been detected in this chromosome (Lahn and Page 1997). Consequently, selection must play an important role in the formation of haplotypic diversity of the Y chromosome in populations. The discovery of numerous polymorphic DNA markers on the Y chromosome has initiated extensive investigation of the possible role of various haplotypes in diseases, particularly their relation to sterility and oncology (Jobling and Tyler-Smith 2000).

**Single Nucleotide Polymorphism (SNP).**    SNP is a polymorphism of a single nucleotide site. As a rule, it is represented by two allelic variants (substitutions) of a single nucleotide in a DNA sequence. At present, due to the improvement and automation of sequencing procedure, the development of DNA microarrays (Gibson 2002), and other analytical methods, these markers are extensively studied in the human genome (Wang et al. 1998) for detecting their association with different complex diseases (Lander 1996; Cargill et al. 1999; Halushka et al. 1999), and for understanding various aspects of genetic differentiation of populations and evolution of humans (Przeworski et al. 2000). According to Cargill el al. (1999), the SNP number per gene in humans ranges from zero to 29, while the coding gene sequences on average contain four polymorphic sites (cSNPs). A typical individual must be heterozygous at about 24,000–40,000 nonsynonymous (i.e., altering an amino acid in an encoded protein) substitutions. According to Halushka el al. (1999), the total human genome contains approximately one million SNPs, of which about 500,000 are noncoding, 200,000 synonymous coding, and 200,000, nonsynonymous coding. Based on SNPs in 75 studied human genes, recalculation of mean heterozygosity for proteins produced the estimate of 17% (Harris and Hopkinson 1972), which exceeds the value summarized by Nevo et al. (1984) from several sources (12.5%).

Below, we briefly discuss other marker types that are of more limited utility.

**Expressed Sequence Tag (EST) and Sequence Tag Sites (STS).**    EST is polymorphism of the expressed, coding genomic sequences, also referred to as ESTR. These are usually fragments or total sequences of complementary DNA, which are obtained with reverse transcriptase from mRNA isolated from various tissues and representing the genes expressed in these tissues (Krutovsky and Neale 2002). Using the EST sequences, PCR primers are generated to amplify ESTs from the individual genomic DNA; polymorphism of this DNA is then examined by means of some methods of amplification product analysis. These markers are very often employed in genetic mapping. In particular, they were used to identify several genes in the human genome (Wolfsberg and Landsman 1997) and to determine many candidate genes for complex diseases (Chakravarti 1999). In plants and animals, these markers are used as candidate genes for analysis of QTL controlling adaptive or productive traits (Krutovsky and Neale 2002). ESTs constitute an important component of the DNA microarray method, which has been currently intensely developed. This method opens unprecedented perspectives in various areas of genomics and in investigation of molecular bases of evolution and ecology (see Gibson 2002; Krutovsky and Neale 2002 for review).

STS markers are simpler and dispersed throughout the genome. They are represented by short, PCR-amplified genomic sequences. These markers are widely used for physical mapping of the human genome and analysis of human polymorphism (Wang et al. 1998).

**Short And Long Interspersed Elements (SINEs and LINEs).**   SINEs and LINEs are repeated, unblocked, and dispersed throughout the genome sequences (Singer 1982). They also can serve as genetic markers. These sequences represent retroposons, i.e., included in the genome transcripts of intracellular RNA. They constitute more than 20% of the genome of humans and other mammals (Rogers 1985). SINEs turned out to be very useful markers for phylogenetic analysis because species exhibit variation in the genomic localization of SINE inserts (Okada 1991).

Some features of the main polymorphic DNA markers are presented in Table 2.10.


## 2.4.4
## Selective Constraints of DNA Variation

As in the case of biochemical population genetics, whose development was based on hereditary protein polymorphism, molecular population genetics is at present ready to tackle the problem of the role of selection in maintenance of polymorphism of various DNA segments in the eukaryotic genome. Indeed, as shown by comparing mtDNA sequences of different species, the amount of variation in mitochondrial genes is selectively constrained by the amino-acid composition and function of their protein products (Kocher et al. 1989; Saccone et al. 1991), and by the gene position relative to the replication origin on the heavy strand (Bielawski and Gold 1996). An analysis of the variation distribution in nucleotide sequences using various neutrality tests has shown deviations from neutrality for mtDNA of genes *ND5* and *cytb* in *Drosophila*, *ND3* in mouse, *ND3* and *CO II* in human and chimpanzee (see Wise et al. 1998). The detection of mutations, duplications, deletions, and inversions in mtDNA

**Table 2.10.** Types of DNA polymorphism (Vogel and Motulsky 1997 with modifications and additions)

| Variation type | Cause of polymorphism | Detection methods | Fields of application |
|---|---|---|---|
| Restriction fragment length polymorphism (RFLP) | Nucleotide differences in restriction sites | Cleavage of double-stranded DNA by restriction enzymes; electrophoresis; visualization of fragments by Southern blotting with a DNA probe or ethidium bromide staining. Used for analysis of genomic DNA, mtDNA, or their individual segments | Population genetics, systematics, phylogeny, genetic mapping, QTL |
| Minisatellites (varying number of tandem repeats, VNTR) | Varying number of tandemly repeated nucleotide DNA sequences with repeat size of 10–100 nucleotides | DNA cleavage by restriction enzymes; Southern blotting with a specific DNA probe complementary to the repeated sequence. Multilocus probes are complementary to repeats commonly occurring in the genome and single-locus probes to rare and unique repeats | Population genetics (single-locus polymorphisms); multilocus polymorphisms are effective for identification of individual genotypes, for estimation of kinship, and pedigree analysis for studying induced mutation process |
| Microsatellites (single tandem repeats, STR; simple nucleotide repeats, SSR) | Varying number of short repeated nucleotide DNA sequences with repeat size of 1–6 nucleotides | PCR amplification with primers complementary to the unique sequences flanking the repeat family; electrophoresis of the amplification products | Population genetics, evolutionary, demographic, and ecological genetics; identification of kinship and population assignment, genetic mapping |
| Randomly amplified polymorphic DNA fragments (RAPD) | Nucleotide differences in sites of binding primers | PCR amplification of random DNA segments using short (10–20 nucleotides) primers with arbitrary nucleotide sequence; electrophoresis of the amplification products | Population genetics, systematics, phylogeny. Identification of plant cultivars and animal breeds, genetic mapping, QTL |

**Table 2.10.** (continued)

| Variation type | Cause of polymorphism | Detection methods | Fields of application |
|---|---|---|---|
| Amplified fragment length polymorphism (AFLP) | Nucleotide differences in restriction sites and flanking sites | Restriction, typically by two restriction enzymes recognizing frequent and rare restriction sites; linking of nucleotide adapters by ligase; selective PCR amplification, electrophoresis | Population genetics, systematics, phylogeny, identification of individual genotypes, analysis of kinship and pedigrees, genetic mapping, QTL |
| Single-nucleotide polymorphism (SNP) | Substitutions of single nucleotides in a DNA sequence | Sequencing of PCR-amplified DNA segments; hybridization of labeled PCR products with microarrays of DNA probes for detection of variants; denaturation of PCR products at critical temperatures and heteroduplex analysis; denaturation fluid chromatography, etc. | Evolutionary and population genetics; genetic mapping; particularly often used in studying SNP associations with diseases |

of humans suffering various maternally inherited diseases also testifies to non-neutrality of mtDNA [139]. Apparently, there are selective constraints on the limiting size of the mtDNA molecules, i.e., on the tandem repeat number in its control region determining size variation of the total molecule (Rand 1993).

MtDNA codes for 13 polypeptides, which, together with the numerous peptides encoded by nuclear genes, constitutes the respiratory chain – that is, the system of electron transport functioning in mitochondria. Because of this, mtDNA variation can significantly affect metabolism, and, consequently, fitness of the organism. The functional association among all these proteins must be maintained by strict selection for the concerted interaction of the mitochondrial and nuclear genomes, i.e., their coadaptation (Blier et al. 2001). The mitochondrial genome must be a leading component of such coadaptation since the mtDNA mutation rate is higher than the nuclear one.

The discovery of numerous SNPs in the human genome has made it possible to identify the effect of selection on this polymorphism. In turn, this permits explanation of molecular differences among species and determination of functional significance of different genomic regions. Let us briefly consider some approaches to detecting selection on the basis of molecular data. Based on the neutrality theory of molecular evolution (Kimura 1968a,

1983; King and Jukes 1969), various statistical tests have been designed that permit one to evaluate deviations from the theory predictions as a possible evidence for selection (see, e.g., Otto 2000). Reviewing these tests, Nielsen (2001) distinguished among them two major groups: (1) tests based on allele distributions or levels of variation and (2) tests based on comparing divergence among different mutation classes within a locus, such as synonymous (i.e., silent, not changing the amino-acid sequence of the encoded protein) and nonsynonymous (changing the amino-acid sequence) nucleotide substitutions. Deviations from the neutrality theory detected by the first group of tests may be explained not only by selection but also by certain demographic factors (a reduction or growth in the population size or population subdivision), but tests of the second group clearly indicate operation of selection (Nielsen 2001). As shown by analyzing SNP frequencies in human gene samples, nucleotide variation in coding gene regions is considerably limited at the sites whose replacements change the amino-acid sequence of the protein molecule. These sites also exhibit a significant excess of rare alleles. In all, this suggests the action of stabilizing (purifying) selection against nonsynonymous SNPs, especially those producing nonconservative amino-acid alterations (Cargill et al. 1999; Halushka et al. 1999; Sunyaev et al. 2000). According to calculations of Halushka et al. (1999), who examined SNPs in 75 human genes, purifying selection, together with genetic drift, eliminates 62% of nonsynonymous SNPs. Some noncoding DNA regions (e.g., nontranslated sequences flanking the coding gene regions) perform important regulatory functions, and their mutations must be eliminated by strong selection. Indeed, variation, i.e., SNP frequency in such sequences, turned out to be several times lower than in degenerate sites of the coding region, where any substitution is synonymous and the level of variation is highest and comparable to the neutral level of variation (Cargill et al. 1999; Sunyaev et al. 2000).

In closing this section, I point to the considerable similarity of the main landmarks in the progress of biochemical and molecular genetics of populations in interpreting protein and DNA polymorphism in a biological context, i.e. in revealing the role of selection for maintaining this giant hereditary diversity.

The more-immediately-above information and the authors' interpretations are reminiscent of the earlier consideration of protein polymorphism (see Sect. 2.2.2). In Chap. 5, devoted to natural selection effects in wild populations, we will discuss another aspect of DNA polymorphism: how the analysis of this variation may find application, taking into account an enormous amount of empirical material as well as theoretical generalizations imposed by the study of biochemical genetics of populations. For this it is necessary to consider carefully the specificity of genetic processes in historically formed population systems of different species.

# 3 Genetic Processes in Natural Population Systems

As emphasized in the preceding chapter, the technique of detecting DNA and protein polymorphisms has opened unique possibilities for studying genetic processes in populations of any species. It should be admitted, however, that the vast majority of works on this subject have restricted the analysis to individual samples, not to populations as historically formed structures. The study of variability at this level, in adhering to long-held concepts of the infinite evolutionary variability of populations, has given rise at the same time to several problems regarding the interpretation of the data obtained. This applies especially to revealing the reasons for the uniformity of the allelic frequencies of several loci on some species ranges, discrepancy in the distribution of allozymes with the usual pictures of chromosomal polymorphism in central and marginal populations, difficulties of demonstrating the effects of overdominance in the analyzing of separate loci, and so forth. All these questions have already been discussed.

Meanwhile the important role that has been assigned to the types and characteristics of population structure is evident even from the internal logic of the development of population genetics theory. This approach finds particularly clear expression in Wright's mathematical models of subdivided populations. It is obvious that the choice of the subject of research and adequate knowledge about it are equally as important to genetics as they are to any other branch of science. Every branch of science has its own subject, whether it be a molecule, a cell, or an individual. Among cytologists, anatomists, or molecular biologists, there is virtually no disagreement about the reality and wholeness of the level that they study. However, if we turn to the biology of populations, we see that there is no such unity: there are over a dozen terms used to designate the subdivisions below a species: "population", "Mendelian population", "deme", "ecotype", "morph", "microtopographical race", "shoal", "stock", "parcel", "subpopulation", etc.

For decades, attempts have been made to find a universal definition for the term *population*. For instance, according to Dobzhansky (1951), population is an array of individuals sharing a common gene pool. In the latest Russian edition of genetic and cytogenetic dictionary by Rieger and Michaelis (1967), population is defined as a community of potentially interbreeding individuals (at a given locality) that share a common gene pool. N.V. Timofeeff–Ressovsky, A.V. Yablokov, and N.V. Glotov (1973, pp. 40–41) give the following definition: "Population is a community of individuals of a given species that for a long period of time (many generations) inhabit a particular location, is characterized by a certain degree of panmixia, lacks noticeable isolation barriers within it, and is separated from neighboring communities of the same species by some form of isolation."

In his monograph *Population Biology* (1986, p. 150), Yablokov defines population as "the smallest self-reproducing group of individuals of the same species inhabiting a particular locality during an evolutionarily long period of time, which forms a distinct genetic system and its own ecological hyperspace".

A universal terminology that allows no ambiguity in its interpretation indicates the maturity reached by any branch of science. So far this cannot be said of population biology, as for a long time it has been passing through the analytical stage of formation representing diverse interests of systematics, morphology, ecology, and other disciplines that have shown an enhanced interest in the populational level of life organization. It is only in recent years that the situation has improved perceptibly (Glotov 1975; Solbrig and Solbrig 1979; Yablokov 1986). At the same time, this terminological diversity also reflects certain fundamental characteristics of the population levels that actually exist but which, unfortunately, are far from always being taken into account. The most important of these is *the multifacetedness of populations* that they present by *virtue of the principle of hierarchical structure internally intrinsic to them.* This *systemic organization* is of major importance both for correctly planning any comparative research and for estimating those factors that determine the specifics of the genetic process at different hierarchical levels of a species' population structure. It is precisely the systemic approach, based on the comprehensive characteristics of the structure of populations, that demonstrates that in many species, despite substantial differences in ecology, *it has been possible to single out at least two qualitative organizational levels of population structure* distinguished by their genetic features. On one hand, there are *fairly stable population systems* that correspond to models of subdivided populations to some degree. On the other, there are more elementary, sometimes extremely variable population units that are *structural components* of these systems and correspond in the highest degree to the Mendelian population model traditionally regarded as an elementary unit of the evolutionary process.

The same approach has enabled us to compare native populations with Wright's island model and to investigate for the first time the so-called evolutionary optimal situation directly in nature and over several generations. As might have been expected, the genetic process in population systems has indeed been found to be of the stationary type because of the reciprocal balance of the interacting random and systematic factors of evolution at a subpopulation level. However, it has been elucidated that the subpopulation structure acts as such a powerful maintaining factor of the genetic stability of an isolated population that it is no longer possible to concur with the view of it, stemming from that advocated by Wagner long ago, that it is the most promising unit of the evolutionary process. Mayr's (1968) well-known pronouncement that "the problem of speciation is ultimately a problem of isolates" has acquired a completely new significance.

The effect of reciprocal balancing of the processes of differentiation and integration of the gene pools was originally described by Serebrovsky (1935) while researching hen populations in isolated Armenian settlements. This was followed by analysis of human populations (Rychkov 1968, 1969, 1973) and of fishes (Altukhov 1969a,b, 1971, 1973a,b, 1974; Altukhov et al. 1969b). Convinced of the universality of the phenomenon of the dynamic stability of a population system, thanks to the variability of its structural components (Altukhov and Rychkov 1970, 1971), we were then able to model the same processes in laboratory populations of *Drosophila melanogaster* (Altukhov and Pobedonostseva 1978, 1979a,b; Altukhov and Bernashevskaya 1978, 1981; Altukhov et al. 1979a). Thus it has been shown that the conclusions from population genetics research may indeed be qualitatively different, depending on the level of the hierarchical structure at which the research is conducted. Let us examine these data in order, analyzing first of all the distributions of biochemical gene markers in native natural populations.

# 3.1
# Natural Populations as Communities
# of Genetically Differentiated Subpopulations

We studied what are called local shoals of fish. These communities are large geographical populations isolated from each other by natural borders at least many thousands of years ago, which still have not been completely destroyed by one or other anthropogenic influence. For decades our group has investigated thousands of fish using a variety of gene markers (blood groups and electrophoretic protein variants). By studying species of different ecologies we hoped to uncover the particular as well as general features of their population genetics organization. New facts derived from this research have enabled us to find a new way of interpreting the characteristics of the genetic process at different levels of population structure.

As representatives of our two principal subjects of research we chose marine fish (American deepwater redfish, *Sebastes mentella* Travin, in the

Newfoundland region of the northwest Atlantic; and the European anchovy, *Engraulis encrasicholus* Linne, inhabiting the Black Sea and the Sea of Azov) and anadromous fish [the Pacific salmons of genus *Oncorhynchus*: the chum salmon *O. keta* (Walb.), the sockeye salmon, *O. nerka* (Walb.), and the pink salmon, *O. gorbuscha* (Walb.)]. These species are primarily of interest because their most important biological features, including those of their internal subdivision into isolated groupings of different ranks, are known well enough from traditional ichthyological works. The unity of area, isolation, and integrity of the morphological and ecological features of the shoals examined make it possible to regard them as historically formed, reasonably stable, self-reproducing communities suitable for studying from the angle of population genetics in the same way that they are usually studied by ecologists and taxonomists.

The anchovy populations were investigated in 1961–1965, the redfish in 1964–1965, and the Pacific salmon have been studied systematically since 1968. In the first two programs, erythrocyte antigens were used as genetic markers tested directly onboard research vessels by methods specially adapted for these purposes (Altukhov et al. 1964). Populations of Pacific salmon were studied during the reproductive period (and in some cases during larval stages) with varied biochemical gene markers, revealed by electrophoresis of proteins in starch and polyacrylamide gels (see Altukhov et al. 1969a,b, 1970, 1975a,b, 1980a; Altukhov 1973b, 1974; Sachko 1973; Salmenkova 1989; Salmenkova and Volokhonskaya 1973; Altukhov and Salmenkova 1981, 1987a,b, 1991, 1994; Salmenkova et al. 1983, 1986, 1992). This has also made it possible to evaluate the levels of polymorphism and heterozygosity of the species studied (Altukhov et al. 1972; Salmenkova and Volokhonskaya 1973; Salmenkova and Omel'chenko 1978).

The most important results have been obtained from genetic–biochemical studies of Pacific salmon. However, there have also been substantial findings derived from redfish and anchovy populations, at least in demonstrating the spatial genetic differentiation of such ecologically varied species. Thus, redfish is a deepwater species with a very settled way of life and a hydrography causing strong isolation of the populations. Anchovy are mobile, pelagic fish which make considerable seasonal migrations, and there are no physico-geographic barriers to panmixia in their areas. However, a marked spatial variability of gene frequencies has been found in both cases during the reproductive period.

Even in our first research into spawning populations of anchovy covering the entire water area of the Sea of Azov (July 1963), we encountered local differentiation in the frequencies of blood groups (Altukhov et al. 1969b). Evidently the discovery of such heterogeneity in a population of what appeared at first sight to be panmictic entails two consequences that predetermine the character of subsequent work.

Firstly, it becomes clear that in order to characterize the population as a whole, one must not restrict oneself to analyzing random samples but one should investigate a reasonably large number of samples which are more or less equally distributed in the area.

Secondly, doubt may be cast on the genetic unity of the investigated population itself if this unity cannot reasonably be adequately inferred from independent sources.

Insofar as the latter hypothesis does not apply, thanks to the works of several scientists who have described the Azov anchovy population as an independent geographical race (= subspecies; for literature on the subject, see Altukhov 1974), it has been possible to concentrate on detailed genogeographical analysis. The results of the most complete study of this kind, which Limansky carried out on the spatial distribution of blood $A_0$ group frequencies during the summer of 1965, are represented on the graph (Fig. 3.1) and on the genogeographical maps (Fig. 3.2).

These data establish the clearly expressed spatial heterogeneity of allelic frequencies, which can be explained in three ways: (1) as a sampling error while investigating a single panmictic population; (2) as a reflection of heterogeneity of the environment in the area of a panmictic population; (3) as evidence of a population's subdivision in a community of subpopulations differing in gene frequencies.

The first explanation is certainly unsatisfactory – the actual variance of the gene frequencies among the samples $V_q = 0.0485$ is far greater than the expected random $V_{\delta q} = 0.038$ ($F = 12.76$; $P < 0.001$).

It is also difficult to accept the premise of heterogeneity of the environment since, as is well known, the water masses of the Sea of Azov are characterized by considerable uniformity of the most important physico-

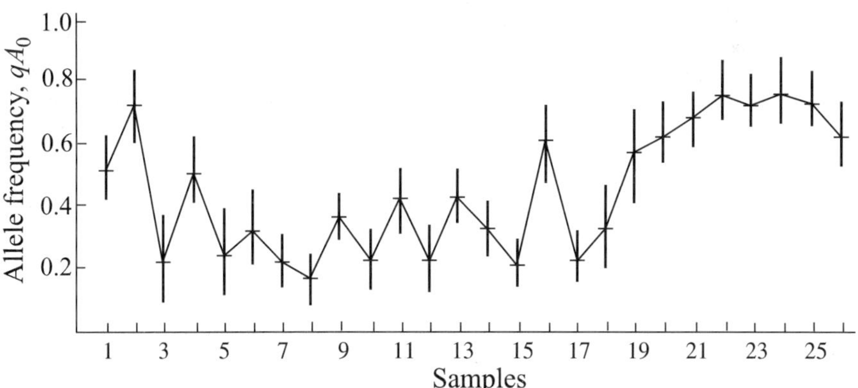


**Fig. 3.1.** The spatial heterogeneity of the frequency of blood group $A_0$ in spawning shoals of anchovy, *Engraulis encrasicholus*, in the Sea of Azov (June 1965). (Altukhov 1974)

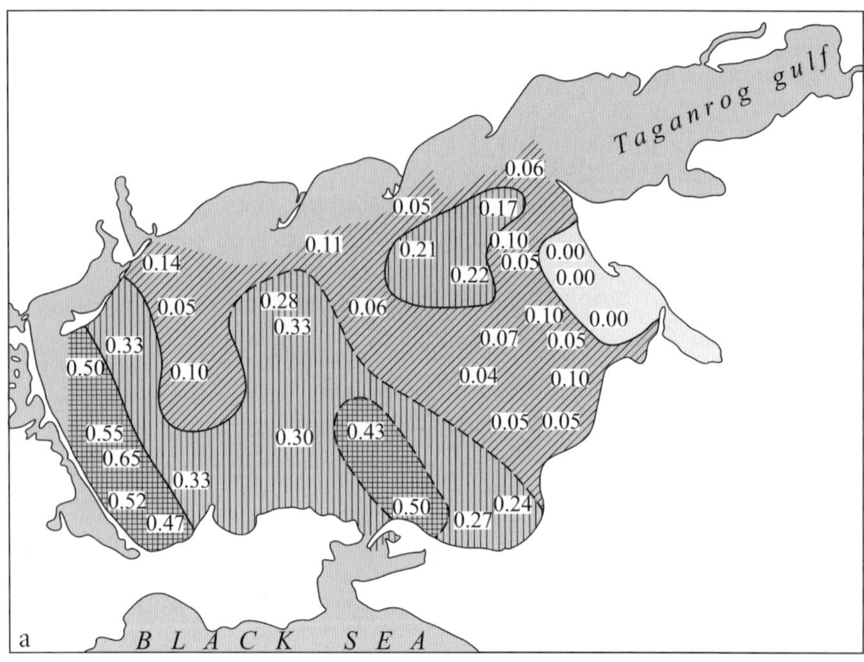
Taganrog gulf
0.06
0.05
0.17
0.11
0.21
0.10
0.14
0.22
0.05
0.00
0.00
0.05
0.28
0.06
0.10
0.00
0.33
0.33
0.07
0.05
0.50
0.10
0.04
0.10
0.55
0.05
0.05
0.65
0.43
0.33
0.30
0.52
0.50
0.24
0.47
0.27
a
B L A C K   S E A

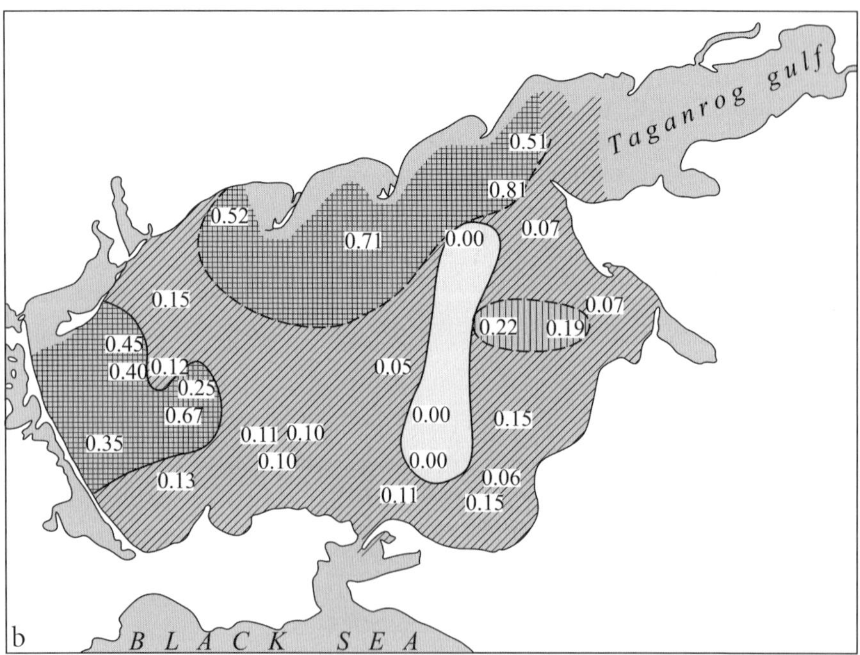
Taganrog gulf
0.51
0.81
0.52
0.07
0.71
0.00
0.07
0.15
0.22
0.19
0.45
0.40
0.12
0.05
0.25
0.00
0.67
0.15
0.35
0.11
0.10
0.00
0.10
0.06
0.13
0.11
0.15
b
B L A C K   S E A

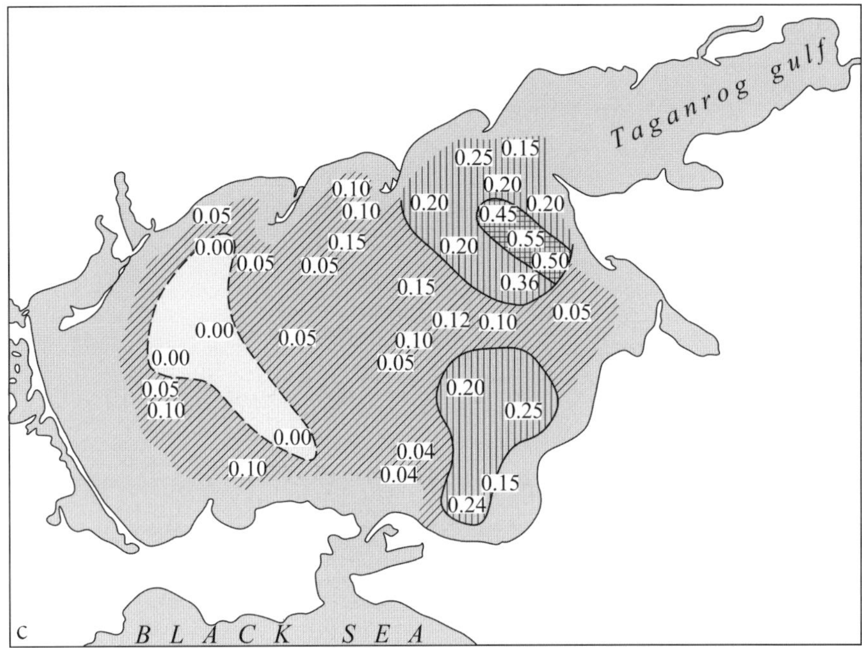


**Fig. 3.2.** The genogeography of the $A_0$ blood group of the anchovy in the Sea of Azov during the summer of 1965 (from Altukhov 1974). Contoured sampling with different frequency values (within 0–0.81 interval). **a** June, **b** August, **c** September

chemical parameters at or near their surface. The time dynamics of areas with different frequencies of genes within them also favor this conclusion: the spatial conjugation of the samples can be seen clearly. Thus, for instance, groups of fish lacking the $A_0$ blood group are found in June to the east in a coastal section of the sea (Fig. 3.2a), in August they are displaced to the west, and in September they are found even further to the west (Fig. 3.2b,c). Nor are changes in the gene $A_1$ frequency observed in the groups. In June it was 72% ($n = 64$), in August 75% ($n = 83$), and in September 72% ($n = 54$).

Consequently, the third hypothesis must be adopted, namely, that smaller, genetically distinctive subpopulations are present within the reproductive part of the Azov population. These are the *elementary populations* discovered by Lebedev (1946, 1967) at the beginning of the 1940s but which he regarded as non-hereditary groupings. Our works have shown the fallacy of this conclusion by revealing the obvious relation of the morphobiological differentiation of anchovy subpopulations with their differentiation by blood group frequencies (Altukhov 1969a, 1974; Altukhov et al. 1969b; Limansky and Payusova 1969). Evidence has also been obtained later of the

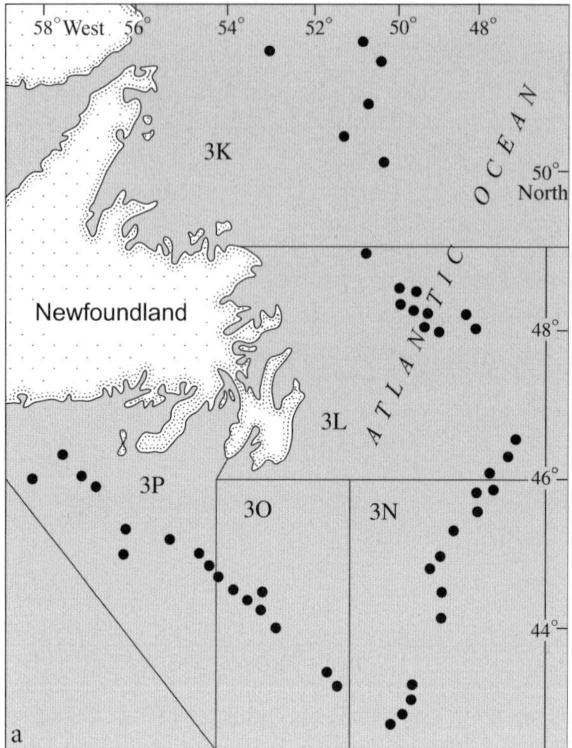


**Fig. 3.3. a** The spatial location **a** of trawl catches of redfish, *Sebastes mentella*, studied on the Newfoundland banks in 1964

spatial heterogeneity of anchovy accumulations on the basis of allozyme frequencies (Kalnin and Kalnina 1982; Kalnina and Kalnin 1984).

In principle, the same picture was obtained when investigating accumulations of redfish on the Grand Banks near the Island of Newfoundland. Figure 3.3a depicts Newfoundland and, to the right of it, with layer upon layer of continental depths where redfish accumulations occur, a continuous belt of fish concealed from our eyes by a mass of seawater several hundred meters deep. The *individual dots* are where the trawling stations are located. Trawling continuously, an expeditionary vessel fishes "from top to bottom", and a fish is selected from each trawl catch so that it can be classified for the characteristics of antigenic differentiation of erythrocytes as well as for biological traits such as body length, sex, age, and stage of sexual maturity. As with the anchovy, genetic heterogeneity is found, which is clearly seen on the genogeographical map (Fig. 3.3b). Simultaneously, the shoal's heterogeneity of biological traits are revealed, providing evidence over two decades of elementary populations (Fig. 3.4); some of these

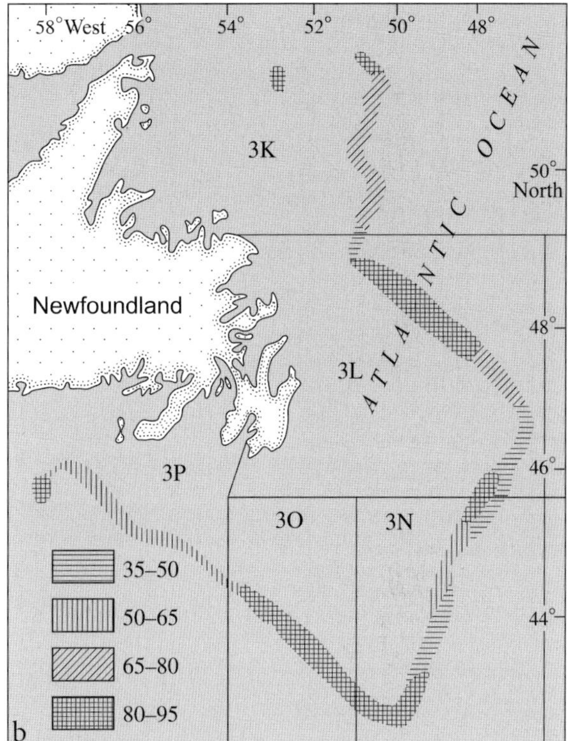


**Fig. 3.3. b** Spatial differentiation of the $A_2$ blood group frequencies in the same redfish accumulations. The *different shadings* indicate the different frequencies (in %) of encountering fish whose erythrocytes carry the $A_2$ antigen. *Capital letters* designate fishing zones allocated in the north-west Atlantic by the International Fisheries Commission

could even be identified two years later during repeated fishing expeditions (Altukhov 1974). Juxtaposition of these two maps shows a high degree of reliable differentiation of subpopulations of the redfish in the $A_2$ blood group frequency (Altukhov 1973a).

Analysis of local shoals of one species of Pacific salmon, the chum salmon, while spawning in rivers on the Asiatic coast (Fig. 3.5), also revealed a high degree of reliability in the heterogeneity of the samples taken for the frequencies of the genes that encode certain proteins (Altukhov 1974). At the same time other markers indicated the *uniformity*, or at least the great similarity, of allelic frequencies of populations known to be isolated and genetically differentiated – a picture identical to that for several *Drosophila* species, as noted in the preceding chapter.

We shall discuss the question of the interlocus differences in the variances of allelic frequencies later. For the present, it is sufficient to show that patent

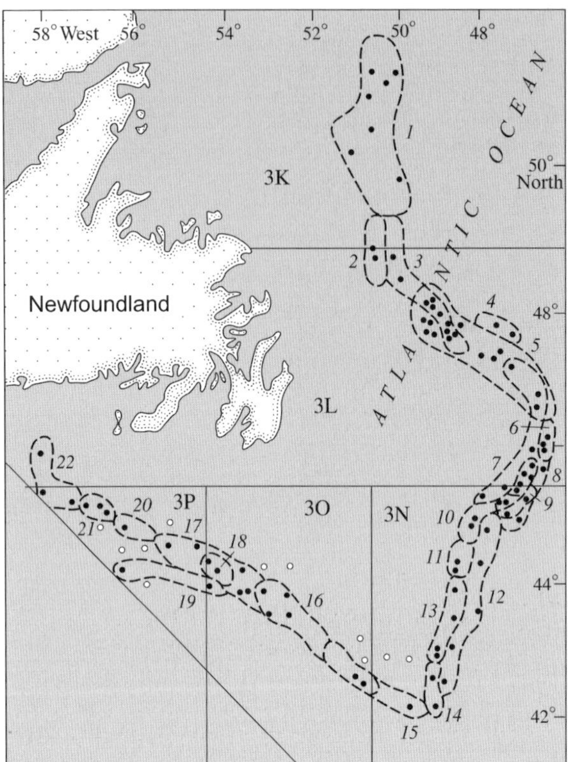


**Fig. 3.4.** Location of elementary populations (nos. *1–22*) of *Sebastes mentella* redfish on the Newfoundland banks. Contoured aggregate samplings in which the fish are characterized by maximal biological uniformity with respect to such parameters as variation of body length, frequency of modal class, sex ratio, etc. (Altukhov 1974)

genetic differences are revealed not only among populations known to be different and isolated, but also within the limits of these populations, that is, in exactly the same way as already discussed for redfish and anchovy populations.

The corresponding evidence is presented in Table 3.1, demonstrating intrapopulation heterogeneity of chum salmon from the Russian Far East and interpopulation differences among chum spawning stocks from the rivers of northeastern Russia and islands of the Japanese archipelago (Figs. 3.6 and 3.7).

The dendrograms are based on the theory of genetic distances/similarities among the populations. Genetic similarity $I$ and genetic distance $D$ among populations $X$ and $Y$ can be found from the formulas proposed by Nei (1975). According to Nei, identity of allelic genes (i.e., theoretically

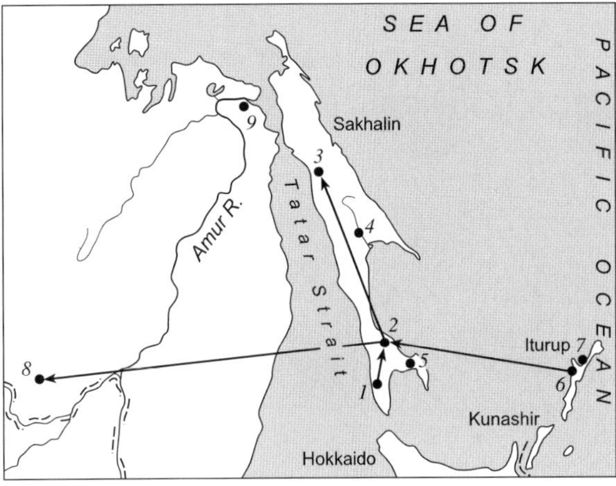


**Fig. 3.5.** River locations of spawning populations of the chum salmon studied. *1* Kalininka; *2* Nayba; *3* Tym; *4* Poronay; *5* Udarnitsa; *6* Kurilka; *7* Reidovaya; *8* Amur basin; and *9* Amur (estuary). Arrows show transplantations

**Table 3.1.** Chi-square tests for homogeneity for allelic frequencies at *Ldh* and *Alb* loci and frequencies of *Mdh* phenotypes among the samples taken in spawning from chum salmon populations of the Rivers Nayba and Kalininka

| Year | Ldh | Alb | Mdh | Year | Ldh | Alb | Mdh |
|---|---|---|---|---|---|---|---|
| | | Nayba | | | | Kalininka | |
| 1969 | 3.22 | 25.97*** | 39.40*** | 1969 | 2.20 | – | 15.00** |
| 1970 | – | 91.31*** | 36.50* | 1970 | – | 10.85* | 5.37 |
| 1971 | 3.52 | 24.61*** | 17.10* | 1971 | 0.61 | 48.25** | 7.99* |
| 1971 | 0.33 | 1.89 | 0.90 | 1974 | 3.00 | 8.85* | 4.40 |
| 1973 | 7.70 | 10.69* | – | 1978 | 5.37 | 24.46*** | 14.40** |
| 1975 | 5.32 | 25.20** | 6.87 | | | | |
| 1976 | 3.22 | 10.68** | 1.98 | | | | |
| 1977 | 4.46 | 9.62** | 6.79 | | | | |
| 1978 | 4.62 | 5.59 | 14.07** | | | | |

*, **, and *** denote $\chi^2$ values significant at the level of $P < 0.05$; $P < 0.01$; $P < 0.001$, respectively

expected homozygosity) in two populations for the *j*th locus is

$$I_j = \sum x_{ji} y_{jj} \Big/ \sqrt{\sum x_{ji}^2 \sum y_{jj}^2},$$ (3.1)

where $x_i$ and $y_i$ are the allele frequencies in populations $X$ and $Y$, respec-

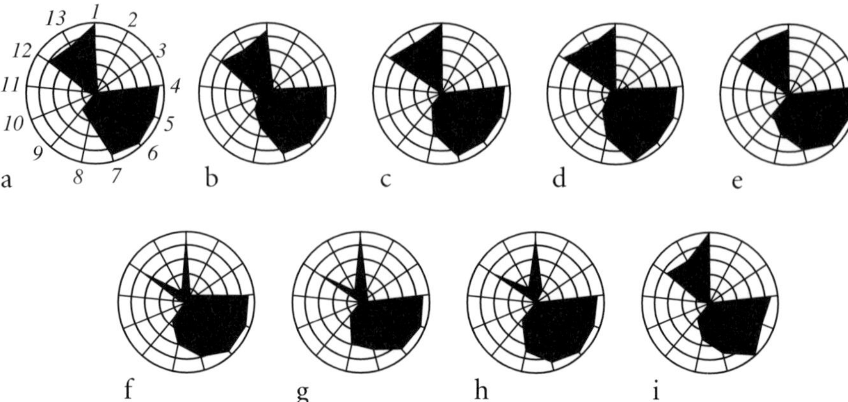

**Fig. 3.6.** The schematic representation of allelic frequencies of allozyme loci in spawning populations of chum salmon in rivers in the Sakhalin–Kuril region (from Altukhov et al. 1980a, with additions). Zero frequency in the center of the circle; frequency 1 on the perimeter. Radii: *1–3* alleles *a*, *b*, and *c* of *Mdh* locus; *4* allele *a* of *Aat* locus; *5* allele *a* of *Ldh-1* locus; *6* allele *a* pf *Pgdh* locus; *7* allele *a* of *Idh-3* locus; *8–11* the corresponding alleles *a*, *b*, *c*, and *d* of *Idh-2* locus; *12* allele *a* of *Me-2* locus; and *13* allele *a* of *Est-D* locus. The system used to designate the alleles is taken from Okazaki (1982) (see also Salmenkova et al. 1986). *a* Udarnitsa; *b* Nayba; *c* Poronay; *d* River Tym; *e* Amur; *f* Kalininka; *g* Zavetinka; *h* Yasnomorka; and *i* Kurilka

tively. For several loci, the generalized genetic identity of the samples is

$$I = \bar{J}_{xy} / \sqrt{\bar{J}_x \bar{J}_y}\,, \tag{3.2}$$

where $\bar{J}_x, \bar{J}_y$ and $\bar{J}_{xy}$ are arithmetic means taken over all loci, i.e.,

$$\bar{J}_x = \frac{1}{L}\sum_{j=1}^{L}\sum_{i=1}^{m} x_{ji}^2;\ \ \bar{J}_y = \frac{1}{L}\sum_{j=1}^{L}\sum_{i=1}^{m} y_{jj}^2;\ \ \bar{J}_{xy} = \frac{1}{L}\sum_{j=1}^{L}\sum_{i=1}^{m} x_{ji}y_{jj}\,,$$

where *m* is the number of alleles and *L* is the number of loci.

$$D = -\ln I\,. \tag{3.3}$$

Based on the values of similarity and difference among the populations, the corresponding samples can be clustered, i.e., combined in hierarchical groups within a *dendrogram* (Zhivotovsky 1991). The methods of clustering are well known; the corresponding procedures can be found in a number of statistical software packages. Here, we use the popular SYSTAT package (Systat Software, Richmond, CA, USA).

The above evidence obtained on chum salmon is also relevant for populations artificially maintained in hatcheries, in which natural reproduction

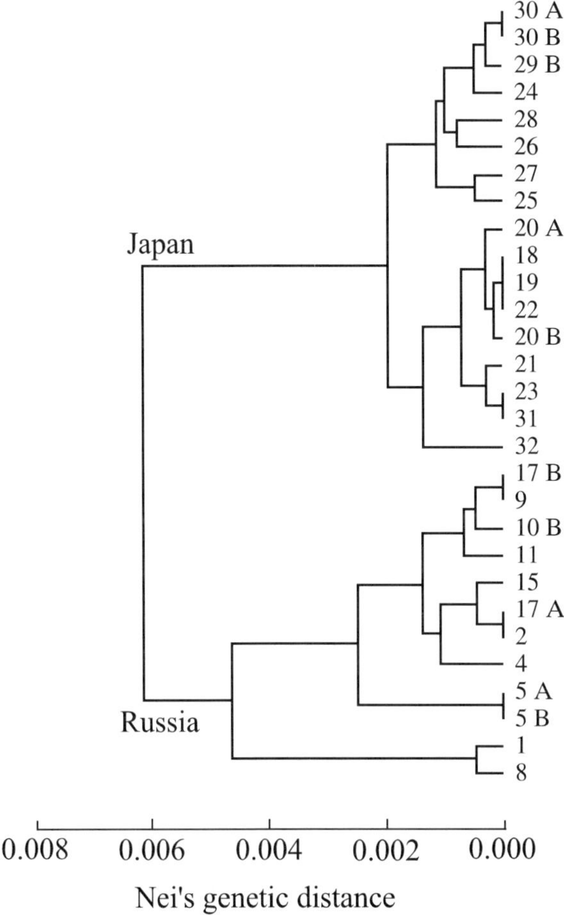


**Fig. 3.7.** Dendrogram based on genetic distance among chum salmon populations in northeastern Russia (*1* Anadyr River; *2, 4, 5A, 5B* east Kamchatka; *8, 9, 10B, 11* west Kamchatka; *15, 17A, 17B* mainland coast of the Okhostk Sea) and islands of the Japanese archipelago (*18, 19, 20A, 20B, 21, 24–29, 30A, 30B* Hokkaido; *22, 23, 31, 32* Honshu). *A* and *B* indicate two different years' samples from the same river. (Winans et al. 1994; 62 loci; reproduced with permission from the National Research Council of Canada)

is to a certain degree absent. Apparently, analysis of such controlled populations virtually cannot yield information on the fine population structure characteristic of this species and known from biological observations (Levanidov 1969). However, we can use for this purpose analogous data obtained by our research team for a closely related naturally reproducing species, sockeye salmon (*Oncorhynchus nerka*). This can provide in-depth analysis of the factors maintaining protein polymorphism and the features of genetic processes in subdivided native populations.

## 3.2
## Genetic Processes in a Natural Population System

### 3.2.1
### Ecology, Demography, and Mating Structure

Isolated populations of the sockeye salmon, spawning in a small area of
9 × 14 km in Lake Azabachye situated in the Kamchatka basin (Fig. 3.8),
have been investigated by the author and his colleagues since 1971, by using
electrophoretically well-detected polymorphism in the autosomal loci of
lactate dehydrogenase and phosphoglucomutase (Fig. 3.9). Simultaneously,
within the overall program, a group of scientific workers at the Institute of
Marine Biology, USSR Academy of Sciences' Far Eastern Scientific Centre,
led by S.M. Konovalov, assessed the numbers and sex ratio in the spawning
grounds, thus enabling us to approach the matter of determining effective
population size (Altukhov 1974; Altukhov et al. 1975a,b; Konovalov 1980).

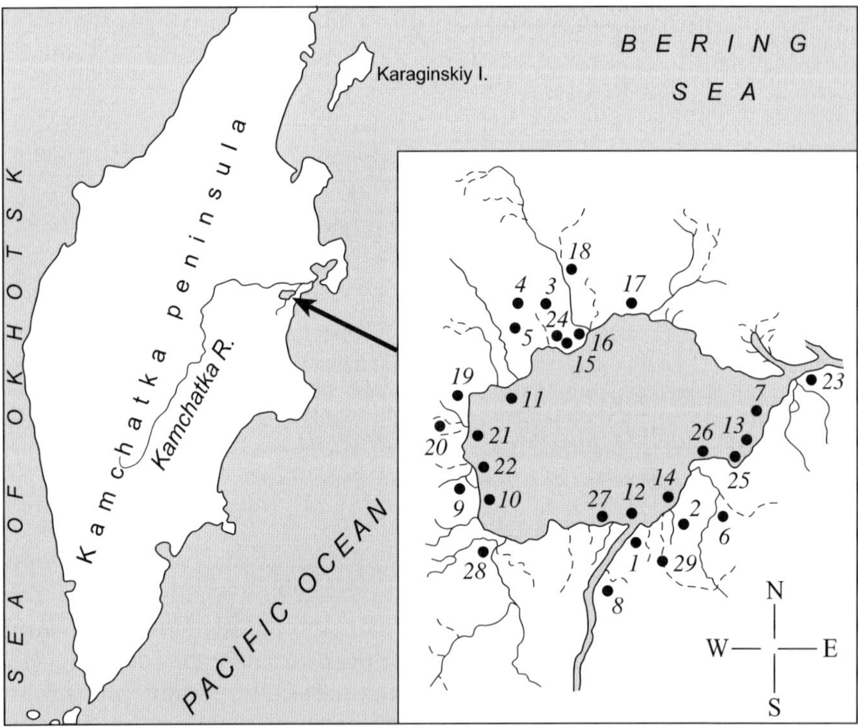


**Fig. 3.8.** Location of Lake Azabachye in the Kamchatka River system (*arrow*) and the spatial
location of spawning subpopulations (*1–29*) of sockeye salmon

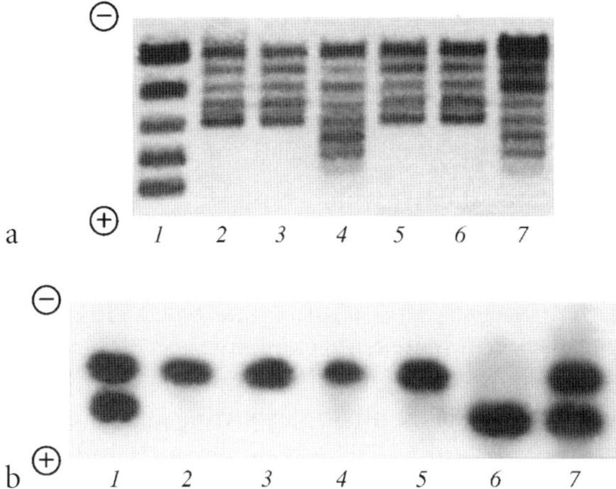

Fig. 3.9. Polymorphism in the sockeye salmon at **a** the lactate dehydrogenase and **b** phosphoglucomutase loci (Altukhov et al. 1975a). **a** *1*: $B'B'$ homozygotes; *2, 3, 5, 6*: *BB* homozygotes; *4, 7*: $B'B'$ heterozygotes; **b** *1, 7*: *AB* heterozygotes; *2–5*: *AA* homozygotes; *6*: *BB* homozygote

Because the results that have been obtained are of major importance for the interpretation of the main factors responsible for maintaining biochemical polymorphism, it is necessary to describe more fully both the subject of our research and the specific features of gathering, processing, and interpreting the material itself, especially that part of it essential for population genetic analysis. Especially important in this connection is that salmon, in general, and Pacific salmon, in particular, are among the best-studied fish species. This is primarily because of their economic value, but also because of several unique biological features of which, drawn from several studies (Neave 1958; Foerster 1968; Konovalov 1971, 1980; Brannon 1972; Altukhov et al. 2000), one must distinguish the following.

**1. Semelparity.** All the species of the *Oncorhynchus* genus are spread out over a huge area of the northern Pacific, reproduce only once, and die soon after spawning. This factor, taken together with the strong predominance (up to 80–100%) of only one age group in the reproductive structure of many stocks (Killik and Clemens 1963; Foerster 1968; Ward and Larkin 1968), makes it possible to regard generations of the sockeye salmon (especially the pink salmon) as not significantly overlapping in time.

**2. The Homing Instinct.** During the very long migrations from spawning rivers to the sea and back, ascending upstream for tens, hundreds, and even thousands of kilometers, the spawning salmon populations display a strict preference for the same spawning reservoir or even separate spawning ground. This homing instinct creates obvious prerequisites for marked intraspecific differentiation, corresponding to the history and geography of a species' habitat. This isolation, reinforced by complex reproductive behavior, promotes the formation of countless reproductive populations or local shoals dispersed over vast areas.

Many years of tagging experiments (Foerster 1968; Malukina 1969; see also Konovalov 1971, 1980) have shown the nearly complete isolation of local sockeye shoals, and even the shoals, which reproduce in different lakes of a river basin, may be regarded to a considerable extent as biologically independent populations – at least during the reproductive period (Altukhov et al. 2000a).

Hartmann and Raleigh (1964) conducted an extensive experiment at Lakes Brooks and Karluk in Alaska involving studies of the fidelity to their own breeding sites of sockeye salmon males during spawning. The authors tagged the fish, transferred them to "foreign" breeding grounds and then traced their return to "their own" spawning sites. With a large amount of experimental data, they conclude that the average "straying" value was little more than 3%. Very similar estimates have been obtained for other salmon species. Thus in rainbow trout, *Salmo irrideus,* about 94% of individuals spawn in natal reservoirs the first time, and 99.6% of the fish return to their usual spawning grounds in the second and third years (Lindsey et al. 1959). It has been found that around 97% of breeding cutthroat salmon *(Salmo clarkii)* carry out their second reproductive season in the same spawning areas where the fish eggs were laid (Cope 1957). And out of 1131 "kokanees" (the resident form of sockeye salmon), less than 3% laid eggs in new spawning grounds (Vernon 1957).

Altogether these data (see also Stabell 1984) show a strong and very similarly developed homing instinct in several species of salmon. If the findings of Hartman and Raleigh (1964) are examined critically, then the average intensity of straying in the populations of sockeye salmon is 2%. Similar experiments conducted on salmon from Lake Azabachye in 1978–1979 produced approximately the same value (Il'in et al. 1983).

As to the doubts expressed by Mina (1977), we would stress that we have never considered the quantitative estimates of homing in the sockeye salmon, or other species closely resembling it in structure, as corresponding exactly to the migration coefficient value in its strictly population-genetic sense. However, this kind of observation is of major interest and provides a natural basis for evaluating the degree of isolation of subpopulations. Furthermore, it is clear that if there are errors in the use of estimates, these tend to produce increases in the migration coefficient, inasmuch as the "mechanical" intensity of the migration of genes is usually higher than their effective migration (Wallace 1966, 1979; Ehrlich and Raven 1969; Raven 1979). However, even an insignificant exchange of genes among subpopulations is sufficient for these groupings to be regarded as "connected populations" (Altukhov 1974).

**3. Subpopulation Subdivision and breeding structure.**   Within the context of an individual spawning reservoir, one discovers that a shoal is subdivided into spawning stocks or subpopulations separated by time and confined to different spawning sites. Systems of these subpopulations (two or more) form geographical populations ("shoals" in our terminology) attached to separate lake basins. Many shoals are subdivided into two ecological races separated by time – a spring race (early migrating subpopulations) and a summer race (late migrating subpopulations). Another "residual" form of the sockeye salmon (kokanee) lives in certain enclosed lakes. The structure of the sockeye shoal in Lake Azabachye studied by us is a community of subpopulations that reproduce both in streams and ponds connected with the lake (the "spring race"), and in its coastal, littoral zone (the "summer race" in Fig. 3.8; for further details see Konovalov 1980).

During the spawning season, there may be 30–40 elementary populations in the lake. In 11 years of research 29 different spawning sites were investigated, but because many of them have been studied repeatedly we shall present data about the biological and genetic structure of 183 such communities, divided by time or space.

The sockeye young live in lakes for 1 to 3, more rarely 4 years, and then set off for the sea where they spend approximately the same time and, on reaching sexual maturity, migrate to river basins for spawning.

Sexual dimorphism in body size and reproductive behavior, including a system of selective mating, are characteristic of the species as a whole. On average, males are larger than females; however, among males one encounters a group of quickly mature small fish which spend not more than 2 years (more often 1 year) in the sea. They are 3–4 years old at time of mating. Canadian authors call sexually mature 3-year-old fish, which have spent 1 year in the sea, "jacks"; Americans call them "grilses"; and on Kamchatka their local name is "kayurki".

It has been shown (Mathisen 1962; Hanson and Smith 1967; McCart 1969) that large males, which have spent over two years in the sea and usually are in the senior age group of 5–7-year-old fish, have a selective advantage in forming mating pairs in spawning grounds. However, it is the group of young small males that are successful in reproducing during years of water shortage and in shallow spawning areas that large males cannot penetrate. The average age of females is 4–6-years.

Despite subdivision, spawning sockeye shoals are characterized by considerable unity and stability of biological parameters such as the sex ratio and mean age. Though there are frequently observed excesses of males or females in different subpopulations, a sex ratio close to one is characteristic of the shoal as a whole. This is clearly seen in the most detailed analyses of generations that arrive at a spawning ground without noticeable traces of the effects of excessive fishing (Altukhov 1974; see also Fig. 3.10).

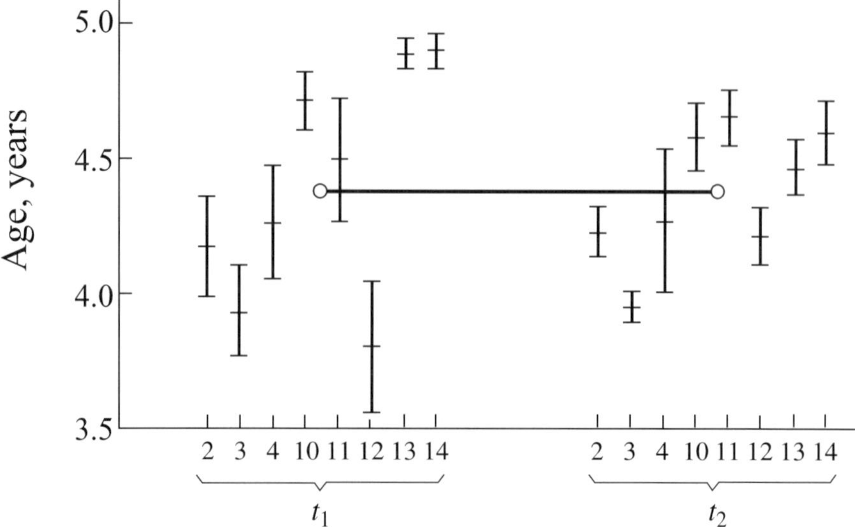


**Fig. 3.10.** The average age of spawning sockeye salmon in separate subpopulations (*dashes with double standard error*) and in the shoal as a whole (*light circles*) for two successive generations, $t_1$ and $t_2$. (Altukhov 1974)

**4. Limits to the $N_e$ Size.** A relatively small number of spawning accumulations is characteristic of all species of the genus *Oncorhynchus*, a feature which is quite strongly marked in the sockeye salmon. These accumulations of fish, which reproduce especially in shallow water streams and basins or lake littorals, are accessible to direct observation; it is possible to calculate the actual number of spawning fish and to determine the sex ratio. Control mechanisms dependent on density may also exercise a certain influence on the $N_e$ value of sockeye shoals in addition to the effect of the sex ratio. Thus, for example, in years of large numbers, the spawning fish that arrive later from the sea remake the spawning nests of fish which had reached the spawning grounds earlier (Krogius et al. 1969). Although the role of this phenomenon has so far not been investigated in detail, we believe that its influence upon the reproductive structure of the Azabachye sockeye salmon is unlikely to be significant as, during the years of our most detailed research, the numbers of the spawning run of the shoal were small because of intensive sea fishing and overcrowding of spawning grounds with mature fish was not observed; large returns were only noticed in 1978 (Shevlyakov 2001). In addition, it was observed that the selective effects of fishing had impaired the sexual structure of several generations that returned to the spawning grounds.

With respect to equilibrium between the numbers of the sexes – an important indicator of the "nativeness" of the population system (Altukhov

1974) – an evaluation of the reproductive portion of the sockeye populations in Lake Azabachye should be made on the basis of an appropriate analysis of only those generations that as a whole are characterized by equilibrium sex ratio.

If one takes this factor into consideration in constructing the distribution of numbers observed throughout the research period, then a series is formed with a mode at an interval of 0 – 500 individuals and with a clearly expressed left-sided asymmetry (Fig. 3.11a). The geometric mean of such a series is 266.5 for spring subpopulations, 350.2 for summer ones, and 332.4 for the entire shoal. The population's effective size, found as a harmonic mean with correction for the sex ratio, is $\sim$ 200 individuals.

A similar processing of data on the number of fish in the stream spawning grounds of the Wood river basin (Alaska; Mariott 1964) leads to a similar distribution (Fig. 3.11b) and an estimated value of $N_e$: 174 individuals; the geometric mean of a series equals 1.504. Throughout their exploitation period, the Wood's sockeye populations suffered less from the fishing than did the populations of Lake Azabachye and other reservoirs in the Kamchatka river basin. Nevertheless, estimates of the genetically effective numbers in both cases are extremely close, differing only in variance which, for the Alaskan populations, is roughly two orders of magnitude higher (2.193 against 0.118 for logarithmic values).

This important fact merely signifies that the fishing, as one might expect, has the greatest effect on large populations, whereas the $N_e$ value is more dependent on groups that contain few individuals, with variability of this parameter in time or space (see Chap. 1).

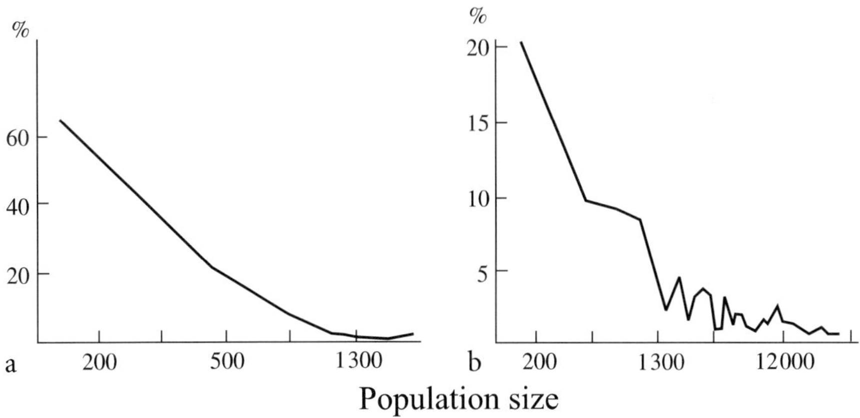


**Fig. 3.11.** The distribution of size of 121 spawning subpopulations in Lake Azabachye in **a** Kamchatka and 179 subpopulations in the river Wood basin in **b** Alaska (Information from Mariott 1964). The abscissa scale is logarithmic

**5. Heterogeneity of Types of Spawning Grounds.**        Three types of spawning grounds, distinguished ecologically, stand out clearly in the Lake Azabachye basin: lake grounds, in which late migrating subpopulations (the summer race) chiefly reproduce, and spawning sections confined to streams and ponds, in both of which grounds spawning of early migrating groups (the spring race) takes place.

The physico-chemical regime of a different type of spawning ground was investigated in July–August 1978 (Novosel'skaya et al. 1982; Table 3.2): measurements were made of pH, oxygen content, and the temperature in the water environment of the upper part of the spawning nests. The two latter parameters are considered the most important in the differentiation of populations of sockeye salmon in types of spawning ground (Ryabova et al. 1978).

The data in Table 3.2 illustrate the heterogeneity of physicochemical characteristics, both among separate spawning sites and among groups of spawning grounds; the difference in temperature, which is maximal in the lake, is particularly marked.

The collection of field material was organized so as to provide maximum coverage of the biological and ecological heterogeneity of the Azabachye sockeye salmon, not only in space but also in time. Standard biological analyses were repeated annually in the same spawning grounds. Simultaneously, in each sample, individual fish were typed for the *Ldh* and *Pgm* loci. As a result, unique information was obtained that makes it possible to approximate the empirical distributions of subpopulations in the gene frequency ranges of both loci, using Wright's mathematical functions for the island model of a population structure (see Chap. 1).

## 3.2.2
## Genetic Dynamics of Population Systems and of Their Structural Components

It follows from the first chapter that Wright's $F_{ST}$-statistics can only be applied if it is known for certain that the test situation is indeed genetically stable. Indirect arguments favoring the structural stability of the sockeye population in Lake Azabachye have been examined above. The stratigraphy of Kamchatkan quaternary deposits attests that Lake Azabachye was formed as a reservoir, differing little from that of today, at least 7,000 years ago (Braitseva 1968; Kuprina 1970). If one estimates the average reproductive age of the Azabachye sockeye salmon as five years, it is clear that over 1,000 generations have passed during the shoal's existence in conditions of isolation, a period sufficient for the development of the stationary genetic process, taking into account effective population number.

**Table 3.2.** Physicochemical characteristics of sockeye salmon spawning grounds of the Lake Azabachye basin

| Type of spawning ground | Number of spawning ground[a] | Date of test sampling | Temperature (°C) | pH | Oxygen content (mg/l) |
|---|---|---|---|---|---|
| Ponds | 2 | 20 August | 8.2 | 7.9 | 9.3 |
| | | 24 August | 8.2 | 7.9 | 10.3 |
| | | 1 September | 8.1 | 7.8 | 9.8 |
| | 3 | 20 August | 9.7 | 7.3 | 8.3 |
| | | 1 September | – | 7.3 | 7.7 |
| | 4 | 20 August | 8.8 | 7.2 | 8.9 |
| | 5 | 20 August | 8.6 | 7.0 | 8.4 |
| | | 24 August | 8.6 | 7.1 | 8.6 |
| | | 1 September | 8.7 | 7.1 | 8.5 |
| | 15 | 20 August | 8.2 | 6.9 | 8.6 |
| | 23 | 20 August | 8.5 | 7.6 | 9.7 |
| | | 1 September | 8.4 | 7.6 | 10.1 |
| | 24 | 20 August | 8.5 | 7.0 | 8.8 |
| Average for ponds | | | 8.54±0.12 | 7.36±0.10 | 9.0±0.22 |
| Streams | 1 | 20 August | 8.3 | 8.1 | 11.3 |
| | | 24 August | 8.2 | 8.2 | 13.3 |
| | 6 | 20 August | 9.1 | 8.1 | 10.4 |
| | | 1 September | 9.1 | 8.1 | 10.9 |
| | 8 | 20 August | 10.0 | 8.0 | 10.0 |
| | | 24 August | 9.8 | 8.1 | 9.1 |
| | | 1 September | 9.9 | 8.0 | 9.3 |
| | 18 | 20 August | 9.0 | 8.1 | 10.6 |
| | 20 | 20 August | 19.1 | 8.1 | 8.6 |
| Average for streams | | | 9.17±0.24 | 8.10±0.06 | 10.43±0.39 |
| Average for the spring race | | | 8.83±0.14 | 7.68±0.09 | 9.65±0.25 |
| Lake | 11 | 20 August | 13.3 | 7.8 | 9.1 |
| | | 24 August | 13.1 | 7.9 | 9.1 |
| | | 1 September | 13.2 | 7.9 | 9.4 |
| | 13 | 24 August | 12.3 | 8.6 | 9.2 |
| | | 1 September | – | – | 10.0 |
| | 14 | 20 August | 10.8 | 8.1 | 9.0 |
| | | 24 August | 10.7 | 8.1 | 9.4 |
| | | 1 September | 10.7 | 8.0 | 10.1 |
| | 22 | 20 August | 13.0 | 7.8 | 8.6 |
| | | 24 August | 13.1 | 7.8 | – |
| | | 1 September | 13.0 | 7.8 | – |
| | 25 | 20 August | 13.2 | 8.4 | 8.2 |
| | | 1 September | 13.2 | 8.6 | 8.0 |
| Average for the summer race | | | 12.47±0.31 | 8.07±0.09 | 9.10±0.20 |

[a] Numeration is the same as for Fig. 3.8

But, of course, one can only obtain direct evidence of a population's genetic stability by comparing the distributions of gene frequencies in appropriate time "segments" – cross-sections through the subpopulation structure of a subdivided population (Altukhov and Rychkov 1970). As already indicated, we have carried out several cross sections of this sort, enabling us to assess distributions of genotypic and allelic frequencies in successive generations (Table 3.3).

Even with cursory analysis of the tabular material one can see an absence of differences among the generations of different years with respect to average values as well as to the intergroup variance of the allelic frequencies of both loci, despite the shoal's constantly revealed genetic heterogeneity.

**Table 3.3.** Stability of the main population-genetic parameters of a shoal of sockeye salmon in Lake Azabachye (from Altukhov et al. 1983a; with supplements)

| Locus | Research year | $K$ | $n$ | $\bar{q}$ | $V_q$ | $\chi^2$ test for variance homogeneity [Bartlet's criterion ($df$)] | Test for the population homogeneity |
|---|---|---|---|---|---|---|---|
| *LDH* | 1971 | 14 | 737 | 0.65 | 0.0074 | $2.69 < \chi^2_{0.05}(10) = 18.30$ | 50.18*** |
| | 1972 | 20 | 1022 | 0.61 | 0.0048 | | 37.53*** |
| | 1973 | 21 | 1007 | 0.65 | 0.0078 | | 65.11*** |
| | 1974 | 14 | 676 | 0.66 | 0.0076 | | 45.59*** |
| | 1977 | 19 | 934 | 0.65 | 0.0085 | | 69.34*** |
| | 1978 | 23 | 1244 | 0.65 | 0.0071 | | 76.72*** |
| | 1979 | 14 | 846 | 0.65 | 0.0091 | | 67.30*** |
| | 1980 | 17 | 843 | 0.67 | 0.0078 | | 63.36*** |
| | 1981 | 11 | 2239 | 0.68 | 0.0073 | | 137.22*** |
| | 1982 | 11 | 1380 | 0.67 | 0.0053 | | 66.74*** |
| | 1984 | 19 | 1408 | 0.64 | 0.0061 | | 79.22*** |
| Totals | | 183 | 12336 | 0.66 | 0.0077 | | 804.49*** |
| *PGM* | 1971 | 14 | 748 | 0.79 | 0.0028 | $6.47 < \chi^2_{0.05}(9) = 16.90$ | 22.57** |
| | 1972 | 20 | 1004 | 0.77 | 0.0019 | | 21.76 |
| | 1973 | 21 | 981 | 0.79 | 0.0041 | | 41.36** |
| | 1974 | 13 | 627 | 0.78 | 0.0039 | | 45.01** |
| | 1977 | 19 | 936 | 0.81 | 0.0032 | | 39.12** |
| | 1978 | 23 | 1241 | 0.77 | 0.0025 | | 35.78* |
| | 1979 | 14 | 846 | 0.78 | 0.0018 | | 17.05 |
| | 1980 | 17 | 831 | 0.77 | 0.0015 | | 24.90 |
| | 1982 | 11 | 1363 | 0.79 | 0.0035 | | 14.53 |
| | 1984 | 19 | 1353 | 0.76 | 0.0019 | | 45.94*** |
| Totals | | 171 | 9950 | 0.78 | 0.0028 | | 317.31*** |

*Remarks:* $K$ is the number of subpopulations; $n$ is the number of fish studied; $\bar{q}$ is the average gene frequency; $V_q$ is the intergroup variance of the gene frequency
* $P < 0.05$; ** $P < 0.01$; *** $P < 0.001$

The chi-square values are calculated by the formula:

$$\chi^2 = \sum_{i=1}^{K} \frac{\left(q_i - \bar{q}\right)^2 2n_i}{\bar{q}\left(1 - \bar{q}\right)},$$

where $i$ is the subpopulation's sequence number; $K$ is the number of subpopulations; $q_i$ is the gene frequency value in the $i$th subpopulation; $\bar{q}$ is the corresponding mean allele frequency for the whole shoal; and $n_i$ is the sampling size of the $i$th population. Heterogeneity is particularly well expressed for the *Ldh* locus. Statistically significant heterogeneity was not observed at the *Pgm* locus in 1972, 1979, 1980, and 1982, but this finding does not influence the conclusion that heterogeneity need only be discovered at any polymorphic locus for it to be regarded as proven. However, where other genes are concerned, especially if polymorphism is not expressed externally and there are no grounds for allowing the differential migration of genotypes, interpopulation variability may be camouflaged by the effects of balanced selection.

In fact, if we turn again to Table 3.3, we see clear differences in interlocus variance: the allelic frequencies at the *Ldh* locus are more variable than those at the *Pgm* locus, and this picture persists throughout the period of study. Moreover, in all analyses of distributions of *Pgm* genotypes one observes either agreement with the Hardy–Weinberg expectations or an excess of heterozygotes, as seen in 1973 and 1974 and in the total distribution for the entire research period (Table 3.4). These facts are very reminiscent of a typical picture of balanced polymorphism maintained through an adaptive advantage of heterozygotes.

Thus, the preliminary stage of research work shows that despite clearly expressed interpopulation differences in gene frequencies, the system of subpopulations as a whole remains genetically stable in successive generations. Furthermore, the substantial interlocus differences in the variance of gene frequencies, repeated from year to year, indicate dissimilarity in the contributions made by random and systematic factors to maintaining the polymorphisms under study.

The role of natural selection in maintaining allozyme polymorphism will be considered in Chap. 5, and here and now we will discuss temporal genetic stability of the population. The example of sockeye salmon from the Lake Azabachye illustrates this stability, which would have been found in any thoroughly studied species if its structure had not been destroyed by anthropogenic impact.

Indeed, the view on stability irrevocably comes out of population-genetic evidence when one passes from simplest populations to higher levels of population hierarchy.

**Table 3.4.** Distributions of genotypes and gene frequencies ($q$) of *Ldh* and *Pgm* loci in the subpopulations of a shoal of sockey salmon in Lake Azabachye

| Year | Ldh | | | | | | Pgm | | | | | |
|---|---|---|---|---|---|---|---|---|---|---|---|---|
| | $B_1B_1$ | $B_1B_1'$ | $B_1'B_1'$ | $n$ | $qB$ | $\chi^2$ | AA | AB | BB | $n$ | $qA$ | $\chi^2$ |
| 1971 | 327 | 319 | 91 | 737 | 0.6601 | 0.92 | 465 | 258 | 25 | 748 | 0.7941 | 2.27 |
| | 321.7 | 330.7 | 85.2 | | | | 471.7 | 244.6 | 31.7 | | | |
| 1972 | 386 | 467 | 169 | 1022 | 0.6061 | 1.88 | 589 | 366 | 49 | | | |
| | 375.4 | 488.0 | 158.6 | | | | 593.6 | 356.8 | 53.6 | 1004 | 0.7689 | 0.67 |
| 1973 | 427 | 461 | 118 | 1007 | 0.6529 | 0.10 | 596 | 351 | 34 | 981 | 0.7864 | 4.14* |
| | 429 | 456.4 | 121.3 | | | | 606.7 | 329.6 | 44.7 | | | |
| 1974 | 296 | 302 | 78 | 676 | 0.6612 | 0.01 | 378 | 253 | 16 | 647 | 0.7797 | 12.38*** |
| | 295.5 | 302.9 | 77.6 | | | | 393.3 | 222.3 | 31.4 | | | |
| 1977 | 409 | 410 | 115 | 934 | 0.6574 | 0.61 | 611 | 298 | 27 | 936 | 0.8119 | 1.70 |
| | 403.6 | 420.8 | 109.6 | | | | 617.0 | 285.9 | 33.1 | | | |
| 1978 | 534 | 538 | 172 | 1244 | 0.6455 | 3.77 | 738 | 443 | 60 | 1241 | 0.7732 | 0.39 |
| | 518.3 | 569.4 | 156.3 | | | | 741.9 | 435.3 | 63.8 | | | |
| 1979 | 373 | 349 | 124 | 846 | 0.6472 | 7.84** | 517 | 284 | 45 | 846 | 0.7790 | 0.54 |
| | 354.3 | 386.4 | 105.3 | | | | 513.3 | 291.4 | 41.3 | | | |
| Totals for seven years | 2752 | 2846 | 868 | 6466 | 0.6457 | 9.35** | 3894 | 2253 | 256 | 6403 | 0.7841 | 9.85** |
| | 2695.7 | 2958.6 | 811.7 | | | | 3936.6 | 2167.9 | 298.5 | | | |

$n$ is the number of fish analyzed. * $P \leq 0.05$; ** $P \leq 0.01$; *** $P \leq 0.001$

In view of this circumstance, the corresponding approach can be applied to other bisexual species. In all cases, this will demonstrate the absence of shifts in the key population-genetic parameter, mean gene frequency, in any number of generations accessible for direct comparison.

Let us examine examples of this stationary state, first turning to widely known studies on genetics of natural populations of *Drosophila* (Dobzhansky 1943; Dobzhansky et al. 1964a). Analyzing frequency dynamics of various chromosomal inversions in *Drosophila pseudoobscura,* these authors came to a conclusion on microevolutionary shifts in the gene pools of the populations examined. However, if this analysis is supplemented with the view of reality of existence of naturally isolated population systems, and with the primary data organized accordingly, another conclusion can be drawn.

The first example (Fig. 3.12) corresponds to ten successive generations of *Drosophila* examined in the late 1940s and early 1950s in the Sierra-Nevada mountains (California) in three localities separated by altitude. Here, we present the data for the Kin Camp locality (about 1,500 m above sea level) studied at five stations situated about 4 km apart.

Since the frequency of the Standard inversion chosen for this discussion exhibits cyclic seasonal variation, we considered the comparable parameters of the May, June, and July samples, shown in Fig. 3.12 as the third (displaying highest variation) level of this population structure. Smaller differences were recorded for stations within a locality (the second hierarchical

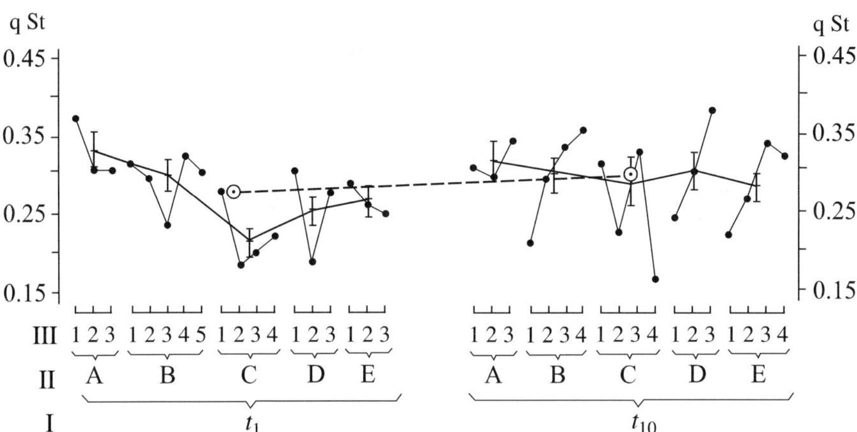

**Fig. 3.12.** Variation in the frequency of chromosome inversion Standard (*qSt*) in the arbitrary generations 1 ($t_1$ 1939) and 10 ($t_{10}$ 1940) in generations of the Kin Camp *Drosophila pseudoobscura* population system as a whole (*I, open circles*) as compared to variation in the stations (*II, A–E*) and in generations (*1–5*) within stations (*III*). *I–III* are the levels of the population structure. (Based on data from Dobzhansky 1943)

level), whereas for the Kin Camp locality as a population system (the first level of the hierarchy), no significant shift in the gene frequency has been found in the generations examined. The values of mean square deviation indicated that the frequency distribution was stable.

The second example (Fig. 3.13), corresponding to the interval of 70 generations in a system of *Drosophila* populations distributed throughout California, provides a possibility of studying the resistance of the system to the powerful selective agent, pesticides.

According to incomplete data (Dobzhansky et al. 1964a), in just the years 1951 and 1955, 17,432 tons of various pesticides were applied on the California landscape. (About one-fifth of all pesticides used in the United States in the examined period was used in California.)

Comparing the sets of the same stations from 1957 to 1963, we see that, in spite of evident frequency fluctuations in the localities (e.g., compare localities 7, 9, and 18 in Fig. 3.13), no statistically significant changes in the frequency distribution of the Standard inversion occurred in California as a whole.

However, apart from these examples and the studies discussed in Chap. 6, I have not found any work in which stability of genetic parameters of a population system was traced over time. Nevertheless, the view on the

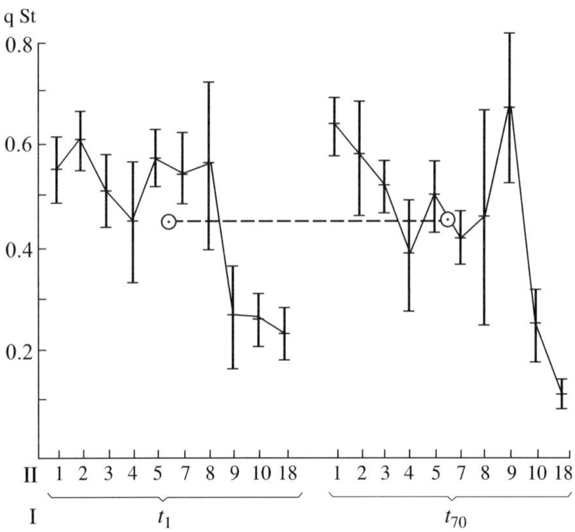


**Fig. 3.13.** Variation of the frequency of chromosome inversion Standard ($qSt$) in the arbitrary generations 1 ($t_1$ 1957) and 70 ($t_{70}$ 1963) in generations of the *Drosophila pseudoobscura* population system of California as a whole (*open circles*) as compared to variation in the stations (localities; *bars* show doubled standard errors; based on data from Dobzhansky 1964a). *I, II* are the hierarchical levels

importance of the approach demonstrating stability of the system despite genetic variation of its parts is shared by a number of authors. For example, Campton and Utter (1987) have shown that local genetic differentiation in trout *Oncorhynchus (Salmo) clarkii clarkii* is as pronounced as in sockeye salmon from Lake Azabachye, and this variation may promote stabilization of the subdivided population as a whole. Substantial subpopulation differentiation has been recently confirmed using DNA markers in fishes [e.g. *Salmo salar* (Tessier et al. 1995, 1997) and *Oncorhynchus kisutch* (Small et al. 1998)], amphibians (Shaffer et al. 2000; Newman and Squire 2001), birds (Abbot and Double 2003; Caizergues et al. 2003), mammals (Goossens et al. 2001), trees (Lacerda et al. 2001), spiders (Pedersen and Loeschcke 2001), and a number of other animal and plant species (see, for example, Duran et al. 2004). Today there is virtually no doubt that most of the genetic variability of a species has a complex hierarchical structure (Avise 2000; Goossens et al. 2001), down to distinct family groups within subpopulations such as colonies, demes, prides, etc. Their primary role is maintaining intrapopulational genetic variation via nonrandom mating structure that prevents close inbreeding.

For example, the settlements of the long-tailed macaque, *Macaca fascicularis*, inhabiting tropical forests of Sumatra, consist of small groups of 8 to 50 animals each. Females of this species exhibit strict philopatry, i.e., they stay at the place of their birth for all of their lives. By contrast, males move from colony to colony; in any colony a reproducing male is thus always an immigrant. Furthermore, there are two types of males: dominant ($\alpha$) and subdominant ($\beta$); the contribution of the dominant males in the gene pool of the subsequent generation attains 80%, and heterozygosity of this group is higher than expected (de Jong et al. 1994; Lawler et al. 2003). A similar mechanism has been recently reported for the Alpine marmot *Marmota marmota* (Goossens et al. 2001) using microsatellite analysis. The authors also cite analogous studies on other social species, and as they rightly note, "Sampling strategy and behavioral information may dramatically affect the validity of the results and their genetic interpretation" (Goossens et al. 2001, p. 41). Pacific salmonids are characterized by a nonrandom system of selective mating: in mating, females prefer large-sized males (Sect. 3.2.1). We can readily add to the number of these examples (see, e.g., the data on cod; Bekkevold et al. 2002). Moreover, it is known now that *interpopulation exchange of individuals is also nonrandom* (Altukhov and Pobedonostseva 1979a), which we examine in detail in Chaps. 4–7.

However, there is an erroneous view often encountered in literature that allele frequencies can be used to describe the population structure of a species. This view is incorrect, first of all, because it leads to the widely spread opinion that random mating is a leading cause of the uniformity of allele frequencies over huge territories of the species ranges, especially in pelagic marine fishes and some invertebrates having a pelagic larval stage (DeWoody and Avis 2000; Ward et al. 1994; see also O'Relly et al. 2004). More than half a century ago, these species, hidden from our eyes in the sea depth, were shown to have a distinct structure consisting of simple subpopulation units, which in Russian literature were termed *elementary populations* (Lebedev 1946, 1967; see Sect. 3.1) and, as shown later, differing in frequencies of selectively neutral alleles of blood group (Altukhov et al. 1969a,b) and allozyme genes (Kalnin and Kalnina 1982; Kalnina and Kalnin 1984). If the proportion of overdominant (or highly mutable) genes prevails in a sample (protein or DNA) of gene markers used, then the actual population structure of the species "disappears" owing to the spatial and/or temporary uniformity of gene frequencies under the pressure of stabilizing selection or mutation (Altukhov 1983; Altukhov et al. 1987a). However, using traditional biological traits (such as linear size, sex, age, stage of sexual maturity, or sex ratio in the sample) and taking into account spatial localization of the samples, one can reliably determine borders of the elementary populations and trace their dynamics in time and space (for details, see Altukhov 1975; Sect. 3.1).

Thus, population structure *is given to us by Nature itself* regardless of genetic markers used to study it. As to other species, for instance our own, only 200 years ago its rural population was represented by ethnic systems consisting of small reproductive groups that at a low rate exchanged genetic material, maintaining balance between inbreeding and outbreeding (see also Chap. 7). At present, such communities occur only at the margins of the *Homo sapiens* range (Neel 1970; Rychkov and Sheremetyeva 1977; see also Sect. 7.5).

The key factor of stability of a population system as a whole is a balance between the processes of differentiation and integration of its gene pool, e.g., random gene drift counterbalanced by migration or selection, as was shown for sockeye salmon from Lake Azabachye.

These effects are readily simulated in computer models. Such simulation has been recently carried out in cooperation with a professional mathematician (Altukhov and Blank 1991, 1992; Blank and Altukhov 1992, 1995). The main results of these studies are examined in the following section.

## 3.3
## Mathematical Modeling of Simplest Population Systems Represented by Small Subpopulations

Many authors (Malécot 1948; Kimura 1953; Kimura and Weiss 1964; Maruyama 1971a; Nagylaky 1989; and others) used mathematical tools to analyze subdivided populations represented by subpopulations connected by migration. The primary aim of these works was to study the effect of subdivision (subpopulation structure) on local genetic differentiation. In our studies, the attention is focused on the phenomenon of genetic stability of the population system as a whole in context of maintaining gene diversity within this population system over many generations. Our efforts have been similar (but nor identical) to those of Maruyama (1972a,b), in the sense of being devoted to rates of loss of genetic variation in structured populations. Moreover, note that in the studies of Malécot and Kimura's school, the population structure is assumed to be homogeneous in time and space, which allowed the authors to obtain precise analytical solutions for stationary states of the models examined. M.L. Blank has developed a mathematical model (Blank and Altukhov 1992) that describes the dynamics of statistical moments of gene frequency distributions in a subdivided population with an arbitrary subpopulation structure and arbitrary (albeit symmetric) migration matrix (Bodmer and Cavalli–Sforza 1968). This approach permits analysis of gene frequency dynamics far from the equilibrium points, which is significant for better understanding of evolution of population systems under various types of anthropogenic pressure, and, consequently, for solving a number of problems of population genetics (Altukhov 1999).

Consider a population comprising of $k$ subpopulations; the $i$th subpopulation consists of $N_i$ individuals. Time $t$ is measured as the number of non-overlapping generations. As the two-allele, one-locus haploid model is considered, each subpopulation consists only of two types of individuals. Consequently, the $i$th subpopulation is described by the number of individuals of one of these types $x_i(t)$ at a given time $t$. The frequency of these individuals (which is equal to the gene frequency) in the $i$th subpopulation at time $t$ is $p_i(t) = x_i(t)/N_i$. Denote by $x_{ij}(t + 1)$ the number of the $t$-generation progeny of these individuals that migrated from the $i$th subpopulation to the $j$th subpopulation. If the number of individuals migrating from the $i$th subpopulation to the $j$th subpopulation is $n_{ij}$, the coefficient of gene migration will be $m_{ij} = n_{ij}/N_j$.

This type of migration will be referred to as local. We also consider the so-called long-range migration (Kimura and Weiss 1964), or migration "from the mainland", with coefficient of migration equal to $\varepsilon$.

Statistical moments of gene frequencies are

$$p(t) = k^{-1} \sum_i M\left[p_i(t)\right],$$

$$C_{ij}(t) = k^{-1} \sum_{ij} M\left[\left(p_i(t) - M\left[p_i(t)\right]\right)\left(p_j(t) - M\left[p_j(t)\right]\right)\right],$$

where $M[\cdot]$ is the mathematical expectation, $p(t)$ is the mean, and $C_{ij}(t)$ is the covariance of gene frequencies in the $i$th and $j$th subpopulations. The temporal dynamics of these moments provides complete information on the behavior of the populations.

Assume for simplicity that local and long-range migrations, selection, and reproduction are separated in time and occur successively.

Considering long-range migration, we obtain

$$p_i(t) \rightarrow (1 - \varepsilon)p_i(t) + \varepsilon p(t) \,,$$
$$C_{ij} \rightarrow (1 - \varepsilon)^2 C_{ij}; \ i,j = 1...k \,.$$

To describe selection, we introduce the following linear model:

$$p_i(t) \rightarrow (1 - s_i)p_i(t) + s_i p_i^{(s)} \,;$$
$$C_{ij}(t) \rightarrow (1 - s_i)(1 - s_j)C_{ij}(t) \,,$$

where $s_i$ is the intensity of selection and $p_i^{(s)}$ is the final equilibrium frequency obtained by selection in the $i$th subpopulation. If $p_i^{(s)} = 0(1)$, this is directional selection; at $0 < p_i^{(s)} < 1$, the model describes stabilizing selection.

For typical initial frequencies ($0.2 < p_0 < 0.8$), our model insignificantly differs from the well-known selection model of Sewall Wright. Analysis of empirical data shows that both models have approximately equal precision in this frequency range. However, for selection against a very frequent allele (e.g., at $p_0 \geq 0.9$), Wright's model predicts an $S$-shaped gene-frequency curve, which contradicts the results of experimental studies (see, e.g., Altukhov 1989b, p. 153; Mettler and Gregg 1969, p. 148 and the review in Curtsinger 1990). Our approach can be used for studying effects of selection in a heterogeneous environment, when selection coefficients $s_i$, and equilibrium allele frequency $p_i^{(s)}$ depend on the number of members of a subpopulation.

Using the system of relationships obtained, we can conduct a temporal analysis of population dynamics taking into consideration various migration rates, types of population structure, and selection modes.

In practice, it is convenient to describe genetic diversity of a population using the proportion of heterozygotes $H(t)$:

$$H(t) = \frac{2}{N} \sum_i N_i \left[ \bar{p}_i \left(1 - \bar{p}_i\right) - C_{ii} \right] \,.$$

The interpopulation ($D_1$) and intrapopulation ($D_0$) components of variance are computed accordingly:

$$D_0 = k^{-1} \sum_i C_{ii} \,,$$

$$D_1 = 2k^{-1} \left\{ \sum_i C_{ii} - k^{-1} \sum_{i,j} C_{ij} \right\}.$$

Earlier we (Blank and Altukhov 1992; Altukhov and Blank 1992), also described statistics $H_S$ and $H_T$ characterizing, respectively, intrapopulation and total gene diversity. As they somewhat differ from Nei's parameters generally accepted in population genetics, we present here their expressions:

$$H_T = M \left[ 1 - \left( \frac{1}{k} \sum_i p_i \right)^2 \right] = 1 - \frac{1}{k^2} \sum_{i,j} M \left[ p_i p_j \right]$$

$$= 1 - \frac{1}{k^2} \sum_{ij} \bar{p}_i \bar{p}_j - \frac{1}{k^2} \sum_{ij} C_{ij};$$

$$H_S = M \left[ 1 - \left( \frac{1}{k} \sum_i p_i^2 \right) \right] = 1 - \frac{1}{k} \sum_i \bar{p}_i^2 - \frac{1}{k} \sum_i C_{ii}.$$

In the following, we review the results obtained by simulation.

### 3.3.1
### Population System Dynamics Upon Interaction Between Random Drift and Local Gene Migration

Simulation of the gene frequency dynamics in panmictic and subdivided populations of equal size confirms the previously established fact of lower genetic stability (in the sense of preservation of gene diversity) of a non-structured panmictic population (Fig. 3.14). A comparison of Fig. 3.14a,b shows that eight out of ten model panmictic populations (each of 500 individuals) became completely homozygous by generation 1,000, whereas none of the subdivided populations of the same total size ($N_i = 20$, $k = 25$, $m = 0.03$) lost their genetic diversity. If we estimate the life span of the panmictic and the subdivided populations of equal sizes expressing the life span as the number of generations that corresponds to the loss of 99% of initial heterozygosity, this value would be 2,301 generations for the panmictic and 5,341 for the subdivided ($N_i = 20$, $k = 25$, $m = 0.005$) populations.

In other words, the existence of a circular subpopulation structure with limited (about 0.5%) local gene migrations causes the same deceleration of the decrease in genetic variation as would doubling of the size of this population. Various comparisons of this kind are given in Table 3.5. In Fig. 3.15, this relationship is presented graphically, demonstrating obvious

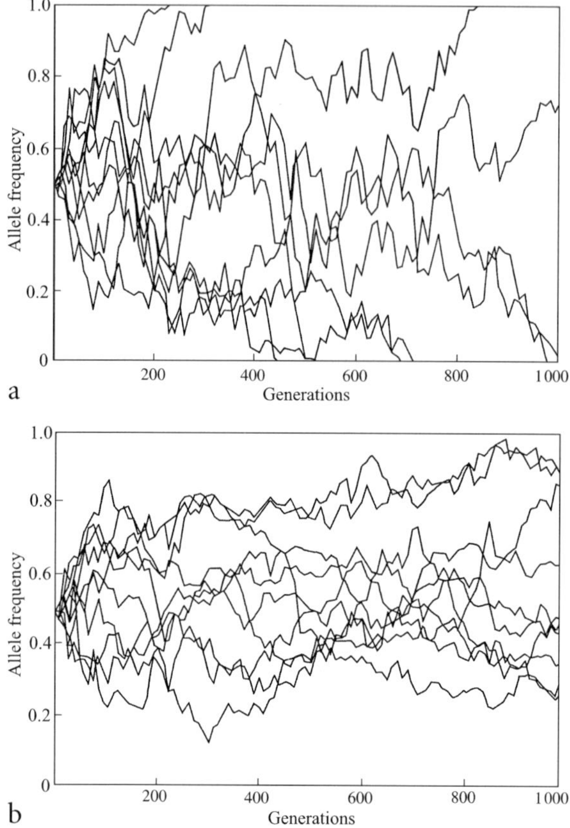

Fig. 3.14. **a** The gene frequency dynamics in 10 panmictic populations of 500 individuals each and **b** in 10 subdivided populations with the circular structure of gene migration. $N_i = 20, k = 25, m = 0.03$

advantages of a population system for continuous preservation of genetic polymorphism. This conclusion is important for conservation biology: it is clear that in subdivided and panmictic populations, the resources required for obtaining the same result are incomparable. In the case of subdivision, the size of a *"minimal viable population"* (Soulé 1987) may be hundreds of individuals because of internal fragmentation, whereas in the case of panmixia, this number may exceed thousands.

From the theoretical viewpoint, it is interesting that a decrease in variation due to genetic drift in a structured population is inversely proportional to the local migration rate; in the first dozens of generations, this decrease is more pronounced than in the panmictic population (Fig. 3.15). However, the situation drastically changes in subsequent generations. A subdivided

**Table 3.5.** "Life span" (in generations) of panmictic and subdivided populations of equal size up to the time of their loss of 95/99% of the initial heterozygosity

| Individual number ($N_i$) | Number of subpopulations ($k$) | Subdivision/gene migration coefficient | | | | | | Panmixia |
| --- | --- | --- | --- | --- | --- | --- | --- | --- |
| | | 0.0005 | 0.001 | 0.005 | 0.010 | 0.030 | 0.000 | |
| 10 | 10 | 33/1119 | 40/1997 | 440/957 | 440/957 | 364/628 | 20/44 | 299/459 |
| 20 | 25 | 81/4977 | 314/8251 | 1731/5341 | 1772/3801 | 1437/2464 | 59/90 | 1497/2301 |

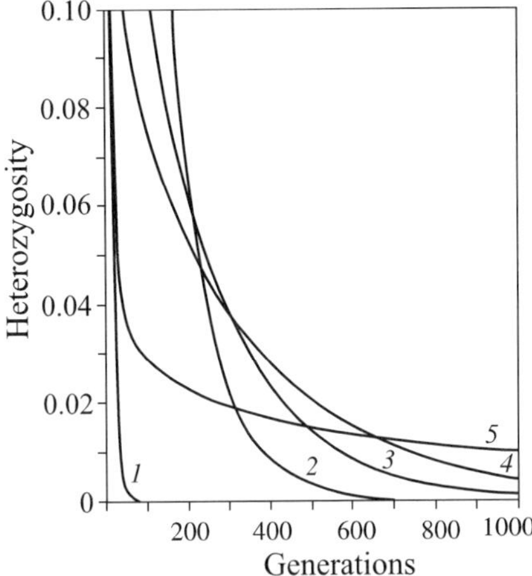

**Fig. 3.15.** The rate of decrease in heterozygosity (*ordinate*) in (*1*) 10 isolates of 10 individuals each, (*2*) a panmictic population of 100 individuals, and (*3–5*) subdivided populations of the same total size ($N_i = 10$, $k = 10$) but with different gene migration rates (*3*, $m = 0.01$; *4*, $m = 0.005$; *5*, $m = 0.001$)

population of 100 individuals ($N_i = 10$, $k = 10$, $m = 0.001$) retains a certain level of heterozygosity for thousands of generations after "degeneration" of the panmictic population (Table 3.5, Fig. 3.15).

This stability of a subdivided population is caused by its nonrandom subpopulation structure. The reorganization of this structure in biological time resembles (at least formally) a stationary or, more precisely, a quasi-stationary process: gene frequency "waves" appear in the linearly organized structure (Fig. 3.16). This is related to the circular structure of the model. In the one-dimensional linear stepping-stone model, the correlation among gene frequencies in subpopulations exponentially declines with distance, since the probability of migration gene exchange rapidly decreases at each step (Kimura and Weiss 1964). However, in the circular system of subpopulations, the correlation attains the minimal value in the middle of the migration axis and then begins to increase in a similar manner.

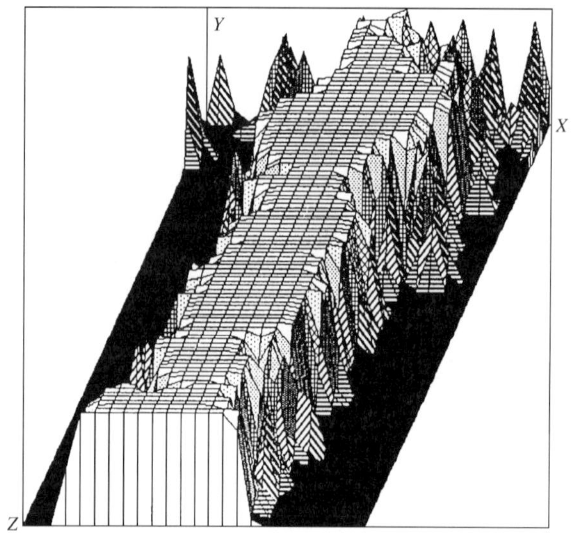


**Fig. 3.16.** The spatial and temporal dynamics of a structured population ($N_i = 20$, $k = 25$, $m = 0.03$). $Z$ time in generations (1–1000); $X$ subpopulation number; $Y$ gene frequency

## 3.3.2
## Genetic Dynamics of a Population System
## with Varying Parameters of Structure and Selection

As in the previous section, here we consider the simplest circular model of 25 subpopulations of 20 individuals each; local gene migration occurs only among adjacent subpopulations. In the case of a more complex population structure, it is difficult to differentiate between the impact of a heterogeneous environment (selection) and intrinsic genetic dynamics (migration and drift).

We have noted above that, in contrast to a panmictic population, the gene frequency dynamics in a structured population is not uniform in time. At the final stage, when the number of elapsed generations $t = kN$ tends to the population size, correlations decrease almost exponentially (Fig. 3.15). In the simplest cases, expressions for the exponent are well known (see, e.g., Moran 1962; Maruyama 1977). However, at the initial stages of this process at $t \leq N$, the gene frequency dynamics is basically different. Assuming that the coefficients of migration are small, we obtain

$$C_{ij}(t+1) = m_{ii}m_{jj}C_{ij}, \text{ at } i \neq j,$$

$$C_{ii}(t+1) = \left[ m_{ii}\left(q_i - q_i^2\right) - C_{ii} \right] / N_i + m_{ii}^2 C_{ii}.$$

The latter expression completely coincides with the well-known equation for an isolated panmictic population. Hence, all subpopulations initially

### 3.3.3
## Nonequilibrium Genetic Dynamics of Population Systems

In previous sections, we have shown how a complex subpopulation structure significantly prolongs the life span of a population by maintaining its genetic diversity for hundreds and thousands of generations (depending on the direction and intensity of gene migration or selection parameters). However, it is well known that at such long time intervals, external conditions (selection pressure) and, more importantly, the population structure itself (e.g., migration flows) can significantly change and thus promote development or, conversely, degradation of population systems. In both cases, we deal with nonequilibrium population dynamics, investigation of which is as interesting as an analysis of nonequilibrium processes in theoretical physics and chemistry.

Main assumptions of the models analyzed above are symmetry and stochasticity of the gene migration matrix, which are equivalent to the equilibrium state. In order to analyze nonequilibrium dynamics, we relax these assumptions in this section. Moreover, we assume that subpopulation size and gene migration rates can change in time. In particular, this corresponds to developing (size increases) or degrading (size decreases) population systems. A more complex situation is possible, when some subpopulations increase in time while other subpopulations decrease or change in a more complicated fashion.

This approach is presented in Blank and Altukhov (1995). The nonequilibrium model differs from the model considered in the first section of this paper in several respects. First, the migration matrix is more complex: migrations among subpopulations can vary in time. Second, heterozygosity is estimated taking into account differences in subpopulation sizes. Third, initial conditions for the system of equations are different in the two models. By assuming that at initial time $t_0$ subpopulations are not connected and their gene frequencies are fixed, we set zero initial conditions for covariances $C_{ij}$ and $C_{ii} = \bar{p}_i(1 - \bar{p}_i)$. Fourth, describing selection, we can take into account the population size dynamics and thus construct a mathematical model of the genetic dynamics of subdivided populations based on their ecological parameters.

To visualize possibilities of this model, we present graphically the dynamics of intra- and interpopulation gene diversity for the simplest circular system consisting of six subpopulations of 20 individuals each. Three situations are considered: (1) local migration is constant (stationary regime); (2) local migration decreases in each generation; (3) local migration increases in each generation. As expected, an increase in the inter-population gene flow results in the higher intrapopulation (Fig. 3.19a, curve 3) and lower interpopulation (Fig. 3.19b, curve 3) components of gene diversity.

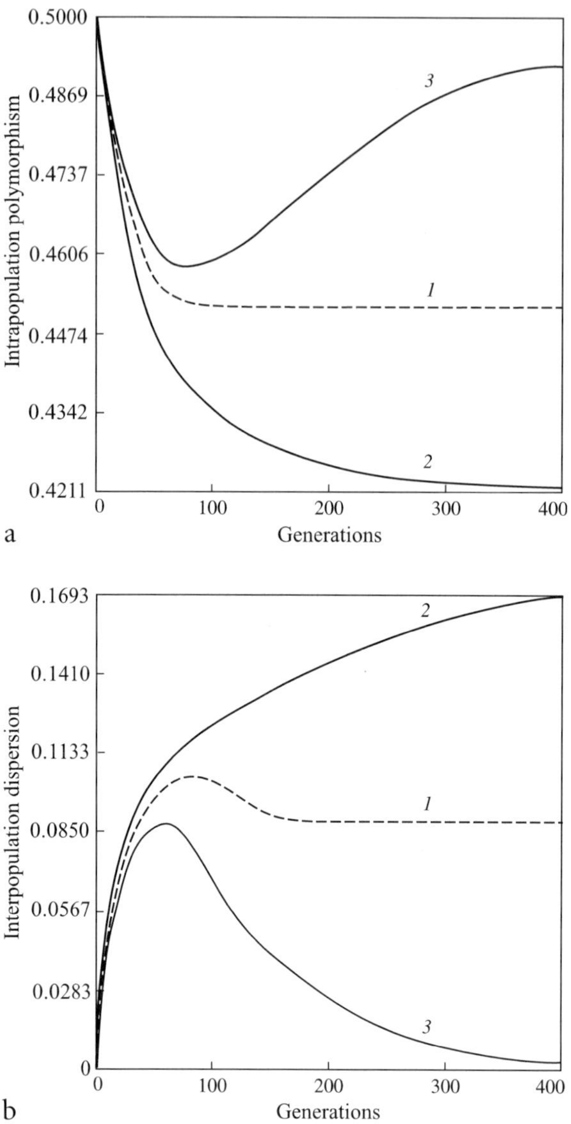

**Fig. 3.19.** The dynamics of **a** intragroup and **b** intergroup genetic variation in a circular subdivided population. *1* Stationary state: $k = 6$, $N = 20$, $m = 0.01$, $\varepsilon = 0.001$; *2* the same structure, but the local migration rate decreases in each generation ($m \rightarrow m \times 0.99$); *3* the same structure, but the local migration rate increases in each generation ($m \rightarrow m \times 1.01$)

Conversely, when the local gene flow decreases, the reverse picture is observed (Fig. 3.19a, curve 2; Fig. 3.19b, curve 2).

Using our approach outlined above, one can model the genetic dynamics of a complex subdivided population in a wide range of environments and subpopulation structure parameters. The importance of this approach for solving various questions related to conservation of gene diversity is obvious. However, we consider this approach as the first approximation of the problem. Clearly, further investigations are required that would deal with genetic consequences of various anthropogenic pressures on populations during time intervals significant for population management (Chap. 7).

## 3.3.4
## A Concept of Population Systems and a Modern View of Subdivision

The examples of studies of genetic processes given in this chapter, be it in natural population systems or in simple computer models, clearly show in what way these population systems differ from panmictic ones and from subpopulations forming the system structure: the former exhibit long-term genetic stability. This conclusion, made over 30 years ago (Altukhov and Rychkov 1970), is currently generally accepted; it is further supported by the new evidence presented here.

Note, however, that the concept of the systemic population organization, though close to the views on the so-called subdivided populations (Sect. 1.3.5), nevertheless, is principally different from them. A population system is always a subdivided population, but the latter may lack the key property of the system, i.e., *hierarchic organization*. The hierarchy of the structural levels within the system is a result of its historical development and a prerequisite of its stable existence in time and space, whereas a population subdivision may be unrelated to its history, being only a transient state observed at a given time. Because of this, when the mathematical apparatus developed for analysis of subdivided populations is used in the context of population systems, it reveals the phenomenon of an increase in genetic stability of the population as the level of its historic and geographical hierarchy increases. This is a principal difference of the conclusion of the population system studies from Sewall Wright's conclusion (see Sects. 1.3.5 and 6.2) on subdivision as a factor of accelerated evolution of populations and species.

The concept of population systems should also be compared to the metapopulation theory, which has been intensely developed by researchers outside Russia. The two theories were formulated independently but almost at the same time, in the late 1960s and early 1970s. In spite of their seeming similarity – in both cases, a population of populations is considered – these concepts differ in at least one important aspect, which requires special consideration.

Metapopulations (Levins 1970; see also Hanski 1991; Gilpin 1991) are supposed to be remnants of formerly prosperous, large panmictic populations that had high genetic diversity. Naturally, anthropogenic fragmentation of a habitat and appearance of "patchiness" instead of continuous distribution leads to a decline in the effective population size "$N_e^{meta}$" (Gilpin 1991) and, hence, to the loss of heterozygosity. In other words, "a metapopulation is an extreme form of spatial structure, in which loosely coupled local populations 'turnover', that is, suffer extinction followed by recolonization from elsewhere within the metapopulation" (Gilpin 1991, p. 165). Consequently, a conclusion is drawn that very few species possess and maintain natural population structure. Adherents of the metapopulation theory admit that the "classic conservation dilemma" is still unresolved; i.e. it is not clear what is better for species preservation: a single large patch or several small patches ("the SLOSS controversy"; Gliddon and Goudet 1994).

Moreover, an extreme conclusion was drawn that aims of conservation genetics do not necessarily require maintenance of subpopulation structure of a species. It was stated that "decrease in $N_e$ with realistic population structures is a common phenomenon, and the increase in $N_e$ with the island model is an artifact of the unnatural assumptions of this model" (Whitlock and Barton 1997).

Evidently, this conclusion follows exactly from the metapopulation ideology focused on the model, in which extinction is caused by stochastic demography and chaos created by excessive anthropogenic pressure that had transformed the natural landscape into an artificial one.

The population system concept is based on cardinally different assumptions.

The term *population system* (Altukhov and Rychkov 1970) was proposed on the basis of the long-term genogeographic studies of populations in various species including humans. These studies showed that historically formed large isolates possessed a subpopulation structure of several hierarchical levels (Rychkov 1968, 1969, 1973; Altukhov 1971, 1973b, 1974; Altukhov et al. 1969b). It was demonstrated above that when these structured populations are not destroyed by anthropogenic activities (slight damage is admissible), they remain genetically stable regardless of strong fluctuations of. gene frequencies in subpopulations. This stability was reported for time periods that are comparable with the span of historic existence of population systems of various species in naturally fluctuating environments (see Altukhov and Rychkov 1970; Altukhov 1971, 1974, 1989b; Rychkov 1973; Altukhov et al. 1983a for details). In population systems that currently represent the result of spatial and temporal differentiation of the ancestral gene pool, effects of extinction and recolonization are counterbalanced. Population size and other demographic parameters remain stable, fluctuating within historically set limits. As shown by computer simulation, the interaction between only two evolutionary factors, migration and genetic drift, in a population system of 500 individuals and 25 subpopulations maintains the same level of genetic diversity as in a panmictic population of double that effective size. Thus, investigation of natural populations and their experimental and simulation models reveals completely different genetic effects from those of the mathematical modeling of artificial constructions termed metapopulations.

Thus, in contrast to the metapopulation concept, the concept of population systems states that many, if not all, biological species (at least before they were subjected to extreme anthropological pressure) possess a fine subpopulation structure, which is a main mechanism of preservation of intraspecific genetic polymorphism. Ignorance of this species structure in the course of their commercial use is one of the main reasons for irreversible changes of genetic diversity in the biosphere.

It would be fair to note, however, that in the recent version of the metapopulation concept (Hanski and Gilpin 1997), elements of the systemic approach to investigation of such communities appear. Some new conclusions of the authors (Hanski and Simberloff 1997; Stacey et al. 1997) are similar to those made in the framework of the population system concept at the initial stage of its development 30 years ago. Presumably one more step in the same direction has also been made in the recent monograph by Hanski (1999).

Nevertheless, we would like to emphasize that, in contrast to metapopulations, population systems are actually existing natural entities with certain intra- and intersystem relationships rather than artificial constructions. They occupy historically formed ranges and have specific levels of gene diversity that are continuously preserved under natural conditions. The evolutionary optimal (according to Sewall Wright) population structure appears to be genetically stable, i.e., more stable than that of a panmictic population of the same size and genetic composition.

Now let us turn to experimental models of population systems, in analysis of which new genetic properties are revealed.

# 4 Genetic Processes in Experimental Population Systems

As far as I know, nobody has yet modeled experimentally the features of the genetic process characteristic of population systems. This approach is significant not only as a way to further verify the qualitatively different level of stability of a population system compared to a panmictic population of similar size, but also as an independent branch of research. It could have particular value as a basis for mathematical models that serve a useful purpose in the theory and practice of the rational utilization of biological resources (see Chaps. 3 and 7). After all, however economical and expressive population computer models may be, they are no substitute for experimental models, through which there is always the possibility of discovering certain features intrinsic to natural communities – features impossible to simulate electronically. In any case, of one thing one can be sure: the incorporation of data from experimental models in a simulation model will enhance its prognostic value.

Fundamental to successful work in experimental population genetics, replication of populations is a cornerstone in our experiments. Particular aspects of the genetic process were investigated in two subdivided (experimental) and two panmictic (control) populations of *D. melanogaster*. However, in order to vary the conditions of the experiment, we had to deviate somewhat from traditional approaches to research of this kind by using two different types of structure and grouping the populations on the basis of different gene pools in each series of experiments, conditions within a series being identical[1]. These variations in experimental conditions had no substantial effect on the basic aim, which was to model the dynamics of the genetic parameters of subdivided and panmictic populations of comparable size.

## 4.1
## The Structure of the Models

Two types of population systems were examined: one of them based on Wright's "island model", the other on the simplest version of the one-dimensional stepping-stone model proposed by Kimura and Weiss (1964).

---

[1]The reader will find more detailed information about these experiments in primary sources published earlier (Altukhov and Pobedonostseva 1978, 1979a,b; Altukhov and Bernashevskaya 1978, 1981).

Special population cages were constructed of Plexiglas. They and the conditions of the experiment must be described briefly so that the results obtained may be better understood. Figure 4.1 represents an "island" model, a cage consisting of nine compartments, one on top and eight below. The exchange of migrants among the lower compartments could only take place through the upper one (no. 1) which was joined to each of the lower compartments by migration openings 2.5 mm in diameter. In such a structure of migration, one would expect a "population core" to form in compartment 1 with a gene frequency corresponding to the average for the peripheral subpopulations. The migration openings could be closed at will, separating the system into nine completely isolated populations.

Genetic drift was determined by limiting the number of flies in each of the compartments, in each of which three test tubes were placed containing about 12 ml altogether of food medium; this made it possible to maintain fly numbers at an average level of about 190 in each compartment. The effective population size was considerably less – some 50–70 individuals.

Flies from a natural population of *D. melanogaster*, caught during the summer of 1971 in the North Caucasus (Shilenko 1974) and kept for a year under standard laboratory conditions, supplied the basic material for establishing the population.

Two lines were obtained: one of them was homozygous at a "fast" allele (FF) of the esterase-6 locus, the other was homozygous at the slow allele (SS) at the same locus. Grossing resulted in heterozygous $F_1$ progeny (i.e., gene frequency 0.5), and three pairs of these flies (FS) were placed in one of the lower compartments of the population cage. After one and a half months the entire cage had been colonized and the system's numbers had become

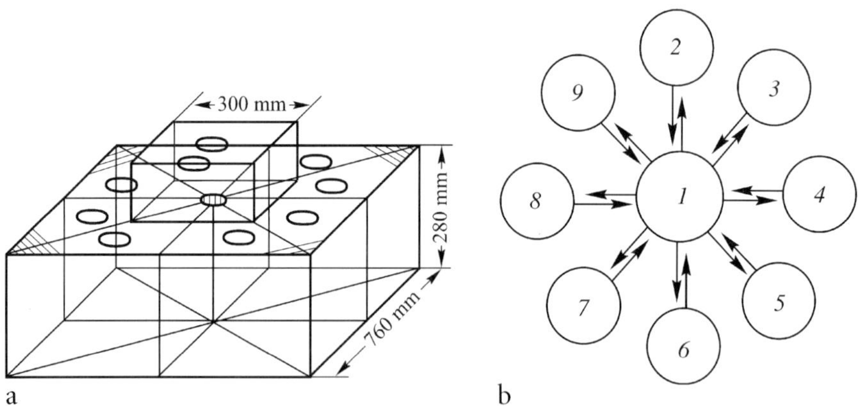


**Fig. 4.1.** The design of a population cage (Altukhov and Pobedonostseva 1979a). **a** General appearance; **b** diagram of migration in the population system

relatively stabilized. At the 17th generation a control panmictic population of a comparable size was organized and having a gene pool extremely close to, if not identical to, the gene pool of the subdivided population (Altukhov and Pobedonostseva 1979a). Both populations were kept under identical conditions – the experiment was conducted at room temperature in the usual kind of laboratory premises, and the generations did not overlap. The research plan comprised the following stages:

1. The number of flies and the sex ratio were determined in each generation and each compartment, genotypes at the Est-6 locus being identified by starch gel electrophoresis.

2. Migration cycles alternated with isolation cycles so that throughout the experiment, which lasted for 84 generations (over 1,600 days), we were able (with three replicates) to examine the behavior of the system and of isolates not connected by migration. This condition of the experiment models certain essential features of the population biology of *D. melanogaster.*

3. A control panmictic population was set up at the 17th generation of the experiment, with subsequent investigation of its biological and genetic parameters.

4. Individual tagging of flies was carried out at one of the last generations (#68) of the third migration cycle so as not to break the continuity of the experiment. This treatment allowed us to estimate migration direction and intensity in males and females separately.

Under modeling conditions a striking migration structure developed: each generation was characterized by intense migration of flies into the upper compartment, part of them subsequently flowing out into peripheral subpopulations.

This phenomenon, caused by the negative geotaxis characteristic of Drosophila and controlled by autosomal genes (Hirsch and Erlenmeyer-Kimling 1962), rapidly led to the formation in compartment 1 of a kind of "ecological optimum zone" – a population core of the system whose average numbers were five times larger than any one of the peripheral subpopulations. Moreover, in compartment 1 there was a marked excess of females in nearly all the generations that were analyzed, caused by a greater outflow of males to peripheral subpopulations.

The noticeable fluctuations in the numbers of males in the subpopulations, against a background of equilibrium in the ratio for the system as a whole, is well illustrated by the changed value of this parameter's inter-group dispersion (Fig. 4.2). Equilibrium of the sex ratio was restored in all the compartments under isolation. This equilibrium of sex ratio also characterized the control panmictic population.

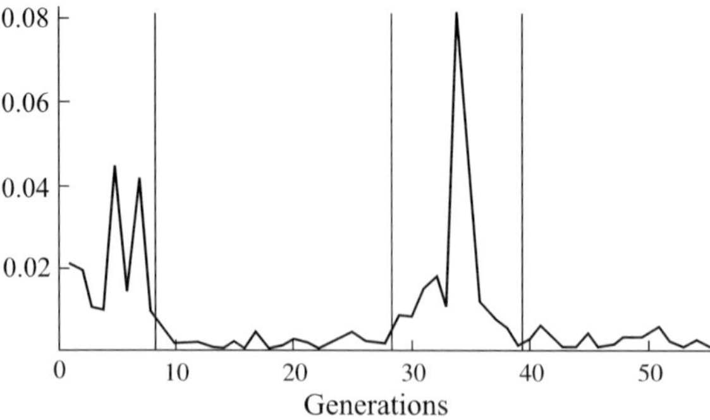



**Fig. 4.2.** Interpopulation variance values of the sex ratio (*ordinate*) in an experimental community of *Drosophila melanogaster* populations (Altukhov and Pobedonostseva 1979a). Generations 1 to 8 and 29 to 39 were run under migration; generations 9 to 28 and 40 to 55 were run under complete isolation among subpopulations

**Table 4.1.** The quantitative evaluation of the migration structure in a *Drosophila melanogaster* population system

| Subpopu-lation number | Subpopulation numbers prior to migration | Number of emigrants | Number of immigrants | Subpopulation numbers after migration | Migration coefficient |
|---|---|---|---|---|---|
| 1 | 89 | 14 | 414 | 489 | 0.847 |
| 2 | 134 | 44 | 11 | 101 | 0.109 |
| 3 | 83 | 32 | 28 | 79 | 0.354 |
| 4 | 129 | 52 | 8 | 85 | 0.094 |
| 5+8 | 263 | 150 | 8 | 121 | 0.067 |
| 6 | 87 | 59 | 7 | 35 | 0.200 |
| 7 | 140 | 126 | 10 | 24 | 0.417 |
| 9 | 75 | 18 | 9 | 66 | 0.136 |

The results of the migration analysis based on individual tagging of flies are given in Table 4.1. Because flies could only migrate through the upper compartment, it was possible to establish how many males and females left the lower compartments, migrated to the upper one, and then moved into other compartments or remained above. Table 4.1 shows that 50% of the population were active in migration. However, this figure may even be lower than the real level as flies returning to "their own" compartment could not be counted. The migration process is represented in stages in Fig. 4.3. It is clear that over half (53%) of the flies migrated from the lower compartments to the upper one, whereas only 14% of individuals moved from the upper compartment to the lower ones. Males comprised 62% of the flies migrating downward. These results confirm the hypothesis that males

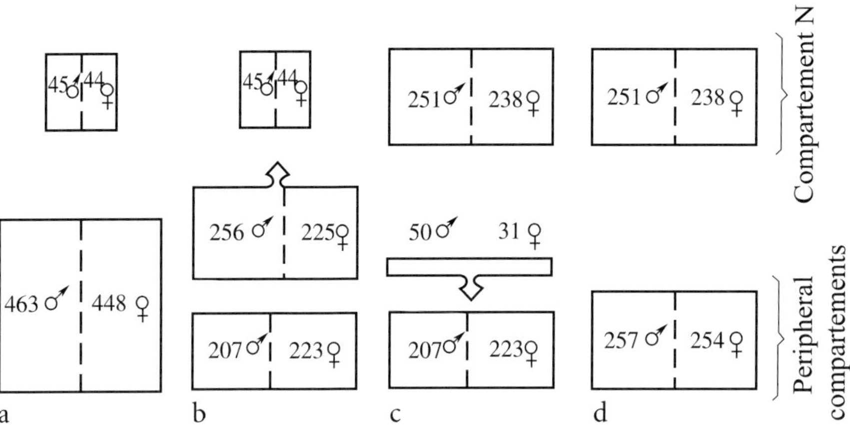



**Fig. 4.3.** The dynamics of numbers and sex ratio in a subdivided population at different stages of the migration process (Altukhov and Pobedonostseva 1979a). **a** Initial numbers and sex ratio in the system core and in the peripheral subpopulations; **b** migration of flies into the core from peripheral subpopulations; **c** migration of flies from the core to the periphery; **d** distribution of numbers and sex ratio at the end of migration. Remaining explanations in text

play the active connecting role in natural population systems (Altukhov 1974). The average coefficient of migration from the system core to the peripheral subpopulations was $0.180 \pm 0.048$. The higher migration activity of males in natural populations is known from a series of works (Blair 1960; Panov 1970; Schwarz and Armitage 1980).

The migration intensity of each generation may also be estimated in another way – the difference between the number of males and females in the system core (compartment 1) divided by the total number of flies in the peripheral subpopulations. This evaluation of the average intensity of males' migration to the periphery gives a value of $0.185 \pm 0.030$, which accords with the migration figure determined in the experiment described above.

It has already been remarked that, as a rule, an excess of females is seen in the system's core, but the amount varies with generations. If our hypothesis about the equally probable migration of flies of both sexes into the upper compartment is correct, then this variability is for one reason only – the different intensity of the males' outflow from the core to peripheral compartments. Moreover, it transpired that the smaller the peripheral subpopulation, the more likely were the males to migrate toward it. Figure 4.4 shows the negative correlation between the number of migrating males and the size of peripheral subpopulations (totals). It follows from these data that the migration structure in our system is indeed non-random,

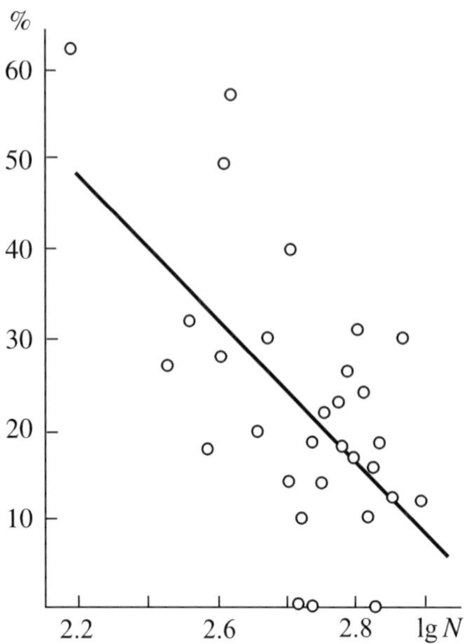



**Fig. 4.4.** Correlation between the size of "island" subpopulations and the proportion of males migrating into them from the "mainland". *x-axis* Logarithm of the size of peripheral subpopulations after the outflow of flies from them into the upper compartment; *y-axis*: the number of males migrating into peripheral subpopulations as percent of their total numbers in the system in a given generation. The theoretical regression line satisfies the equation $y = 1.34 - 0.39x$; $r = -0.57$; $P < 0.05$

and that the smaller the size of the "island" subpopulations, the greater the number of males that migrate in every generation. In other words, we see how the system's structure is divided, as it were, into two parts – one regulates ("mainland") and the other is regulated ("island").

The numbers of flies in the subdivided and panmictic populations fluctuated over generations (Fig. 4.5), evidently because of environmental factors not accounted for. It is interesting to note, however, that if the fluctuation of the numbers of both populations during the first half of the experiment was virtually synphasic, and numbers were practically the same in the interval between the 36th and 48th generations, then, beginning with the 50th generation, the number of flies in the experimental system exceeded the number in the control panmictic population by 293 specimens on average (the level of significance is $P < 0.01$; see Fig. 4.5). On comparing the average values of numbers of flies in the experiment in cycles with and without migration, no significant differences were found ($P > 0.05$).

Figure 4.6 shows the dynamics of subpopulation numbers in separate compartments. It can be clearly seen that whereas the numbers of all subpopulations are approximately the same under isolation conditions, during

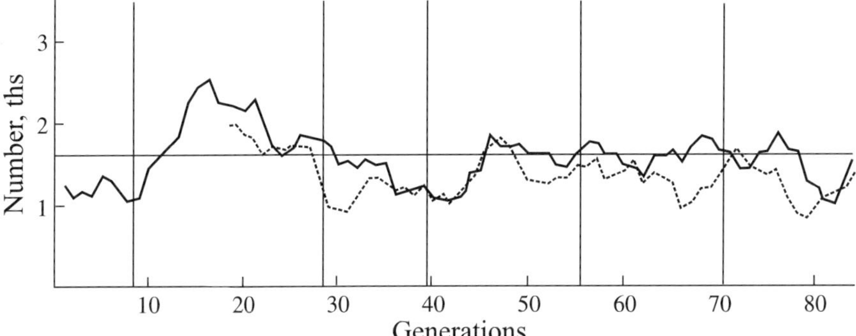

**Fig. 4.5.** The dynamics of total numbers in a subdivided (*continuous line*) and a panmictic (*dotted line*) population over several consecutive generations. Generations 1 to 8, 29 to 39, and 56 to 70 were run under migration conditions; generations 9 to 28, 40 to 55, and 71 to 84 were run under conditions of complete isolation among subpopulations. (Altukhov and Pobedonostseva 1979a)

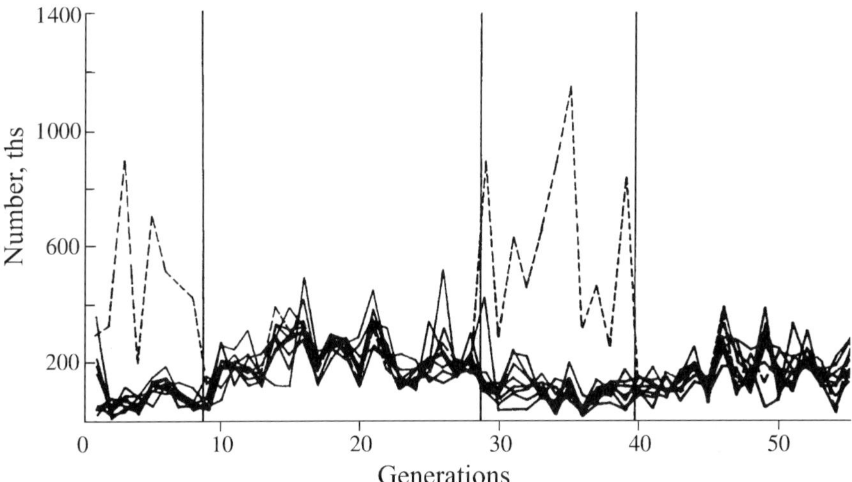

**Fig. 4.6.** The dynamics of numbers in separate subpopulations of a population system in several consecutive generations (Altukhov and Pobedonostseva 1979a). The notation is the same as in Fig. 4.5

migration the system core is characterized by considerable fluctuation – from 20 to 60% of the total population number. One would have thought that such sharp fluctuations of individual numbers in compartment 1, caused by intensive migration, would lead to increased variation in the size of the peripheral subpopulations. However, this was not the case: the value of the coefficient of variation of the total numbers in the peripheral compartments remained at a level typical of the experiment as a whole (Table 4.2).

**Table 4.2.** Certain parameters of population numbers of *Drosophila melanogaster* under different experimental conditions

| Experiment conditions | $n$ | $N\pm SE.$ | $\sigma$ | $C.V., \%$ |
|---|---|---|---|---|
| Subdivided population (1–84)[a] | 83 | 1621.22±56.78 | 517.32 | 31.91 |
| Subdivided population with migration (1–8, 29–39, 56–70) | 34 | 1501.41±75.57 | 440.64 | 29.35 |
| Subdivided population with isolation (9–28, 40–55, 71–84) | 49 | 1704.35±79.09 | 553.66 | 32.49 |
| Panmictic population (17–84) | 66 | 1403.67±57.98 | 471.02 | 33.56 |
| Peripheral populations with migration (1–8, 29–39, 56–70) | 33 | 941.79±61.37 | 352.54 | 37.43 |
| Population core (1–8, 29–39, 56–70) | 33 | 555.06±45.14 | 259.30 | 46.72 |

Conventional designations: $n$, number of observations; $N\pm SE$, average number value and standard error; $\sigma$, dispersion of number; $C.V.$, coefficient of variation of numbers
[a] Brackets give generation numbers

Furthermore, the characteristic migration structure resulted in the redistribution of the total number of the population system, about 40% of which on average was found in compartment 1. This induced a fall in the mean size of the population and its dispersion in the peripheral subpopulations, as compared with observations under conditions of complete isolation.

It can be seen that migration in the population system resulting from increased numbers of *Drosophila* in the uppermost compartment, shouldering the brunt of the fluctuations, kept the peripheral subpopulation numbers at a relatively stable level.

Thus, the dynamics of the biological parameters studied differ qualitatively according to whether a panmictic population, a community of isolates not linked in any way, or a system of subpopulations are involved. In the latter case the most essential features of the system are: (1) a clearly expressed dispersion of the sex ratio in peripheral subisolates when a subdivided population as a whole is in equilibrium at this characteristic; (2) a non-random migration structure, reflecting ecological features of the system core; and (3) a core regulatory role in the migration process and the maintenance of a number of island subpopulations at a relatively stable level.

Furthermore there was evidence of greater numbers in the subdivided system (which increased with time) than in the control panmictic population. This can plausibly be explained by the heterotic effect of outbreeding (migration) replacing inbreeding cycles (isolation). A similar alternation of separation and interaction among subpopulations, caused by pulsed, seasonal changes in numbers, is characteristic of many animal species. For instance, it was shown for *D. melanogaster* (Danieli and Costa 1977) that migration among elementary subpopulations is disrupted during the winter/spring season because of the very low numbers of flies. In the summer and autumn, population numbers increase, their areas partially overlap, and a population structure is formed, which with the advent of cold weather is disrupted again.

The structure of the stepping-stone model, and correspondingly the character of migration, were different. An experiment was conducted in a cage consisting of 30 communicating compartments joined together by vinyl chloride tubes 35 mm long and 4 mm in section. This ensured an average m value of about 0.03 (Fig. 4.7). In accordance with theory, each population could only exchange individuals directly with two adjacent subpopulations (although this does not mean at all that this, in fact, is the only possible kind of migration).

Migration was not interrupted, enabling the behavior of the experimental subdivided population to be compared with that of the control panmictic population for several dozens of generations. The stepping-stone model also differs from the island model in another important respect, namely, the considerably greater heterogeneity of the original gene pool of the

**Fig. 4.7.** The structural characteristics of an experimental *D. melanogaster* population system corresponding to the circular stepping-stone model (Altukhov and Bernashevskaya 1978). *Arrows* indicate possible migration direction

population. Two autosomal diallelic loci: *Est*-6 and $\alpha$-glycerophosphate dehydrogenase, $\alpha$-*Gdh* were studied.

Genetic heterogeneity of the original lines was assured by the fact that they were taken from natural populations of the Crimea and the North Caucasus. In addition, the colonization of the experimental "area" was effected by means of 150 pairs of double heterozygotes, placed in the population cage at the same time. The whole cage was colonized within 24 h (the number of individuals in different compartments fluctuated from 1 to 37), after which tubes were inserted in the migration openings ensuring subsequent exchange of migrants at a level of approximately 3%. Each compartment always had two test tubes containing standard yeast medium supporting an average subpopulation number at the level of 135 individuals and $N_e \sim 50$. As in the island model, the generations were discrete.

A parallel experiment was carried out with the control population cage. This panmictic population was set up as follows: an extra test tube containing fresh medium was placed in each compartment of the subdivided cage before the first generation individuals emerged. Later, at the end of the first generation, all the additional test tubes were taken out and used to establish a panmictic population in the undivided cage. Virtually complete identity of the genetic composition of subdivided and panmictic populations was thus achieved, and the conditions observed for the "evolution" of the populations from the initial frequency of 0.5 for the alleles at both the loci followed.

The numbers of flies were determined in the control and experiment populations for 50 generations. These numbers turned out to be almost identical. The average number in the subdivided population was 4,036, against 4,097 in the panmictic population (Fig. 4.8). The minor changes in numbers in the subdivided and panmictic populations, caused by negligible uncontrolled fluctuations of the environment and probably by individual competition for food (Noguès 1977), were also synchronous. This fact reflects the identity of the populations' genetic structure and thus the parallel effect of the environment. Evaluation of population size revealed not only its temporal fluctuations (Fig. 4.8), but also differences in numbers of flies in the subpopulations. Hence, variability of the sizes of the population system's structural components was observed in both time and space.

The sex ratio was recorded for 30 generations. Significant fluctuations in time and space were observed for the sex ratio with an average variance of 0.0035 (Fig. 4.9). However, the mean sex ratio taken for the system of subpopulations as a whole was virtually unchanged in time, at 1.0 ($\sigma = .00025$). This agrees completely with the pattern described earlier for natural population systems (Altukhov 1974).

Throughout the entire experiment, which lasted for about 800 days (over 60 generations), the population cages were kept in a controlled room with

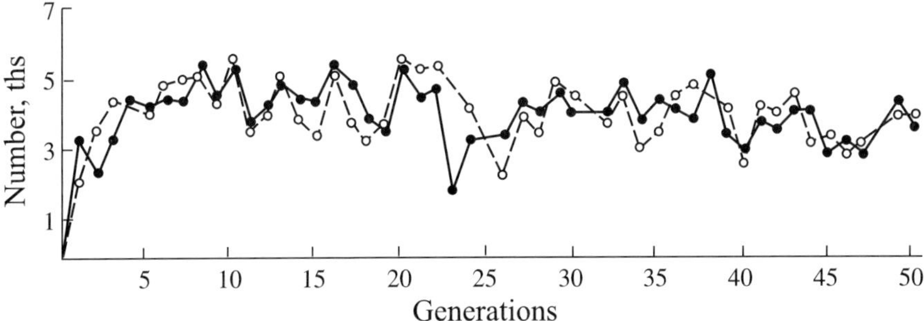

**Fig. 4.8.** The dynamics of the numbers of flies in successive non-overlapping generations of subdivided (*continuous line*) and panmictic (*dashes*) populations. (Altukhov et al. 1979b)

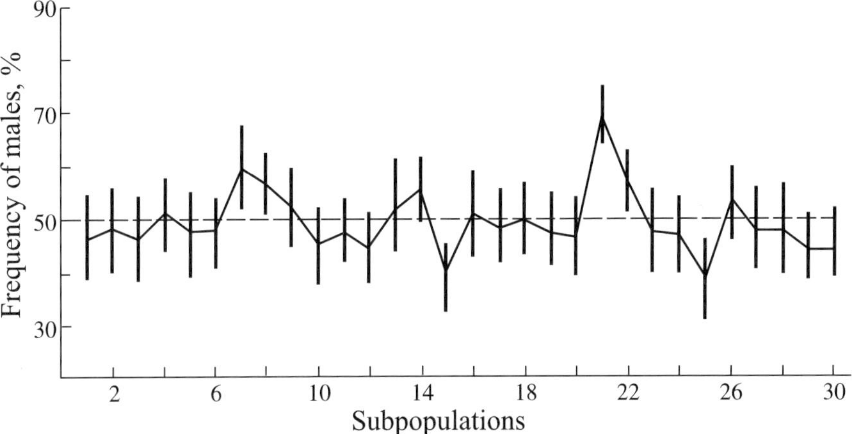

**Fig. 4.9.** Fluctuations in the sex ratio within subpopulations of an experimental *Drosophila melanogaster* population with stepping-stone migration structure

constant humidity and at a temperature of 24 °C±1 °C. The subdivided cage was evenly illuminated from top to bottom through a transparent lid so that light rays fell perpendicular to the direction of migration. Despite positive phototaxis in the *D. melanogaster*, this ensured similar migration throughout the whole system of subpopulations. In order to avoid unevenness of illumination in the different compartments, the population cage was placed on a revolving base, and turned through arbitrary angles at preset time intervals. This allowed each compartment to occupy a random position in the space surrounding it.

Electrophoretic analysis of enzymes was conducted for 5–40 generations. The sample size for each subpopulation averaged 40 individuals – about 30% of the number in each subpopulation; that for a panmictic subpopulation averaged 100 individuals.

Thus, there were fundamental differences between the two series of experiments both in the types of population structures compared, and with the characteristics of the ancestral populations: one was represented at the start by a maximum of three pairs of flies, the second by several dozens of founders.

Let us now examine specific aspects of the genetic processes in these two experiments.

## 4.2
## The Genetic Process in the "Island" Population Model

As already indicated, the main purpose of these experiments was to compare the dynamics of gene frequencies over time in: (1) a population system, (2) a panmictic population, and (3) a community of completely separated isolates.

In two variants of an experiment (migration I and II, and isolation I and II) the genetic structure of a subdivided *D. melanogaster* population was analyzed in each generation. We examined the dynamics of characteristics such as the average gene frequency and interpopulation variance of gene frequencies in migration cycles and in isolation cycles. In a third variant of the experiment (migration III and isolation III) an analysis of genotypes was made only at the beginning and end of cycles in order to verify the results obtained previously.

Figure 4.10 represents the frequency dynamics of a fast F allele of an esterase locus in an experimental subdivided population and a control panmictic population. Just after formation of the control population we see a sharp fall in its allele frequency from 0.63 (generation #17) to 0.36 by the 30th generation of the experiment. Some 20 generations later the frequency remained close to 0.3. Then, beginning with the 50th generation, it began to fall again and by generation 83, the last to be analyzed, it reached 0.13. Consequently, during the time interval of 66 generations, a significant shift downwards occurred in the frequency of the fast allele in the control population as a result of frequency-dependent selection effects or some similar process. Let us now look at the dynamics of gene frequencies at a population-system level and in a community of isolates not connected by migration interaction. In the latter case only two factors of evolutionary dynamics are effective – selection and random genetic drift.

Figure 4.11 shows the variability of gene frequencies in separate sections of a population system at the initial and final generation in each cycle. The black circles represent the mean values of a given generation, light circles over an entire cycle. In the migration cycles, despite the considerable fluctuations in frequency in subpopulations, it will be seen that the original

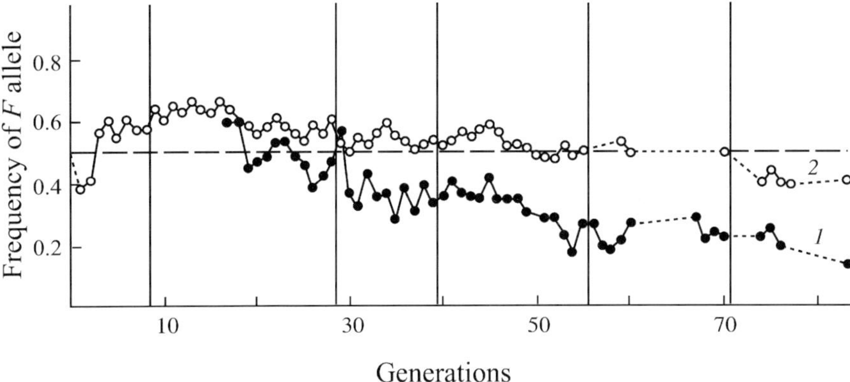

**Fig. 4.10.** The dynamics of the frequency of *Est-6*$^F$ in successive generations of a control panmictic (*dark circles*) and experimental subdivided (*light circles*) *Drosophila melanogaster* populations (Altukhov and Pobedonostseva 1979a). The *vertical lines* on the graph separate migration cycles (generations 1–8, 29–39, and 56–71) from isolation cycles (generations 9–28, 40–55, and 71–83). The *dotted line* represents the starting frequency of 0.5

mean value of 0.5 is preserved from generation to generation, as well as from cycle to cycle. Stemming from three pairs of heterozygous flies, the population system has kept the frequency set at the start. It is clear from Table 4.3 that there is essentially no difference between the frequency in compartment 1 (the system core) and the average for the peripheral sub-populations. This distribution of gene frequencies corresponds to Wright's "island model".

When the migration factor is plotted there is no such stability. In Fig. 4.11 one can clearly trace a fall in the average gene frequency within a community of completely isolated populations, although this process also proceeds at a slower rate than in the control panmictic population. It is evident that in the isolates, each of which is subjected to random genetic drift, selection effectiveness is substantially lower (by about a half) than in the control panmictic population: the reduction in gene frequency per generation ($\Delta p$) in the community of isolates equals 0.0031 while in the control $\Delta p = 0.0071$. Hence, as one might expect, the experimental system of populations is characterized by a stability level of average gene frequency which is qualitatively different from that of the panmictic population or the community of independent isolates with the same total size.

Stability of the experimental system is also observed when analyzing interpopulation variance of gene frequencies: its average values in all migration cycles are extremely close to each other, and corresponding shifts in several successive generations are also small (Table 4.3). There is a different picture in isolation cycles. The intergroup variance behavior is virtually unpredictable, as the example of the first cycle shows most clearly. In this

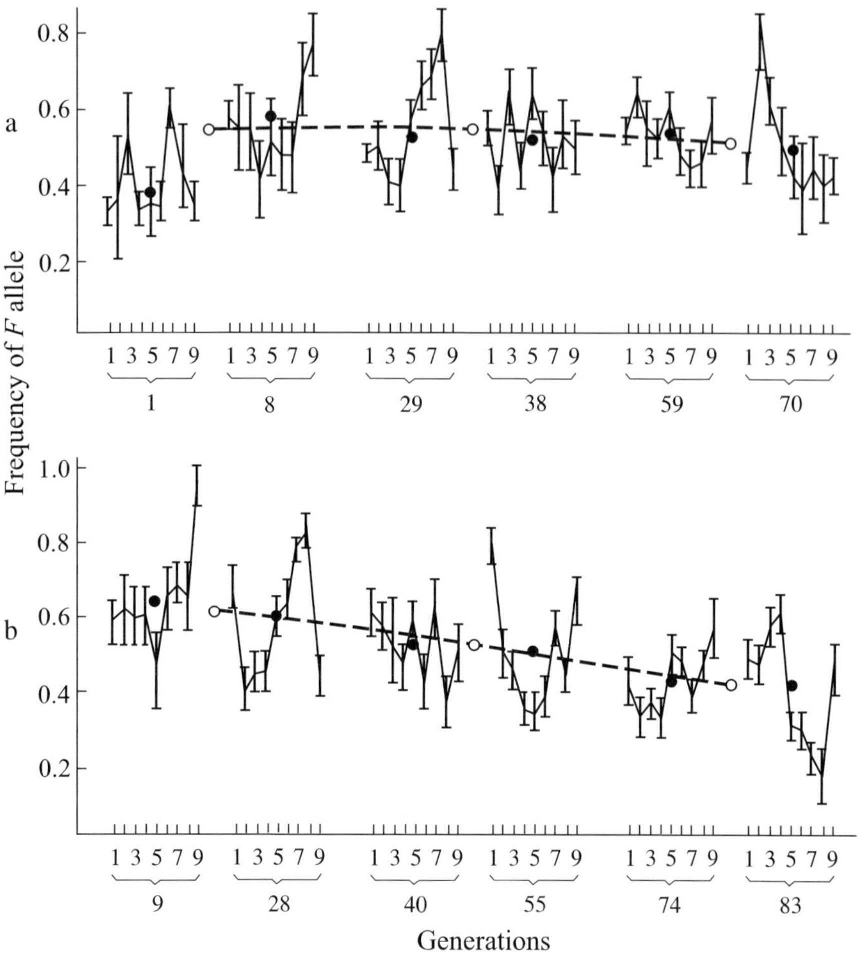

**Fig. 4.11.** The frequency dynamics of *Est*-6$^F$ **a** at different levels of a population system structure and **b** in a community of isolates (Altukhov and Pobedonostseva 1979b). *Dots joined by a continuous line* within the limits of one generation represent the frequencies in separate subpopulations (**a**) and isolates (**b**). *Vertical lines* represent twice the standard errors of estimated gene frequencies. The *dotted line* connects the average frequencies of the cycles. The remainder of the explanations are in the text

instance, the range of fluctuation is particularly large – from 0.001 in the 15th generation of the experiment to 0.034 in the 25th and 27th generations. Under the effects of selection, this parameter no longer displays such marked differences between generation in the second and third isolation cycles, although the average variance value is reduced from 0.018 in cycle 1 to 0.01 in cycle 3 (Table 4.3). Hence, stochastic effects alone are not sufficient to counteract the directed effects of selection. Presumably, migration determines the system's genetic stability. This results primarily from the constant change in direction and/or the intensity of selection at a subpopulation level that it elicits. Furthermore, it should be emphasized that interpopulation heterogeneity of allelic frequencies (Table 4.3) is maintained under the conditions of the population system despite a naturally evolved high level of migration ($m{\sim}0.18$). This means that under the conditions of our experiment, migration does not prevent the differentiation of the gene pool of a subdivided population, which might have been expected from theoretical formulations (Kimura and Weiss 1964; Kimura and Maruyama 1971; Moran 1973). We shall discuss the possible reasons for this phenomenon later.

Thus, clear differences have been revealed between the genetics of a subdivided and a panmictic population, as investigated in relation to the frequency dynamics of alleles at the *Est*-6 locus.

Using the example of this locus, Kojima and Yarbrough (1967) were the first to demonstrate the effects of frequency-dependent selection in a series of experimental panmictic populations of *D. melanogaster*. After 15 generations the populations under study reached the equilibrium frequency of the F allele, close to 0.3, at which all three genotypes were selectively neutral. The populations were observed for 30 generations. Over this same time interval the dynamics of the allelic frequencies in our panmictic population coincided completely with the results obtained by Kojima and Yarbrough.

However, not all workers observed this picture of the kinetics of genotypic composition at the *Est*-6 locus of populations. In several cases – usually studying populations with a heterogeneous initial gene pool (McIntyre and Wright 1966; Altukhov and Bernashevskaya 1978) – the *Est*-6 locus acted as if overdominance was present. Moreover, detailed analysis of the effect of environmental conditions on the allelic composition of populations revealed the fitness dependence of *Est*-6 genotypes not only on their frequency but also on temperature and population density (Birley and Beardmore 1972). It is interesting to note the works of French researchers (Anxolabéhère 1976) who discovered the effect of frequency-dependent selection, which phenomenon brought *D. melanogaster* populations through several generations to the same equilibrium frequency (0.3) at the "sepia" locus. This gene is also located in the third chromosome almost adjacent to locus *Est*-6 (sepia III – 26.0; *Est*-6 III – 36.8).

The above data may suggest that frequency-dependent or another type of selection does not influence an esterase locus directly, but acts upon a cluster of closely linked genes. However, the direction and strength of selection can change depending on the qualitative composition of this supergene and the initial genetic background. We cannot exclude some effects of linkage disequilibrium here.

**Table 4.3.** Dynamic of *Est-6*$^F$ allele and its intergroup variances in "island" system of subpopulations and under condition of complete isolation in *Drosophila melanogaster*

| Experimental conditions | No. of genera-tions | $P_1$ | $\tilde{p}_{2-9}$ | $\tilde{p}_{1-9}$ | $V_p$ | $\chi^2$-test for homogeneity of dispersion (*df* in paren-theses) | $\chi^2$-test for population homogeneity |
|---|---|---|---|---|---|---|---|
| 1 | 2 | 3 | 4 | 5 | 6 | 7 | 8 |
| Migration | 1 | 0.321 | 0.393 | 0.378 | 0.008 | | 100.02 |
| | 2 | 0.340 | 0.473 | 0.411 | 0.010 | | 61.15 |
| | 3 | 0.583 | 0.561 | 0.575 | 0.007 | | 84.47 |
| | 4 | 0.586 | 0.610 | 0.602 | 0.002 | | 12.36 |
| | 5 | 0.579 | 0.528 | 0.549 | 0.008 | 6.88 (7) | 115.79 |
| | 6 | 0.624 | 0.590 | 0.601 | 0.009 | | 117.92 |
| | 7 | 0.620 | 0.560 | 0.585 | 0.003 | | 26.01 |
| | 8 | 0.593 | 0.558 | 0.575 | 0.007 | | 47.95 |
| Average per 1st cycle | | 0.553 | 0.525 | 0.536 | 0.007 | | |
| | 29 | 0.483 | 0.558 | 0.527 | 0.009 | | 163.50 |
| | 30 | 0.481 | 0.508 | 0.500 | 0.011 | | 87.04 |
| | 31 | 0.552 | 0.529 | 0.537 | 0.005 | | 74.98 |
| | 32 | 0.558 | 0.522 | 0.534 | 0.006 | | 66.01 |
| | 33 | 0.520 | 0.584 | 0.556 | 0.009 | | 106.65 |
| | 34 | 0.617 | 0.529 | 0.586 | 0.006 | 8.31 (10) | 73.28 |
| | 35 | 0.527 | 0.585 | 0.500 | 0.007 | | 114.34 |
| | 36 | 0.500 | 0.559 | 0.529 | 0.005 | | 26.48 |
| | 37 | 0.500 | 0.524 | 0.515 | 0.003 | | 31.29 |
| | 38 | 0,556 | 0,521 | 0,518 | 0,007 | | 61,37 |
| | 39 | 0.546 | 0.519 | 0.533 | 0.002 | | 22.85 |
| Average per 2nd cycle | | 0.535 | 0.542 | 0.538 | 0.006 | | |
| | 59 | 0.536 | 0.527 | 0.530 | 0.003 | | 37.45 |
| | 60 | 0.430 | 0.514 | 0.505 | 0.006 | 4.47 (2) | 78.63 |
| | 70 | 0.442 | 0.508 | 0.488 | 0.013 | | 169.72 |
| Average per 3rd cycle | | 0.481 | 0.516 | 0.508 | 0.007 | | |
| Isolation | 9 | 0.587 | 0.654 | 0.642 | 0.010 | | 65.37 |
| | 11 | 0.593 | 0.657 | 0.648 | 0.014 | | 216.87 |
| | 13 | 0.764 | 0.644 | 0.658 | 0.014 | | 184.07 |
| | 15 | 0.604 | 0.631 | 0.627 | 0.001 | | 25.79 |
| | 17 | 0.686 | 0.633 | 0.638 | 0.017 | 31.83 (19) | 244.67 |
| | 19 | 0.679 | 0.583 | 0.593 | 0.024 | | 459.96 |
| | 21 | 0.574 | 0.549 | 0.577 | 0.024 | | 423.80 |
| | 23 | 0.457 | 0.600 | 0.588 | 0.022 | | 236.28 |
| | 25 | 0.521 | 0.546 | 0.543 | 0.034 | | 491.84 |
| | 27 | 0.592 | 0.555 | 0.559 | 0.034 | | 442.16 |

**Table 4.3.** (continued)

| Experimental conditions | No. of genera- tions | $P_1$ | $\tilde{p}_{2-9}$ | $\tilde{p}_{1-9}$ | $V_p$ | $\chi^2$-test for homogeneity of dispersion (*df* in paren- theses) | $\chi^2$-test for population homogeneity |
|---|---|---|---|---|---|---|---|
| 1 | 2 | 3 | 4 | 5 | 6 | 7 | 8 |
| Average per 1st cycle | | 0.618 | 0.605 | 0.608 | 0.018 | | |
| | 41 | 0.542 | 0.530 | 0.532 | 0.006 | | 50.05 |
| | 43 | – | 0.548 | 0.548 | 0.013 | | 102.04 |
| | 45 | 0.587 | 0.570 | 0.572 | 0.017 | | 89.48 |
| | 47 | 0.653 | 0.510 | 0.521 | 0.010 | | 122.59 |
| | 49 | 0.636 | 0.503 | 0.511 | 0.007 | 8.12 (15) | 129.13 |
| | 51 | 0.684 | 0.442 | 0.475 | 0.014 | | 175.75 |
| | 53 | 0.730 | 0.487 | 0.523 | 0.011 | | 124.39 |
| | 55 | 0.786 | 0.458 | 0.503 | 0.021 | | 313.84 |
| Average per 2nd cycle | | 0.647 | 0.506 | 0.523 | 0.011 | | |
| | 74 | 0.417 | 0.413 | 0.414 | 0.005 | | 76.43 |
| | 75 | 0.550 | 0.410 | 0.427 | 0.004 | | 48.96 |
| | 76 | 0.458 | 0.393 | 0.398 | 0.012 | 6.76 (4) | 217.05 |
| | 77 | 0.367 | 0.394 | 0.393 | 0.008 | | 116.51 |
| | 83 | 0.482 | 0.398 | 0.409 | 0.020 | | 237.04 |
| Average per 3rd cycle | | 0.476 | 0.402 | 0.407 | 0.010 | | |

Conventional designations: $P_1$, allele *F* frequency in compartment no. 1 (population system core under migration conditions); $\tilde{p}_{2-9}$ and $\tilde{p}_{1-9}$, weighted average allele *F* frequency, respectively, in peripheral subpopulations (compartments 2–9) and in the entire subdivided population (compartments 1–9); $V_p$, interpopulation variance of gene frequencies
Only odd generations are given for cycles 1 and 2 under isolation conditions. The calculation of all the average values and the estimate of dispersion homogeneity based on Bartlet's criterion have been made, taking into account all the analyzed generations. The standard $\chi^2$ values for eight degrees of freedom at (*P*) value levels 0.05, 0.01, and 0.001 equal, respectively, 15.51, 20.09, and 26.13

It is evident that in the present context these features are not of major significance, as our task involved only a comparison of the dynamics of the gene frequencies in different types of population structure with similar gene pools and under identical environmental conditions. Irrespective of whether evolutionary factors directly affect the *Est*-6 locus or a supergene for which it acts as a marker, the results of our investigation clearly show that no substantial deviations from the initial gene frequency of 0.5 occurred during many generations of the experimental population system, despite the variability of its structural components. Note that during exactly the same time interval the frequency of *Est*-6$^F$ fell to 0.13 in the control panmictic population.

One would have expected that the stability of the average gene frequency in the subdivided population to correspond to the simplest random drift model when a population, differentiated in time, starts from one or several heterozygous pairs of parents. However, in that case one should observe an increased intergroup variance in the populations (see Chap. 1). In fact the $V_p$ values remain stable in time, which indicates the integrating role of migration.

The role of migration as the chief factor in maintaining the genetic stability of the population system is revealed no less distinctly when the population system is compared with the community of isolates. In plotting the migration factor we observe a fall in the average frequency of $Est\text{-}6^F$, although the rate of this is only half that seen in the control panmictic population. This difference is caused by the diminished effectiveness of selection resulting from the action of genetic drift. In addition, a comparison of the stability of intergroup genetic variance in the migration cycles and variability of this parameter in the isolation cycles also provides evidence of the fundamental difference between the population system and the mechanical community of the elementary populations.

In population genetics theory it is usually believed that the principal role of gene flow consists in the leveling and smoothing off of differences among populations, shaping clinal variability or even in establishing the uniformity of gene frequencies over wide areas (see Chap. 2). However, all models of this kind are based on the hypothesis that migration has a random character.

Insofar as this is not the case, the maintenance of a high level of heterogeneity of allelic frequencies in our experimental system with a migration coefficient of about 0.2 may serve to indicate either the extremely low genetic effectiveness of migration or its non-random nature. This latter hypothesis is supported by the increase of the intergroup variance of gene frequencies obtained in the experiment, accompanied by an increase in the intensity of the migration flow from the core to the system's periphery of as much as $m = 0.35$. Only when this exceptionally high level is exceeded is there a gradual fall in the intergroup variance values (Fig. 4.12).

Since the feedback between "mainland" and "islands" in our population system is formed by males, the correlation that has been found may mean it is this sex that makes a decisive contribution toward maintaining hereditary heterogeneity at a subpopulation level.

Note that in the recent years, owing to the use of genetic markers, ample new evidence for nonrandom structure of gene migrations and intrapopulation mating has been obtained for various animal species (Tregenza and Weddel 2000; Winters and Waser 2003; Stow and Sunnucks 2004). These observations from a new viewpoint support the conclusion on the key importance of systemic organization of populations in maintaining balance

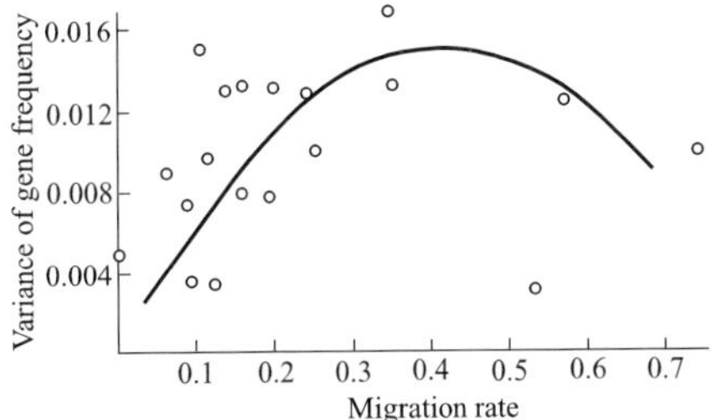

Fig. 4.12. The relation between the intensity of migration of males from a system core to the periphery and the variance values of gene frequencies at the level of peripheral subpopulations nos. 2–9. $y = 0.0128 \pm 0.0011 \times lnx$; $r = 0.56$; $P < 0.05$

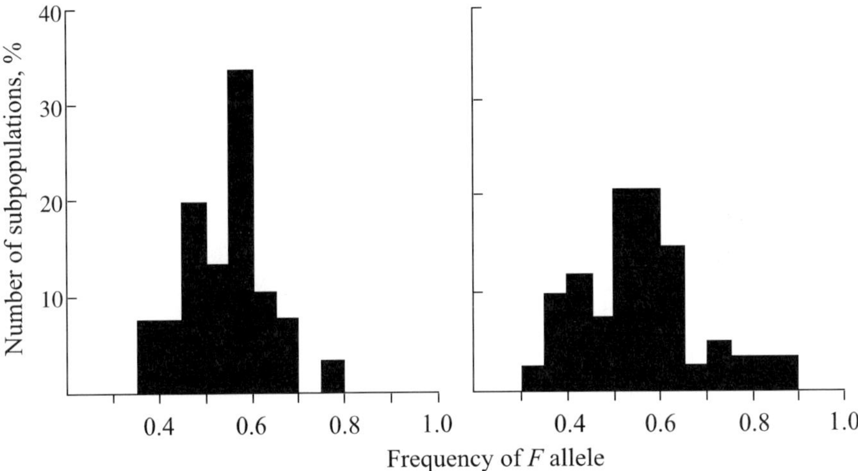

Fig. 4.13. Distributions of frequency of $Est$-$6^F$ in subpopulations of $Drosophila$ $melanogaster$ with an excess of a females ($K = 60; \overline{N} = 95.7 \pm 15.0; pF = 0.54; V_p = 0.008$) and b males ($K = 61; \overline{N} = 94.8 \pm 6.7; pF = 0.56; V_p = 0.017$) under conditions of an island-type population system ($F = 2.13; P < 0.05$). Corresponding distributions are constructed for subpopulations in which the proportion of males is over 60% and less than 50%

between inbreeding and outbreeding, i.e., the optimal genetic diversity ("inbreeding optimization" in terms of Stow and Sunnucks 2004).

Similar data have already been examined in Chap. 3 during the analysis of the genetic processes in a natural population system of the sockeye salmon, where the intergroup variance of allelic frequencies was significantly higher in subpopulations with an excess of males. Now exactly the same result represented in Fig. 4.13 has been obtained for the experimental system of populations belonging to a completely different biological species.

It should be pointed out that in estimating migration intensity, we have hitherto operated with coefficients which reflect the value of the "mechanical" flow of genes; and the value, as has been noted, may be very far removed from the genetically effective migration measured by the migrants' contribution to the gene pool of subsequent generations. If the real intensity of gene migration in our experimental system of populations is less than our estimates of it, then the conditions for local differentiation become even more favorable. However, of course, other explanations may also be suggested for mechanisms that maintain the genetic differentiation of "island" subpopulations.

What is far more important to us is that while investigating the experimental system of partially isolated populations, we established that the same effects of stability as those noted in the preceding chapter for natural population systems are seen. Key characteristics, such as average gene frequency and intergroup genetic variance, displayed no essential changes in several consecutive generations.

Nor can one ascribe the stability observed in the experimental population system, set against a background of the variability of the subpopulations that form it, to differences in numbers at various hierarchical levels. If the stability of the average values of gene frequencies depended solely on the size of the population, the control panmictic population, with numbers comparable to those in the population system, would also have to be fairly stable. However, this was not observed, either within the context of the experiment that has just been reviewed, or in the experiments with a *D. melanogaster* population system based on the type of one-dimensional circular stepping-stone model examined below. Moving ahead of the subject, I shall point out that in the latter case, the dynamics of gene frequencies at the $\alpha$-glycerophosphate-dehydrogenase locus in the control panmictic population was subjected to the effects of powerful directional selection, whereas in the population system the allelic frequencies, to all intents and purposes, did not change when the system reached a stationary regime.

When the concept of the genetic stability of population systems was first formulated (Altukhov and Rychkov 1970), some authors attributed the phenomenon to the "averaging effect". The data that have been examined enable us to answer this question. The experimental material presented here shows that averaging does indeed take place, but this is no bare statistical effect resulting from the method of processing the material, but an intrinsic characteristic of the population system itself deriving from the non-random structure of gene migration. As we have seen, it is the system core which assumes an integrating role remaining stable through a long series of generations despite the fact that its effective numbers are considerably less than in the control panmictic population. Virtually the same effects of the stability of the genetic composition in time were revealed in the course of research on a different experimental system having a stepping-stone structure of gene migrations.

## 4.3
## The Genetic Process in the Stepping-Stone Population Model

Assay of genotypes at the *Est-6* and *α-Gdh* loci was undertaken in generations 5, 10, 16, 30, 36, 41, 51, and 61. The results of three of these "cross-sections" are shown as sector diagrams of the $Est\text{-}6^F$ and $\alpha\text{-}Gdh^F$ frequencies in generations 16, 36, and 61 pictured in Fig. 4.14. The large dispersion of allelic frequencies at both loci signifies local differentiation which had become established at the 5th generation, becoming highly significant by generation 36. In generations 41, 51, and 61 where the gene frequencies show practically no change, comparisons were made of actual and expected genotype distributions. As one would expect, a considerable deficit of heterozygotes was characteristic of the system as a whole (the $\chi^2$ values in the above generations were $\chi^2 > 53.26$ for *α-Gdh* and $\chi^2 > 18.24$ for *Est-6* at d.f. = 1).

Table 4.4 gives the population-genetics parameters of the system studied, enabling evaluation of the observed variability. We see that at the *Est-6* locus

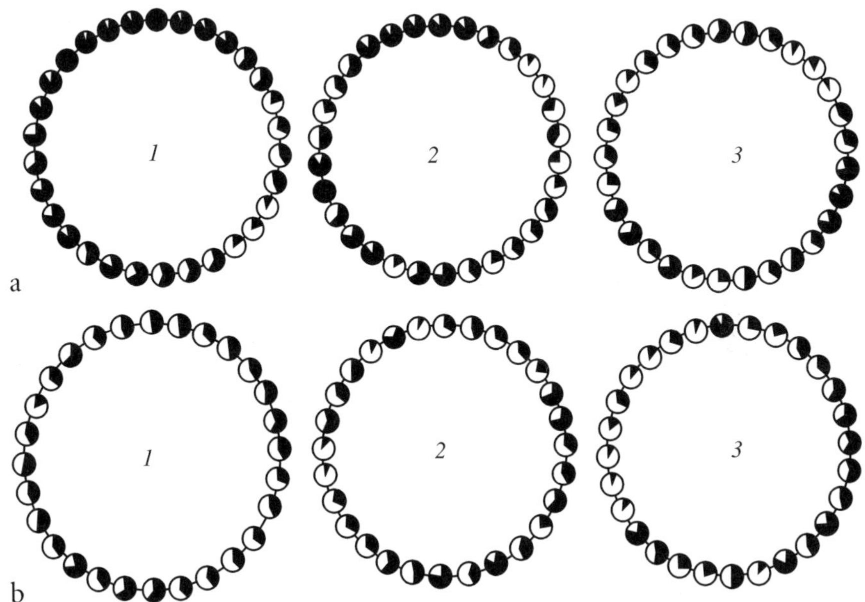

**Fig. 4.14.** Local differentiation of allelic frequencies at the **a** *α-Gdh* and **b** *Est-6* loci in successive generations of a *D. melanogaster* experimental population system correspond to the stepping-stone model (from Altukhov et al. 1979a). *1, 2, 3* Samples at 16, 36, and 61 generations, respectively. The *darkened sections* of the *circles* indicate the values of the estimated frequencies of fast alleles in each of the subpopulations

no directional changes in time occur and the frequency of *Est-6*[F] fluctuates around 0.43. The average frequency of $\alpha$-*Gdh*[F], however, increases between generations 5 and 41, indicating directional selection on this locus; we shall return to the subject later. Now, however, we must pay special attention to the clearly defined heterogeneity of the gene pool of the subdivided population at both loci, evidently reflecting the strong influence of genetic drift. To the extent that this effect directly relates to small population sizes, such a marked spatial differentiation where the subpopulation has $\overline{N} = 135$ individuals might seem somewhat unexpected. It should be borne in mind, however, that as a rule effective population size $(N_e)$ is less than total population size (Chap. 1). Taking this fact into account along with some ecological properties intrinsic to our system[1] we may conclude that the subpopulation $N_e$ should be at least half the average total population size. This is confirmed by a comparative analysis of the theoretical and observed intergroup variances of the allelic values calculated for generations 51 and 61, based on the hypothesis that $N_e = 50$ (Table 4.5). The calculations were derived from the formula obtained by Kimura and Weiss (1964) for a stationary population in the absence of selection (formula (42), Chap. 1). As Table 4.5 shows, both the experimental and theoretical values of variance of $N_e = 50$ are of the same order.

As was pointed out in Chap. 1, the local differentiation in a subdivided population depends on the parameter $N_e m$ which represents the number of migrants in a subpopulation during one generation. The prerequisite for local differentiation to arise in a one-dimensional stepping-stone model (Maruyama 1970a,b) is

$$N_e m < k/\pi^2 \, ,$$

where $m$ is the migration coefficient and $k$ the number of subpopulations. In our experiment $m = 0.03$ and $k = 30$. $N_e = 50$ fulfills the condition for differentiation ($N_e m < 3.03$).

The results we obtained raise once again the question of the reasons for the geographical variability of gene frequencies, casting doubt on conclusions both about selection in a heterogeneous environment (Spieth 1974) and the smoothing-out effects of even minimal migration pressures. Moreover, as previous work indicates (Yarbrough and Kojima 1967; Berger 1971),

---

[1] In Buri's (1956) experiment on small *D. melanogaster* populations, $N_e$ comprised 56 and 72% of the total numbers in populations with greater and lesser density, respectively. It must be noted that in Buri's work all females were placed on the medium simultaneously. Under these conditions, females that laid eggs first were at an advantage in the process of reproduction (Sang 1949a,b). In our experiment, when oviposition takes place throughout the period of emergence, the difference between $N_e$ and total numbers should increase, since not all the individuals succeed in reproduction.

**Table 4.4.** Main statistical parameters of allelic frequency distribution of $\alpha$-$Gdh$ and $Est$-6 loci in successive generations of an experimental system of subpopulations with a stepping-stone pattern of gene migration

| Locus | Generation | $K$ | $N$ | $N'$ | $\bar{p}$ | $V_p$ | Test for population homogeneity $\chi^2_{0.001}$ (*df* in parentheses) |
|---|---|---|---|---|---|---|---|
| $\alpha$-$Gdh$ | 5 | 27 | 4203 | 1109 | 0.36 | 0.0122 | 616.19>(26)=54.1 |
| | 10 | 23 | 5315 | 943 | 0.41 | 0.0340 | 1250.17>(22)=48.3 |
| | 16 | 29 | 5447 | 1215 | 0.40 | 0.0485 | 2000.03>(28)=56.9 |
| | 30 | 29 | 4163 | 1227 | 0.51 | 0.0819 | 2629.00>(28)=56.9 |
| | 36 | 30 | 4525 | 1198 | 0.57 | 0.0726 | 2683.00>(29)=58.3 |
| | 41 | 29 | 3835 | 1149 | 0.62 | 0.0491 | 1550.99>(28)=56.9 |
| | 51 | 30 | 3571 | 1097 | 0.62 | 0.0549 | 1659.84>(29)=58.3 |
| | 61 | 30 | 3605 | 949 | 0.65 | 0.0717 | 1650.39>(29)=58.3 |
| $Est$-6 | 5 | 28 | 4203 | 1196 | 0.50 | 0.0121 | 569.61>(27)=55.5 |
| | 10 | 29 | 5315 | 1367 | 0.42 | 0.0146 | 629.53>(28)=56.9 |
| | 16 | 28 | 5447 | 1149 | 0.46 | 0.0119 | 488.48>(27)=55.5 |
| | 30 | 27 | 4163 | 1228 | 0.48 | 0.0453 | 1353.11>(26)=54.1 |
| | 36 | 28 | 4525 | 1166 | 0.44 | 0.0409 | 1373.85>(27)=55.5 |
| | 41 | 29 | 3835 | 1212 | 0.42 | 0.0440 | 1360.42>(28)=56.9 |
| | 51 | 25 | 3571 | 757 | 0.34 | 0.0299 | 739.85>(24)=51.2 |
| | 61 | 28 | 3605 | 853 | 0.38 | 0.0610 | 1273.99>(27)=55.5 |

Conventional designations: $K$, number of subpopulations investigated; $N$, total number of individuals in the system; $N'$, number of analyzed individuals; $\bar{p}$, average allele frequency; $V_p$, interpopulation variance of allele frequencies

**Table 4.5.** Theoretically expected and actual values of gene frequency variance in a population system with a stepping-stone structure of gene migration

| Locus | Generation | $\bar{p}$ | $r(1)$ | $V_{p(exp)}$ | $V_{p(obs)}$ |
|---|---|---|---|---|---|
| $\alpha$-$Gdh$ | 51 | 0.62 | 0.72 | 0.0879 | 0.0549 |
| | 61 | 0.65 | 0.81 | 0.1063 | 0.0717 |
| $Est$-6 | 51 | 0.34 | 0.31 | 0.0437 | 0.0229 |
| | 61 | 0.38 | 0.24 | 0.0424 | 0.0610 |

Conventional designations: $\bar{p}$, average allele $F$ frequency; $r(1)$, correlation of gene frequencies among adjacent subpopulations; $V_{p(exp)}$, theoretically expected variance value; $V_{p(obs)}$, actual variance value. Calculations were made for $N_e = 50$ and the migration coefficient $m = 0.03$

both loci that we examined are subject to selection (frequency-dependent or stabilizing for $Est$-6 and directional for $\alpha$-$Gdh$) that should lead to a homogenization of gene frequencies in subpopulations. In our experiment, however, a high level of local differentiation is seen, particularly in terms of variance estimates (in generation 61 $V_p = 0.0610$ for *Est*-6 and $V_p = 0.0717$ for $\alpha$-*Gdh;* Table 4.4).

These findings accord with the results of research on natural populations of animal species and in human populations in which studies were not confined to the analysis of random samples, but were also aimed at obtaining reliable evidence of the effect of isolation on spatial and time distribution of gene frequencies. (We examined this question in the previous chapter.)

Considerable heterogeneity prevails in the gene frequency distribution of the MN and ABO blood group systems of human populations indigenous to Siberia (Rychkov et al. 1973). In research on the shell pattern polymorphism *Chondrus bidens,* a land snail (Altukhov and Livshits 1978), quantitative assessment of spatial differentiation gives a value of approximately the same order as in the present experiment. It should be noted that certain of the loci studied in the above work are not selectively neutral, but local differentiation of gene frequencies in population systems is persistently maintained despite the selection pressure. Furthermore, it has been found that this differentiation is periodic in character, evidently reflecting the stationary process of genetic reorganizations in a population system (Rychkov and Sheremet'yeva 1976).

The periodic nature of gene frequency dynamics is best seen by figures that demonstrate "time cross-sections" through the same subpopulations in several successive generations (Figs. 4.15 and 4.16). The figures along the abscissa ("migration axis") designate the migration distance in steps of any subpopulation, from the initial population (in our case no. 1, see Fig. 4.7); one step is thus equivalent to the move from one population to the next.

For convenience of analysis, the generations in Figs. 4.15 and  4.16 are compared in pairs, and the dash line connects allelic frequencies in each subpopulation of the previous generations, while a continuous line unites the same for the subsequent generations.

To begin with, let us examine the dynamics of the $\alpha$-*Gdh* gene frequencies (Fig. 4.15). It is seen that whereas variability assumes a random appearance in the earlier generations (Fig. 4.15a), a cyclic tendency becomes evident after the 10th generation: a series of "peaks" and "valleys" appears, alternating with zones of intermediate frequencies. In other words, there is a periodicity in the dependence of the allelic frequencies on distance, and this periodicity becomes more marked with the generations.

The changes in gene frequencies at the $\alpha$-*Gdh* locus within the population system can be approximated by the simplest periodic function in the form

$$pF = a + b\left(\cos l/\lambda + 1\right),$$

where $a$ and $b$ are the numeric coefficients, $l$ is the distance in "steps" along the migration axis, and $\lambda$ is the period of change in frequencies along the "migration" axis (Fig. 4.16). An increase in the period of change in gene

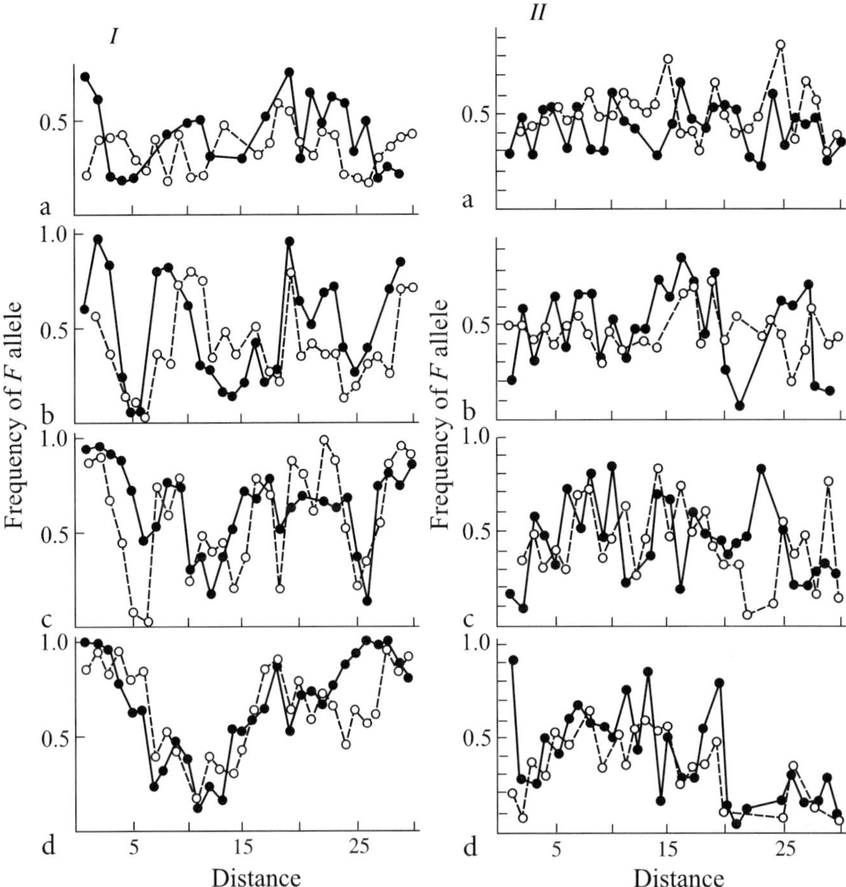

**Fig. 4.15.** Distribution of the allelic frequencies at the $\alpha$-Gdh locus (**I**) and Est-6 locus (**II**) in subpopulations in consecutive generations of an experimental population system having a stepping-stone migration structure. **a–d** Generations 5–10, 15–30, 36–41, 51–61 respectively. See text for further explanations

frequencies, with several periods merging into one, is a general tendency of the dynamics observed, though this process had not been fully completed by the 61st generation. For convenience, and taking into consideration the special features of the spatial differentiation of the allelic frequencies at the $\alpha$-Gdh locus, we can divide the "history" of our population system into three stages: (1) the formation of the periodic structure (generations 1–16); (2) a period of increased frequency change (generations 30–51); (3) a stationary (or quasi-stationary) phase (generations 51–61). Further research is necessary in order to define the tendency towards increase in period length. However, the mere fact that the system forms a plateau in relation

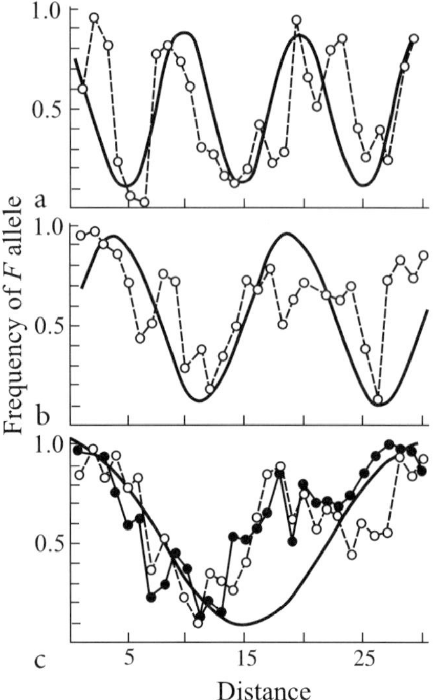

**Fig. 4.16.** Approximation of the periodic distribution function of $\alpha$-*Gdh* gene frequencies at a subpopulation level in successive generations of a *D. melanogaster* experimental population system (Altukhov and Bernashevskaya 1981). **a, b,** and **c** Generations 30, 41, and 51–61, respectively. The *dashed lines* are the actual distributions; *continuous lines* are the theoretical distributions: $\mathbf{a} - pF_{30} = 0.1 + 0.40\left(\cos 2\pi\frac{1}{10} + 1\right)$; $\mathbf{b} - pF_{41} = 0.1 + 0.42\left(\sin 2\pi\frac{l-3.5}{15} + 1\right)$; $\mathbf{c} - pF_{51} = 0.1 + 0.45\left(\cos 2\pi\frac{l}{30} + 1\right)$

to the average allele frequencies and their variance for a generation interval corresponding to the genetically effective size of the subpopulation, justifies the thought that our interpretation is not so remote from reality. A more revealing picture would probably appear only in a system constructed to incorporate long range migration as the stabilizing factor (see Chap. 1). However, I make the proviso that the main objective of the experiment has already been reached: *even the analysis of the simplest population system provides ample evidence that the genetic stability of such a system is greater than that of a panmictic population.*

Now let us turn to the spatial variability of gene frequencies at the second locus investigated – *Est-6* (Fig. 4.15). Variability at this locus, as for $\alpha$-*Gdh,* is purely random in the early generations of the experiment. However, by the 50th generation, there are signs of relatively stable valleys (in subpopulations 20–30) and peaks (subpopulations 7–18) on the curve, which describes the dependence of gene frequencies on distance; the peaks and valleys were maintained over 10 generations. This indicates that spatial genetic variability at the *Est-6* locus also displays a tendency to form a stable non-random structure, although it is expressed considerably less clearly than in the case of the $\alpha$-*Gdh* locus.

Analysis of the correlations with the data provides a further argument in favor of increased stability of the population in time. The correlation coefficients of the gene frequencies for different generations within the same subpopulations increase in time at both loci, testifying to the increased stability of the system structure with time (Table 4.6).

Theoretical analysis by Kimura and Weiss (1964) of the linear one-dimensional stepping-stone model has shown that if the system is in equilibrium, the correlation among gene frequencies decreases with distance from the exponent (see Chap. 1). This dependence is perfectly understandable because in proportion to an increase of the distance between subpopulations, the possibility of their exchanging genes diminishes rapidly in such a model. We have studied this correlation of the allele $\alpha$-*Gdh*$^F$ with distance in the 61st generation. The results of our analysis are represented on the graph in Fig. 4.17. One observes an initial fall in the value of the correlation to a minimum in the middle of the migration axis, followed by a symmetrical rise. In the instance given, the symmetrical character of the correlation change results from the closed state of the system investigated.

"Moving" along the circumference, we "move away" from the subpopulation no. 1 taken as the reference point (with diminishing probability of gene exchange with it), and then from the middle of the path (population no. 15) we begin to approach the original subpopulation from the opposite side, and the probability of gene exchange begins to increase in the same way.

Hence, periodicity is the feature that is characteristic of the dynamics of gene frequencies in a system of subpopulations connected with each other by the enclosed one-dimensional stepping-stone type of model. Independently, Yu.G. Rychkov and V.A. Sheremet'yeva (1976) discovered periodic changes of gene frequencies at several loci with geographical distance while carrying out research on human populations indigenous to the Eurasian

**Table 4.6.** Correlation of allelic frequencies at subpopulation level between pairs of generations of an experimental system having a stepping-stone gene migration structure

| Locus, allele | Generations | | | | |
| --- | --- | --- | --- | --- | --- |
| | 5–16 | 16–30 | 30–41 | 41–51 | 51–61 |
| $\alpha$-*Gdh* | 0.10 | 0.48 | 0.38 | 0.64 | 0.77 |
| *Est-6* | 0.07 | 0.21 | 0.16 | 0.53 | 0.52 |

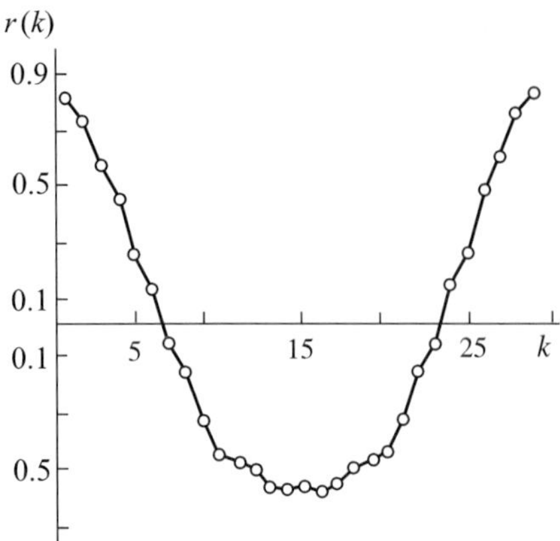

**Fig. 4.17.** Relation of the correlation between the frequencies of $\alpha$-*Gdh* alleles in different subpopulations and the distance in "steps" between the subpopulations at generation 61 (Altukhov and Bernashevskaya 1981). *x-axis* The distance between subpopulations in "steps" ($k$); *y-axis* correlation of gene frequencies of subpopulations $k$ steps behind each other

circumpolar zone. The system studied by Rychkov is approximately equivalent to a one-dimensional linear stepping-stone model with a migration coefficient of about 0.03 and an effective subpopulation size equal to 45, which means that the structural parameters are very close to those of our experimental population.

Transferred to a plane, the periodic dependency of frequency on distance corresponds to a relief "surface" with alternating peaks and valleys – maximum and minimum zones. Selander (1970) found a similar pattern for populations of *Mus musculus,* the house mouse, whose subpopulation structure agrees with the two-dimensional stepping-stone model in its general features. In these populations the spatial distribution of the allozyme frequencies was mosaic, with smooth transition between maximum and minimum zones. The colony (or tribe) size of the house mouse averaged 20 individuals and, because of behavioral characteristics, migration among the tribes was extremely low. Selander interpreted the origin of this type of local genetic differentiation to be a result of the population's subdivision since the conditions in the barn in which the mice lived were uniform.

Such a mosaic character of genetic variability was also found in computer simulations of a population whose structure corresponded to "isolation by distance" (Rohlf and Schnell 1971). It is interesting to note that in this work the population structure that arose proved to be stable; the areas of peaks and valleys persisted for tens of generations. We also observed a similar persistence of maximum and minimum areas of gene frequencies in our experiment and in simulation experiments in which the quasi-stationary phase lasted for hundreds of generations (Altukhov et al. 1984).

Using general computer principles for simulating genetic processes in populations (Fraser and Burnell 1971), we investigated the simplest case of a diallelic autosomal gene in a circular stepping-stone model. There were 25 subpopulations each of which represented a panmictic unity ($N_e = 40$). A parental pair was selected randomly using pseudo-random numbers. The model was based on the hypothesis of no limitation in the free segregation of genes, no selection, no overlapping of generations, and minimal variation in the subpopulation size in each subsequent generation. Migration among neighboring subpopulations was possible at an intensity of m ($m/2$ – the probability that an individual progeny migrates to one of two neighboring populations). The total population size was the same in all the experiments – 1,000 individuals. Four simulation experiments were conducted: in two of them the number of generations was 3,000 and the migration coefficient 0.01 and 0.03; in the third experiment there were 1,000 generations with $m = 0.1$; in the fourth, 325 generations with $m = 0$. Panmictic populations of the same total size served as the control.

The data obtained can be summed up as follows:

1. Highly significant local differentiation of allelic frequencies was observed in the whole interval of generations, with simultaneous relative stability of their average values and variance, at least in time intervals, measured by hundreds of generations (Table 4.7, generations 150–400; 600–1,000).

2. There was evidence of less genetic stability in panmictic populations than in those subdivided of the same total size (Fig. 4.18). The gene frequency in subdivided populations with a migration coefficient among

**Table 4.7.** Statistical parameters of distributions of allelic frequencies in successive generations of a subdivided "computer population" (3,000 generations, $m = 0.03$)

| Generations | Average gene frequency | Gene frequency dispersion | Generations | Average gene frequency | Gene frequency dispersion |
|---|---|---|---|---|---|
| 50 | 0.53 | 0.22 | 600 | 0.59 | 0.37 |
| 100 | 0.45 | 0.27 | 700 | 0.63 | 0.38 |
| 150 | 0.56 | 0.28 | 1,000 | 0.56 | 0.36 |
| 200 | 0.53 | 0.30 | 1,250 | 0.42 | 0.36 |
| 250 | 0.58 | 0.31 | 1,500 | 0.65 | 0.35 |
| 300 | 0.61 | 0.33 | 1,625 | 0.65 | 0.23 |
| 350 | 0.59 | 0.28 | 2,200 | 0.74 | 0.31 |
| 400 | 0.56 | 0.34 | 2,600 | 0.51 | 0.23 |
| 500 | 0.49 | 0.32 | 3,000 | 0.66 | 0.32 |

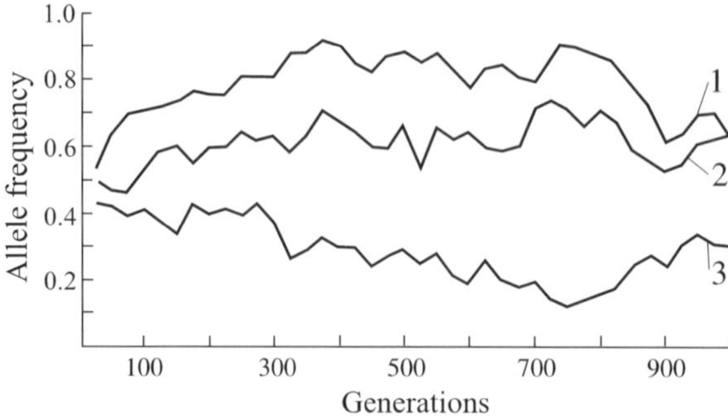

**Fig. 4.18.** Frequency dynamics of an allele (*ordinate*) over generations (*abscissa*) of three model panmictic populations

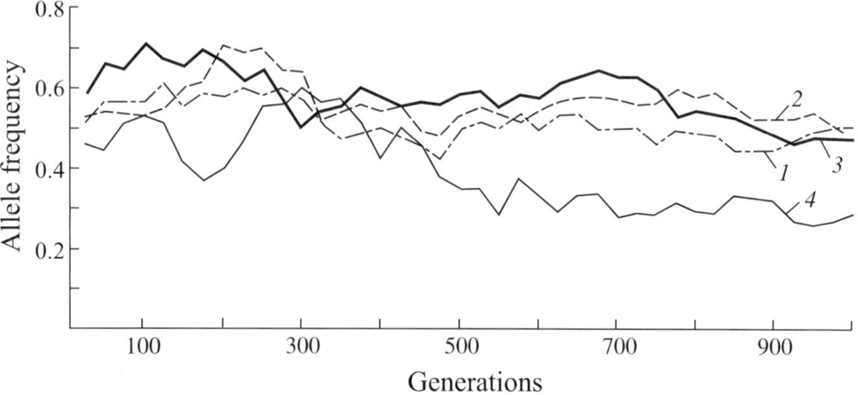

**Fig. 4.19.** Frequency dynamics of an allele (*ordinate*) over generations (*abscissa*) of four model subdivided populations with a gene migration coefficient of $m$ = 0.01 (*1–3*) and 0.1 (*4*)

subpopulations of $m$ = 0.01 remained practically at the starting level of 0.5 (Fig. 4.19). Simultaneously, the panmictic populations diverged considerably from each other, as they did from the ancestral population which had an initial gene frequency of 0.5. If these dynamics are evaluated quantitatively, even if only for the difference in frequencies between the first and last generations, then the $\Delta q$ value per generation for populations 1 and 2 is 0.00015, and for population 3 is 0.00018. It is interesting that the system already acts as a panmictic unity when the intensity of gene exchange among subpopulations is increased to a level of 0.1 (Fig. 4.19, curve 4).

It is understandable that even in a community of completely isolated populations, taken as a whole, random drift alone is capable of delaying the directional evolution of genetic structure resulting from fluctuations of allelic frequencies with different directional thrust. However, the picture is different when genes are exchanged among subpopulations. When, as in the case of an experimentally subdivided *Drosophila* population, we also investigate genetic variability at a subpopulation level in a series of "time sections", then we reveal the process of the stable interval structure which is formed along the axis of gene migration (Fig. 4.20). An analysis of the intrapair correlation of gene frequencies in the same subpopulations examined in previous and subsequent generations shows an intensification of this link over time, peaking at a maximum in the interval between the 150th and 1,000th generations (Table 4.8). At the same time, we see the transition from apparent chaotic variability at the beginning of the process (20th–50th generations) to an ordered structure thereafter. By the 350th generation, peaks and valleys appear united by zones of intermediate frequencies, in other words, a periodicity is observed in the dependency of allelic frequencies on distance which persists till at least the 400th generation. Following this, the structure is rearranged, assuming the shape of a half-wave with one maximum, and it maintains this form until the 1,100th generation when it degenerates.

Possibly, further simulation would show the formation of some kind of different structure, but the interval of generations over which the subpopulation structure and genetic parameters of the situation remain constant, is exceptionally large in biological terms. In discussing this conclusion, another important factor should be taken into account. The fact of the matter is that in our experiments we used a modified variant of Kimura and Weiss' model, lacking the stabilizing factor of long-range migration ($m_\infty$; see Chaps. 1 and 3). In our opinion, the introduction of this parameter to the model would only increase the stability of its structure in time.

Obviously, the genetic stability found in similar cases reflects the stationary (or at least, quasi-stationary) process of reorganization of generations of the system of partially isolated populations.

At the same time, analysis of the subdivided experimental *Drosophila* population reveals a difference in the picture of the spatial variability of

Table 4.8. Coefficients of rank ($r_s$) correlation among gene frequencies in the same "computer subpopulations"

| Generations | Values $r\pm$SE | Generations | Values $r\pm$SE |
|---|---|---|---|
| 40– 50 | –0.028 | 500– 600 | 0.719±0.145 |
| 50–100 | 0.347±0.196 | 700–1,000 | 0.785±0.129 |
| 150–200 | 0.720±0.145 | 1,250–1,500 | 0.266±0.201 |
| 250–300 | 0.871±0.102 | 1,625–2,200 | 0.422±0.189 |
| 350–400 | 0.613±0.165 | 2,600–3,000 | 0.059±0.205 |

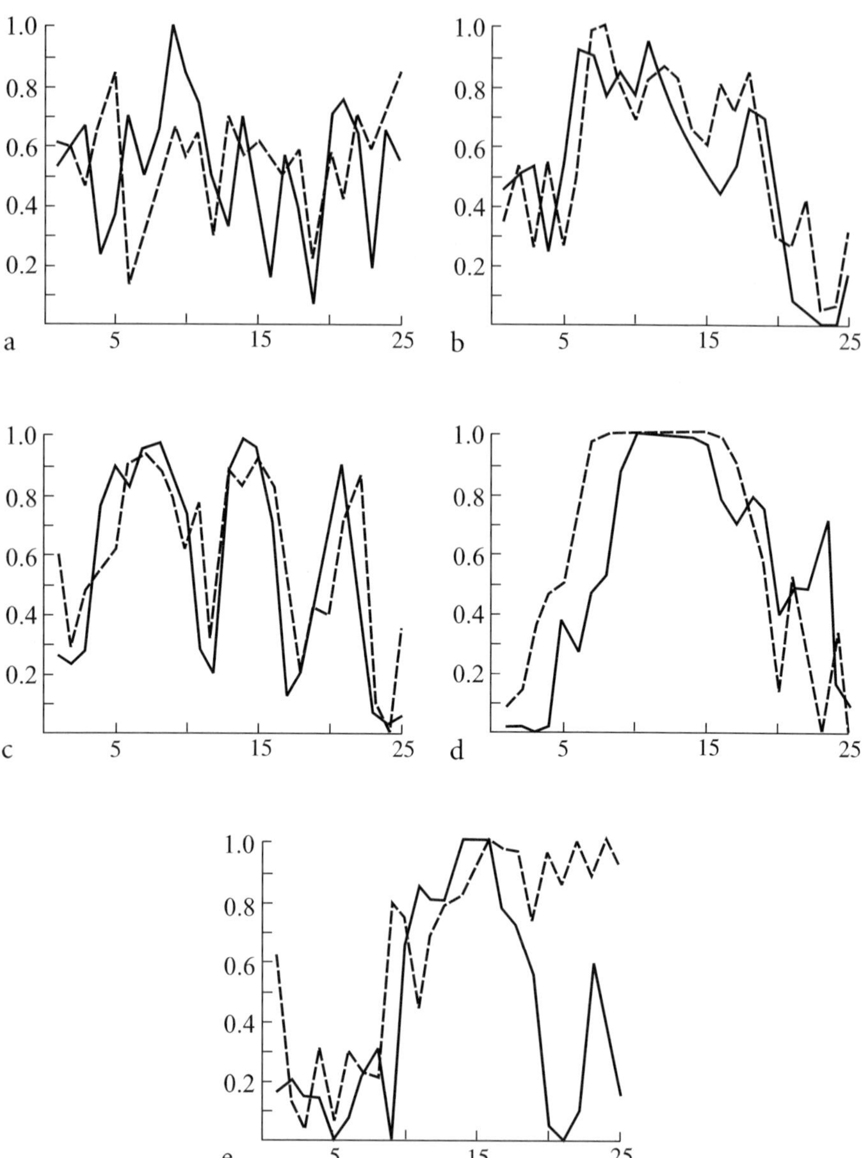

**Fig. 4.20.** Distributions of allelic frequencies (*y-axis*) at the level of separate subpopulations (1–25; *x-axis*) in consecutive generations of a computer simulation of a population system. The gene migration coefficient $m = 0.03$. *Continuous lines* Previous generations; *interrupted lines* subsequent ones. **a** Generations 20 and 50; **b** 150 and 200; **c** 350 and 400; **d** 700–1,000; **e**: 1,250–1,500

allele frequencies at the $\alpha$-*Gdh* and *Est*-6 loci, expressed by a clear spatial structure in $\alpha$-*Gdh* and the "erosion" of this structure in *Est*-6. This may be explained by the difference of selection types, which are directional for $\alpha$-*Gdh* and balancing (or frequency-dependent) for *Est*-6, with an equilibrium point of about 0.45.

Clearly, if balancing selection considerably exceeds the effects of random genetic drift, it will override the subpopulation structure (and shift the allelic frequencies in all subpopulations towards the equilibrium point); this was very evident in analyzing the distribution of *Pgm* gene frequencies of sockeye salmon populations (Chap. 3).

However, the contribution of genetic drift is more marked in the directional selection that leads to the fixation of the allele $\alpha$-*Gdh*^F in *Drosophila melanogaster*. It results in the balance of the two forces, one them tending to reduce the system's genetic diversity, the other to increase it. This is also reflected in the emergence of periodic local differentiation of allele frequencies, maintained for a long time, with a corresponding structure of gene migration.

It is important to stress that this structure and the average gene frequency that characterizes it are maintained, despite the effects of selection registered in the panmictic population: in 60 generations, allele $\alpha$-*Gdh*^F went from a starting frequency of 0.5 to a frequency of 0.98 ($\Delta q = 0.008$ per generation); the relevant data are given in Fig. 4.21.

In computer simulations, Zhivotovsky (unpublished) attempted to find more adequate values for the fitness of three genotypes at the $\alpha$-*Gdh* locus. The best correlation between experimental and theoretical "time-frequency" curves was obtained for a panmictic population with fitness

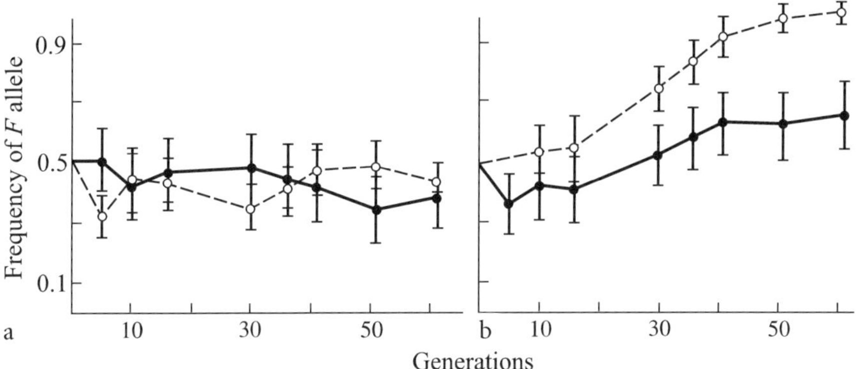


**Fig. 4.21.** The dynamics of allelic frequencies of **a** *Est*-6 and **b** $\alpha$-*Gdh* at the loci in successive generations of subdivided (*continuous line*) and panmictic (*interrupted line*) populations of *D. melanogaster*. (Altukhov and Bernashevskaya 1978)

coefficients of 1.05, 1.00, and 0.80 for the $\alpha$-*Gdh* genotypes FF, FS, and SS, respectively.

However, if these same *W* values are used for computer simulation of the genetic process in a subdivided population, the theoretical curve is somewhat higher than the experimental one, and the effects of selection can be traced virtually to the end (Fig. 4.22, curve 2). It is obvious that at these coefficients of directional selection, delay in the evolution of a sub-divided population occurs when stochastic processes are a considerable contributory factor.

In the computer experiments indicated above, the $N_e$ value was reduced to five individuals, but this did not induce a substantial agreement of the theoretical and experimental curves. It is only when $N_e = 5$ and the migration coefficient $m = 0.01$ that the empirical and modeled curves draw together, although it is evident that both these population structure param-eters, especially the $N_e$ value, clearly do not correspond to the experimental data. An artificial increase of genetic drift, accompanied by the reduced effects of migration, leads to a rise in the order of magnitude of gene frequency variance in the simulation, compared with what is observed in the experiment. Under these conditions the fixation of allele F would be modeled by the 60th generation – which does not happen in reality.

Hence, the computer simulation gives us the possibility of saying that effects of random genetic drift are not sufficient to explain the substantial delay (or even termination) of evolution seen in a subdivided, compared with a panmictic population.

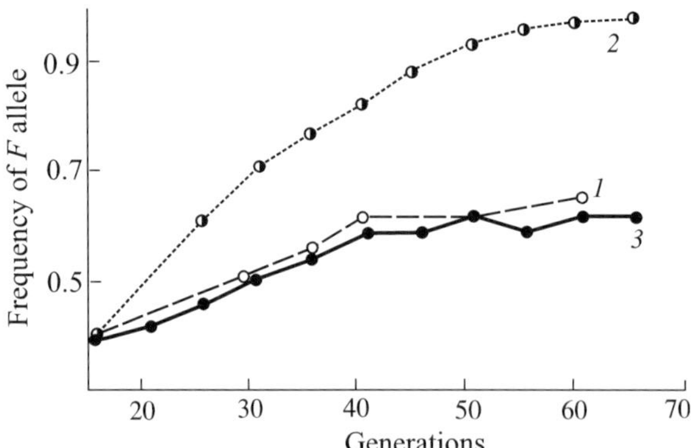

**Fig. 4.22.** Comparison of the dynamics of the allelic frequencies of the $\alpha$-*Gdh* locus in a subdivided population with the results of a computer simulation. *1* Experimental curve; *2* theoretical curve at *W* values for genotypes FF, FS, and SS of 1.05, 1.00, and 0.80, respectively; *3* theoretical curve at corresponding *W* values for the same genotypes: 0.95, 1.00, and 0.90. The initial frequencies in computer simulation correspond to the frequencies of the genotypes in the experiment at the 16th generation

A satisfactory approximation is attained only with fitness values of 0.95, 1.00, and 0.90, respectively for genotypes FF, FS, and SS. When these co-efficients and the values $N_e = 25$ and $m = 0.03$ are introduced into the simulation, the distribution of expected frequencies comes extremely close to the experimental; expected and observed variances are also similar (Fig. 4.22, curve 3).

Although it is difficult from a formal viewpoint to visualize how both the intensity and direction of selection can change in this way under subdivision (when the practical identity of environmental conditions has been established in the experiment and control), it is a fact that a populational system remains genetically stable for an interval of 20–25 generations against the rapid evolution of a structureless panmictic population prior to the attainment of a stability phase. Of course, in a species having a longer generation time than Drosophila, this stability phase may last for hundreds of years.

Which factor in a population system is so effective in balancing the effects of directional selection? Insofar as it is not random genetic drift, the possibility of migration should be considered which, whether in pure form or together with drift, evidently also changes the fitness of genotypes that remain different and constant in a panmictic population. It is interesting to compare our data with the results of Narise (1968, 1969, 1974) who studied the effects of migration on the fate of a semilethal *vestigial* (vg) gene in populations of *Drosophila melanogaster*. Although the aims of her research and our experiments were different, both cases had in common the utilization of subdivision with migration, a panmictic population serving as the control.

In Narise's experiments the subdivided population consisted of 10 test tubes containing medium and joined by migration pipes so that one central test tube connected with three peripheral ones (Fig. 4.23). At the beginning of the experiment 120 heterozygous flies (*vg*/+) were placed in a central test tube, and simultaneously 120 individuals of the same genotype were inserted in a standard population cage. Migration links were interrupted every fortnight and a count made of the number of flies in the test tubes. Then the flies were transferred to test tubes containing fresh food, after which the migration structure was re-established. The experiment lasted for 20 generations, and was repeated three times.

The fitness of the vestigial homozygotes is greatly reduced in comparison with wild-type flies (Dubinin et al. 1937; Zurabyan and Timofeyev–Resovsky 1967). Accordingly, a sharp reduction of the *vg* gene concentration was observed both in subdivided and panmictic populations. However, substantial differences between the two population types were already noticeable in the dynamics of this process at the 4th generation. The gene frequency in the subdivided population was significantly higher than in the panmictic case and this difference persisted to the end of the exper-

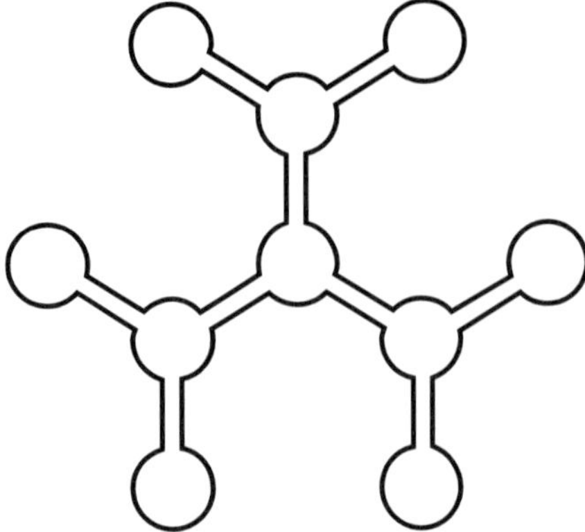

**Fig. 4.23.** The structure of an experimental population during the study of the effect of migration on the fate of a semilethal *vestigial* gene. (Narise 1968)

iment. By the 20th generation, $vg$ homozygotes had almost completely vanished from the panmictic population whereas their frequency in the subpopulation community had stabilized at a level of 1%, remaining constant despite the strong effects of negative selection. Narise explained this result by the increased migration activity of the less adapted individuals under conditions of migration. While migrating, the flies had sought out more favorable environmental conditions, thereby freeing themselves, albeit/partially, from the powerful effects of selection. It is very likely that under the terms of our experiment the differential (non-random) migration of genotypes may also have a determining effect upon the dynamics of allelic frequencies in the systems of interacting populations.

Thus our findings have established that the dynamics of gene frequencies are different in principle in panmictic and subdivided populations: under the same conditions of the environment the directional shift of allelic frequencies are incomparably more marked in the former than in the latter. In the subdivided population, the frequency of allele $\alpha\text{-}Gdh^F$ was virtually stabilized at a level of 0.62 by the 41st generation, remaining there for 20 generations. Meanwhile, in the 51st and 61st generations of the panmictic population, the SS homozygotes were virtually absent and the FS heterozygotes were encountered extremely rarely. This means that the local differentiation of gene frequencies, which had already emerged under subdivision in the first generations, persisted until the end of the experiment. The $F_{ST}$ values for the $\alpha$-glycerodephosphate dehydrogenase and esterase-6 loci remained 0.297 and 0.245, respectively.

When the environment is uniform, the only reason for maintenance of heterogeneity of this kind is the interaction of all the factors of population dynamics present during subdivision and absent in panmixia. Of course, it is not always possible to obtain quantitative evaluations of the interaction of these factors. Nevertheless, a new feature emerges at this level of population organization even under experimental conditions, which cannot be replicated by computer simulation or be resolved through analytical methods – the genetic stability of a population system in time. But such stable polymorphism results from the interaction of different factors of evolutionary dynamics, not from any one of them as is frequently suggested by the works of "selectionists" and "neutralists". Hence, the results of the present chapter are in agreement with the main conclusions drawn from the study of natural population systems. Moreover, investigation of the genetic process in an experimental population of the island type has revealed a nonrandom structure of migration between the "continent" and the "islands": the smaller size of the peripheral populations the more intense the inflow of individuals from the "continent". This migration plays an important self-regulating role, maintaining the integrity of the system and temporal stability of the gene frequencies of the ancestral population. Bases on the results obtained for the natural and experimental subdivided populations, let us now turn to broader examination of the biological significance of hereditary protein and DNA polymorphism, i.e., the role of natural selection in its maintenance.

# 5 The Role of Natural Selection in the Maintenance of Protein and DNA Polymorphism

The materials presented in the two last chapters have presumably convinced us that the intra- and interpopulation dynamics of allozyme gene frequencies cannot be regarded as merely a transient phase of molecular evolution, that, to the contrary, biochemical polymorphism is an extremely stable condition which a fine subpopulation structure with a limited flow of genes alone is capable of maintaining for lengthy time intervals. This structure can maintain a high level of genetic diversity even under conditions of directional selection, as is the case, for instance, with the locus $\alpha$-*Gdh* in an experimental population of *Drosophila melanogaster* (Chap. 4). However, as with sockeye salmon natural populations, so with experimental *Drosophila* populations, it would be wrong to exclude selection favoring heterozygotes or other forms of balancing selection as additional factors maintaining and promoting the greater stability of protein polymorphism.

Yet, it must be admitted that we have very few clear facts about this aspect in biochemical population genetics. One can only agree with Lewontin (1978a), that more often than not the "tired old Bucephalus", that is, sickle-cell anemia, linked with hemoglobin polymorphism in certain ethnic groups, is advanced to demonstrate single-locus heterosis in natural populations. But what causes this kind of difficulty? Is selection favoring heterozygotes at enzyme loci truly a rarity in nature or is it a defect of research?

From my point of view, the results of the present work favor the second alternative. As has already been remarked, in many works on population genetics, the authors are not dealing with *whole populations but with random samples,* which does not permit one to pass judgment on the specifics of the genetic processes taking place in native population systems. We have already drawn attention to this circumstance in previous publications (Altukhov 1974, 1977; Altukhov et al. 1975).

Although the above studies have convincingly demonstrated the effects of selection, the adaptive significance of protein polymorphism is still under discussion (see, for instance, Hey 1999; Allendorf and Seeb 2000; Ohta 2000). Only as late as in 1997, a monograph by Jeffrey Mitton (1997) appeared in which the author, in parallel to our studies (Altukhov 1990, 1991) stated a clear view on the role of natural selection in the maintenance of biochemical polymorphism in populations. Mitton lists a number of such examples, including obvious cases of overdominance.

Thus, some researchers consider that protein polymorphism is maintained by different forms of balancing (stabilizing) selection (e.g., Ayala et al. 1974a,b). According to another viewpoint, this variation is selectively neutral (Kimura 1983), i.e., whatever is the genotype for any protein variant, its carrier gains no adaptive advantage. The objective resolution of the controversy between "selectionists" and "neutralists" is not only of purely academic interest, but also of prime significance for practice. Indeed, if protein polymorphism is selectively neutral and, according to Kimura (1983), represents a transitional phase of molecular evolution, it may only be used for solving the problems of phylogeny and microsystematics. Conversely, if protein polymorphism is functionally loaded and represents a stable (rather than transitory) stage of evolution, it offers a unique opportunity for use in breeding programs, and the necessity of preserving this form of genetic variation becomes apparent.

In recent years, an approach has been developed for assessing the adaptive significance of protein polymorphism, which is based on the use of five methods:

1. The analysis of stationary distribution of gene frequencies in natural population systems that can be reliably characterized with respect to parameters of their breeding structure, such as effective size $N_e$ and the coefficient of gene migration $m$ (Rychkov et al. 1973; Altukhov 1974; Altukhov et al. 1975a; Rychkov 1975; Rychkov and Sheremetyeva 1976)

2. The analysis of genotypic distributions and correlations of gene frequencies in consecutive generations of the same subpopulation (Altukhov 1989a,b).

3. The analysis of distributions of genotypes for allozyme loci at early and late ontogenetic stages (Altukhov 1983a, 1989a,b; Altukhov et al. 1991)

4. The analysis of joint variability of adaptively significant polygenic morphophysiological and monogenic biochemical traits (Altukhov et al. 1979b; Altukhov 1983, 1985a,b; Altukhov and Kurbatova 1990; Singh and Zouros 1978; Koehn and Gaffney 1984; Garton et al. 1986; Mitton 1997)

5. The comparison of observed and demographically predicted levels of spatial genetic differentiation among populations (Rychkov and Sheremet'yeva 1976; Rychkov and Balanovskaya 1990a,b; Balanovskaya and Rychkov 1990)

Regretfully, such approaches for analyzing variation at the level of DNA polymorphisms are now only emerging (see Sects. 2.4.4 and 5.7) but they evidently hold great promise.

## 5.1
# Analysis of Stationary Distributions of Gene Frequencies

Let us approximate observed distributions of subpopulations of sockeye salmon population system of Lake Azabachye in corresponding ranges of allele frequencies of *LDH-B2** and *PGM-2** loci by using Wright's stationary functions [(33) and (34), Chap. 1]. The values $N_e$, $\bar{q}$, and $m$ must serve to construct the theoretical curves. Our material (Sect. 3.2) enables us to estimate the first two parameters, whereas the migration coefficient value is assessed by the average value of 2%, derived from the data cited above (Sects. 2.3.1 and 4.1.1).

It is evident from the theory of stationary distributions that in the case of unsatisfactory approximations of empirical distributions to those expected, which take into account only migration and random genetic drift, one is justified in making allowance for the effects of selection, measured by $\overline{W}$, the value of the mean population fitness.

By comparing the values of function [(33), Chap. 1] and the empirical distribution using the same values of $q$, we can estimate the $\overline{W}$ value and then calculate the $W_i$ values for each separate genotype. This can be done by several methods. Here, with some small changes, we have applied the method used earlier by A.I. Pudovkin (Altukhov et al. 1975b).

Dividing the distributions of gene frequencies into three intervals ($\leq$ 0.60, $\leq$ 0.70, and $\geq$ 0.70 for lactate dehydrogenase and $\leq$ 0.750, $\leq$ 0.800, and $\geq$ 0.800 for phosphoglucomutase), we determine the observed and expected (according to the neutral model) probabilities of finding the subpopulation (point) in each of these intervals and the ratio of these probabilities. Taking the $2N$ roots of resulting values, we obtain $w$ for each interval. Using the formula (18; Sect. 1.5.1), the system of three linear equations in three unknowns is obtained:

$$\overline{w}_1 = w_{AA}\varphi_{11} + w_{AB}\varphi_{12} + w_{BB}\varphi_{13} \,,$$
$$\overline{w}_2 = w_{AA}\varphi_{21} + w_{AB}\varphi_{22} + w_{BB}\varphi_{23} \,,$$
$$\overline{w}_3 = w_{AA}\varphi_{31} + w_{AB}\varphi_{32} + w_{BB}\varphi_{33} \,,$$

where $\overline{w}_1$, $\overline{w}_2$, $\overline{w}_3$ are the values of intralocus fitness for 1st, 2nd and 3rd intervals, respectively; $\varphi_{ij}$ is the expected proportion, according to neutral model, of $j$ genotype in $i$ interval, i.e.,

$$\varphi_{11} = \sum_{1}^{n} q_i^2 W_i, \quad \varphi_{12} = \sum_{1}^{n} 2p_i q_i W_i, \quad \varphi_{13} = \sum_{1}^{n} p_i^2 W_i$$

and so on, where $n$ is the number of points ($q_i$) in the interval; $W_i$ is weighted within the $i$ interval value $\Phi(q)$. By solving the above equation system we will find the estimates for the fitness of genotypes.

Corresponding calculations for the entire community of spawning subpopulations have shown that when $N_e = 200$ and $m = 0.02$ in the case of lactate dehydrogenase locus, there is agreement between the empirical distribution of subpopulations and the expected stationary one in the presupposed selective neutrality of this polymorphism ($\chi^2 = 10.1, P > 0.1; df = 6$; Fig. 5.1a). However, the observed distribution of subpopulations for the phosphoglucomutase locus differs significantly from the expected stationary distribution: $\chi^2 = 35.9, P < 0.001; df = 3$ (Fig. 5.1b).

In dealing with the lack of agreement with expectation as an effect of selection, we have defined the absolute values of the fitness of genotypes as $W_{AA} = 0.987; W_{AB} = 1.048; W_{BB} = 0.861$.

If the fitness of the heterozygote is set at 1.0, the normalized fitnesses become $W_{AA} = 0.942; W_{AB} = 1.000;$ and $W_{BB} = 0.822$.

These estimates attest that the genotypes of *PGM-2** locus evolve under strong pressure of balancing selection. Introducing the selection factor to the model also improves the approximation for the *LDH-B2** gene frequencies distribution. In this case $W_{BB} = 0.98, W_{BB'} = 1.00,$ and $W_{B'B'} = 0.97$. These estimates give $\chi^2 = 3.04, P > 0.5; df = 6$ (Fig. 5.1a).

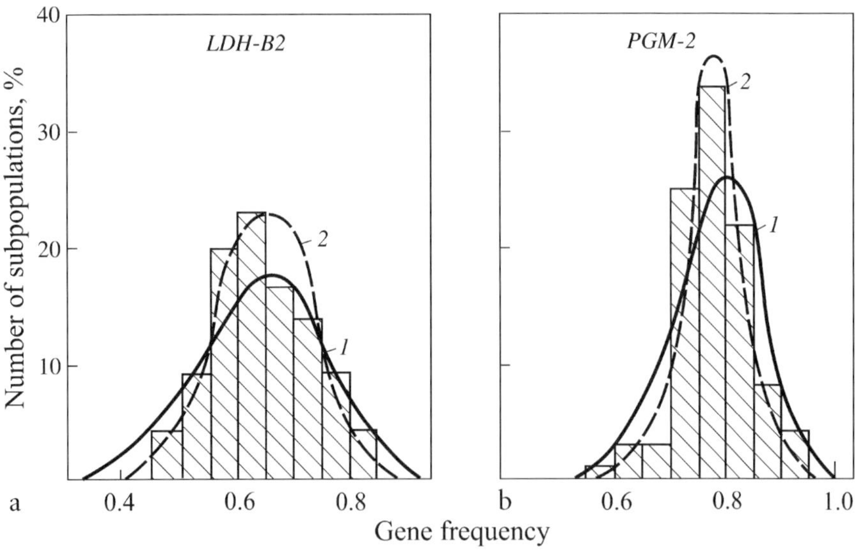


**Fig. 5.1.** Distributions of spawning subpopulations of sockeye salmon of Lake Azabachye in corresponding ranges of the allelic frequencies of *LDH-B2** (**a**, 183 samples) and *PGM-2** (**b**, 171 samples). Empirical distributions: the *curves* are the expected distributions, *1* on the basis of genetic drift equilibrated by migration; *2* under the combined effects of drift, migration and selection. See text for further details

Insofar as a precise definition of gene migration intensity of the natural populations of most species (the human species is the only exception) is unusually complex and difficult to resolve, the estimates obtained above for the fitness values should be tested by any appropriately independent method. There are two possibly methods of analysis. Firstly, stable differences in the interpopulation variances of the allelic frequencies of the two loci studied may be used to hypothesize greater or lesser selective neutrality of genotypes, and accordingly, to evaluate the parameter of structure $N_e m$ through the intergroup gene frequency variance [see (26), chap. 1] at the most variable *LDH-B2** locus; then this value can be used in formula (34). Secondly, inasmuch as the population is characterized as a whole by constancy of the distributions of the *PGM-2** gene frequencies in generations with a clearly expressed excess of heterozygotes, direct comparison can be made of the actual and expected numbers of genotypes by the Hardy–Weinberg equation.

We utilized both of these methods, and both revealed virtually complete coincidence of new $W_i$ estimates with what has been indicated above (Table 5.1). At the same time we also see a close identity of estimates of the $\hat{q}$ equilibrium value with the allele's average frequency in the subdivided population. Facts of this kind are indicative of the effect of overdominance in maintaining polymorphism at the *PGM-2** locus.

It is clear that we only discovered this because of a system for collecting material that had been planned ahead, envisaging the lengthy observation of the genetic structure of a well-defined natural population. However,

**Table 5.1.** The $W_i$ values of genotypes at the locus *PGM-2** determined by different methods

| Method of determining $W$ | Genotype | $W_i$ values | Equilibrium (stationary) gene frequency $\hat{q}$ | Observed average gene frequency $\bar{q}$ |
|---|---|---|---|---|
| Using Wright's stationary functions. Parameter *Nm* found from the ecological data | AA | 0.942 | | |
| | AB | 1.000 | 0.756 | 0.782 |
| | BB | 0.822 | | |
| Using Wright's stationary functions. Parameter *Nm* found from the variance of allelic frequencies at *LDH-B2** locus | AA | 0.937 | | |
| | AB | 1.000 | 0.766 | 0.782 |
| | BB | 0.799 | | |
| Using the ratio of observed and expected numbers of genotypes | AA | 0.943 | | |
| | AB | 1.000 | 0.782 | 0.782 |
| | BB | 0.796 | | |

Here and further see note under Fig. 3.9 for genotype designation

the same approach allows one to proceed further and to conduct a more detailed analysis of the distribution of genotypes in successive generations of a separate subpopulations, taking into account the microheterogeneity of an environment when different types of spawning grounds are available (Ryabova et al. 1978; Novosel'skaya et al. 1982).

## 5.2
## Analysis of the Genotype Distributions and the Correlations of Gene Frequencies in Successive Generations of the Exact Same Subpopulations

Table 5.2 presents the genetic parameters of subpopulations attached to different spawning sites. It follows from this that whereas there are, for the *PGM-2** allele frequency, no significant differences among groups of spawning grounds, the picture is different for the *LDH-B2** locus. When the populations of ponds and streams are compared with lake populations, local differentiation is evident.

In comparing the actual and expected distributions of *LDH-B2** genotypes, the significant deficit of heterozygotes is seen in the population as a whole. The character of the distribution of *PGM-2** genotypes is different: an excess of heterozygotes is typical for the system as a whole and for the subpopulations of ponds, whereas the distribution of genotypes in the lake subpopulations is not different from what is expected from the Hardy–Weinberg equation. These findings justify an assumption that *PGM-2** genotypes are subject to considerably greater selection pressure than *LDH-B2** – at least in ponds. The effects of selection should, however, not be so great for the distribution of *PGM-2** genotypes in lake subpopulations. This kind of deduction also accords with estimates of the genetic variance at the *PGM-2** locus (Table 5.3), which is minimal for subpopulations in ponds and streams and maximal for those in lakes, although the differences are not large. Comparison of the allele frequency variances for the *LDH-B2** locus reveals a more salient, statistically significant differentiation among the types of spawning grounds: maximal variability typifies the lake populations, the streams are intermediate, and the ponds have least variability (Table 5.3).

The genetic differentiation of populations thus revealed permits one to specify the previous conclusion about the intensity of selection at the *LDH-B2** locus and the assumption that in ponds, at least, the effects may be more considerable. The same surmise suggests itself for estimates of the $W_i$ genotypes of the phosphoglucomutase locus. There are two possible ways of checking this hypothesis:

**Table 5.2.** Observed and expected distribution of genotypes and gene frequencies of *LDH-B2** and *PGM-2** loci in sockeye salmon subpopulations preferring different types of spawning ground

| Types of spawning grounds | LDH-B2* | | | | | |
| --- | --- | --- | --- | --- | --- | --- |
| | Genotype | | | $n$ | $qB$ | $\chi^2$ |
| | *BB* | *B′B* | *B′B* | | | |
| Ponds | 827 | 1080 | 413 | 2320 | 0.5892 | 3.41 |
| | 805.40 | 1123.08 | 391.52 | | | |
| Streams | 654 | 778 | 238 | 1670 | 0.6246 | 0.07 |
| | 651.51 | 783.15 | 235.34 | | | |
| Totals (spring race) | 1481 | 1858 | 651 | 3990 | 0.6040 | 2.81 |
| | 1455.62 | 1908.68 | 625.70 | | | |
| Lake (summer race) | 1253 | 957 | 200 | 2410 | 0.7185 | 0.82 |
| | 1244.14 | 974.88 | 190.98 | | | |
| Population system as a whole | 2734 | 2815 | 851 | 6400 | 0.6523 | 10.43* |
| | 2723.2 | 2903.1 | 773.7 | | | |
| Ponds | 1350 | 886 | 80 | 2316 | 0.7742 | 20.51** |
| | 1388.18 | 809 .74 | 118.08 | | | |
| Streams | 934 | 618 | 80 | 1632 | 0.7616 | 3.01 |
| | 946.62 | 592.63 | 92.75 | | | |
| Totals (spring race) | 2284 | 1504 | 160 | 3948 | 0.7690 | 20.61** |
| | 2334.69 | 1402.64 | 210.67 | | | |
| Lake (summer race) | 1535 | 767 | 95 | 2397 | 0.8004 | 0.00 |
| | 1535.61 | 765.89 | 95.50 | | | |
| Population system as a whole | 3819 | 2271 | 255 | 6345 | 0.7809 | 13.32** |
| | 3869.2 | 2171.2 | 304.6 | | | |

* $P < 0.01$; ** $P < 0.001$

1. By analyzing the correlations of gene frequencies in the same spawning grounds in successive generations

2. By analyzing stationary distributions

Taking into account the 5-year length of generation of Azabachye sockeye salmon (Konovalov 1980), we correlated the allele frequencies of *LDH-B2** and *PGM-2** loci for the subpopulations of the exact same spawning sites in successive generations (1972–1977; 1973–1978; 1974–1979).

The significant correlation of gene frequencies in successive generations of one and the same subpopulation provides an indirect evaluation of the vector of selection, because it is obvious that the more considerable the effect of stabilizing selection, the greater should be the similarity in time of gene frequencies estimated repetitively in the same subpopulations. And

**Table 5.3.** The main genetic parameters of a subdivided population of sockeye salmon in Lake Azabachye characteristic of different types of spawning grounds

| Locus | Type of spawning ground | $K$ | $n$ | $\bar{q}$ | $F\varphi$ | | $V_q$ | $\chi^2$ test for homogeneity of variances ($df$) |
|---|---|---|---|---|---|---|---|---|
| | Pond | 44 | 2320 | 0.5892 | Pond-stream | 4.98* | 0.0026 | |
| $LDH$-$B2^*$ | Stream | 32 | 1670 | 0.6246 | Pond-lake | 87.58*** | 0.0043 | $8.62 > \chi^2_{0.05}(2) = 5.99$ |
| | Lake | 48 | 2410 | 0.7185 | Stream-lake | 39.69*** | 0.0058 | |
| | Pond | 44 | 2316 | 0.7742 | Pond-stream | 0.78 | 0.0026 | |
| $PGM$-$2^*$ | Stream | 31 | 1632 | 0.7616 | Pond-lake | 4.75* | 0.0029 | $1.46 < \chi^2_{0.05}(2) = 5.99$ |
| | Lake | 48 | 2397 | 0.8004 | Stream-lake | 8.22** | 0.0037 | |

$F_\varphi$, The estimated differences of gene frequencies using Fisher's criteria. The remaining designations are as for Table 3.3

* $P < 0.05$; ** $P < 0.01$; *** $P < 0.001$

in fact the results obtained do not contradict this model (Fig. 5.2): the pair correlation coefficients at the $PGM$-$2^*$ locus proved to be adequately high and reliable for the pond and stream spawning grounds ($r = 0.51$; $P < 0.05$ and $r = -0.51$; $P < 0.01$, respectively) and non-significant for the lake ones ($r = -0.13$); at the $LDH$-$B2^*$ locus a significant positive correlation was found only for subpopulations reproducing in ponds ($r = 0.47$; $P < 0.05$).

It is curious that the correlation for the $PGM$-$2^*$ locus is positive for the pond subpopulations, but is negative for the stream subpopulations. This may be explained as follows: evidently *selection is of a universal nature and shifts the allelic frequencies in each subpopulation and in every generation to the equilibrium point* (see Chap. 1), which is the same for total population or even for such an integral component of its structure as represented, for example, by a community of early spawning populations (the spring race). If this model is correct, then one would expect the differences among allelic frequencies in any subpopulation in two successive generations to be the more marked the further the subpopulation is from the equilibrium point in the preceding generation. *In this case, if an allele's frequency is lower than the equilibrium frequency in the preceding generation, it should increase in the next generation; if higher than the equilibrium, it should decrease.*

Indeed, the data represented in Fig. 5.3 are in accordance with this expectation, also satisfying all the hypotheses about potential differences in the

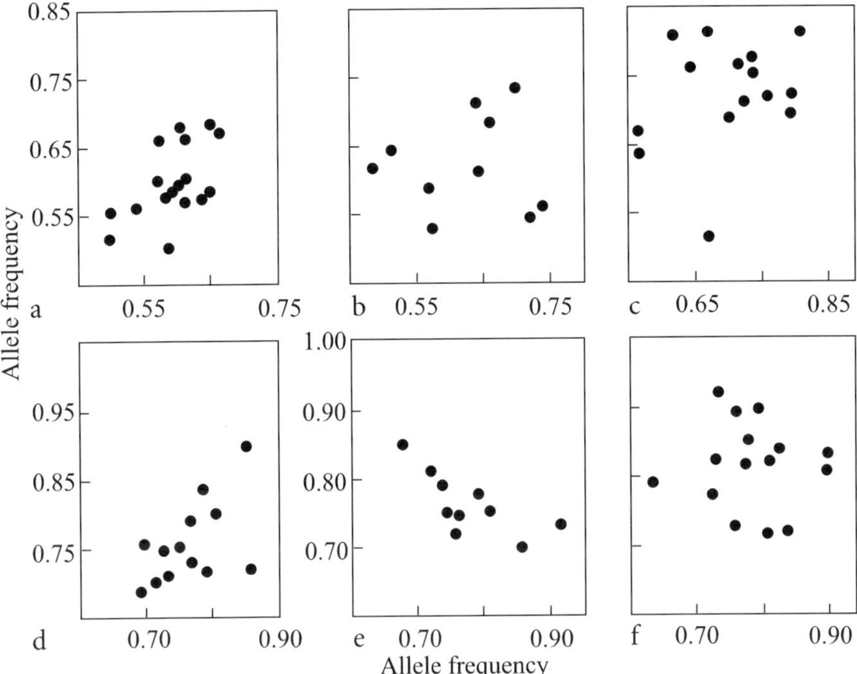

**Fig. 5.2.** The relationship between gene frequencies in consecutive generations of individual subpopulations of sockeye salmon in one and the same spawning site. **a–c** *LDH-B2** allelic frequencies for subpopulations of ponds, streams, and lake, respectively. **d–f** The same for the *PGM-2** allelic frequencies. Remaining explanations in the text

effects of selection on both loci in relation to the different types of spawning grounds. We clearly see that a correlation can be traced, for both races, for the *PGM-2** locus, which is more subject to selection, and the strength of the link at both loci is maximal in the case of the spring-spawning subpopulations that undergo the most significant selective pressures.

Let us now define again the appropriate coefficients of genotype fitness through approximation of empirical distributions of allelic frequencies to the expected stationary distributions; we can estimate the $Nm$ parameter from the $F_{ST}$ statistics of the distribution of *LDH-B2** gene frequencies in lake subpopulations of the sockeye salmon (Table 5.4). The estimates of the $W_i$ values obtained are practically indistinguishable from those above.

Thus, the methods we have used have enabled us to obtain very similar estimates of fitness coefficients of the genotypes at both loci and, moreover, to reveal finer genetic differentiation of populations by types of spawning grounds. This differentiation (and the corresponding differences in $W_i$) also appear to harmonize very well with the state of the environment.

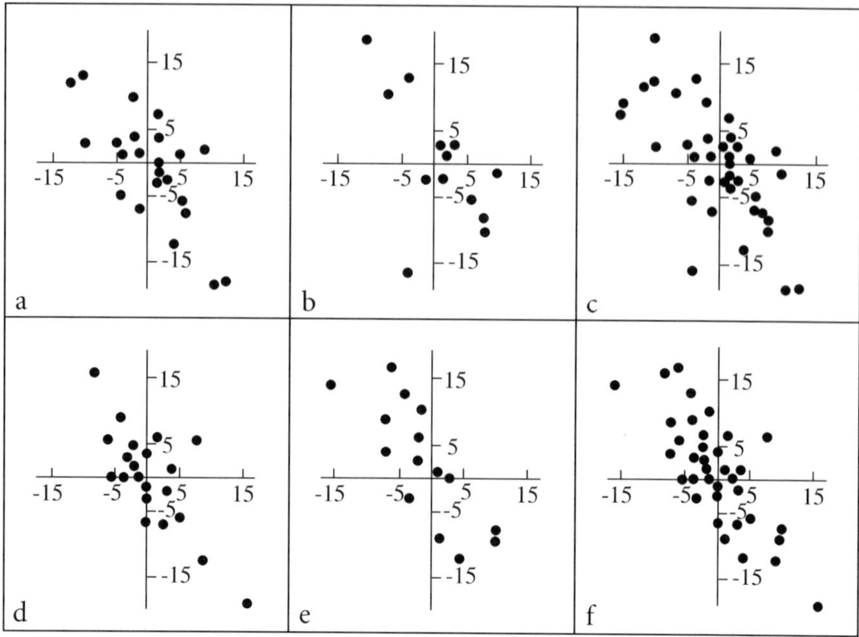

**Fig. 5.3.** The relationship between the deviation of gene frequency $q_i$ from equilibrium $\hat{q}$ in a given generation ($q_i - \hat{q}$; $x$-coordinate), and the difference between gene frequencies in preceding and following generations of the same subpopulation ($q_{i+1} - q_i$; $y$-coordinate). **a–c** Spring and summer populations and the population system as a whole, locus *LDH-B2**. **d–f** The same for locus *PGM-2**. **a** $r = -0.70$***; **b** $r = -0.50$; **c** $r = -0.62$**; **d** $r = -0.87$***; **e** $r = -0.77$***; **f** $r = -0.75$***. ** $P < 0.01$; *** $P < 0.001$

**Table 5.4.** $W_i$ values of sockeye salmon genotypes in different kinds of spawning grounds

| The studied group of subpopulations (type of spawning grounds) | $W_i$ values of genotypes at a locus | | | | | |
|---|---|---|---|---|---|---|
| | *LDH-B2** | | | *PGM-2** | | |
| | *BB* | *B′B′* | *B′B′* | *AA* | *AB* | *BB* |
| Ponds | 0.90 | 1.00 | 0.86 | 0.94 | 1.00 | 0.80 |
| Streams | 0.98 | 1.00 | 0.97 | 0.91 | 1.00 | 0.72 |
| Spring race total | 0.94 | 1.00 | 0.91 | 0.92 | 1.00 | 0.74 |
| Summer race (lake) | | | | 0.99 | 1.00 | 0.97 |
| Population system as a whole[a] | 0.98* | 1.00* | 0.97* | 0.94 | 1.00 | 0.79 |

[a] The parameter $Nm$ is a determined from the ecological observations

It is clear, at least at a general level, that ecological conditions in a lake – a "lake environment" – are more stable than those in streams and ponds, with their varying water regimes and the accompanying fluctuations of temperature and other physico-chemical parameters at different seasons of the year, especially during the period coinciding with early ontogenesis of the sockeye salmon (the embryonic period and early postnatal stages).

Naturally, the question arises as to which parameters of the environment can act as the selective agents on genotypes of the two loci studied. There are several ways of approaching this problem, and one of them is to explore links between parameters of environment and frequencies of alleles at different stages of ontogenesis. Although the number of these characteristics may be large and their influence indirect (see, for example, Nevo et al. 1984; Philipp et al. 1985), let us examine the connection between the allelic frequencies at both loci and such physicochemical characteristics of spawning sites as concentration of oxygen, pH, and temperature in spawning nests of the sockeye salmon.

Appropriate estimates were made in one year of the observations (Fig. 5.4). It appeared that the allelic frequencies for both loci do not relate at all to concentration of oxygen, although a reliable correlation with temperature exists: for *PGM-2** $r = 0.60$ ($P < 0.05$) and for *LDH-B2** $r = -0.85$ ($P < 0.001$). This association is similar to a clinal one – the allelic fre-

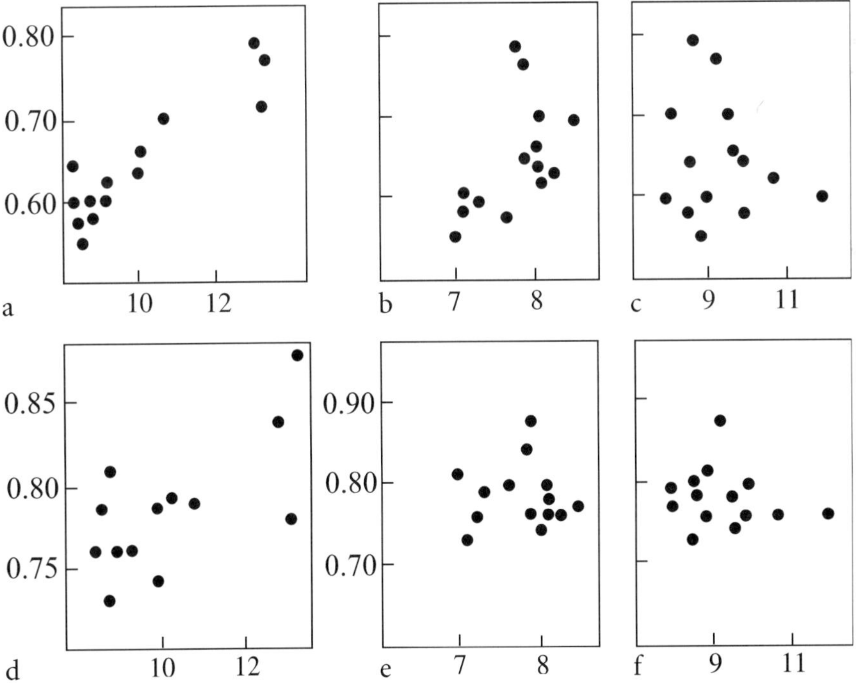

**Fig. 5.4.** The correlation between allelic frequencies at lactate dehydrogenase (**a–c**) and phosphoglucomutase (**d–f**) loci and several physicochemical parameters of spawning ground areas. **a, d** Temperature; **b, e** concentration of hydrogen ions; **c, f** oxygen content

quencies both at the *PGM-2** locus and (particularly) the *LDH-B2** locus
gradually change in the spawning grounds from lakes to ponds in accordance with decreasing water temperature. A significant correlation between
the frequency of *PGM-2** and pH of the spawning grounds has not been
found, but for *LDH-B2**, the correlation coefficient is 0.54, and at the given
value of sample size ($n = 14$) the probability approaches 0.05. Clearly, additional study is merited here. It would also be interesting to partition the
time over which selection may be acting. Evidently, it is most effective in
the very earliest stages of development, because during the catadromous
migration of the young, the genotype distributions within the young cohorts are virtually indistinguishable from the corresponding distributions
in the spawning fish (Novosel'skaya et al. 1982). A similar picture applies for
another species of Pacific salmon, the pink salmon (Altukhov et al. 1987).

We do not exclude the possibility that the heterozygote excess at *PGM-
2** locus observed in the spring race of sockeye salmon may be the result
of a selective fishing impact, as to be considered in detail in Chap. 7.
Nonetheless, in this case also, if the selection vector is directed in favor of
heterozygotes, the comparison of the late ontogenetic stages with the early
ones may be an appropriate method of revealing the selection effect.

## 5.3
## Analysis of Genotype Distributions at Early and Late Stages of Ontogenesis

This study was carried out on successive generations of pink salmon population reproducing in the Naiba River basin (southeast coast of Sakhalin
Island; Fig. 6.8). It consisted of a comparative analysis of the distributions
of allozyme genotypes (at *MDH-B1,2**, *PGM-2**, *G3PDH-1**, *PGDH**, *MEP-
2** loci) in adult fish (spawners) and alevins. In all the polymorphic loci
studied the proportion of heterozygotes was higher for the spawners than
the alevins (Fig. 5.5).

Although the differences between spawners and alevins are small, they
are persistently repeated from generation to generation and are observed
in different populations, both artificially propagated in hatcheries and in
nature (for details see Altukhov et al. 1987).

In order to estimate the selection intensity one can, as a first approximation, calculate the fitness coefficients (viability) of the genotypes at
individual loci ($W_i$), or their different paired combinations ($W_{ij}$), by dividing the frequencies of one- or two-locus genotypes in the spawners into
corresponding frequencies in the alevins. Here, we shall use the data for
the generations of odd years, for which sample sizes are larger (Table 5.5).
It can easily be seen that the fitness of the heterozygotes is higher than that

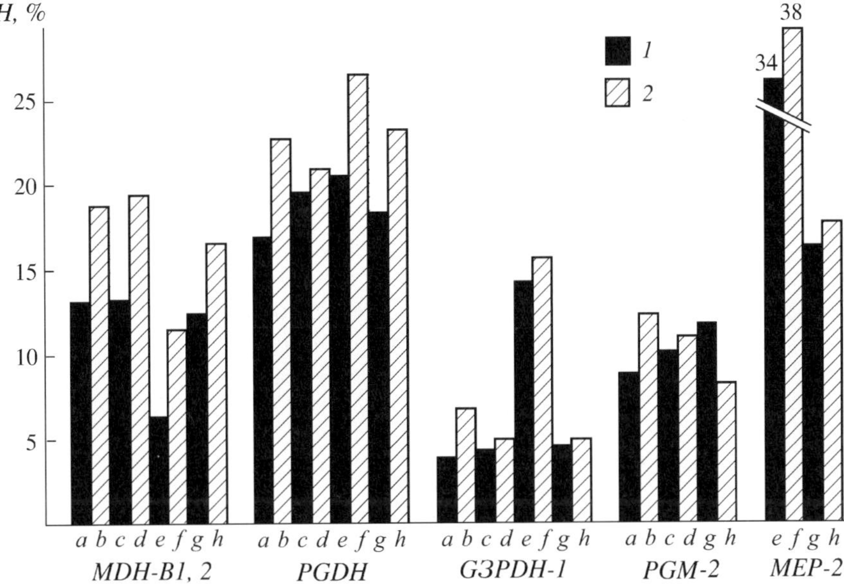

**Fig. 5.5.** Heterozygosity (%) observed at several allozyme loci in alevins (*black columns*) and spawners (*hatched columns*) in successive generations within one and the same population of pink salmon. *a, c, e, g* Alevins of 1979, 1983, 1984, and 1985 spawning years, respectively; *b, d, f, h* spawners of 1979, 1981, 1984 and 1985, respectively. The differences between the spawners and alevins are highly significant by the sing test. The total sample sizes are 1,614 spawners and 2,316 alevins

**Table 5.5.** Fitness values of genotypes of allozyme loci ($W_i$) in the pink salmon and their normalized values ($W_i'$) in adult fish in relation to alevins

| Geno-types | MDH-B1,2* | | | G3PDH* | | | PGDH* | | | PGM-2* | | | MEP-2* | | |
|---|---|---|---|---|---|---|---|---|---|---|---|---|---|---|---|
| | $W_i$ SE | | $W_i'$ | $W_i$ SE | | $W_i'$ | $W_i$ SE | | $W_i'$ | $W_i$ SE | | $W_i'$ | $W_i$ SE | | $W_i'$ |
| AA | 0.92 0.02 | | 0.63 | 0.98 0.01 | | 0.58 | 0.94 0.03 | | 0.71 | 0.98 0.02 | | 0.65 | 0.98 0.03 | | 0.92 |
| AB | 1.45 0.16 | | 1.00 | 0.69 0.39 | | 1.00 | 1.32 0.14 | | 1.00 | 1.20 0.18 | | 0.78 | 1.07 0.21 | | 1.00 |
| BB | | | | | | | 0.57 0.26 | | 0.43 | 1.53 2.00 | | 1.00 | 0.93 0.32 | | 0.93 |

Designations for genotypes are standard: *AA*: predominating homozygote; *AB*: total heterozygotes at all alleles; *BB*: alternative homozygote or following a predominating one. Average sample size per locus: adult fish: 542, alevins 910

of the homozygotes at nearly all loci, except the *PGM-2** locus. However, in latter case, the deviation has simply a statistical character because of the extreme rareness of one homozygote genotype.

The fitness of genotypes in spawners in relation to alevins was calculated by comparing actual distributions of individual heterozygosity – by taking

**Table 5.6.** Distributions of individual heterozygosity at allozyme loci in pink salmon spawners and alevins

| Fish groups | Frequencies of classes with a different number of heterozygous loci per individual | | | Sample size, $n$ | $\chi^2$ test for differences between distributions in spawners and alevins |
| --- | --- | --- | --- | --- | --- |
| | 0 | 1 | $\geq 2$ | | $(df = 3)$ |
| Spawners | 0.495±0.016 | 0.391±0.016 | 0.114±0.015 | 606 | 38.6* |
| Alevins | 0.620±0.020 | 0.335±0.020 | 0.044±0.007 | 901 | |

*Notes*: The "fitness" of frequency classes in spawners in relation to alevins equals 0.80 ± 0.04; 1.17 ± 0.08 and 2.57 ± 0.51 tending towards increased individual heterozygosity. * $P < 0.001$

into account the frequencies of fish that were heterozygous at 0, 1, 2, and more loci (Table 5.6). The estimate for this fitness also increases as the class heterozygosity grows. This means that one cannot exclude the possibility of overdominance as the mechanism maintaining variability at the loci investigated in the pink salmon.

As follows from the data in Table 5.7, the fitness of the interlocus paired genotypic combinations $W_{ij}$ is extremely close to the products of the fitness of the genotypes of individual loci, that is $W_{ij} \cong W_j \times W_j$. This corresponds to the multiplicative model, although additional research is required to reach a final conclusion.

*Thus, comparing genotype distribution at various stages of ontogenesis for several generations within a population indicates, at least in pink salmon, the important role of stabilizing selection in the maintenance of protein polymorphism.*

We now turn to detection of the adaptive nature of protein polymorphism through analysis of the interrelated variability of monogenic and polygenic traits.

## 5.4
## Interrelated Variability of Monogenic and Polygenic Traits

This approach is characterized by the analysis of the distribution of individual heterozygosity of various genotypic combinations along the variability curve of a quantitative character, or of a set of such characters that appear to be adaptively significant. Here, a very important fact is taken into account: those individuals that approximate the population mean value are shown to have the highest fitness, particularly at early stages of ontogenesis (Bumpus 1899; McAtee 1937; Dubinin 1948; Karn and Penrose 1951, etc.).

**Table 5.7.** The fitness of the interlocus paired genotypic combinations in spawners in relation to alevins

| Interlocus genotypic combinations | MDH-B1,2×G3PDH | | MDH-B1,2×PGDH | | MDH-B1,2×PGM-2 | | G3PDH×PGDH | | G3PDH×PGM-2 | | PGM-2×PGDH | |
|---|---|---|---|---|---|---|---|---|---|---|---|---|
| | $W_{ij}$ | $W_i \cdot W_j$ | $W_{ij}$ | $W_i \cdot W_j$ | $W_{ij}$ | $W_i \cdot W_j$ | $W_{ij}$ | $W_i \cdot W_j$ | $W_{ij}$ | $W_i \cdot W_j$ | $W_{ij}$ | $W_i \cdot W_j$ |
| $AA \times AA$ | 0.89 | 0.90 | 0.86 | 0.86 | 0.89 | 0.90 | 0.93 | 0.93 | 0.96 | 0.95 | 0.95 | 0.94 |
| $AB \times AA$ | 1.43 | 1.43 | 1.38 | 1.35 | 1.44 | 1.39 | 1.57 | 1.61 | 1.53 | 1.81 | 1.19 | 1.26 |
| $AA \times AB$ | 1.53 | 1.55 | 1.26 | 1.28 | 1.16 | 1.10 | 1.27 | 1.28 | 1.13 | 1.17 | 1.00 | 1.13 |
| $AB \times AB$ | 5.44 | 2.47 | 2.02 | 2.01 | 1.29 | 1.69 | 2.71 | 2.23 | 4.90 | 2.24 | 2.49 | 1.54 |
| $AA \times BB$ | – | – | 0.69 | 0.55 | – | – | 0.58 | 0.54 | 1.45 | 1.42 | 0.41 | 0.46 |
| $\chi^2$ ( $df$ ) | 2.0 (3) | | 0.4 (4) | | 1.6 (3) | | 3.5 (4) | | 3.7 (3) | | 5.7 (4) | |

It is supposed that if allozyme loci are involved, in one degree or another, in the adaptive process, the levels of individual heterozygosity must not be distributed randomly in the field of variability of adaptively significant qualitative traits. We can state at present that in the case of heterotic selection (overdominance), the maximum heterozygosity must be concentrated in the tails of the distribution of adaptively significant polygenic characters or in the center of a such distribution (under a simple additive effects of genes). This approach turned out especially effective in the study of sockeye salmon populations (Altukhov 1983a; Altukhov and Varnavskaya 1983). We have already given examples that provide evidence for a strong heterotic selection at the *PGM-2** locus of sockeye spawning in Lake Azabachye (Fig. 5.6). The differences between the fitness coefficients of the different genotypes were highly significant, creating the question of how to explain this differentiation in relation to the characteristics that normally mark the effect of heterosis (e.g., increased rates of growth and maturation). And really, we can reveal a clear relationship between intrapopulation differentiation of the species for these characters and allozyme heterozygosity at *LDH-B2** and, especially, *PGM-2** loci.

Sockeye salmon is characterized by sexual dimorphism for body size and complicated reproductive behavior including a system of selective matings (see Sect. 3.2.1). On average, males are larger than females. As a rule, these large males have spent more than 2 years in the sea and are 5 to 7 years old by spawning time. Besides these large males, a second group of males exists that grow rapidly and mature as small, 3-year-old individuals who have spent only 1 year in the sea (they are called "kayurki" on Kamchatka, and "jacks" in the USA and Canada). The mean age of females at spawning varies between 4 and 5 years.

As was shown by Hanson and Smith (1967) and McCart (1969), large males possess a selective advantage during the formation of mating pairs by holding a territory in the spawning grounds. However, in years with a water level lower than normal, and in the shallow areas of the spawning grounds, small males will have a reproductive advantage over large ones.

Interestingly, variances of gene frequencies were shown to differ among population groups of sockeye salmon with unequal sex ratios: at the lactate dehydrogenase locus, the intergroup variation was maximum for subpopulations with a marked excess of males (Fig. 5.7). The same situation was also found in an experimental populations system of *D. melanogaster* (Chap. 4). Analysis of heterozygosity simultaneously at two loci reveals characteristic differences among sockeye individuals of different sexes and ages (Fig. 5.8).

Three groups of fish, which can be easily distinguished during spawning, were compared with regard to their level of heterozygosity for both the *LDH-B2** and *PGM-2** loci. The estimates were made using the sockeye salmon of Lake Azabachye. The data show that females exert an intermedi-

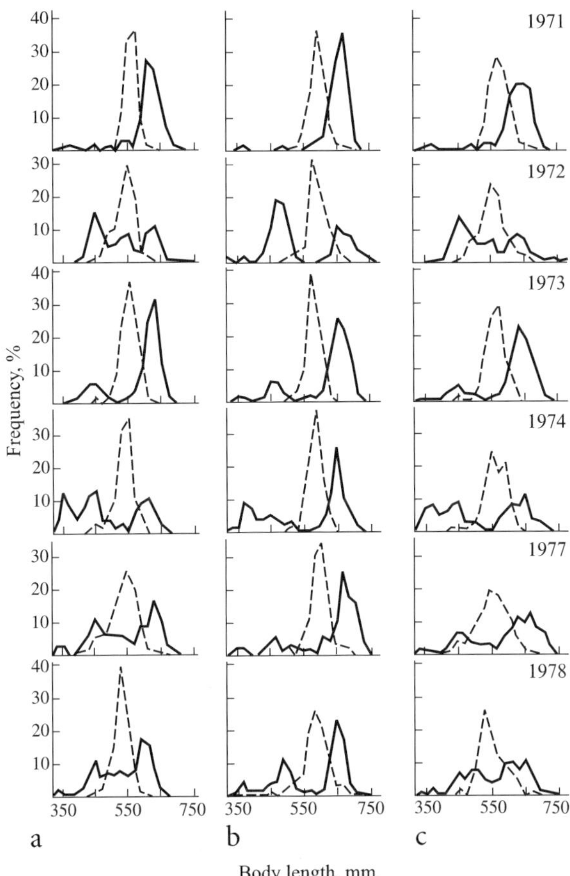

Fig. 5.6. Distribution of body length of males (*continuous line*) and females (*dotted line*) in the reproductive part of a population system of sockeye salmon in Lake Azabachye. *x-coordinate* Body length, mm; *y-coordinate* frequency as %. **a** Spring race; **b** summer race; **c** population system as a whole. Samples sizes: 1971: 542; 1972: 654; 1973: 793; 1974: 676; 1977: 800; 1978: 1,090

ate level of heterozygosity, while small and large males are characterized by maximal and minimal levels, respectively (Table 5.8). Thus, sockeye males and females represent two adaptive systems with the maximal variance of fitness in males. The high degree of heterozygosity of small males correlates (Altukhov and Varnavskaya 1983) with their increased rate of growth and early sexual maturation (Krogius 1960, 1975).

The relationship between increased levels of heterozygosity for the two allozyme genes and the morphophysiological traits, reflecting heterosis for growth and maturation rates, has become clear. The heterozygosity level is especially high in the so-called dwarf males which do not migrate to sea,

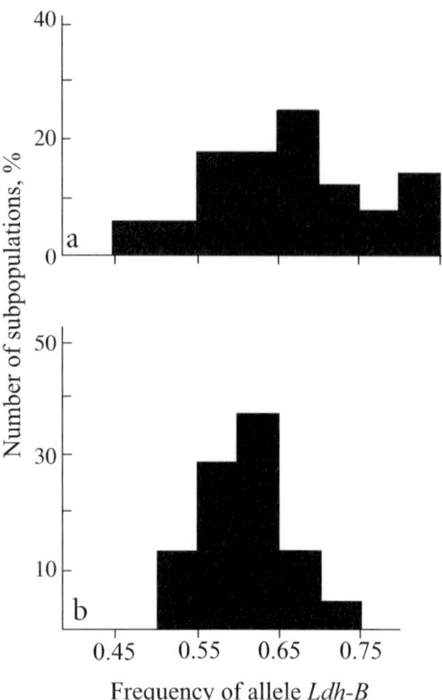


**Fig. 5.7.** Distribution of frequency of allele $B_1$ at locus *Ldh-B2** in *Oncorhynchus nerka* of Lake Azabachye subpopulations (*abscissa*) with different sex ratio (*ordinate*). **a** The subpopulations of excess of males; $n = 33$; $V_q=0.0089$. **b** The subpopulations of excess of females; $n = 22$; $V_q = 0.0028. F = 3.18; P < 0.001$

but remain in freshwater throughout life (Kirpichnikov 1981; Altukhov and Varnavskaya 1983).

Similar relationships, although not certainly interpreted, between multilocus individual allozyme heterozygosity and growth rate and sexual maturation have been found for *Salvelinus fontinalis* (Liskauskas and Ferguson 1991) and *Oncorhynchus mykiss* (Danzman et al. 1988; Ferguson 1992). At least a negative correlation between multilocus heterozygosity and body length for sexually mature males in *S. fontinalis* is evident, as well as the fact that mature males of *O. mykiss* turn out to be more heterozygous than young males, simultaneously showing a slowdown in growth rate (Ferguson 1992). Clearer data completely conforming to the observations for sockeye salmon were obtained for *Salmo trutta* (Makhrov et al. 1997).

The results of different authors concerning the elimination of homo- and heterozygous allozyme genotypes in the fry also agree. It can be assumed proven that a considerably higher mortality among multilocus homozygotes at early stages of ontogenesis, first discovered in pink salmon (Altukhov 1983a; Altukhov et al. 1987), is also typical for some other species

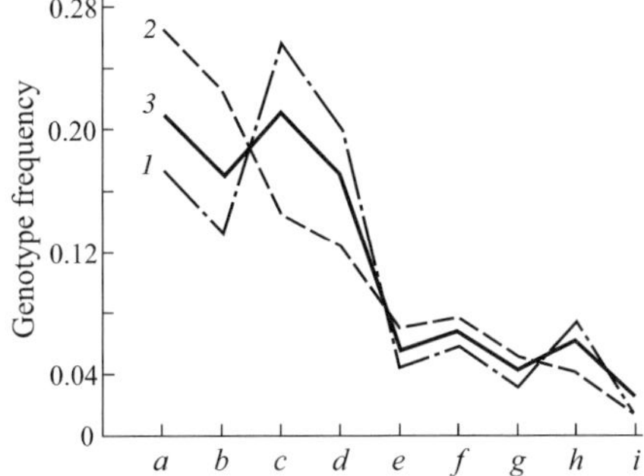

**Fig. 5.8.** The distribution of frequencies of paired genotype combinations at the *LDH-B2** and *PGM-2** loci in the Lake Nachikinskoe spawning sockeye salmon. *1* Small males, *2* large males, and *3* females. *a* 100/100/100/100; *b* 100/120/100/100; *c* 100/120/100/135; *d* 100/100/135/135; *e* 100/120/100/100; *f* 100/120/135/135; *g* 120/120/100/135; *h* 120/120/ 100/100; *i* 120/120/135/135. The probability of random differences between large and small males, and also between small males and females is *P* < 0.05. The genotypic combinations are ranged in the order in which their theoretical frequencies diminish

**Table 5.8.** Total levels of heterozygosity for *LDH-B2** and *PGM-2** in males and females of sockeye salmon populations (in accordance with distributions in Fig. 2.6)

| Research years | Males | | | | Females | |
|---|---|---|---|---|---|---|
| | Small | | Large | | Sample size | Heterozy-gosity |
| | Sample size | Heterozy-gosity | Sample size | Heterozy-gosity | | |
| 1971 | 62 | 0.74 | 205 | 0.62 | 275 | 0.63 |
| 1972 | 209 | 0.66 | 121 | 0.63 | 324 | 0.66 |
| 1973 | 71 | 0.69 | 331 | 0.65 | 391 | 0.67 |
| 1978 | 202 | 0.61 | 317 | 0.60 | 571 | 0.65 |
| 1979 | 90 | 0.65 | 226 | 0.63 | 318 | 0.65 |
| Totals | 705 | 0.67 | 1455 | 0.61 | 2353 | 0.65 |

Statistical analysis shows here the comparison of heterozygosity in pairs of small and large males (large males and females or small males and females) by means of the chi-square test: small and large males 8.45 (*P* < 0.01); large males and females 6.86 (*P* < 0.01); small males and females 1.00 (*P* > 0.05)

(Mitton and Grant 1984; Allendorf and Leary 1986; Zouros and Foltz 1987; Quatro and Vrijenhock 1989).

We studied eight combinations of parent pairs characterized with respect to the loci *MDH-B1,2**, *GPDH**, *PGM-2**, and *MEP-2**, allowing us to rank the families by the average level of heterozygosity. Artificially fertilized eggs were incubated at a hatchery in southern Sakhalin, and mortality was registered throughout the period of incubation and rearing to the smolt stage. The level of individual heterozygosity at the above allozyme genes in alevins of the same age demonstrated a significant positive correlation with body length, i.e., heterosis in growth rate was observed. As expected, a positive correlation was also revealed between the variance of alevin body length and the level of allozyme heterozygosity in the family (Fig. 5.9). Concerning mortality in the progeny (total over the period of incubation and rearing), it was minimal among alevins from intermediately heterozygous parents and maximal in the progeny of pairs with both high and low levels of heterozygosity (Altukhov et al. 1991). One can see in Fig. 5.9 that the mortality of offspring was rather high under these experimental conditions. These results agree with all the previous data confirming that the genotypes of the studied enzyme loci are under the strong pressure of stabilizing selection. In recent years, several authors have presented new facts on the important role of balancing selection for maintaining allozyme polymorphism in natural populations of various animal species (Karl and Avise 1992; Pogson et al. 1995; Wilding et al. 1997).

Our next experiment was performed with 20 pink salmon families differing in the level of heterozygosity at a set of eight allozyme loci: *G3PDH*, MDH-3*, MDH-4*, MEP-2*, PGDH, PGM-2*, GPJ-3*,* and *FDHG**. The results of this experiment, in which the rate of progeny mortality was less than in the previous one, showed the correlation between heterozygosity and body length (at the same pre-smolt stage). More detailed analysis showed that this relationship is conditioned by the effect of heterozygosity on the interfamilial component of variation. In other words, the average body length in the progeny of a family largely depends on the level of family heterozygosity. Moreover, if the families are ranked with respect to average

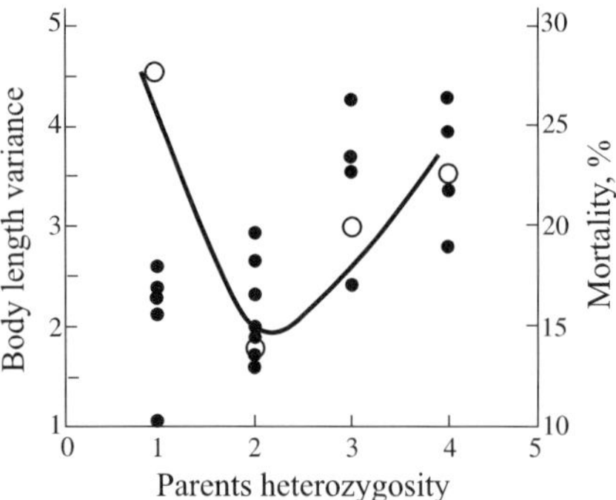


**Fig. 5.9.** The relationship between the level of individual allozyme heterozygosity of parental pairs (*x*-axis), variance of alevin body length (*y*-axis, *dark circles* $r = 0.72; P < 0.001; n = 843$), and death rate (*ordinate to the right, light circles* $n = 5,184$) throughout incubation of fish eggs and development to the premigration stage in the pink salmon (*Oncorhynchus gorbuscha*)

heterozygosity, the largest average body size is observed in the progeny of intermediately heterozygous families, and the smaller progeny is characteristic of families with the highest and the lowest levels of heterozygosity (Dubrova et al. 1995; Fig. 5.10).

Especially significant differences are observed among progeny of families with intermediate and high levels of heterozygosity: average body length in the latter is markedly smaller, and fry from the families with the intermediate level of heterozygosity apparently have a higher growth rate. This result contradicts the predictions based on the heterotic model and does not exclude the possibility that we are dealing with the case of epistatic gene interaction. The latter is characteristic of hybrid and rare genotypic combinations and can have an effect on the type of dependence of variability in quantitative traits on individual heterozygosity (Dubrova and Gavrilets 1989).

We found that parent pairs similar to those giving rise to families with a high heterozygosity rank occur in Sakhalin pink salmon populations with a frequency of approximately 0.05. In other words, highly heterozygous pink salmon families artificially created in our experiments are rare in nature. A unique combination of alleles of the studied polymorphic loci can provide the basis for the epistatic interaction affecting variation in quantitative traits, including body length in young fish. However, the problem is that mortality at early embryonic stages in this experiment was

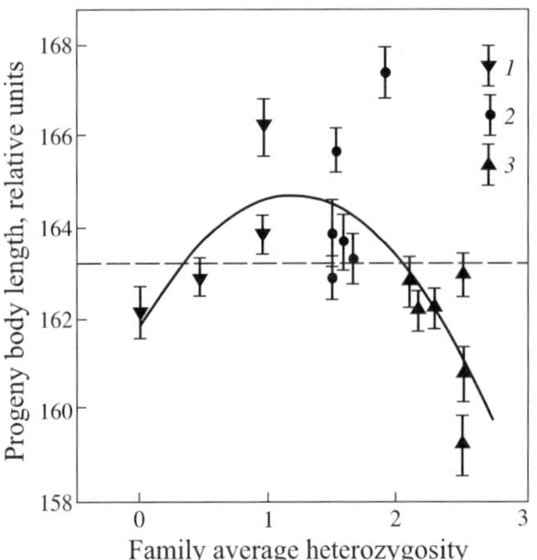


**Fig. 5.10.** The relationship between mean family heterozygosity and progeny body length of pink salmon. *1* Families with low heterozygosity; *2* families with intermediate heterozygosity; *3* families with high heterozygosity. *Broken line* indicates the mean body length value for all the progeny examined. Quadratic fit $y = 161.7 + 5.51x - 2.27x^2$; $F_{2,2562} = 189.23$, $P < 0.0001$, $R^2 = 0.1290$

relatively low, whereas the effects of intrapopulation heterosis (identified by allozyme genes) manifest themselves under suboptimal environmental conditions (Altukhov 1990, 1991; Altukhov et al. 1991); hence, the relationship between a heterotic model and a model involving epistasis requires further investigation.

Nevertheless, the data reviewed corroborate experimental observations providing evidence for the hypothesis that a medium (rather than maximum) level of gene diversity, which is maintained by stabilizing selection, is optimal for the population (Altukhov et al. 1987; Altukhov 1990, 1991; Dubrova et al. 1990, 1991). Body length in young fish has adaptive significance, and individuals with low values of this quantitative trait are characterized by decreased viability (Danzmann et al. 1988; Beacham 1988). According to our results, a decrease in body length of pink salmon fry is associated with a high heterozygosity of the progeny. Therefore, natural selection, eliminating young fish with a small body length from the population, simultaneously has a strong effect on its genotypic diversity and eventually leads to a decrease in heterozygosity. This agrees with the results of our previous experiments in which minimal mortality was observed in families with intermediate levels of genetic diversity (Altukhov 1989b; Kartavtsev et al. 1990; Altukhov et al. 1991). From this point of view, one may try to explain a group of interesting facts noticed by Lynch and Walsh (1998). These authors found, in analyzing data in the literature on the correlation of multiple-locus heterozygosity with fitness-related characters, that increases in heterozygosity with cohort aging only appear at loci that initially have heterozygote deficiencies. If we are to interpret this as a result of selection directed against multiple heterozygotes at the earliest stages of ontogenesis, but in favor of heterozygotes at the later stages, then it is not necessary to resort to the hypothesis of associative overdominance, and the intrinsic overdominance mechanism appears to be applicable in the framework of a heterozygosity-adaptive optimum.

It should be noted, however, that great potential for assessing the adaptive value of polymorphism at many biochemical loci simultaneously is also offered by the approach based on the comparison of their standardized genetic variances with variance expected in the case of neutral evolution, provided the latter is determined on the basis of the parameter of demographic structure $N_e m$ estimated with sufficient accuracy (Rychkov and Sheremetyeva 1976; Rychkov and Balanovskaya 1990a, b; Balanovskaya and Rychkov 1990).

## 5.5
## Analysis of Empirical and Expected Interlocus Genetic Variances as a Method of Estimating Selective Significance of Biochemical Polymorphism

It was shown in Sect. 5.1 that different polymorphic loci are characterized by different intergroup variances of allelic frequencies in one and the same community of subpopulations combined into a whole by migration links.

Insofar as allozyme polymorphism is not expressed externally at all ("hidden" variability) and, hence, there are no grounds for assuming non-uniformity of migration with regard to the genotypes of different loci, persistent interlocus differences of standardized genetic variance can only be explained by these loci (for instance, *Ldh* and *Pgm* in sockeye salmon) being subjected to different effects of selection (Cavalli–Sforza 1966; Rychkov 1969). And indeed, these effects have been shown above. However, are the loci studied involved in the adaptation process in themselves, or do they act as the markers of chromosome segments having other vitally important genes, or supergenes, located in them?

To answer this question, the sampling of genes should be extended. The first analysis in this attempt, unfortunately not successful, was conducted in some interesting research by Lewontin and Krakauer (1973). Their work entailed a method of estimating the possible effects of selection by analyzing the average values of standardized genetic variance for a set of loci studied. If the $F_{ST}$ value (the authors use the symbol $f$) is assigned to each individual locus, their dispersion in the case of neutral evolution should only be determined by sampling error. However, if the level of genetic differentiation differs significantly for different loci, the reason for this may lie in their non-identical proneness to selection pressures differing in direction and intensity.

Lewontin and Krakauer applied this method to estimating protein polymorphism in *D. pseudoobscura* using the data of several authors, and in analyzing the polymorphism of dozens of human blood group loci, using Cavalli–Sforza's (1966) material. They discovered that the variance of the $f_i$ values, characteristic of human blood group genes and of Drosophila protein loci, is several times higher than what would be expected under the hypothesis of random differentiation of populations. Simultaneously, the authors showed very large differences in the values of the $f$ statistics of individual loci; however, this analysis was not pursued to its logical conclusion as "the statistical test for heterogeneity of $f$ values is not very powerful unless a large number of loci and populations have been examined, so it cannot be applied to the data so far gathered for other species or for the electrophoretic variation in man" (Lewontin 1978b, p. 216).

This conclusion is justified if there is no unknown element in the reproductive structure of a subdivided population. But if the parameters of such a situation are known and we know that the genetic process in a population is stationary (that is, the individual $F_{ST}$ values are stable in time), then even two loci are sufficient for extracting important information on selection – as we have shown above.

Naturally, the potentialities of this approach are extended even more when the number of genetic markers studied simultaneously increases.

If genetic drift and gene migration are in equilibrium, the corresponding coefficient of local genetic differentiation expected for a selectively neutral process and described by Eq. (35; Chap. 1) may be designated as $\widehat{F}_e$ and compared with the value $\overline{F}_0$ empirically determined from allele frequencies of different allozyme or DNA loci (Eq. 30; Chap. 1):

$$F_0 = 1 \Big/ K \sum F_i = \frac{1/K \sum V_p}{\overline{p}_i(1 - \overline{p}_i)} \, ,$$

where $K$ is number of loci, $\overline{p}_i$ and $(1 - \overline{p}_i)$ are mean allele frequencies of the $i$th locus in the total subdivided population consisting of $n$ subpopulations, and $V_p = 1/n \sum (p_i - \overline{p}_i)^2$ is Wahlund's variance (Eq. 28; Chap. 1).

If a locus is selectively neutral in a given moment and in a certain site of the species range, $F_{0i} \approx \widehat{F}_e$; if a polymorphic gene evolves under the effect of stabilizing selection, $F_{0i} < \widehat{F}_e$; and in conditions of disruptive or oppositely directed local selection, $F_{0i} > \widehat{F}_e$.

Studies on the indigenous Mongoloid populations of the circumpolar zone of Russia showed that, as estimated according to the set of polymorphic loci, the genetic process occurring in them actually is selectively neutral, and, hence, $\overline{F}_0 \approx \widehat{F}_e$ (Rychkov and Sheremet'yeva 1976).

On the other hand, the species *Homo sapiens* at different phases of its genetic differentiation was exposed to different types of environmental selection pressure, and the ratio of different gene groups in its gene pool changed with time: some genes that behaved as adaptively important or, conversely, selectively neutral at the early stages of historical development of the human population system changed their rank at later stages, whereas the rank of other genes remained unchanged (Balanovskaya and Rychkov 1990). It is noteworthy that none of polymorphic gene loci proved to remain selectively neutral throughout the period of time considered in this study, i.e., since the Upper Paleolithic Period (Rychkov and Balanovskaya 1990b; Balanovskaya and Rychkov 1990).

The analysis performed in our studies has provided evidence that local genetic differentiation value (average per locus) confirms to selectively neutral not only in the case of human populations, but also in population systems of biological species in general (Altukhov 1995), including salmonid fishes (Altukhov and Salmenkova 1994).

Estimations of $N_e$ and coefficients of gene migration $m$ were obtained from publications and on the basis of our own field studies (Altukhov 1974; Altukhov et al. 1975a; Altukhov and Salmenkova 1994). They may be considered approximate, but correspondence between the expected values ($\widehat{F_e}$) and those observed from allele frequencies of allozyme loci ($\overline{F_0}$) are too close to be accidental.

In Fig. 5.11, the correlation between values $\widehat{F_e}$ and $\overline{F_0}$ are shown for several biological species to demonstrate universality of these relationships.

This method obviously needs further improvement because it may be applied only if:

1. The number of loci is sufficiently large.

2. The limit of the estimated polymorphic state of the gene is not lower than 5%.

3. The asymmetrical distribution of $F_{ST}$ values for $i$th loci is normalized.

4. Estimated $F_0$ values for $i$th locus, and comparison of them with multilocus $\overline{F_0}$, is made by using the bootstrap analysis.

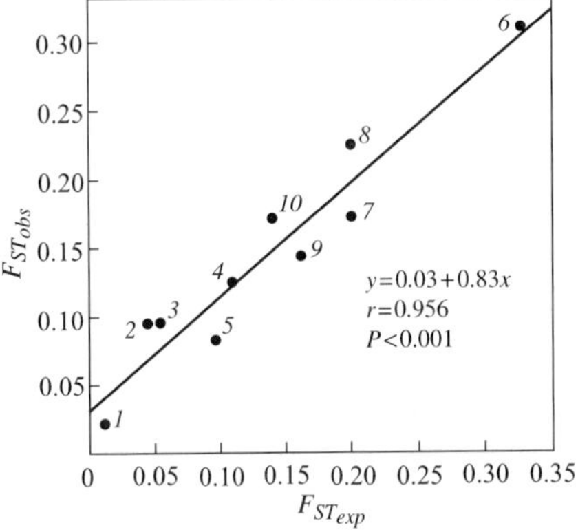


**Fig. 5.11.** Relationship between observed $F_{ST_{obs}}$ and expected $F_{ST_{exp}}$ values of spatial genetic differentiation in different biological species. *1* Pink salmon, *Oncorhynchus gorbuscha*; *2* sockeye salmon, *Oncorhynchus nerka*; *3* chinook salmon, *Oncorhynchus tshawytscha*; *4* Atlantic salmon, *Salmo salar* (data for these four fish species from Altukhov and Salmenkova 1994); *5 Homo sapiens*, Siberian isolate (Rychkov and Yashchuk 1985); *6* black-tailed prairie dog, *Thomomys bottae* (Lidicker and Patton 1987); *7* mouse, *Mus musculus* (Petras 1967; Lidicker and Patton 1987); *8, 9* red howler monkey, *Alouata seniculus* (Pope 1992); *10* green turtle, *Chelonia mydas* (Carr et al. 1978; Bowen et al. 1992; Karl et al. 1992). Reproduced from Altukhov (1995)

Nevertheless, when gene pool reproduction in a population system is a steady-state or at least quasi-stationary process (i.e., when the processes of its differentiation and integration are in equilibrium), the selective status of a gene may be determined reliably enough even if the number of loci is small – so long as number of subpopulations is large. To illustrate this conclusion, let us consider the data on the sockeye population system of Lake Azabachye. Although the necessary data are available for only two polymorphic genes, this obstacle can be overcome owing to the mathematically equivalent increase in the number of subpopulations characterized genetically (over the entire period of studies, a total of 183 and 171 populations for loci $LDH-132^*$ and $PGM-2^*$, respectively). At the corresponding sample size, estimations of $F_0$ by individual loci are fairly stable. The same applies to $N_e$ determined on the basis of direct ecological observations, which value proved to be almost the same as that obtained by analyzing the data of world scientific literature (see 3.2.1; additional evidence in Altukhov et al. 1997).

Comparing empirical estimates of $F_0$ with the expected level of selectively neutral local genetic differentiation $F_e$ calculated from demographic parameters, we can approximately determine the intensity of environmental selection pressure on the loci under study using the following formula:

$$S = \frac{\widehat{F}_e - F_0}{4N\widehat{F}_e F_0}$$

where $\widehat{F}_e$ is the expected genetic variance under conditions of selectively neutral polymorphism, $F_0$ is empirically determined variance of the gene frequency (Eqs. 30, 34, and 41, Chap. 1), and $N$ is effective population size (see Kimura and Weiss 1964; Cavalli–Sforza and Bodmer 1971; Rychkov and Sheremet'yeva 1976).

In the sockeye salmon population system of Lake Azabachye, the $\widehat{F}_e$ value, is expected for a neutral stationary process and calculated on the basis of demographic parameters ($N_e = 200, m = 0.02$), is 0.059. In the case of $PGM-2^*, F_0 = 0.008$. As shown above, polymorphism in this gene is maintained by a strong pressure of stabilizing selection.

Using the above formula to estimate selective environmental pressure on the locus $PGM-2^*$, we obtain $S = 0.135$. At gene frequencies in the point of equilibrium, this value is approximately similar to standard deviation $\sigma$ in genotype fitness $w_i$. Calculating this deviation on the basis of different estimates of $w_i$ values (from the comparison of observed and expected genotype distribution for PGM-2* locus in spring race (Table 5.2) and from solution of stationary distributions, we obtain $\sigma_W = 0.118 \pm 0.048$, which does not differ considerable from $S = 0.135$ estimated above.

For the locus $LDH-B2^*$, which is selectively more neutral, $F_0 = 0.027$, and estimated $S = 0.025$ and $\sigma_W = 0.032$ are virtually identical.

Thus, the method of comparing standardized variances of gene frequencies for individual loci ($F_{0i}$) with that expected for a selectively neutral process ($F_e$), with demographic parameters of population structure $N_e$ and $m$ taken into account, produced the results reliably characterizing the selective status of corresponding allozyme loci and fully confirmed the initial conclusion that the locus *PGM-2** is evolved under strong stabilizing selection, whereas polymorphism at the locus *LDH-B2** is nearly selectively neutral. Remember that this conclusion was also drawn after solving Wright's stationary functions, which allowed us to obtain accurate estimates of genotype fitness for these allozyme loci (Sect. 5.1).

Although the methods of analysis involved were different, they revealed the important role of one form of natural selection, i.e., stabilizing selection. We succeeded in revealing the selective advantage of heterozygotes not only for genotypes of single loci (*PGM-2** or *LDH-B2**) in sockeye salmon, but also for a set of unlinked genes when genotypic distributions were compared in alevin and spawners of pink salmon. Such data are interesting in the context of theoretical discussions on the biological significance of protein polymorphism, and they also are very important for practice.

However, at least three questions remain unsolved:

1. How important is the role of balancing selection on a macrogeographical scale?

2. To what degree can the effects of diversifying selection be detected?

3. What is the ratio of different gene groups with respect to their selective status throughout a species?

This analysis was undertaken for a Pacific salmonid species, chum salmon. The populations of this species were simultaneously examined at 16 allozyme loci in rivers of Alaska, British Columbia, and the northeastern Russian Far East (Konzela et al. 1994; Wilmot et al. 1994; Winans et al. 1994; Table 5.9). The differences between $F_{ST_i}$ and $F_{ST_0}$ in each region were estimated statistically based on conditions listed above in Sect. 5.5. Clear variation of adaptive significance of genes over the species range has been found. For instance, *mAAT-1** behaves as a selectively neutral locus in the Alaskan populations and as a locus subject to diversifying selection in the populations from rivers of British Columbia and the northeastern coast of Asian Russia. The *ESTD* locus is under diversifying selection in Alaska and Russia regions but remains selectively neutral in British Columbia. Interestingly, the proportion of different gene classes grouped according to their selective rank is about the same in Alaska and northeastern Russia: 38–50% of loci under stabilizing selection, 36–38% under disruptive selection, and 14–23% of selectively neutral loci. At the same time, these three gene groups occur in equal proportions in the region of British Columbia, which

**Table 5.9.** Values of mean population heterozygosity ($H_S$) and interpopulation differentiation ($G_{ST}$) at shared polymorphic loci in chum salmon of three regions

| Loci | Alaska including Yukon R. basin | | British Columbia | | Northeastern Russia | |
|---|---|---|---|---|---|---|
| | $H_S$ | $G_{ST}$ | $H_S$ | $G_{ST}$ | $H_S$ | $G_{ST}$ |
| *mAAT-1** | 0.176 | 0.036 | 0.373 | 0.062[b] | 0.080 | 0.087[b] |
| *sAAT-1, 2** | 0.192 | 0.021[a] | 0.182 | 0.018[a] | 0.253 | 0.030[a] |
| *ALAT** | 0.169 | 0.023[a] | 0.276 | 0.036[b] | 0.271 | 0.018[a] |
| *ESTD** | 0.452 | 0.101[b] | 0.109 | 0.033 | 0.249 | 0.067[b] |
| *mIDHP-1** | 0.063 | 0.057[b] | 0.124 | 0.052[b] | 0.237 | 0.097[b] |
| *sIDHP-2** | 0.561 | 0.028[a] | 0.687 | 0.041[b] | 0.618 | 0.034 |
| *LDH-A1** | 0.332 | 0.049[b] | 0.062 | 0.023[a] | 0.218 | 0.033[a] |
| *sMDH-B1, 2** | 0.034 | 0.024[a] | 0.092 | 0.026[a] | 0.031 | 0.021[a] |
| *mMEP-2** | 0.264 | 0.052[b] | 0.222 | 0.028[a] | 0.317 | 0.055[b] |
| *MPI** | 0.211 | 0.024[a] | 0.150 | 0.022[a] | 0.180 | 0.015[a] |
| *PGDH** | 0.561 | 0.011[a] | 0.002 | 0.044[b] | 0.103 | 0.033[a] |
| *PEPB-1** | 0.198 | 0.033 | 0.438 | 0.031 | 0.183 | 0.034 |
| *PEPLT** | 0.051 | 0.044[b] | | | 0.027 | 0.034 |
| *PEPA** | 0.001 | 0.019[a] | 0.001 | 0.010[a] | 0.022 | 0.072[b] |
| $\overline{G}_{ST}$, average per loci | | 0.032 | | 0.032 | | 0.039 |

$H_S$ and $G_{ST_i}$ were calculated from data of Wilmot et al. (1994); Kondzela et al. (1994); Winans et al. (1994). Average $\overline{G}_{ST}$ value and statistical estimates of differences between $G_{ST_i}$ and average $\overline{G}_{ST}$ in each region are obtained by means of bootstrap analysis (1,000 permutations) of log-transformed $G_{ST_i}$ values. [a] indicates that $G_{ST_i}$ is less than $\overline{G}_{ST}$ b indicates that $G_{ST_i}$ is more than $\overline{G}_{ST}$; $G_{ST_i}$ without marks does not differ significantly from $\overline{G}_{ST}$. In all cases probability level $P < 0.01$

seems to have milder environmental conditions. The proportions averaged over all of the populations examined are as follows: 40, 36, and 23% of genes of the first, second, and third groups, respectively. A comparison of these values with the corresponding data for the total ancient humankind (33 genes) of the Upper Paleolithic (the epoch of division and spreading of the original population; Balanovskaya and Rychkov 1990) reveals a striking similarity among these proportions: 48.5 and 24.2% of genes of *Homo sapiens* were respectively under stabilizing and diversifying selection while 27.3% were selectively neutral. In more recent times, when peoples were colonizing new territories and new regional population systems emerged, selection intensity decreased. This resulted in a change in the above ratio so that the proportion of selectively neutral genes increased (27.3, 24.2, and 48.5%). Interestingly, using a different approach, Motoo Kimura (1983) obtained very similar estimates of the proportion of selectively neutral electrophoretic alleles for such distant species as *Drosophila willistoni* (13%) and man (21%; to wit, a tribe of Yanomama American Natives). In a recent study on SNP polymorphisms in man, the proportion of neutral mutations was estimated as approximately 20% (Fay et al. 2001).

Thus, we clearly see that over a long time interval of 15,000 to 30,000 years the selective status of human genes has been changing in accordance with the cultural evolution. The general outcome of this change came to be weaker selection pressure.

Unfortunately, similar chronological analysis is impossible for the fish species studied. Nevertheless, as was in part shown earlier, the spatial dimension can be taken into account. A change in selective rank of a gene can be more graphically demonstrated for a group of loci in another Pacific salmonid species, pink salmon, using a known algorithm (Slatkin 1993).

Geographically and historically related populations are known to demonstrate the effect of isolation by distance with respect to selectively neutral genes, i.e. the genetic divergence of populations increases proportionally to the geographic distance between them. Slatkin (1993) suggested such effects could be revealed by the method in which the logarithm of gene flow $N_e m$ between pairs of populations (calculated from $F_{ST}$ values for these pairs) is regressed on the logarithm of corresponding geographic distance. A significant negative regression means isolation by distance. If the loci evolve under stabilizing selection, the effect of isolation by distance should be absent. We applied this approach to examine the relationship between $N_e m$ (estimated according to different groups of loci) and geographic distance in pairs of more or less linearly connected populations of pink and Chinook salmon using published data.

In pink salmon populations from coastal rivers of eastern Kamchatka, which were studied for 12 loci (data from Varnavskaya and Beacham 1992), no isolation by distance was revealed by this method (Fig. 5.12a). This fact provides evidence for the prevalence of genes exposed to stabilizing selection among the 12 genes studied, confirming our data that such selection at several allozyme loci does exist in pink salmon populations (Altukhov 1983a, 1989a; Altukhov et al. 1987a). Different results were obtained on the basis of data by Shaklee et al. (1991) for pink salmon populations from the North American coast (southern Canada): isolation by distance was traced when the parameter $N_e m$ was calculated according to all 20 loci studied (Fig. 5.12b) or to the set of loci used in the cited work on populations of eastern Kamchatka (Fig. 5.12c). Isolation by distance appeared to be absent, however, when the group of genes characterized by minimal $F_{ST}$ values were used for calculating $N_e m$ (Fig. 5.12d). According to these data, stabilizing selection actually plays a key role in the spatial variation of some allozyme loci, and their selective rank varies in different parts of the species range. However, it was shown that the intra- and interpopulation variation is maintained at an optimal average level, which minimizes the genetic (segregational) load.

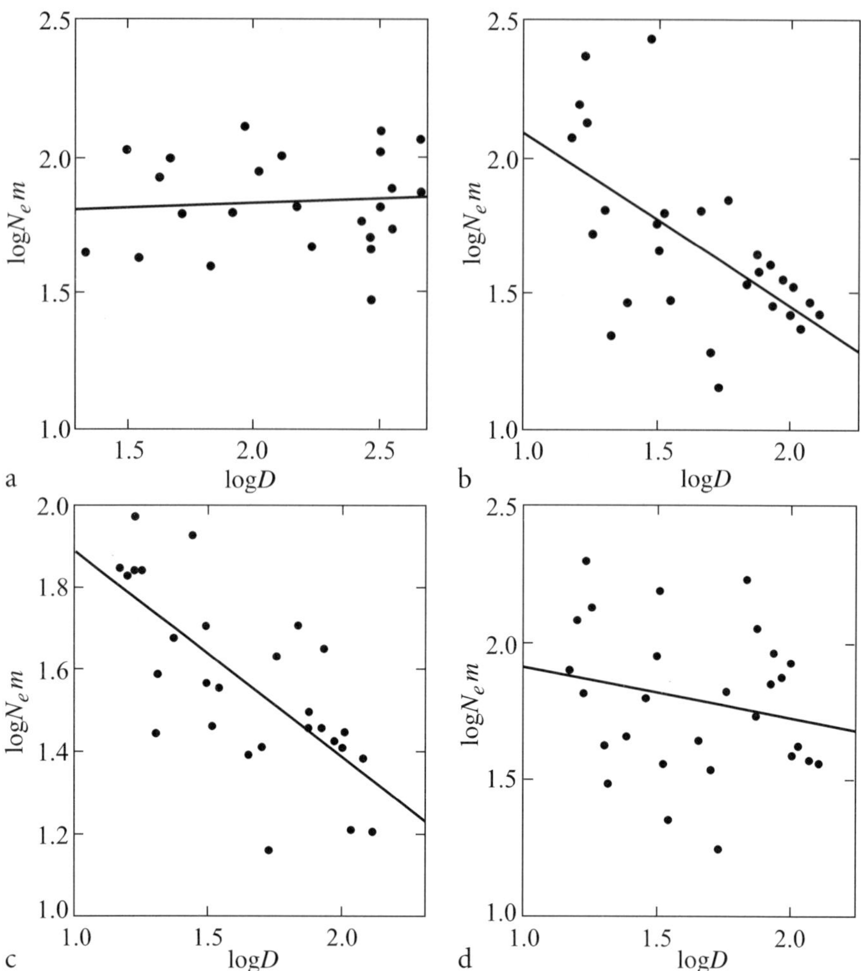

**Fig. 5.12.** Relationship between geographic and genetic distances for pink salmon populations. Calculations of the authors (Altukhov et al. 2000) based on published allele frequency data. **a** Correlation between log $D$ and log $N_em$ for eight populations from east Kamchatka; $N_em$ is estimated using 12 allozyme loci (by data from Varnavskaya and Beacham 1992); correlation is non-significant, $y = 1.75 + 0.05x, r^2 = 0.128$. **b** Correlation between log $D$ and log $N_em$ for eight populations from British Columbia rivers (20 allozyme loci, by data from Shaklee et al. 1991); correlation is significant, $y = 2.40 - 0.51x, r^2 = 0.511$. **c** Correlation between log $D$ and log $N_em$ for the same eight populations from British Columbia, but $N_em$ is estimated using the same 12 loci used in **a**; correlation is significant, $y = 2.55 - 0.60x, r^2 = 0.391$. **d** Correlation between log $D$ and log $N_em$ for the same eight populations, but $N_em$ is estimated using four loci with minimal $F_{ST}$ values; correlation is non-significant, $y = 2.08 - 0.18x, r^2 = 0.043$

# 5.6
# Optimal Genetic Diversity of a Population as a Measure of its Adaptive Maximum

As shown experimentally, *optimal* (intermediate) *levels of gene diversity* play an important role in population survival. This is explained by the fact that high heterozygosity for a number of allozyme loci, which confers maximum viability to an individual, may be disadvantageous for the progeny of this individual and thus for the population as a whole, since many of the segregating genotypes would not be adaptive in suboptimal environments (Altukhov 1989b; Altukhov et al. 1996, 1997). High levels of polymorphism and heterozygosity can lead to a drastic rise in genetic recombination rate, which would disrupt harmonious interlocus interactions. The probability of disturbance of cooperative interrelationships among loci must be particularly high precisely at the early stages of development since at these stages, because of immaturity of many functional systems of the developing organism, its viability primarily depends on the most sensitive links of the metabolic processes (Livshits and Kobyliansky 1985).

If our assumption is correct, we would expect that the individual heterozygosity is specifically related to the viability of the progeny (or to its variance). Indeed, this has been found in at least one experiment on pink salmon (Sect. 5.4).

The conifers are of exceptional interest for the experimental approach to this problem. As with all Gymnosperms, they have a unique reproductive system, in addition to numerous polymorphic enzyme systems, which can be investigated at the earliest stages of ontogenesis, including the gametic phase (Adams 1983). As distinct from Angiosperms, the endosperm (the female gametophyte or megagametophyte) of the Gymnosperms constitutes a haploid reserve food tissue with its origin a single haploid megaspore. The latter produces all the other cells within an ovule, including the egg cell, and after fertilization the embryo begins to develop and seeds are formed. Megaspores are products of meiosis, and as a rule allozymes in the megagametophytes of heterozygous trees segregate in a ratio of 1:1. This makes it possible to determine the maternal genotype without crossing (Fig. 5.13).

While studying allozyme polymorphism in a natural population of the Norway spruce, *Picea abies* (Altukhov et al. 1986a,b), we, as expected, discovered a positive correlation ($r = 0.52; P < 0.001$) between individual heterozygosity levels at six allozyme loci (*Idh-1*, *Pgd-1*, *Pgd-2*, *Gpd*, *Got-3*, *Gdh*) and the total fraction of sterile and dead seed (Fig. 5.14). It is important to note that the correlation is significantly non-linear ($P < 0.05$): when the heterozygosity of the maternal trees increases, the proportion of nonviable

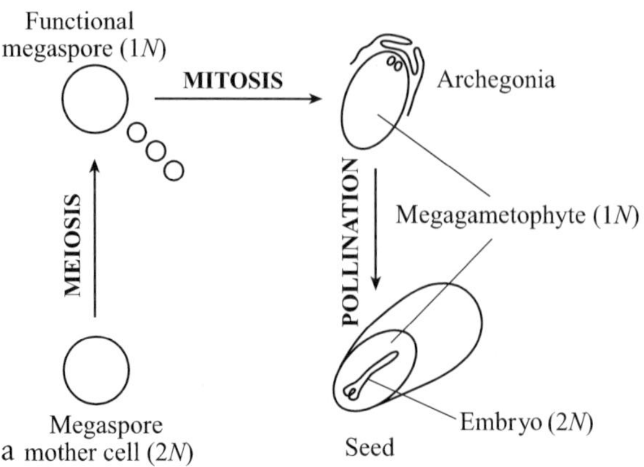


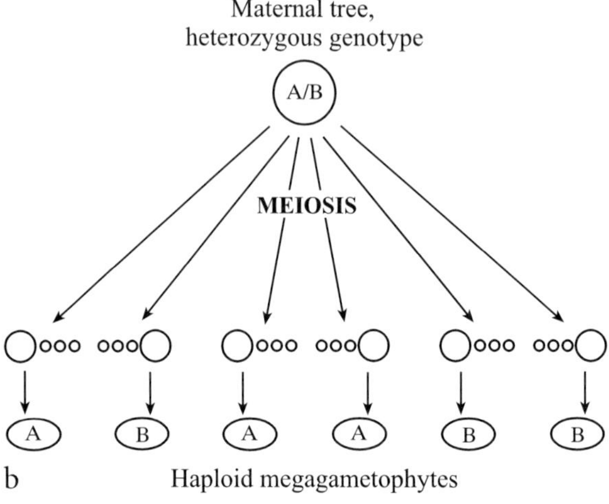


**Fig. 5.13.** Genetic basis of reproduction in gymnosperms. **a** The endosperm (female megaga-metophyte) in coniferous plants is represented by storage of nutritious haploid tissue stem-ming from a single functional megaspore. The functional megaspore undergoes mitosis to produce the megagametophyte, which becomes an archegonium containing female gametes. Pollen can fertilize several gametes but only one of them produces an embryo in a mature seed. **b** Segregation of a diallelic locus of the maternal plant genome. Megagametophytes of individual seeds genetically correspond to single haploid offspring, the ratio of which does not deviate from the equilibrium (1A:1B). If the tree is heterozygous for a null allele, 50% of the offspring will lack the corresponding DNA-fragment or protein (empty gel in electrophoresis)

seed is reduced initially and then, after reaching a certain level virtually identical to the population mean, it begins to rise.

Conifer seed formation, as is well known, depends on many factors but, first and foremost, on the interaction of the developing embryo and the endosperm. Disturbance of this development can suppress endosperm development, leading to the formation of sterile and dead seed which, as is known, form a substantial part of the harvest of conifers – in many cases as much as 50–70%. Homozygosity of recessive embryonic lethals during self-fertilization is believed to cause this (Fowler 1965; Koski 1971; Fowler and Park 1983; Griffin and Lindgren 1985).

Presumably, trees that are more heterozygous at biochemical loci have a larger amount of recessive lethals or semilethals. If heterozygosity is maintained by a system of balanced lethality and is accompanied by linkage of corresponding genes with allozyme loci, then we shall observe the picture represented in Fig. 5.14: the higher the individual biochemical heterozygosity, the greater the portion of nonviable seed. However, without denying the possibility that such a mechanism exists, I would suggest that the results from the spruce as well as from the pink salmon (see Sect. 5.4) are connected with levels of individual biochemical heterozygosity per se.

In this case, the "cost" of high heterozygosity is segregation of numerous inadaptive genotypes. These genotypes, if we assign to them not only less adaptive homozygotes but any genes or their combinations that negatively affect fitness, constitute the segregational load of the population (Altukhov 1985b).

However, of greatest interest seems to be the evidence on strong balancing selection at the early development stages and on many polymorphic

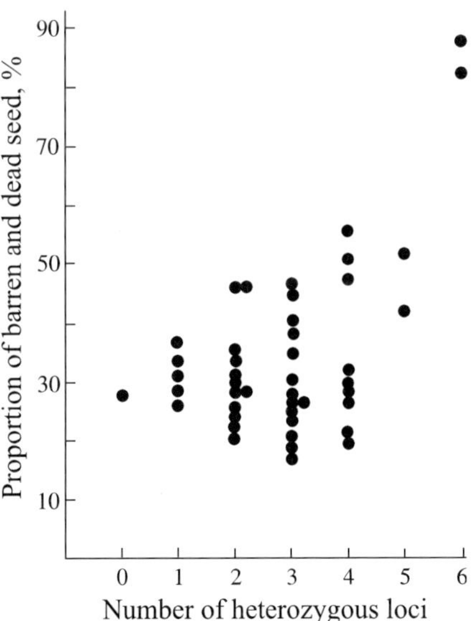


**Fig. 5.14.** The correlation between allozyme heterozygosity of Norway spruce trees (*x*-coordinates) and the total fraction of barren and dead seed produced by them. (Altukhov et al. 1986b)

genes participating in this process (Altukhov 1989a,b, 1991; Mitton 1997). Apparently, the intensity of selective elimination is positively related to the probability of detecting joint variation of an adaptive polygenic trait and a discrete genetic marker. Upon multiplicative interactions, the sought associations can be found using far smaller samples of genes and individuals than in the case of neutral allozyme loci.

In any case, studies of the association between individual tree heterozygosity and seed production in the reproductive part of gymnosperm populations yielded striking results. More heterozygous trees were generally more productive, but seed production of trees with intermediate heterozygosities proved to be maximum on account of a higher proportion of empty and dead seeds in trees with high and low heterozygosities (Malyuchenko and Altukhov 2002). Thus, these data suggest adaptive significance of exactly the average (optimal) levels of allozyme polymorphism important for successful reproduction of the population.

During the last decade, many papers have been published attempting to analyze the effects of heterosis by investigating allozyme polymorphism in a wide variety of plant and animal species. The results have been reviewed by Zouros and Foltz (1987), Lynch and Walsh (1998), and in J. Mitton's book (1997). Zouros and Foltz note that the study of a correlation between the individual allozyme heterozygosity and adaptively loaded characters has led both to positive and negative results, making it reasonable to summarize these somewhat contradictory findings. One should, however, address the following questions: Why are the connections between the variability of quantitative characters and allozyme polymorphism not always discovered? And why are these links, when discovered, associated with a small number of allozyme loci?

When answering the first question it is very important to regard the "genotype × environment" interactions. As was noted earlier (Altukhov 1983a, 1991), under the same environmental conditions various gene loci are subject to different forms of selection, or are sometimes selectively neutral. Correlations can be estimated only when the conditions are strict enough, i.e., when the uniformly vectorized selection simultaneously influences a great number of genes. The early stages of ontogenesis are the most appropriate for revealing these correlations. This conclusion can be illustrated by the aforementioned data for European spruce. In contrast to the situation depicted in Fig. 5.14, where correlation between the individual heterozygosity of maternal trees and the proportion of non-viable seeds was estimated for the total population, we will now examine the same relationship for two isolated plots with contrasting environments: in one case, trees occupy a zone of ecological optimum (Fig. 5.15a), whereas in the other, they grow under adverse conditions, on swampy peat soil. The correlation was stronger in the case of a harsh environment. Similar results

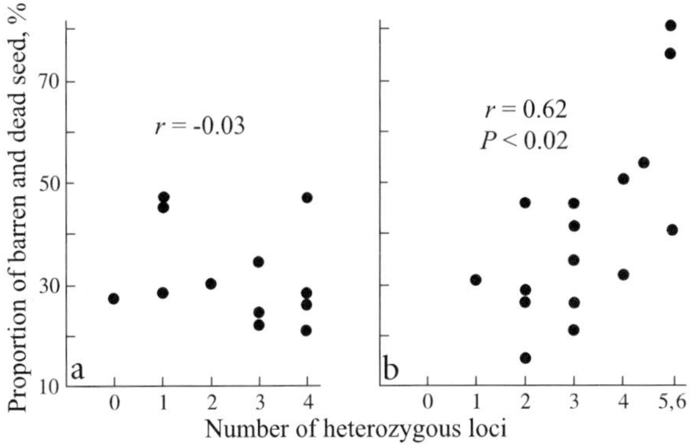

**Fig. 5.15.** The connection between the level of heterozygosity of maternal Norway spruce trees and the proportion of the inviable seed they produce. **a** Trees from an ecologically optimal zone (a dry valley) and **b** trees from suboptimal conditions ( a marshy area)

were obtained for other animal and plant species (Govindaraju and Dancik 1987a,b; Mitton 1997). These observations are in good agreement with the conclusion on an important role of genotype × environment interaction in the manifestation of heterosis (Griffing and Zsiros 1971; Barlow 1981).

Answering the second question, we should take into account several issues in addition to those already addressed. These issues include, for instance, yet unstudied but possible pleiotropic effects of allozyme genes or the possibility that the actual functional variability of the genome may be substantially lower than what we currently believe (Sect. 2.3; see also Sect. 6.3).

The evidence on pink salmon obtained under virtually selectively neutral conditions raises the issue of the role of epistatic gene interaction in the expression of a polygenic trait (Sect. 5.4; Dubrova et al. 1994, 1995).

The effects of intrapopulation heterosis of allozyme genes are expressed in suboptimal environments (Sect. 5.4; Altukhov 1989a; Altukhov et al. 1991). Hence, the issue of the relative contribution of the heterosis model and the model with epistasis requires further consideration.

Nevertheless, the above data confirm a number of experimental observations that testify in favor of the hypothesis that the average (but not maximum) genetic diversity maintained by stabilizing selection is optimal for populations (Altukhov et al. 1987a; Altukhov 1989a,b; Altukhov and Kurbatova 1990; Dubrova 1990; Kurbatova 1996). Body length of juvenile fish is an adaptive trait; individuals with low values of this trait have low viability (Danzmann et al. 1988; Beacham 1988). Our results indicate that reduced body length in juvenile pink salmon is correlated with high heterozygosity of the offspring. Therefore, natural selection eliminating low-body-length juveniles from the fish population substantially affects genotypic diversity of the population, which ultimately results in a reduced heterozygosity level. This is in good agreement with the results of our previous experiment, in which families with intermediate levels of genetic diversity were characterized by minimum mortality (Altukhov 1989a; Kartavtsev et al. 1990; Altukhov et al. 1991).

Finally, a question arises as to whether natural selection acts directly on genotypes of enzyme loci or affects some functionally significant chromosome segments that are marked by the corresponding genes. In this context, of key importance is the evidence presented in Sect. 5.3, where we compared early and late developmental stages in consecutive generations of the same population of pink salmon.

Given that the chromosome number in this species is relatively high ($2n = 52, NF = 104$), and the genetic differences between juveniles and spawners were found at all examined and completely independent allozyme genes, we can conclusively state that *selection acts precisely on the genotypes of these loci rather than on some other genes or supergenes that are marked by these loci.*

Likewise, the above data obtained by estimating selective gene values via Wright's $F_{ST}$ statistics solve the issue of linkage disequilibrium: the gene pools of the species examined are viewed as flexible, mobile systems, in which proportions of different genes rapidly change in time and space according to the character of the genotype × environment interaction.

Thus, it is not surprising that when the intact structured gene pool reproduces in the optimal environment, the coefficient of local genetic differentiation averaged over an array of allozyme loci ($F_{ST_0}$) corresponds to the expected value ($F_{ST_e}$) for a selectively neutral process, and is practically the same as the value of ($F_{ST_0}$) inferred from polymorphisms of random DNA fraction samples from the eukaryotic genome (excepting microsatellites; Chap. 2). This relationship was first found by Rychkov and Balanovskaya (1990a,b) in human populations examined for numerous DNA markers (Cooper and Schmidtke 1984; Bowcock et al. 1987) and for 20 systems of immunobiochemical polymorphism (Yu.G. Rychkov and associates). Much later Allendorf and Seeb (2000) tabulated similar results for a vide variety of animal and plant species, but they did not give appropriate population-genetic interpretation to these facts. In 12 of 16 species examined, the levels of local genetic differentiation for protein and DNA markers fully coincided, and in only 4 cases were discrepancies observed.

For instance, $F_{ST}$ values for proteins and DNA coincided in Atlantic salmon *Salmo salar* (proteins = VNTRs), maize *Zea mais* (RFLPs = proteins), in Douglas fir *Pseudotsuga menziezii* (RAPDs = proteins), red ant *Isolenopsis invicta* (RAPDs = VNTRs = proteins), marine beet *Beta vulgaris* (SCN = RFLPs = proteins), etc. Analogous results have been recently obtained for a number of ethnic groups from Eastern Europe (Limborska et al. 2002, p. 169).

As to the discrepancies, they are probably caused by sampling errors (unequal coverage of the population structure examined by different marker groups) or, which seems more common, prevalence of one of the above groups of gene loci in the analyzed sample of markers.

If we deal with a prevailing incidence of overdominant protein loci, the genetic differentiation of the species would involve substantial spatial similarity, or even uniformity of the frequencies of the corresponding protein-coding genes, together with marked local differentiation of the populations at DNA loci. This was indeed observed in populations of cod *Gadus morhua* (Pogson et al. 1995; RFLPs > proteins), pine *Pinus flexilis* (Latta and Mitton 1997; RAPDs > proteins), and oyster *Crassostrea virginica* (Karl and Avis 1992; scn DNA > proteins). Only in one case (gray toad *Bufo bufo*) was the situation different (proteins > VNTRs; Scribner et al. 1994). In view of the population structure characteristic for the gray toad, its results may be explained by diversifying selection. The uniformity of allozyme gene frequencies in the Pacific pink salmon, as a result of the potent effect of stabilizing selection, has been discussed by us earlier (Altukhov 1983; Altukhov 1989a,b; Altukhov et al. 2000).

As a whole, if our concepts of adaptation as a compromise between the individual and population's components of this process are correct, one would expect there to be several mechanisms in nature aimed at maintaining optimal gene diversity and, correspondingly, at minimizing the genetic load of populations. And indeed it is possible to deduce these mechanisms as well as to show real situations that correspond to them.

**1. Selective Mating.**    We have already discussed this in a previous section in relation to the Pacific sockeye salmon. The following is supplementary: the mating of some females with large males and of others with small ones having distinctive genetic features preserves the optimal (average) level of heterozygosity characteristic of a population (Altukhov and Varnavskaya 1983; Varnavskaya 1984; see also Sect. 7.1). Other types of analogous reproductive behavior are also known – see, for instance, Schröder (1983).

**2. Subdivision of Populations into Subpopulations.**    This mechanism is of interest because a balance is formed between the processes of differentiation and integration of genetic variability, thanks to a fine subpopulation structure. This phenomenon, well known in many species, has been examined in detail in Chap. 3.

One must add that in all probability the parameter $N_e m$ is autoregulated in population systems: the greater $N_e$ is, the less $m$ is, and vice versa. This association was discovered in an experiment (Sect. 4.1, Fig. 4.4), but it is also seen in natural population systems. For example, for the circumpolar human populations of the North Asiatic Mongoloids: $N_e = 45$ and $m = 0.034$ ($N_e m = 1.53$), while for northern Asia as a whole $N_e = 200$ and $m = 0.010$ ($N_e m = 2.00$) (Rychkov and Sheremet'yeva 1976).

**3. Restriction of Free Recombination.** This is a well-known mechanism associated with the pressure in a population of inversions, linkage, or other genetic phenomena that induce the formation of stable gene clusters (supergenes; Altukhov and Rychkov 1972; Carson 1982). But, of course, findings on parthenogenetic species and populations having maximally high levels of biochemical heterozygosity are unique (see the review by Nevo et al. 1984). In many such cases the individual and population components of fitness coincide completely, and there is 100% facultative or obligatory heterozygosity in all the protein loci studied. A strong positive link has been established between individual heterozygosity for a set of genes and a tendency to artificial ameiotic parthenogenesis (Altukhov and Klimenko 1978; Barauskene et al. 1985).

**4. Selection that Varies in Direction at Different Stages of Ontogenesis or in Individuals of Opposite Sex; Other Types of Similar Selection.** Data of this kind have been given above for the pink and sockeye salmon (see also Sect. 7.1). In addition, it should be remembered that a theoretical analysis of this question by Lewontin et al. (1978) has highlighted the difficulties encountered by a heterotic model when there are more than two alleles at a locus. Of greater interest in this connection are models based on frequency-dependent selection and, in particular, a model of the selection that varies in both time and space. As Gillespie (1977) has shown, if a population occupies a large number of microniches, and if the homozygotes differ slightly in their fitness from niche to niche (even if the fitness of the heterozygotes is somewhat higher than that of the homozygotes), this creates a favorable condition for maintaining stable allozyme polymorphism at many loci. One should only add that each locality ought to have its own level of heterozygosity as adequate as possible to its specific living conditions. Each of these populations, if historically formed, should be regarded as having reached maximum adaptation to a specific environment.

**5. Physiological Homeostasis Aimed at Preserving a Population's Optimal Genotypic Structure.** Unfortunately, genetic literature sheds very little light on this mechanism discovered by B.P. Ushakov (1982a,b, 1984, 1989) and studied in detail on the basis of temperature adaptations in poikilothermal animals. The main conclusion of these original works is that each population responds to changes in the environment, not exceeding the ecological optimum, by a complex of individual physiological reactions that preserve hereditary diversity. Universal reactions, such as the *phenotypic masking of genotypic differences* in thermostability among individuals and *the change* *in their selective rank* were discovered during broad comparative research on many animal species – which lends special authority to Ushakov's conclusions. From this point of view it is not difficult to explain, for example, why the interpopulation variance of heterozygosity at allozyme loci is lower in species that are active in their reproductive behavior than in not very mobile, attached species that are fertilized externally and have a passive larval stage (Nevo et al. 1984).

More examples could be given of the control of the optimal level of gene diversity in natural populations, although there is clearly a need for further research. However, even now one can say that in the case of a heterotic model, heterozygosity should be concentrated in the tails of the distribution curve of adaptively significant polygenic traits, whereas in other cases morphologically average phenotypes may have maximal heterozygosity (see, for instance, Beardmore and Shami 1979). Chakraborty (1987) has examined theoretically the unresolved questions of this kind.

In general, the pattern of natural selection operating in a system of subpopulations is quite specific. It includes variation in selection direction and intensity in different loci and sexes, as well as differences of the same type at different developmental stages, under different conditions, etc. The balance of these forces together with gene drift in the population system can make a false general impression of neutrality of the given polymorphism, which is probably the case for the lactate dehydrogenase locus in sockeye salmon from Azabachye Lake. Under these conditions, the gene frequencies may not be in equilibrium, but the population system will consistently maintain the mean frequency of this gene and its intergroup variance.

The "neoclassical" and "balance" views on spatial homogeneity of frequencies of many independent genes and exceeding segregational population load are given new meaning on the basis of the evidence presented in Chaps. 3–5; asymmetry of balancing selection and effective population size assume key significance.

## 5.7
## Theory of Neutrality in the Light of Recent Data

In fact, if we rely on estimates of population numbers given in works of representatives of both the neoclassical and balanced schools (amounting to thousands, if not millions of individuals), then as we have seen (Sect. 1.3), the volume of the segregational load is indeed inordinately large, even at very insignificant selection coefficients. But the whole point of the matter is that there are no such gigantic populations in nature: by virtue of subdivision they are composed of pluralities of subpopulations limited in effective and total size. In the course of their natural history these subpopulations may die out, become reestablished through recolonization, or experience sharp fluctuations of numbers in the reproductive period. Under these conditions, and even for a species as a whole, the genetically effective number (total) is much lower than direct estimates would lead one to expect (Maruyama and Kimura 1980).

There is also the problematic approach whereby, in estimating heterozygosity levels, the total number of a species is usually taken into account without allowing for its subdivision into isolated subpopulations. This concept would appear to be feasible if the rate of the migration exchange of genes among separate subdivisions of a species is known to be higher than the rate of the spontaneous mutation process. But in the event of these values becoming commensurable (see, for example, Larson et al. 1984), then estimates of the heterozygosity or effective number of alleles, anticipated by making allowance for their selective neutrality, should be made taking into account the numbers of actual isolated populations and not of a species as a whole. In these cases, which are not so rare in nature, the segregational load problem no longer becomes so acute. It is true that the absence of spatial differentiation is often explained away as migration in discussions between "selectionists" and "neutralists", who consider that literally one migrant per generation is sufficient to delete the genetic differences at a subpopulation level. The next logical step is taken from here and it is claimed that if migration is so effective, spatial differences in gene frequencies among separate samples only reflect the *heterogeneity of the environment* over an area of large panmictic populations. It should be pointed out, however, that this conclusion is valid only for Moran's model of population structure (Moran 1959, 1973) in which the intensity of gene exchange increases in proportion to the number of subpopulations. The picture is different in natural population systems: in a one- or two-dimensional area having an "island" or a "stepping-stone" structure of gene migration, such intensity is either independent of the number of subpopulations or inversely dependent on it. Of course, heterogeneity of environment must not be denied, but nor should one forget that in nature there are no random mechanical accumulations of individuals that interact with each other, but real subpopulation units characterized by the unity of their behavior and morphophysiological features. This perception emerges both from the results of our own research and from other works in which the biology of populations is given due attention (see Lebedev 1967; Petras 1967; Selander 1970; Shilov 1977; Yablokov 1986).

On the other hand, even now it is clear that in a wide circle of organisms – from invertebrates to humans – the $N_e$ estimates (or "neighborhoods") come within an interval of 10 to $10^4$ individuals with clearly expressed left-handed asymmetry of distribution and a mode in the range of 10–200

**Table 5.10.** Estimates of effective population size in different species

| Researched species | Effective size | Author, year |
|---|---|---|
| Plants | | |
| *Latris aspera* | 14–27 | Levin and Kerster (1968, 1971) |
| *Linanthus parryae* | 75–282 | Wright (1943b) |
| *Lithospermum caroliniense* | 205–547 | Kerster and Levin (1968) |
| *Phlox pilosa* | 167 | Levin and Kerster (1968) |
| *Viola blande* | 310 | Beattie and Culver (1979) |
| *V. pedata* | 176 | Beattie and Culver (1979) |
| *V. pensylvanica* | 4 | Beattie and Culver (1979) |
| *V. rostrata* | | Beattie and Culver (1979) |
| Animals | | |
| *Aedes aegypti* | 500–1000 | Tabachnik and Powell (1978) |
| *Capaea nemoralis* | 236–8440 | Lamotte (1951) |
| | 190–12,000 | Greenwood (1975, 1976) |
| *Chondrus bidens* | 50 | Altukhov and Livshits (1978) |
| *Drosophila pseudoobscura* | 500–1000 | Dobzhansky (1970) |
| *D. subobscura* | 400 | Begon (1977) |
| *Euphydryas editha* | 10–3700 | Ehrlich (1965) |
| *Mus musculus* | 10 | Anderson (1970) |
| *Michigan populations* | 10–75 | Rasmussen (1964) |
| *Arizona populations* | 30–130 | Rasmussen (1970) |
| *Oncorhynchus nerka* | 200 | Altukhov et al. (1975a, b) |
| *Ovis canadiensis* | 98 | Geist (1971) |
| *Platycercus eximius* | 31–83 | Brererton (1962) |
| *Porcellio scaber* | 19–180 | Brererton (1962) |
| *Scelopus olivaceus* | 250 | Kerster (1964) |
| *Thais lamellosa* | 10–700 | Spight (1974) |
| *Triturus cristatus* | 4.0–49.3 | Jehle et al. (2001) |
| *T. marmoratus* | 4.5–21.4 | Jehle et al. (2001) |
| *Uta stansburiana* | 14 | Tinkle (1965) |
| Native *Homo sapiens* populations | | |
| Parma valley | 214–266 | Cavalli–Sforza et al. (1964) |
| North Asiatic Mongoloids | 45–218 | Rychkov (1968); Rychkov and Sheremet'yeva (1976) |
| South American Native tribes | 288–14,400 | Neel and Rothman (1978) |

individuals (Table 5.10; see also Lande 1979). Furthermore, recent estimates by Frankham (1995), inferred from literature data for 102 animal and plant species, clearly indicate that the effective sizes of natural populations are far smaller than previously thought. Taking into account size fluctuations, unequal sex ratio, and family size variance, the $N_e/N_r$ ratio (averaged over 192 estimates) constituted only 0.10–0.11.

Even allowing for the legitimately exaggerated figure that $N_e$ comprises 75% of a population's reproductive numbers and approximately 30–40% of the total numbers, the population size that should be used in formula (28; see Sect. 1.3) for determining the segregational load, at least insofar as many species are concerned, must be measured not in hundreds of thousands of individuals but by a considerably smaller figure. Clearly, in this case, the segregational load is sharply reduced while the number of polymorphic loci, maintained by the selective advantage of the heterozygotes, increases significantly.

Table 5.11 presents the results of corresponding determinations of the $e^L$ value based on realistic estimates of $N$ and coefficients of selection in favor of heterozygotes in a different number of diallelic loci whose effects

**Table 5.11.** Excess fertility $e^{\tilde{L}}$ depending on population size, number of overdominant loci and selection coefficients against homozygotes

| Number of over-dominant loci | Population size, number of individuals | | | | | | | |
|---|---|---|---|---|---|---|---|---|
| | 100 | 200 | 500 | 1000 | 5000 | 10,000 | 50,000 | 100,000 |
| | $s_1 = s_2 = 0.05$; $\widehat{p}* = 0.5$ | | | | | | | |
| 100 | 1.97 | 2.10 | 2.27 | 2.37 | 2.65 | 2.77 | 3.04 | 3.16 |
| 500 | 4.56 | 5.23 | 6.17 | 6.92 | 8.84 | 9.74 | 12.03 | 13.11 |
| 1000 | 8.56 | 10.38 | 13.11 | 15.43 | 21.81 | 25.02 | 33.73 | 39.07 |
| 5000 | 121.71 | 187.41 | 315.61 | 454.58 | 984.79 | 1,339.23 | 2,610.90 | 3,422.47 |
| | $s_1 = 0.01$; $s_2 = 0.09$; $\widehat{p} = 0.1$ | | | | | | | |
| 100 | 1.28 | 1.30 | 1.34 | 1.36 | 1.42 | 1.44 | 1.49 | 1.51 |
| 500 | 1.73 | 1.81 | 1.92 | 2.01 | 2.19 | 2.27 | 2.45 | 2.52 |
| 1000 | 2.17 | 2.32 | 2.52 | 2.68 | 3.03 | 3.19 | 3.55 | 3.71 |
| 5000 | 5.63 | 6.58 | 7.94 | 9.05 | 11.96 | 13.35 | 16.98 | 18.72 |
| | $s_1 = s_2 = 0.1$; $\widehat{p} = 0.5$ | | | | | | | |
| 100 | 3.89 | 4.39 | 5.09 | 5.64 | 7.02 | 7.66 | 9.25 | 9.99 |
| 500 | 20.84 | 27.38 | 38.07 | 48.95 | 78.19 | 94.97 | 144.87 | 171.81 |
| 1000 | 73.31 | 107.86 | 171.91 | 238.25 | 475.70 | 626.24 | 1137.81 | 1449.47 |
| 5000 | 14,812 | 35,124 | 99,611 | 206,640 | 969,821 | 1,793,529 | 6,816,819 | 11,713,309 |
| | $s_1 = 0.01$; $s_2 = 0.19$; $\widehat{p} = 0.05$ | | | | | | | |
| 100 | 1.29 | 1.32 | 1.36 | 1.39 | 1.45 | 1.45 | 1.53 | 1.55 |
| 500 | 1.78 | 1.87 | 2.00 | 2.09 | 2.29 | 2.37 | 2.57 | 2.66 |
| 1000 | 2.26 | 2.43 | 2.66 | 2.83 | 3.23 | 3.40 | 3.81 | 3.99 |
| 5000 | 6.20 | 7.30 | 8.90 | 10.23 | 13.72 | 15.42 | 19.88 | 22.03 |

*Notes:* * $\widehat{p}$ is the equilibrium frequency of one of the alleles

on fitness are presumed to be multiplicative. Estimates have also been made for the equilibrium frequencies of alleles of hypothetical "symmetrical" ($\hat{p} = 0.5$) and "asymmetrical" ($\hat{p} \neq 0.5$) selection. We see that when the task is formulated in this way, made possible by many years' field research experience, the best adapted individual's excess fertility is perfectly "meaningful" in the biological sense and hence, the hypothesis of overdominance's important role as the support mechanism for the hereditary variability of populations cannot be refuted. This conjecture becomes even more valid if one realizes that the segregational load is markedly reduced due to asymmetry of selection effects, in other words, at equilibrium frequencies significantly differing from 0.5 (Table 5.11). This conclusion holds at least for diallelic loci, which prevail in random gene samples from various species; besides, in calculations multiallelic systems can be reduced to diallelic ones.

It is stabilizing selection acting simultaneously on many loci that is responsible for uniformity (or close similarity) of allelic frequencies over enormous ranges of diverse species: Pacific salmon (Altukhov 1983; Altukhov 1989a,b; Altukhov et al. 1997), *Drosophila pseudoobscura* (Lewontin 1978a), cod *Gadus morhua* (Pogson et al. 1995), marine beet *Beta vulgaris* (Raybould et al. 1996) and a number of other species, particularly conifers (see for review Politov et al. 1992; Politov and Krutovskii 2001). In all cases, the gene frequencies are shifted; that is, the most common and very slightly spatially differentiated allozyme alleles occur at frequencies very far from intermediate ones, where the segregational load at overdominant loci reaches maximum. Apparently, in such cases the possibility of excessive migration as a mechanism of allele-frequency uniformity is completely excluded (cf. Glubokovsky and Zhivotovsky 1986; Zhivotovsky et al. 1987, 1989; Glubokovsky et al. 1989) since selection only masks the subpopulation structure of the species. This structure is easily revealed by analyzing morphobiological traits or enlarging the sample of markers with selectively neutral loci (see Altukhov et al. 1997 for details).

As an example, we can present the studies on cod and marine beet, in which pronounced spatial heterogeneity of allele frequencies was detected using RFLP analysis of nuclear DNA. The picture obtained by approximation of the experimental DNA and allozyme data in *Beta vulgaris* is very expressive (Fig. 5.16). Likewise, a long-term study of mtDNA polymorphism in pink salmon showed not only interpopulation but also intrapopulation heterogeneity of haplotype frequencies (Brykov et al. 1996, 1999).

In general, new molecular genetic evidence confirm the conclusions inferred from studies of immunobiochemical polymorphisms (Altukhov 1974; Altukhov 1983; Altukhov 1989a,b; Rychkov and Sheremetyeva 1976; Rychkov and Balanovskaya 1990a; Balanovskaya and Rychkov 1990; Altukhov 1991; Altukhov 1995). These conclusions have made an important

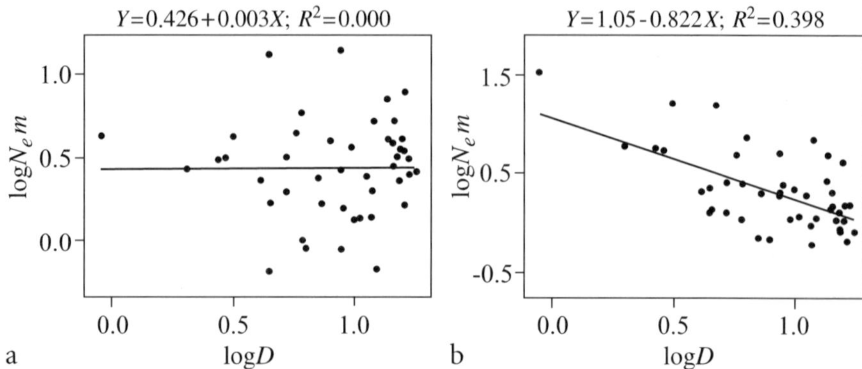

**Fig. 5.16.** Linear regression of logarithm (log) of gene flow *Nm* (estimated from isozyme of RFLP data) against log of distance between all population pairs in sea beet *Beta vulgaris*. *P* (probability of the null hypothesis, Mantel's test) = 0.4913 for **a** isozymes and 0.0006 for **b** RFLP. (Altukhov and Salmenkova 2000)

contribution toward solving the problem of the role of balancing selection in maintaining this variation in humans, plants, and animals. These issues have only recently become clear for our Western colleagues. A paper published in PNAS USA in late January of 2001 has a telltale title "Order Emerging from Chaos in Human Evolutionary Genetics" (Rogers 2001). This study and the evidence presented in Sect. 2.4.4 suggest that polymorphisms maintained by balancing selection also occur at the DNA level. We believe that only on average does genetic differentiation of the species correspond to the selectively neutral model (and here M. Kimura is right) but is accompanied by selective processes acting in different directions and ensuring its stabilization. The process of reproduction of the species gene pool, in which the ratio of its intragroup ($H_S$) and intergroup ($F_{ST}$, $G_{ST}$) variation components remains constant in generations, was termed normal. It provides a natural reference point for estimating population-genetic dynamics under various external influences (Altukhov 1995). These issues are considered in detail in Chap. 7.

Concluding this chapter, I would like to emphasize that the absence of clarity with regard to selective values of various gene groups may lead to gross errors in different phylogenetic reconstructions. In this case the estimates of genetic distances (times of divergence) may be biased: the prevalence in the sample of analyzed loci of overdominant genes and genes under disruptive selection would respectively "shrink" and "extend" these distances.

The results discussed above show that in native populations, genetic differentiation at an array of loci can model selective neutrality if individual loci are subject to different types of selection. Thus, our study of genetic processes in natural populations of various species corroborates our earlier conclusion that the views of "neutralists" and "selectionists" should not be in sharp opposition. The genetic differentiation of a species, as well as evolution at the level of amino acid or nucleotide substitutions in biological macromolecules cannot be explained by either selection or genetic drift. In nature, these two factors are in constant interaction, which results in unique distribution patterns of genotype and gene frequencies. However, to detect these effects, we need reliable information on the history of a population and its internal structure; here, analysis of early developmental stages is of utmost importance. I have endeavored to show this.

On the basis of this evidence, we can envision the relative contribution of different factors of genetic dynamics to the current pattern of protein and DNA polymorphisms and draw the important conclusion that this pattern is a stationary rather than transient phase of molecular evolution, at least in native populations. If this is true, then we should thoroughly reexamine the generally accepted concept of the species and speciation based on the view on evolution as a gradual change in population gene frequencies.

In the following, based on the results of our studies and works by other authors (many of which only recently appeared in print), we discuss this problem in more detail.

# 6 Population Genetics and Evolution

To forestall possible censure, I qualify my position by saying that although the term "evolution" figures in the chapter heading, I shall restrict analysis to the problem of species or speciation, that is, microevolution, and the shift of gene frequencies at a population level, vectorized in time (or space) will be considered as an evidence of evolutionary transformation in complete accordance with the traditional approach.

Taking these remarks into consideration, I shall attempt to answer the three following questions which are fundamental to the interpretation of genetic bases of microevolution: (1) is it possible to accept the idea of gradual evolution by analyzing genetic processes at the population level? (2) Is it sufficient to limit oneself to the phenomenon of genetic polymorphism in order to understand the speciation process, or is it necessary to examine simultaneously a species' monomorphic genetic system? (3) Does speciation represent a gradual process of a population's gene pool rearrangement, or does it result from a qualitative reorganization of a species' functionally most significant genetic system?

The discussion of these issues is preceded by a brief summary of the current key views on speciation and the species.

## 6.1
## The Species and Speciation

The concept of the species as one of the hierarchical levels of organization of living matter, its primary taxonomic unit, does not raise objections in most biologists. In logics, species (from the Greek *eidos* and the Latin *species*) corresponds to the category occupying the intermediate position between the more general category of a genus and more specific category of an individual. In this sense, the meaning of species may be very conventional and not correspond to any entity existing in reality. However, even in ancient Greece the notion of "the species" was intended to be fixed by giving it an ontological meaning. According to Plato, the observed variability of the organic world is nothing more than a distortion of some constant, ideal essences of diverse forms of life. Aristotle said that species is "that in everything which it essentially is". The typological concept of the biological species postulating the existence of true specific traits is based on this definition.

The English botanist, John Ray, was probably the first to define the biological species: "Specific identity of a bull and a cow, of a man and a woman is caused by their descent from the identical parents, often from the same mother; in plants in the same manner the truest sign of belonging to the same species is a descent from the same plant. The forms belonging to different species preserve the constant character of their species; one species never arises from seeds of another, and vice versa" (*Historia plantarum*, 1686). Apparently, the essential aspect of Ray's definition is *reproductive isolation* of the species, although this requirement is not precisely implied in it.

This approach was used by Carl Linnaeus as the foundation of his system of nature based on the binary nomenclature, which has been employed by taxonomists ever since. Linnaeus united most similar species into genera and assigned a binary name to each of the species.

This approach replaced the previous long definitions of species and thus revolutionized biology, revealing hitherto unknown perspectives in systematization of plant and animal kingdoms in studies of both neonto-logical and paleontological material. However, as to the origin of species, Linnaeus, like his predecessors, adhered to the Bible's stand. He believed that all individuals of any species had descended from a single pair created by God and no new species arose after the act of Creation ("*tot numeramus species, quot ab initio creavit infinitum Ens*").

In his late years, Linnaeus was not so rigid in his views admitting that new species could arise, e.g., by hybridization. However, in the 18th and early 19th centuries, the view on constancy of species was still predomi-nant in biology, which theory was to a great extent promoted by Georges Cuvier. He linked the idea of constant species with the theory of geological cataclysms that had occurred in different epochs resulting in disappear-ance of previously existing animals and plants and their substitution by new species not related to their predecessors. A special act of creation was postulated for the plant and animals kingdoms of each of these epochs.

Meanwhile, evolutionary views have been developed in biology since the late 18th century. These views found their embodiment in the materialistic theory of the origin of species by natural selection, advanced by Charles Darwin. Since that time, the typological view of the biological species as an immutable entity has been gradually replaced by the population principle. This process led to the development of the so-called synthetic theory of evolution (STE) accepted today by many scientists. The very name STE reflects the synthesis of the Darwinian concept of natural selection and principles of population genetics, which latter concept was started in 1926 by the well-known article of S.S. Chetverikov and reached its final definition in the 1940s (see, e.g., Vorontsov 1984). The chief feature of STE (also called Neodarwinism) is its reliance on probability. Both Darwin's theory of natural selection and the most successful models of population genetics are based on random genetic variability. However, this variability, approached either as a *phenomenon* or as a *process*, can only be expressed in probabilistic terms. It is not surprising, therefore, that the synthesis of these two theories happened so naturally. As the American biologist Ernst Mayr said, this replacement of typological thought by the populational was "maybe the greatest conceptual revolution in biology." Mayr (1968, p. 20) noted that the reasoning of populationists and typologists is opposite. The populationist emphasizes the uniqueness of any event in the living world. As no two humans are alike, neither are any other two plants or animals. All organisms and life events have individual features and, as a group, can be described only in terms of statistics. Individuals or other living entities form populations, for which an arithmetic mean of a trait and its variability can be estimated. Mean values are statistical abstractions; only individuals comprising the population are real. The ultimate conclusions of a typologists and populationists are directly opposite. For the typologist, the species (*eidos*) is real and variation is an illusion, while for the populationist the type (average) is an abstraction and only variation is real. It is hard to imagine views on nature that could be more different.

Indeed, these differences are evident with regard to criteria of the species and, consequently, to the species and speciation definitions. According to the typological concept *true traits of the species exist, which are identical within the species and differ from the corresponding traits in any other species.* According to the population concept, there are no such traits since "the difference between a species and a variety is in degree rather than in essence. The variety is an arising species and a species is a markedly expressed variety". Mayr (1968, p. 270) rephrased this Darwinian thesis saying that all traits used for distinguishing species exhibit geographical variation. Accordingly, based on the typological thought, the species can be defined as "a community of individuals identical in the specific trait" while based on the population thought, species is defined as "groups of actually or potentially mating populations reproductively isolated from other such groups" (Mayr 1942).

This principal difference is characteristic for interpreting speciation mechanisms. If we abstract ourselves from creation of species by a higher mind, which cannot be rationally explained, and stay within the framework of scientific methodology, speciation for a typologist would be a saltatory process involving variability of exactly species-specific traits that would lead to reproductive isolation. From the viewpoint of a populationist, speciation is a gradual process based on intraspecific hereditary variability causing small but accumulating phenotypic changes. In relation to the population adaptation to the environment, the theory of gradual evolution is based on the *uniformitarianism principle* formulated by a geologist, Charles Layel. According to this principle, the slow and small natural environmental changes that occur today, also occurred many thousands of years ago, resulting in step-by-step differentiation of populations up to their complete isolation as species. In terms of population genetics, speciation is a long adaptive process of allele replacement, with the result of temporally transforming the species as a whole (*anagenesis, phyletic evolution*), or in splitting the species into daughter species due to geographical isolation of the corresponding populations (*clagensenesis, true speciation*).

These are the fundamental differences between the two concepts of the species. These concepts overlap, albeit partially, only in one point: they both regard reproductive isolation as a key criterion of the species. However, even here, they only seem similar. In the typological concept of species, reproductive isolation is thought to be the primary condition of speciation, whereas in the population concept, it is only a side effect of pronounced adaptive divergence of spatially isolated populations. It is no coincidence that the model of *allopatric* (geographic) speciation was for a long time very popular with Neodarwinists who often excluded the possibility of *sympatric* speciation, i.e., speciation without a spatial separation of the original population, as, for example, in the case of polyploidy. Since the latter speciation mode is widespread in nature (e.g., in plants), many efforts were made to substantiate qualitative differences in speciation mechanisms between animals and plants. For instance, it was stated that although polyploid series are rather common in parthenogenetic or hermaphroditic animals, saltatory speciation is not feasible for the overwhelming majority of them, since it would disrupt chromosome balance in sex determination (H. Muller 1925): a polyploid individual would have too many genetically imbalanced and thus sterile progeny. (If the mother carries chromosomes XXXX, and the father, XXYY, the progeny will have too many non-viable XXXY individuals, and the species would not be able to survive.) As pointed out by Mayr (1968 p. 359), "even if polyploidy is confirmed in some of these groups [i.e., crustaceans, lepidopterans, fishes, etc., Yu. A.], this is an exceptional phenomenon in animals, as opposed to plants. *We lack evidence for the polyploid origin of groups of species or genera in animals, in contrast to what we often encounter in the plant kingdom* [emphasis is mine, Yu. A]."

Because of this, the category of species in the population concept can be applied only to bisexual species with cross-reproduction. The existence of unisexual species in nature creates insurmountable difficulties for STE in the context of establishing their species status, since the criterion of the absence of intercrossing for them is not relevant. It was even suggested to use a neutral term (e.g., *binome*) for such forms, employing the term *species* only for bisexual organisms.

In the typological concept, the problem of species identification is easily resolved irrespective of a unisexual or bisexual nature of the organisms in question. It requires only the typological definition of the species presented above and acceptance of the existence of diagnostic characters that are invariant within the species but differentiating within the genus. This approach is effective in the case of "good" species that differ in morphological traits. However, in the case of so-called sibling species, which do not noticeably differ in morphology, species identification turns into an unsolvable task.

Nevertheless, a search for interspecific differences at other (e.g., molecular or cellular) structural levels of organization sooner or later yields characters providing the possibility of successful specific identification.

In the following, we examine new evidence on the problem of species obtained in recent studies.

## 6.2
## Do Population-Genetic Studies Suggest the Idea of Evolution?

As noted above, the modern theory of evolution must be primarily regarded as a population-genetic theory. Because of this, it lacks methods of quantitative estimation and of causal analysis of genetic reorganizations underlying speciation: *population genetics is in essence confined within the framework of the species*. Exactly for this reason, as also noted above, reproductive isolation is discussed in STE only as a side product of population divergence to the species status based on gradual adaptive substitutions of polymorphic gene alleles.

This process is represented very graphically in a generalized diagram by T. Dobzhansky (1955b), reflecting the so-called cladogenesis model – "true speciation" – when an original species splits into two (Fig. 6.1). Nevertheless, population genetics can say nothing about *what degree of difference among populations accounts for their belonging to one and the same species or to different species* after the bifurcation point has been reached. It is clear that the traditional populational genetics analysis must be supplemented by the *comparative genetics of a species*, requiring first and foremost a comparison between the character of intra- and interspecies variability based on strictly determined genetic traits. If the difference between a species and variety is not in "essence" but in "degree" or, in other words, the variety is an "incipient species", the nature of genetic differences among species within a genus should be *commensurable* with that among populations within a species. We shall carry out this form of

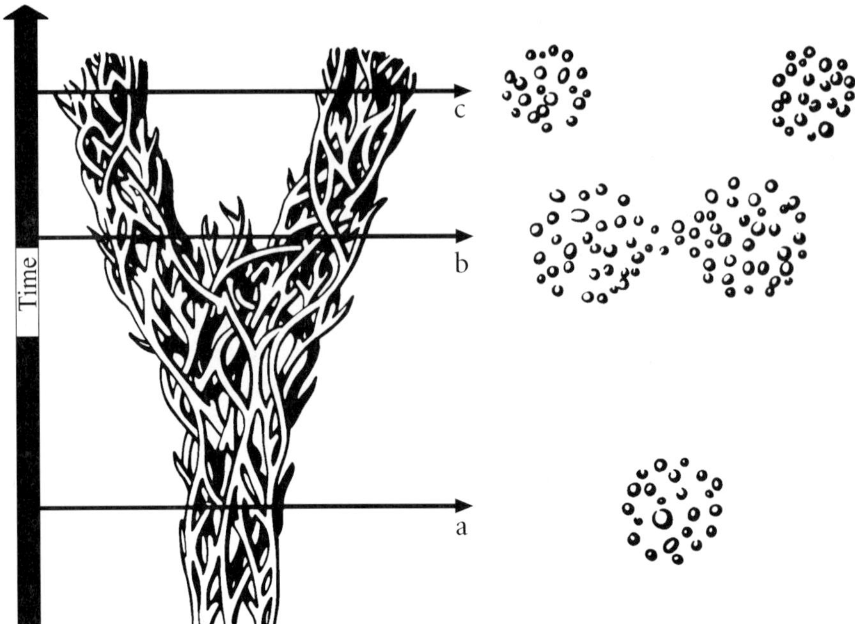

**Fig. 6.1.** Diagram of the splitting of an original species (*a*) into two (*b* and *c*) in time (Dobzhansky 1955b). The individual, more or less *interconnecting branches*, are the "Mendelian populations"

analysis in Sect. 6.3, but let us return to Dobzhansky's diagram (Fig. 6.1) to stress yet another fundamental matter: it is evident that according to the ideas and concepts of population genetics developed in this diagram, it is not the individual Mendelian populations, but their communities that diverge, i.e., the so-called subdivided populations or, in our terminology, population systems. As has already been pointed out (in Sect. 1.5), it is only in such a population – Sewall Wright often equates it with a species – that the conditions exist for maintaining maximum genetic diversity, enabling the populations to react more effectively to changes of the environment and to alter their genetic structure accordingly (see also Dubinin 1966; Dobzhansky 1970; Dobzhansky et al. 1977). It is assumed that the new founder population bears only a limited part of the genetic variation of the original species (bottleneck effect), which promotes a rapid (so-called *peripatric*) speciation in the new environment (Mayr 1999). However, it is for some reason overlooked that the fate of this isolate is either extinction or transformation into a new subpopulation system, to which, as noted above, Sewall Wright attributed the property of "evolutionary optimality".

The fact of the matter is that this purely deductive conclusion has not been corroborated adequately either by experimental or by field research.

In the meantime, the results examined in previous chapters show that if research is not confined to an elementary population level, but natural subdivided populations are studied within their own historically formed boundaries, or appropriate experimental models are used, it is then possible to find the important intrinsic systemic quality. This quality, not obviously derived from the properties of the constituent components, is the *genetic stability* of the population system in time and space.

It is important to answer the following question: how long can this stability of subdivided populations be maintained? Evidently, if the inner characteristics of the process are revealed and the stability of structure thus disclosed is shown over time intervals measured in hundreds of generations, then this mechanism is presumably effective for longer periods of history. Indeed, effectively organized research permits one to observe stability of a population's genetic composition commensurate with the length of its historical existence under specific conditions of the environment (Rychkov and Movsesyan 1972; Altukhov and Kalabyshkin 1974). As data of this kind have been discussed frequently (Altukhov and Rychkov 1970; Rychkov 1973; Altukhov 1974), we shall confine ourselves to examining the results of two of the most systematic investigations.

The first example illustrates the stability of the genetic structure of the system of ancient isolates of indigenous North Asian mongoloid peoples, measured in hundreds of generations. Yu.G. Rychkov and his colleagues have studied these populations for over 20 years, using a wide range of immunological and biochemical gene markers, and also certain hereditarily determined anomalies of the bone system (Movsesyan 1973; Rychkov and Movsesyan 1972). Having examined the frequency distributions of several uncorrelated anomalies in modern and Neolithic populations, the authors concluded that there were no reliable differences in the given character between the system of modern populations taken as a whole and the aggregate of ancient populations, also regarded as a whole (Fig. 6.2). This means that despite the strong and lengthy isolation of the ancient North Asian populations, their system has preserved the original genetic information and hence, its continuity of development for at least 5,000 years, that is, $\sim 200$ generations.

The second example reproduces the results of analyzing shell pattern polymorphism in modern and fossil *Littorina squalida* mollusc populations in the isolated Busse lagoon, South Sakhalin (Altukhov and Kalabushkin 1974; Kalabushkin 1976, Kalabushkin and Zhivotovsky 1979). This mollusc species has well-expressed polymorphism of shell pattern, controlled by a diallelic system with incomplete dominance (Fig. 6.3). As with *Cepaea nemoralis*, the land snail, which has been the object of exceptionally complete studies, these differences in genotypes are also preserved in fossil forms, making it possible to examine their distribution in samples

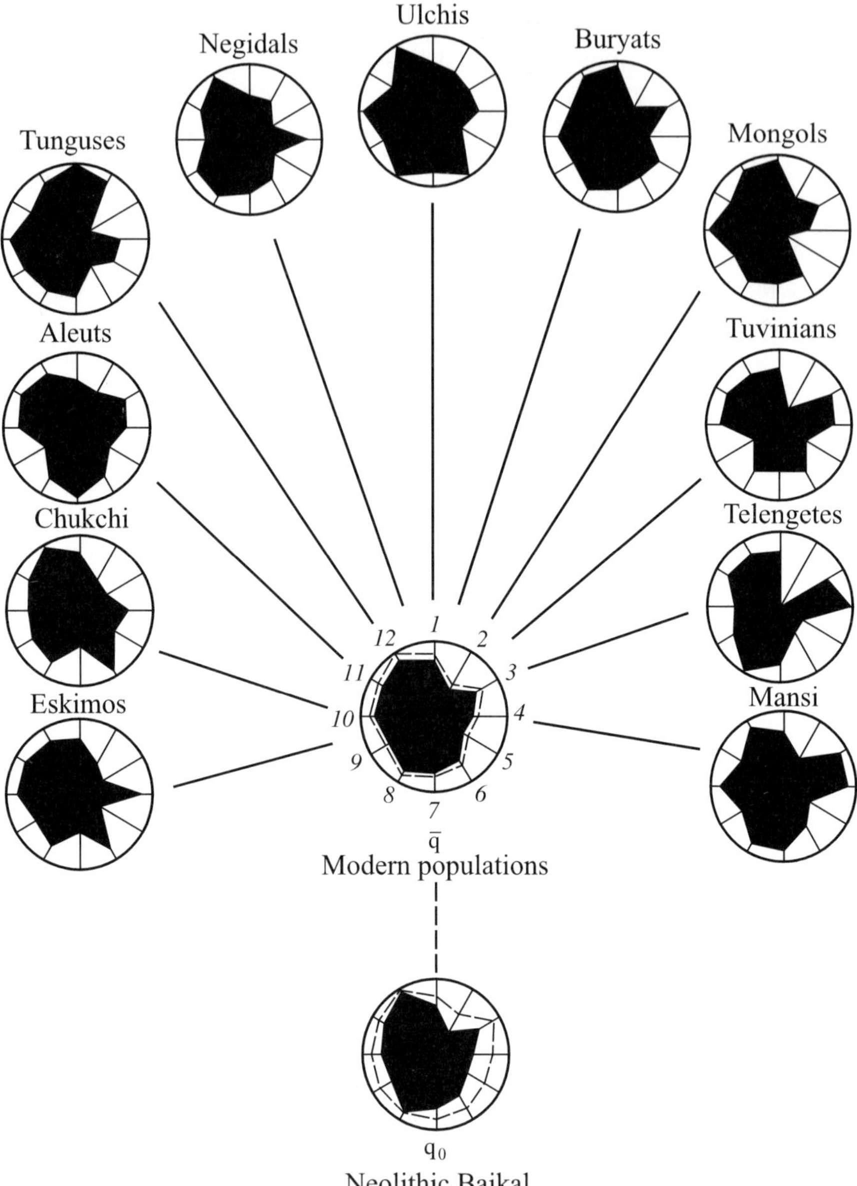


**Fig. 6.2.** The reconstruction of an ancestral population's genetic structure from the frequencies of 12 independent characters ($\overline{q}$) in a modern population system of indigenous Siberian peoples and the comparison of this distribution with a corresponding one for Neolithic Baikal people ($q_0$; from Rychkov 1973). *Dotted lines* Minimal confidential limits; 0 frequency at the perimeter, 0.3 frequency at the center of the circle

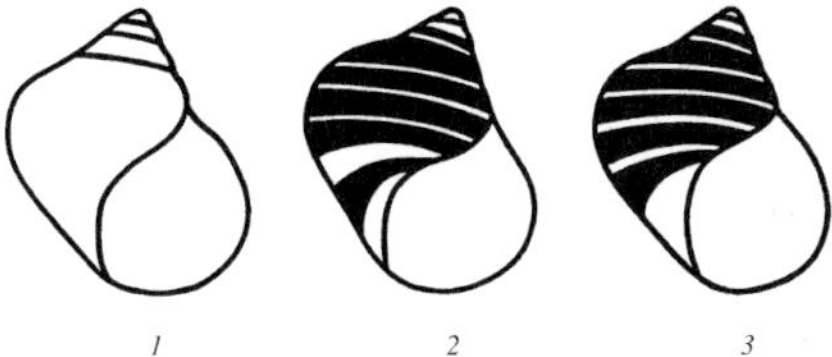



**Fig. 6.3.** Shell pattern polymorphism in Sakhalin populations of *Littorina squalida*, the gastropod mollusc (Altukhov and Kalabushkin 1974). *1, 3* Presumed homozygotes, AA and BB, respectively; *2* heterozygote AB

from living and ancient populations separated by a time interval on the order of 4,500–5,000 years; this corresponds approximately to 2,000–2,500 successive generations (Fig. 6.4). This evaluation, based on the character of deposits, and their accompanying thermophilic fauna, and taking into account periwinkle age structure, is highly reliable (see Zhuze 1959; Golikov and Skarlato 1967; Kalabuskin 1976). The distributions of morphs was examined in three samples of modern material and five fossil samples between 1969 and 1974. The gene frequency estimates presented in Fig. 6.5 show that, by comparing separate modern samples with separate fossil ones, erroneous conclusions may be reached when signs are found of genetic similarity as well as of differences in time and space.

However, if one relies on the clear concepts of the systemic organization of populations and carries out the collection and analysis of primary material for all the elements of a population structure, the only conclusion that can be reached is that, despite variability in parts, the system as a whole preserved the genetic composition inherited from the ancestral population. This stability, as was shown in Chap. 4, persists despite the effects of directional selection.

It should be pointed out that the detailed study of molluscs in Busse lagoon also revealed the effects of strong selection, disruptive in the early stages and stabilizing in the late stages of ontogenesis (Kalabushkin 1976). Although the periwinkle's habitat changed substantially as the lagoon evolved (Kalabushkin and Zhivotovsky 1979), the average gene frequency remained almost unchanged for thousands of generations.

Because time and space are linked in the stationary genetic process, the same factor of subdivision, reinforced in the developmental process of population structure, also plays a decisive role in limiting the genetic differentiation of spatially separated population systems. The genetic distances between and within two long-since diverged Mongoloid population systems of North Asia and America, evaluated by a wide range of biochemical markers, very clearly exemplify this (Rychkov and Sheremet'yeva 1977).

Corresponding estimates made for 28 allelic genes of 11 loci controlling erythrocytic antigens and blood proteins show that the differentiation level

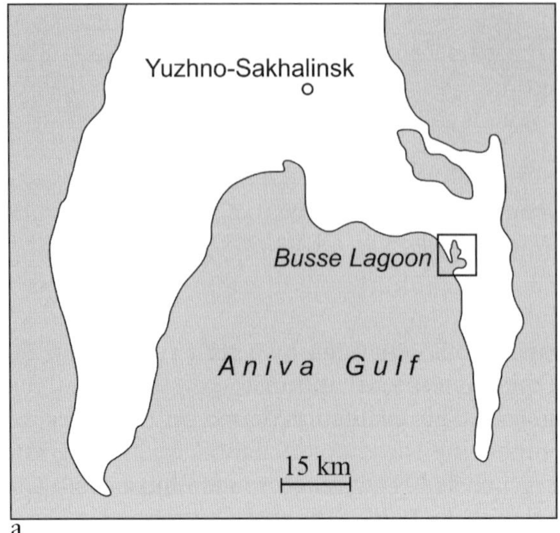



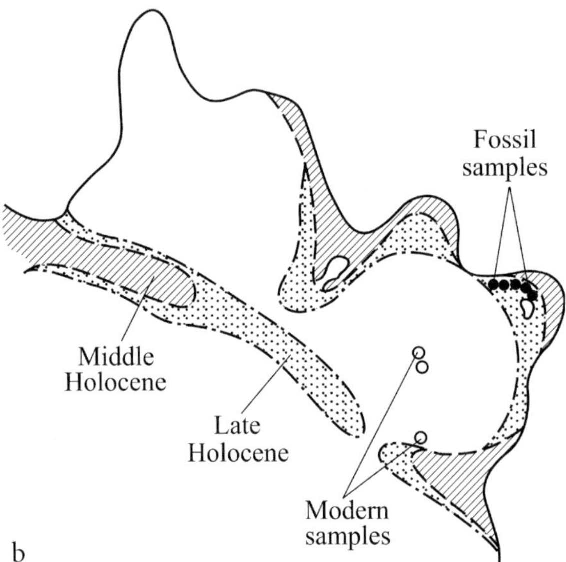



**Fig. 6.4.** Scheme of modern localization of **a** Busse lagoon on Sakhalin Island and **b** approximate historic reconstruction of its coast lines

of population systems, reached in hundreds of generations, is hardly in excess of the differences among separate elementary populations now living and isolated from each other for some four or five generations (Fig. 6.6). Thus, it is clear that if we study genetic differentiation of fine-structured

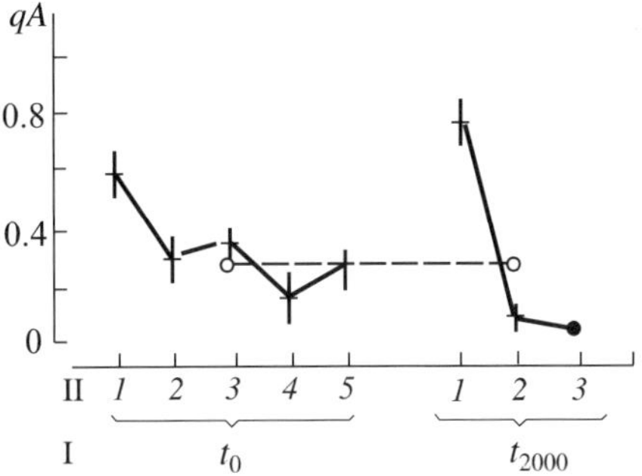

**Fig. 6.5.** Gene frequency variability based on conventional zero ($t_0$; $n$ = 479 specimens; $q_0$ = 0.293 ± 0.015) and approximated 2,000 ($t_{2000}$; $n$ = 1.252 specimens; $\bar{q}$ = 0.280 ± 0.010) generations of a *Littorina squalida* population system as a whole (*I*), compared with variability at separate locations (*II*)

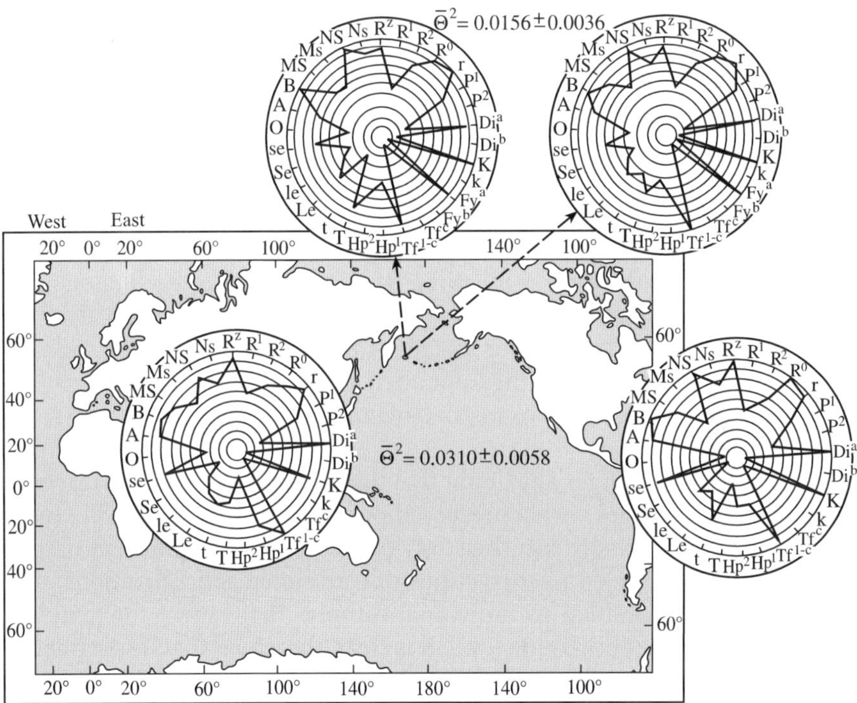

**Fig. 6.6.** Schematic representation of allele frequencies at 11 loci that characterize the differentiation of North Asian and American population systems and two populations of Aleuts in the Komandorskiye Islands (from Rychkov and Sheremet'yeva 1977). $\overline{\Theta}^2$ is the mean square of the genetic distance between populations in radians. Frequency 0 at the perimeter and frequency 1 at the center of the circle

populations in space or in time by comparing only separate random samples, we may draw evolutionary conclusions where we should not or, conversely, we shall reach a conclusion about stability where really a vectorized genetic process has taken place. It would be interesting to examine the whole series of examples that often feature in reports on evolution from this viewpoint. With the possible exception of industrial melanism, where the directional shift in the corresponding gene's frequency under conditions of catastrophic environmental changes is in no doubt, all these examples need to be re-examined in this way.

At any rate, in the case of the best-known example of protracted polymorphism stability from the Pleistocene period to the present, which Diver (1929) described in his day for *Cepaea nemoralis,* we have been able to establish that this stability was revealed through a system of observations analogous to that outlined above. Diver, of course, does not specify this at all, and he gives no quantitative data, but his approach itself leaves no doubt. In the living colonies studied by me the various phenotypes, as might be expected, occur with very different frequencies; and the frequencies of any one phenotype may show a significant difference in samples separated by only a few yards. But on the whole series of colonies, banded types, taken together, are more prevalent than the unbanded form.

Fortunately, the banding of the shell is easily seen in fossil material; detailed data, therefore, of the frequency of the different types in fossil samples could be obtained and would be of considerable interest from the point of view of genetics and natural selection theory. Kennard has supplied me with the figures for three fair-sized samples, each containing both species (*Cepaea nemoralis* and *C. hortensis,* Yu.A.), from deposits near Goodwood ... "In all three samples the banded types taken together are numerically superior to the unbanded shells and *the frequencies of the different types are just such as might be found today*" (Diver 1929 p. 83; our italics, Yu.A.).

Now there is no longer any doubt that this kind of legitimacy is traced in many bisexual species with a preserved population structure and is easily modeled experimentally, but the study of the genetics of different natural populations has not led to corrections of traditional generalizations: field research is still confined to individual samples that reflect the simplest, extremely variable population level. It is clear, therefore, that the important problem of population genetic stability has been reduced merely to studying the mechanisms of homeostasis (Lerner 1954), while main attention has been focused on evolutionary aspects in general, and in particular, on the relative contribution of random genetic drift and natural selection to reorganizing the gene pools of elementary populations. Variability at this level is indeed great, which also helps to promote views about the unlimited evolutionary lability of populations. Apparently, it was believed, and still is believed, purely by virtue of the tradition of considerable interest in the problem of evolution, that if changes at the simplest population level serve as evidence of evolution taking place before our eyes, then the evolutionary possibilities are even greater for the communities of such populations and even more so for widely distributed species – these true "pioneers" of evolution which are constantly experimenting with it (Mayr 1968, 1974).

Meanwhile, the discovery of the systemic organization of native populations and their detailed analysis provide evidence of a species' non-adaptive divergence being averted by migration links, and even the effects of directional selection under subdivision conditions, as we have seen, have little effect. At the same time the existence of fine-structured population systems that are stable in their genetic characteristics, as the example of a natural sockeye salmon population clearly shows, testifies to the conservativeness and uniqueness of local adaptations which the powerful environmental changes that accompany different attempts at acclimatization have not succeeded in altering. Examples of the large number of these unsuccessful efforts with Pacific salmon have been well documented (Ricker 1972; Altukhov 1974; Altukhov et al. 1980a; Salmenkova et al. 1983; Altukhov and Salmenkova 1987b; Altukhov et al. 2000). We shall return to these questions in the last chapter of the book when we discuss the significance of population genetics for the rational management of biological resources and the conservation of the biosphere's gene pools.

Now, however, it should be re-emphasized that a population system – both in nature and under experimental conditions – preserves the genetic characteristics of an ancestral population for many generations, thanks to the reciprocal balance of all the known factors of evolution, although in the absence of a system's "core" these features may not be at all characteristic of populations now living and only reconstructed in the averaging process for all the structural components (Rychkov 1969, 1973). However, not only the stochastic balance of the centripetal and centrifugal forces in an "island" population system is responsible for its genetic stability. There is another, more important mechanism of stabilization. This was discussed in Chap. 4 when a negative feedback association between the number of peripheral subpopulations of *D. melanogaster* and immigration in these subpopulations from the core of the system (Sect. 4.1, Fig. 4.4) considered: the lower the number of the island subpopulations, the greater the inflow of flies from the "continent", and vice versa. Later, the same relationship was found in natural populations of other species and even in humans (Altukhov 1989a,b; Altukhov and Salmenkova 1994; Altukhov et al. 2000).

For instance, in a system of circumpolar populations of native inhabitants of North Eurasia, $N_e = 45$ and $m = 0.034$ (Rychkov and Sheremet'yeva 1976), whereas in the Siberian isolate as a whole, $N_e = 200$ and $m = 0.01$

(Rychkov 1969). The butterfly *Euphydrias editha* exhibited very high dispersal of individuals (up to 100%) when the colony size was about 100 and only 0–7.3% from colonies of about 1,500 individuals (Gilbert and Singer 1973; cited from Yablokov 1986).

In the Pacific Chinook salmon a negative correlation was found between the level of straying and the number of spawners returning for reproduction (Quinn and Fresh 1984). In the same fashion, high straying of tagged fish was found in the years of low numbers whereas low straying was characteristic of the abundant years (Zolotareva 1980).

Very recently, new evidence has been obtained concerning statistically significant negative correlation between effective size and immigration rate for the system of rural human populations (Evsyukov et al. 1999). These new findings strongly support our conclusion about autoregulation of gene diversity levels within native population systems.

This self-regulation suggests *the maintenance of stable (optimal) proportions of the intra- and interpopulation components of gene diversity in the population system*, i.e., a balance between inbreeding and outbreeding. The results of a disturbance of this balance will be considered later, when evaluating the consequences of various anthropogenic impacts on populations (Chap. 7).

Here, we would like to note the following important point: based on the above results, genetic differentiation of populations may be interpreted not as a Markov chain, in which the evolution cannot be predicted for more than one generation ahead, but as a self-regulating, branching process of reorganization of the gene pool of the ancestral population during its differentiation into subpopulations in generations and over the area. With a normally fluctuating environment, the mean gene frequency, and after reaching the stable equilibrium, also the variance, remain steady with regard to mutually compensating microevolutionary transformations. In other words, *if an isolated population does not disappear, it deploys into itself maintaining a dynamic equilibrium with the environment.*

Thus, the evolutionary optimal population structure proves to be genetically stable: more stable than a panmictic population of the same size and genetic composition.

Apparently, we are dealing here with a population superorganism that is able to *maintain genetic diversity as a memory of the preceding stages of its development* over tens, hundreds, and thousands of generations (Rychkov 1969, 1973; Altukhov and Rychkov 1970; Altukhov 1974; Altukhov and Kalabushkin 1974; Kalabushkin 1976; Altukhov and Pobedonostseva 1978, 1979a,b).

As shown in studies of native human populations (Rychkov 1969, 1984; Rychkov and Movsesyan 1972; Rychkov and Sheremet'yeva 1977; Sheremet'yeva and Rychkov 1978; Rychkov et al. 1982; Rychkov and Yashchuk

1985), this depth of genetic memory opens a unique possibility of reconstructing its structure and volume.

The *structure of genetic memory* is related to the constancy of the total gene diversity owing to the constant mean gene frequencies in a population system; they can be restored by averaging corresponding frequencies over all currently existing subpopulations. The *volume of genetic memory* is determined by the process of accumulation of interpopulational gene diversity during the formation of the population system structure in time and space, and can be analyzed using Wright' $F_{ST}$.

The possibility of performing reconstructions of the first type (the structure of genetic memory) has been repeatedly demonstrated for population systems of different species, including man (Serebrovsky 1935; Rychkov 1969; Altukhov and Rychkov 1970; Rychkov and Movsesyan 1972; Altukhov and Kalabushkin 1974; Altukhov and Pobedonostseva 1978; see also Chap. 4). As to reconstructions of the second type (the volume of memory), such studies developed, until recently, only as applied to human populations. This is certainly not accidental, as the history of human populations is the domain shared by many sciences, and its chronology may be reconstructed by study independent of genetics. It is noteworthy that the genetic chronology of historical events as the method for determining the age of populations consistently gives estimates agreeing with the data on radiocarbon dating of ancient campsites or evidence provided by historical chronicles (Rychkov 1984).

Moreover, the dynamics of accumulation of interpopulation gene diversity demonstrated the previously unknown trend, indicating the equivalence of its volumes at different stages of population structure formation. This phenomenon was named the *genetic equidistance of stages of ethnogenesis* (Rychkov and Yashchuk 1980, 1985).

The analysis of data on frequencies of many genes in indigenous population systems of Europe, northern Asia, and America showed that the same value of gene differentiation coefficient, expressed in $G_{ST}$ units, is characteristic for every level of population hierarchy, irrespective of the historical time of its establishment. Note that this principle can only be revealed if the total genetic differentiation of the system ($G_{STi-t}$) is subdivided into regional components ($G_{STi}$) on the basis of ethnic classifications that are independent of genetics but that, at the same time, disclose the actual hierarchical levels of unity historically developed in the process of microevolutionary subdivision of a population (e.g., linguistic, ethnoconfessional, etc., processes; Rychkov and Yashchuk 1980). The equivalence of $F_{ST}$ ($G_{ST}$) values at different levels of the system hierarchy means the existence of a certain limit in the volume of total gene diversity accumulated by the system in course of its historical development. Above this limit, a new stage of microevolution of the system begins.

In studies on other species, it is difficult to treat the initial material in the same way because similarly reliable phyletic classifications are usually absent. However, it is sometimes possible to recognize natural–historical features of species' range formation and to identify the corresponding "levels of antiquity" (i.e., stages in the development of population structure). In such cases, the results can be comparable to those obtained for human populations (Altukhov 1995). Such a method of the replacement of historic time of population system formation by mathematically equivalent space of the same process may be illustrated by the example of sockeye salmon, *Oncorhynchus nerka*, as a species with a complex subpopulation structure (Fig 6.7). $G_{ST}$ estimates are based on our own results and published data on allele frequencies of several allozyme loci in 54 samples from native populations spawning in the basins of 20 rivers of northern Asia and North America (Table 6.1). These data clearly demonstrate that the degree of local genetic differentiation remains the same at any of the three hierarchical levels distinguished by us.

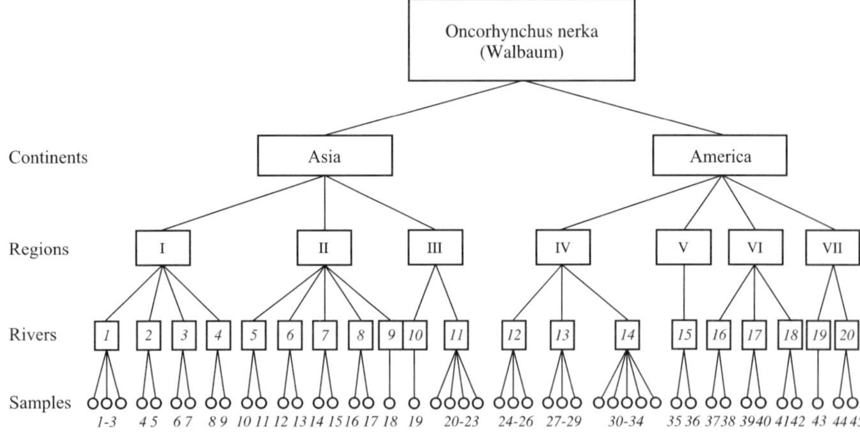


**Fig. 6.7.** Scheme of hierarchic intraspecies subdivision in sockeye salmon (by data from: Grant et al. 1980; Utter et al. 1984; Pustovoit 1994; Altukhov et al. 2000). Regions: *I* West Kamchatka, *II* Kamchatka River basin, *III* East Kamchatka, *IV* Cook inlet (Alaska), *V* Skeena River basin (northern British Columbia), *VI* Fraser River basin (British Columbia), *VII* southern part of American range. Rivers: *1* Palana (3 samples 1992), *2* Khairuzova (2 samples 1988, 1992), *3* Bol'shaya (2 samples 1988, 1990), *4* Ozernaya (2 samples 1988, 1989), *5* Kamchatka (2 samples 1988, 1990), *6* Andrianovka (2 samples 1988, 1990), *7* Kirganik (2 samples 1988, 1990), *8* Kimitina (2 samples 1988, 1990), *9* Kitil'gina (1 sample 1988), *10* Avacha (1 sample 1987), *11* Pakhacha (2 samples 1988 and 2 samples 1990), *12* Kenai (3 tributaries 1975), *13* Susitna (3 tributaries 1975), *14* Kasilof (5 tributaries 1975, 1976), *15* Skeena (2 samples 1980, 1981), *16* Fraser, upper part (2 lakes 1982), *17* Fraser, middle part (lake and tributary 1982), *18* Fraser, lower part (lake and tributary 1982) *19* Quinault (1 sample 1982), *20* Columbia (lake and tributary 1981)

**Table 6.1.** Subdivision of interpopulation genetic diversity in sockeye salmon *Oncorhynchus nerka* (at *LDH-B2**, *PGM-2**, and *ALAT* loci)

| Hierarchy level | The value of local genetic differentiation | |
| --- | --- | --- |
| | $G_{ST_i}$ | Proportion of total $G_{ST_{i-t}}$ % |
| Between continents, $G_{ST_{3-4}}$ | 0.0254 | 34 |
| Among regions within continents, $G_{ST_{2-3}}$ | 0.0237 | 32 |
| Among rivers within regions, $G_{ST_{1-2}}$ | 0.0249 | 34 |

*$G_{ST_{i-t}}$ = 0.074

The same regularity was observed for another Pacific salmon, Chinook (*O. tshawytscha*) from Alaska, in which estimates of the interpopulation component of gene diversity at the levels of regional population groups and individual river populations within these regions proved to be very similar. In Atlantic salmon *S. salar* from northeastern Europe, genetic differentiation among population complexes of individual sea basins and among populations within these complexes was also characterized by similar values. In chum salmon (*O. keta*) from northern regions of the Asian and American ranges, almost equal $G_{ST}$ values were obtained at the levels of regional population groups and river populations within regions (Altukhov et al. 1997, 2000; see also Chap. 7).

The analysis of published data shows that the same trend is also characteristic of other biological species whose spatial differentiation has been studied with respect to many allozyme genes simultaneously. Thus, in the barramundi perch (*Lates calcalifer*) inhabiting ocean waters near Queensland (Australia), $G_{ST}$ values for different hierarchical levels were quite similar: 0.008 for localities within subregions, 0.012 for subregions within regions, and 0.009 for regions within Queensland (calculated from the data obtained by Shaklee and Salini 1993). In the ant *Myrmica rubra,* the values of $F_{ST}$ estimated among the sites within localities and among localities were 0.205 and 0.199, respectively (Seppa and Pumilo 1995). The same trend is observed for the rodent, prairie dog *Cynomys ludovicianus,* if we consider differentiation of subpopulations ("wards") within two populations studied in greater detail and designated CAPU and PORT by Chesser (1983): in the first case, $F_{ST}$ = 0.0554; in the second, $F_{ST}$ = 0.0541. $F_{ST}$ values for the Atlantic–Mediterranean and Indian–Pacific groups of green turtle populations (*Chelonia mydas*), estimated from haplotype frequencies revealed by RFLP analysis of nuclear DNA, are also identical: $F_{ST}$ = 0.13 in both cases (Karl et al. 1992).

The analysis of literature from this point of view shows that similar examples are numerous, but, unfortunately, none of the known publications contains the data allowing us to make a comprehensive and independent investigation of genetics' hierarchical classification of populations and to identify the actual levels of unity established in the process of species microevolution. This problem is likely to attract the attention of researchers, and the required data will probably appear in scientific publications.

In this context, we should emphasize an important property of Wright's $F_{ST}$ coefficient. Being a measure of local genetic differentiation and, simultaneously, a parameter of structural inbreeding in a subdivided population, this parameter has great biological significance: it reflects the ratio of homo- and heterozygous genotypes in a population system and, hence, a balance between inbreeding and outbreeding (Chap. 5). In general, at the level of population systems, the stable ratio between intra- and intergroup components of gene diversity provides evidence that the processes of differentiation and integration of the species gene pool counterbalance one another. This balance can only be revealed in specifically planned studies of natural population systems that have not yet been completely destroyed by anthropogenic impact and still can reproduce their gene pools at a natural rate. Now it is necessary to address the reason for the similarity of the $F_{ST}$ ($G_{ST}$) values at different hierarchic levels in the population systems of different biological species, on one hand, and of human populations, on the other. The question is whether this similarity is purely superficial or is based on a certain universal regulatory mechanism?

Rychkov and Yashchuk (1985) studied evolution of the indigenous human population system in Siberia and concluded that such a mechanism actually exists: its role is played by gene migration that decreases with time as each new hierarchic level is formed in the system. They argued that the increasing numerical size of the population system is associated only with an increase in the number of subpopulations, whereas the average effective size of a local population remains unchanged throughout the history of differentiation of the vast Siberian isolate. A comparison of corresponding $F_{ST}$ values – empirical, calculated for a large set of loci ($\overline{F}_0 = 0.0810$), and expected, for a selectively neutral microevolutionary process ($F_e = 0.0974$) – allowed the authors to estimate the time during which the system approached the existing level of gene diversity (according to the model of its exponential growth):

$$\overline{F}_0 = F_e \left(1 - e^{-t/2N_e}\right) .$$

The result was 19,500 years (approximately 780 generations; Rychkov and Yashchuk 1980), which agrees well with archeological data on radiocarbon dating and recent estimates based on mtDNA polymorphism (Torroni et al. 1993; Starikovskaya et al. 1998).

Therefore, the equality in proportions of the total $F_{ST}$ ($G_{ST}$) at different hierarchic levels in subdivided human populations reflects a continuous process of genetic microdifferentiation directed toward a certain limit, which gradually recedes during historical development of the human population as a system (Rychkov and Yashchuk 1980). As for zoological species, they probably achieved ecological–genetic balance thousands of years ago (Altukhov 1983a). Since then, the structure of intra- and interpopulation gene diversity has been maintained at an optimal level (reflecting the maximal adaptation of the species to the historically established conditions of its natural reproduction), owing to self-regulation of the parameter $N_e m$ (Altukhov 1989a). It is interesting that such regulation is apparently found even in plant species pollinated by insects or birds. In any case, the term *optimal outcrossing distance*, which reflects a higher fitness of the progeny from a certain combination of parents in the herbaceous plant *Ipomopsis aggregata*, has long been successfully used in botanical literature (for details, see Waser and Price 1989, 1991, 1994; Waser et al. 2000). There is more direct evidence that in plants, as in animals, the spatial genetic differentiation at comparable levels of natural hierarchy is virtually identical. Thus, studies on the same perennial plant, *I. aggregata*, showed that, at the level of its geographic races (subspecies), $N_e m$ values estimated from $G_{ST}$ statistics for 18 allozyme loci are similar (if not identical): *I. aggregata*: 2.62; *I. a. formosissima*: 2.46; *I. a. attenuata*: 2.44; etc. (the mean for eight subspecies is 2.49; Wolf and Soltis 1992).

Our assumption that genetic processes in natural populations, in contrast to human populations, are of the steady-state type is supported by one more fact: geochronologically dating via parameters of the subpopulation structure in wild species is ineffective. The age of the population estimated by the above-mentioned formula is always less than the actual value. Thus, the age of a subdivided sockeye salmon population from Lake Azabachye (Kamchatka) was estimated at approximately 200 generations or 900–1,000 years. However, this lake, as a water reservoir with virtually the same characteristics as today, was formed at least 5,000–7,000 years ago, and fish populated it during approximately the same period. The "age" determined in this case most probably corresponds only to the period during which the population system approaches genetic equilibrium. It also suggests that the natural populations occupied and assimilated their modern ranges in a short period of time. Thus, in addition to a decrease in the intensity of gene migration with time, our model postulates that both the number of subpopulations and their effective size increase simultaneously until the system attains a stationary reproductive state. However, we cannot exclude the possibility that, during exponential growth of the size of a natural population system, the mean $N_e$ value for the constituent local subpopulations remains constant, as it does in human populations. *In any case, when the hierarchy of structural levels in the system is finally established, the subdivided population exists as if beyond the time scale of natural history.* Under such conditions, the only way to estimate the age of this population (of course, roughly) is to apply the concept of the molecular clock and the theory of genetic distances (Altukhov et al. 1997).

Although this model needs further proof, it still makes evident a following important circumstance: because the $F_{ST}$ ($G_{ST}$) coefficient for each hierarchic level of the population structure is constant, it can be regarded as the criterion of the norm (optimum). This approach permits us to estimate the degree and type of deviations from the optimum in the regional gene pool and, on this basis, to predict and prevent negative consequences of such deviations. In other words, when organizing programs of population monitoring we should take into account the unique possibility of "overcoming time through space" by exchanging long-term surveillance over the genetic structure of the same group of populations for an analysis of samples taken uniformly over the whole species range, or separate geographical region, with due regard to its natural history. The plausibility of this approach is discussed in the final chapter of this book.

In view of the totality of the facts considered above, we should like to stress an important, often neglected, point. The theory of gradual evolution, upon evaluation of the state of the environment to which the evolving population should adapt, is based, as already noted, on the uniformitarianism principle. As our experience shows, these environmental changes together with genetic drift are evidently sufficient for generating differently directed fluctuations at the subpopulation level. However, the same experience shows that these same changes are not sufficient to produce a directional shift of genetic parameters of the whole system. Note that the time required for forming the intrinsic subpopulation structure of the ancient isolated population and its reaching the modern level of differentiation is equal to several generations (Rychkov 1984), being thus negligibly small compared to the time required in STE for the hypothetical gradual development of post-reproductive isolation.

Altogether, our results suggest that there are no strictly directional microevolutionary processes in native population systems. Rather, these systems are characterized by various stabilization mechanisms preventing random fixation of genes. The directional reorganizations of the genetic population structure primarily result from potent anthropogenic pressures to be discussed in the next chapter.

As for processes of normal reproduction, the more complex the internal organization of a population system and the more considerable its internal diversity, the greater its resistance to different kinds of external influence. Seen from this vantage point, it is precisely a widely distributed species that should have maximum stability in time and space, for the shift in the level of adaptation achieved by it necessitates external influences that it is unable to tolerate. But if one remains within the framework of the principle of uniformitarianism mentioned above, then is it difficult to visualize how the processes that determine the maximal stabilization of species as integral population systems can simultaneously cause their evolutionary transformation (anagenesis) and lead to the emergence of new species. The same applies to cladogenesis. Evidently, the traditional population genetic approach, based exclusively on polymorphism, provides no adequate answer to the question. With this in mind, it is essential to turn to the last stage in the development of genetics, which is directly connected with the molecular aspects of the genome organization of eukaryotes, and to the comparative analysis of variability in different genetic markers; and attention should be paid both to polymorphic and monomorphic proteins.

## 6.3
## Genetic Monomorphism of Species as a Real Natural Phenomenon

In the preceding pages, we have chiefly focused on the phenomenon of genetic polymorphism and have examined its significance for analyzing genetic processes in populations and evaluating the total level of genome variability. It would now seem to be possible to visualize, with greater clarity, the natural mechanisms involved in supporting this kind of genetic variability and to conclude that analysis of it is inadequate to interpret the genetic bases of evolution. In addition, however, the findings of Chap. 2 establish that the genetic contents of a species cannot be reduced to variability alone and that monomorphic invariant proteins are always to be found in populations, along with polymorphic genetic markers, when a biochemical approach is adopted.

It is probably because these characters do not allow one to study the genetic divergence of populations that they remain outside the field of vision of most researchers. Moreover, each protein plays one or another functionally important role in the body, and for this very reason the fact of the genetic invariance of the greater part of a genome requires special analysis. Several years ago we undertook such an analysis (Altukhov 1969a, 1970, 1974; Altukhov and Rychkov 1972; Altukhov et al. 1972), on the basis of which a hypothesis was formed that explains in a satisfactory way the genetic diversity of natural populations for loci coding for both enzymatic and non-enzymatic proteins. Our approach combines the views of the "balanced" and "neoclassical" schools and postulates duality in the structural–functional organization of the eukaryote genome. In this system protein polymorphism is regarded as relatively neutral variability associated with the secondary adaptive properties of a species, and genetically monomorphic proteins as markers of cardinal functions whose normal variability is biologically impermissible ("conditio sine qua non");

mutations in this section of the genome, which lead to pathology, should be immediately eliminated by selection, especially in the early stages of ontogenesis.

Genetic monomorphism may be defined as *the absence of variability in a trait known to be hereditary for a whole species' area, or as the presence within a species of rare discrete variants having a frequency which does not exclude their being maintained by repeated mutations.* This definition is hypothetical, being the converse of Ford's definition of polymorphism. However, I consider it important because it emphasizes the reality of genetic monomorphism as a natural phenomenon, characterizing a species as a whole, and presupposes the existence of this invariance at any structural level of a living organism.

There is no doubt about the phenomenology of biochemical polymorphism and its reality. However, as far as genetic monomorphism is concerned, recognition of its reality requires special, reasoned argument. One cannot exclude the possibility of invariance simply reflecting inadequate sampling, a characteristic of an individual population, or the result of a method's insufficient resolving capacity. Finally, the thesis of the absolute connection between monomorphic proteins and viability needs substantiating.

Thanks to the work carried out by our group and by other researchers in recent years, these difficulties have now been surmounted.

1. It may be regarded as an established fact that the genetic invariance recorded during protein electrophoresis cannot be ascribed to an inadequate solution of the method in all cases. It has been known for a long time that complete analysis of the primary structure of enzymes reveals their catalytic center often to be invariable, whereas a molecule's functionally less significant areas vary. For instance, lactate dehydrogenase has a different amino acid composition in different vertebrate species, but the enzyme's catalytic center, composed of 12 amino acids, is identical in all the cases researched (Kaplan 1965). Histidine residues, located respectively in the functionally important 58th and 87th, and the 63rd and 92nd positions, respectively, of the $\alpha$- and $\beta$-chains of the human hemoglobin molecule, are completely invariant; in many groups of vertebrates this monomorphism has been conserved for the 500 million years of their natural history (Jukes 1971). The substitution in humans of tyrosine for any of these histidines causes hemoglobinopathy (see Kimura and Ohta 1973; Berger and Weber 1974; and Sect. 7.5 of the present work).

   In addition, the reality of the monomorphic state of several protein loci has been reaffirmed recently despite the discovery of heat sensitive alleles. Lewontin, for example, showed at the 14th International Genetics

Congress that further refinements of methods for detecting hereditary variations at a molecular level do not alter the conclusion about the existence of two groups of gene loci that are responsible for protein synthesis. The rule that emerges is that loci which are highly variable under one method of analysis, also show increased allelic diversity when more improved methods are employed, whereas loci that appear to be monomorphic in usual electrophoresis remain monomorphic even when finer analysis is carried out (Lewontin 1978b, p. 465).

Singh (1979, p. 1014), when analyzing *Drosophila pseudoobscura* proteins electrophoretically, concluded that monomorphism was a reality. The author indicates that the so-called monomorphic loci "form a separate group and are not just the tail of the same distribution covering the polymorphic loci."

Studies of tissue proteins of *Drosophila*, *Mus*, and *Homo sapiens* by means of two-dimensional electrophoresis discovered such low genetic variability that it was concluded that previous estimates of the levels of polymorphism and heterozygosity in natural populations should be revised (Brown and Langley 1979; Lee et al. 1979; McConkey et al. 1979; Racine and Langley 1980). According to these new data, the proportion of polymorphic genes does not exceed 11%, while average heterozygosity per locus is only 1–4%.

However, it should be recalled that a very similar result was obtained in 1971 (Altukhov et al. 1972) when we deliberately included several groups of proteins not previously investigated by electrophoretic analysis, not confining ourselves to enzymes alone. Lewontin, unfortunately, considered this a defect of our work, whereas in fact this method enabled us to reveal for the first time protein systems characterized by extremely low levels of genetic variability.

2. It may be regarded as proved that genetic monomorphism, as had been postulated, is not only a population phenomenon but also a phenomenon that characterizes a species as a whole. This has been established with particular reliability for *Homo sapiens*, the representatives of all the main races of which have been studied up to the present from an adequately large number of gene protein markers, some monomorphic throughout the species' range (Nei and Roychoudhury 1974; Rychkov et al. 2000). Taking into account the findings of American, Canadian, and Japanese colleagues, our laboratory has obtained evidence of monomorphism of certain protein systems in Pacific salmon. This conclusion is based on electrophoretic analysis of several thousand specimens gathered from large areas of the Asian as well as American coasts of the North Pacific (Altukhov 1969b, 1974, 1977; Altukhov and Rychkov 1972; Altukhov et al. 1972; Omel'chenko 1974; Salmenkova et al. 1986). As research into

different species using varied gene protein markers is extended, the reality of the existence of genetically monomorphic systems becomes increasingly obvious (see also Coulthart et al. 1984).

3. Direct evidence has been obtained of the link between genetically monomorphic systems and viability from electrophoretic research of blood protein in children with anomalies of development and of different tissues in spontaneous human abortions (see Sect. 7.5). It has been found that only a small proportion of the total number of gene mutations arising in a given generation come within the reproductive section of a population and are transmitted as relatively neutral. The direct evidence that has been obtained of the negative selective value of rare electrophoretic protein variants makes it possible to explain in a new way the excess in populations of different species of rare alleles when approximating the empirical distributions of their frequencies by means of a mathematical function that hypothesizes the neutrality of protein polymorphism. Yu.E. Dubrova (1980) performed the requisite analysis, and the approximation proved to be more satisfactory when our model was adopted, according to which the existence of rare alleles in a population reflects to a considerable degree the balance between selection and the spontaneous mutation process.

In fact, it is possible for any polymorphic locus to be regarded as monomorphic with respect to rare electrophoretic protein variants and hence, to differentiate an entire pool of alleles into two groups which are divided by a clear hiatus into "young" alleles, chiefly maintained in a population by a balance between the mutation process and selection, and "old" alleles that have already entered and been incorporated into a species' gene pool and are represented in "polymorphic" frequencies. From this viewpoint the excess of rare alleles (see Sect. 2.3) can be perfectly well explained by taking into account the effects of selection and revising all previous estimates of the mutation rates at protein loci. The "bottleneck" effect (Ohta 1975; Nei 1976; Chakraborty 1977) and intragenic recombination (Koehn and Eanes 1976; Strobeck and Morgan 1978) can hardly be excluded as sources of the excess of rare alleles in theoretical approximations of empirical distributions that presuppose polymorphism neutrality (Kimura and Crow 1964; Kimura and Ohta 1975).

If one uses Ohta's method to analyze the findings of Harris' group (Harris et al. 1974) on the variability of 42 human protein loci and the material of Ayala and his coauthors (Ayala et al. 1974b) on three *Drosophila* species, dividing the gene loci preliminarily into polymorphic and monomorphic, then one sees a considerably improved agreement of theoretical neutral distributions with observations of polymorphic loci (Altukhov and Dubrova 1981; Table 6.1). This approximation of the empirical distributions was implemented theoretically for the number of $n_a$ alleles falling in the frequency range of $x, x + \Delta x$:

$$n_a(x, x + \Delta x) = \int\limits_{x}^{x+\Delta x} \Phi(y)dy$$

where $\Phi(x) = 4N_e u_0 p^{-1}(1 - p)^{4N_e u_0 - 1}$ – Kimura and Crow's distribution for an infinite number of neutral alleles, in which $N_e$ is the effective population size and $u_0$ the rate of neutral mutations at a locus in a generation.

The $N_e u_0$ product was derived from data for average heterozygosity, using the equation $\overline{H} = 4N_e u_0/(4N_e u_0 + 1)$ (Kimura and Ohta 1971).

This improved approximation only for polymorphic loci may be regarded as a factor of definite importance. However, although such an approach is justified from a biological point of view, it must not be considered effective in a purely statistical sense; further research is needed in this direction. Nevertheless there is no doubt that the data that have been obtained recently not only corroborate the conclusion of a species' genetic monomorphism as a real natural phenomenon, but also shed light on the mechanisms maintaining the phenotypic stability at a protein level. The following are among the most important mechanisms.

**The Effects of Purifying Selection.**    If a protein has a quaternary structure and/ or is responsible for the most essential functions, then the vast majority of mutations should be categorized as "forbidden", which means that they are eliminated by natural selection at the earliest stages of ontogenesis. These are mutations, impinging on the active center of an enzyme or else sections of a molecule important to the process of the association of subunits.

**Organizational Characteristics of Genetic Material.**    If there is a mechanism for correcting and removing mutations in protein synthesis, then it is evident that its particular function lies in maintaining the stability of primary protein structure. This would have been expected even in the case of Kellan's master–slave hypothesis, but the discovery of "mosaic" (or "split") genes (see summaries in Dubinin 1978, 1979; Crick 1979) have made this possibility all the more obvious if one regards introns as a kind of "entrapment" for mutations (see Sect. 2.2). Although this is only speculation, there is no doubt that the analysis of the connection between the fine structure of eukaryotic genes and their levels of variability may lead to an important area of research. At this point, however, attention should be paid to yet another factor that, from our viewpoint, is of major importance for elucidating the biological significance of the genetic monomorphism of a species: considerable invariance often reveals protein systems characterized by exceptional multiplicity of structural components. By way of example one can point to several electrophoregrams of monomorphic protein systems described by us for Pacific salmon, which are allotetraploid (Fig. 6.8). It is still not clear what mechanism is involved in such tolerance of spontaneous mutations of duplicated genes that encode the synthesis of isofunctional protein families.

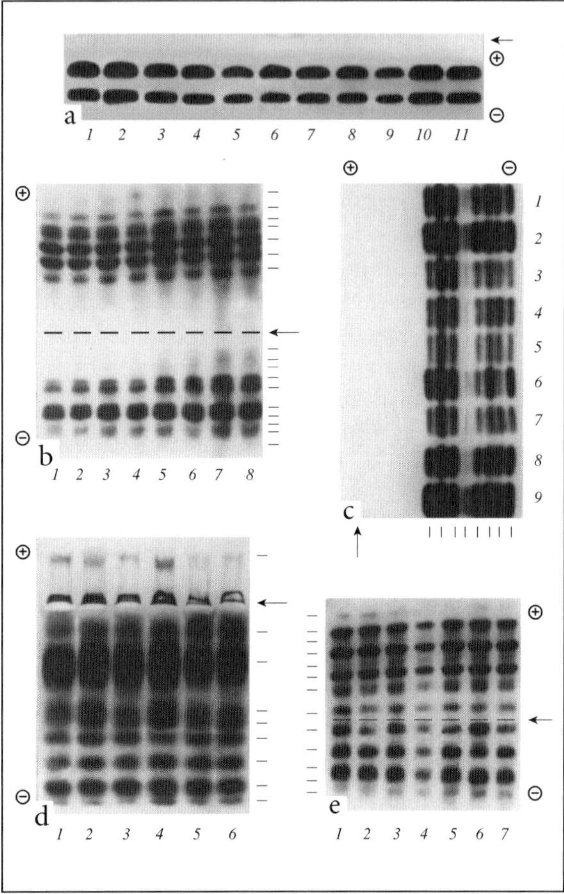

**Fig. 6.8.** Multiple molecular forms of proteins in Pacific salmon (Altukhov 1974). **a** hemoglobin of 11 specimens of chum salmon, *Oncorhynchus keta.* **b** hemoglobin of eight chum salmon specimens. **c, e** the same for several specimens of pink salmon, *O. gorbuscha.* The multiplicity of hemoglobins in salmons is determined by presence of seven to eight different subunits in them which form tetrameric molecules (Wilkins 1970; Tsuyuki and Ronald 1970); **d** water-soluble proteins of the eye lens (crystalline) in the chum salmon. All the photographs show the absence of individual variability in the number or position of separate protein components: monomorphism. **a, c, d** electrophoresis in agar gel, medinalveronal buffer, pH 8.6; **b, e** electrophoresis in a starch gel, borate buffer system, pH 8.5

This phenomenon has been explained on a purely functional plane: the maximum duplication of genetic information in eukaryotes in proportion to its biological significance. In other words, the most important functions may be encoded by multiple cistrons, ensuring for an organism the advantages of broad adaptation in a normally fluctuating environment (Altukhov and Rychkov 1972; Altukhov 1974, 1977).

In fact, if Pacific salmon are amphidiploid (Altukhov 1974; Altukhov et al. 1972; Bushuyev 1973; Bushuyev et al. 1975; Omel'chenko and Gerasimenko 1981), then with regard to homologous genes that differ in primary structure and are obtained from different ancestors, they, as other groups of like origins, are effectively fixed heterozygotes. Correspondingly, in the case of multimer proteins, their formation is effected by the interaction of noticeably diverged subunits, encoded by *different gene loci* and not by *alleles of the same gene.*

It is possible that selection places particularly harsh demands on precisely these complexes, in consequence of which polymorphism levels should be small in allopolyploids (Altukhov et al. 1972; Salmenkova and Volokhonskaya 1973; Salmenkova and Omel'chenko 1978).

Harris and his colleagues (Harris et al. 1977) have noted reduced variability of multimer proteins in man, being concerned not so much with multimer proteins in themselves as with such of them as are encoded by different loci and form hybrid molecules. D. Comings' (1972) work points out the possible role played by allotetraploidy in evolution of humans.

It is undeniable that the multiplicity of isofunctional proteins, betokening the monomorphism of a species as a whole, results from the duplication of genetic material caused by unequal crossing over and polyploidy. These mechanisms were discussed in Sect. 2.2, and as a whole the concept of "multigene families" (Hood et al. 1975) and their "concerted" evolution has been formulated on a theoretical plane with a fair degree of adequacy (Ohta 1978, 1980, 1983, 1984, 1987a,b; see also Kimura 1983; Nei 1987). However, does this not mean that the corresponding genetic mechanisms also form the basis of speciation that, if so, should be regarded not as a gradual, step-by-step process of allelic substitutions but as a phenomenon associated with rapid genome reorganization? A review of the variability characteristics of polymorphic and monomorphic genetic systems at a species level is necessary in order to answer this question.

## 6.4
## Interspecific Variability Characteristics of Polymorphic and Monomorphic Traits

It has not been long since questions about the character of genetic differences among species were solved on the basis of crosses that permitted the thesis of the "internal fertility and external sterility of a species" (Dobzhansky). However, this old and time-honored procedure says nothing about the actual nature of these differences. However, with the development of molecular genetics we now have the direct possibility of investigating the differences among species in homologous genes themselves, for one could scarcely doubt the true functional homology of cistrons that encode the same proteins in representatives of systematically close as well as distant species. So far, such comparisons have been made without dividing the loci into polymorphic and monomorphic ones by applying different measures of genetic distance or indexes of similarity. According to one such method proposed by Nei (1972), the normalized identity of genes between two populations for the $j$th locus is

$$l_j = \frac{\sum x_i y_i}{\sqrt{\sum x_i^2 \sum y_i^2}} \, ,$$

where $x_i$ and $y_i$ are the frequencies of the $i$th allele in $x$ and $y$ populations, respectively. The genetic identity of the samples for all the loci studied is

$$l = \frac{J_{xy}}{\sqrt{J_x J_y}} \, , \tag{6.1}$$

where $J_x$, $J_y$ and $J_{xy}$ are the arithmetic means for all the loci.

Using this criterion, Ayala and his colleagues (Ayala et al. 1974b) undertook the analysis of the degree of differentiation within a *Drosophila willistoni* group at different taxonomic levels and showed the parallelism between the differentiation revealed and that anticipated by the notion of gradual evolution. Maximal and minimal genetic similarity was found, respectively, for populations of the same species and for species within a genus (Fig. 6.9). Evidently, if we had conducted an analysis of the rate of amino acid substitutions at the level of a single protein, as is widely practiced in molecular taxonomy constructions (Nei 1975, 1987; Ayala 1976,

**Fig. 6.9.** Distributions of gene loci with regard to the identity of allelic frequencies at different stages of evolutionary divergence in a group of forms of *Drosophila willistoni*. (Ayala et al. 1974b) (with permission from Sinauer Associates and Author)

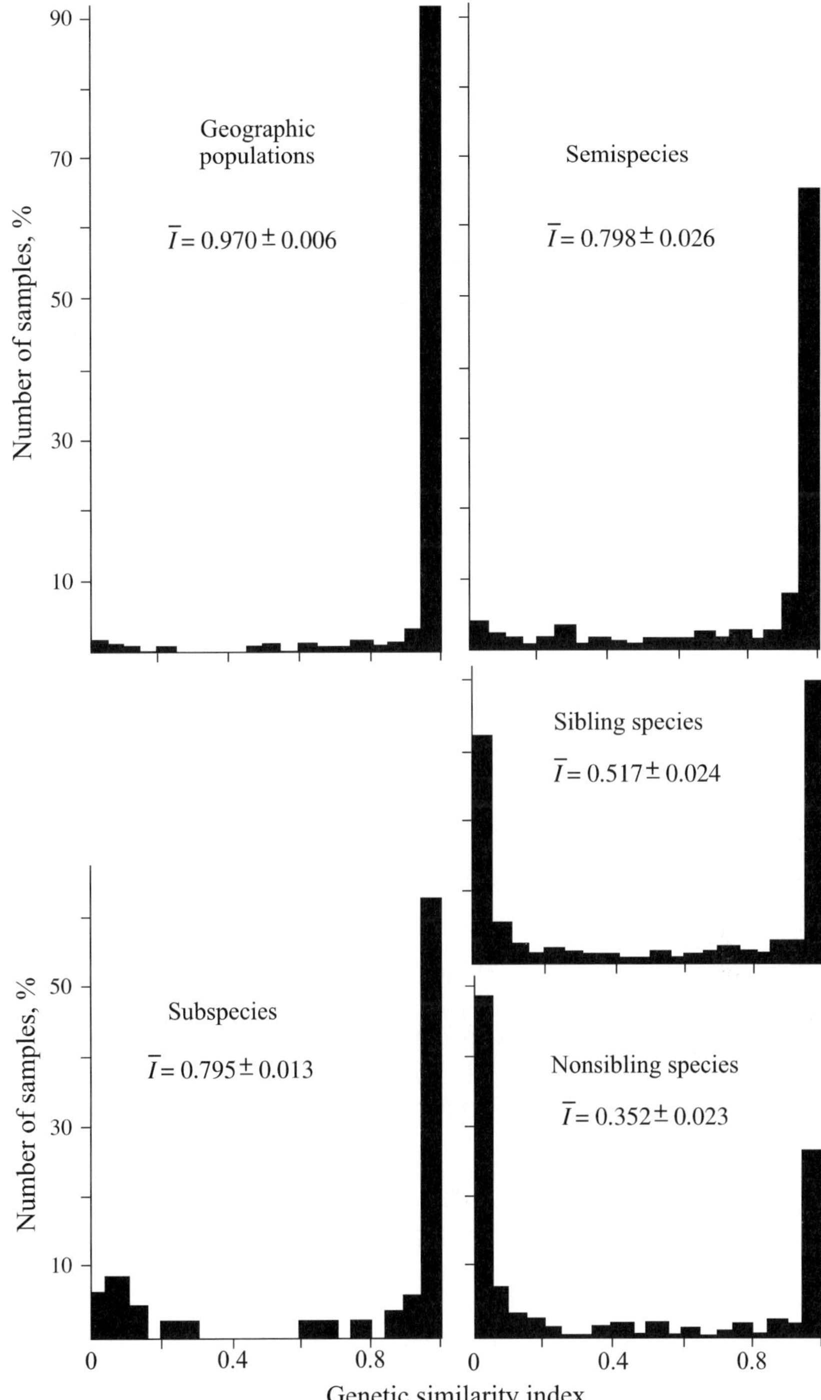


Number of samples, %
90
70
50
30
10
Geographic
populations
$\bar{I} = 0.970 \pm 0.006$
Semispecies
$\bar{I} = 0.798 \pm 0.026$
Sibling species
$\bar{I} = 0.517 \pm 0.024$
Number of samples, %
50
30
10
Subspecies
$\bar{I} = 0.795 \pm 0.013$
Nonsibling species
$\bar{I} = 0.352 \pm 0.023$
0
0.4
0.8
0
0.4
0.8
Genetic similarity index

1983), we would have obtained the same picture in principle. The same relationships must be found in studies of genetic phylogeography, which have become very popular in recent years owing to DNA techniques (see Hewitt 2001 for review).

However, do these differences reflect the real character of the genetic reorganizations on which speciation is founded?

Is it not more logical to admit that the amino acid substitutions, recognized by analyzing primary protein structure or based upon its electrophoretic revelations, are only a reflection of the antiquity of the compared groups, marking the time of their evolutionary division and subsequent relatively independent existence? It is clear that younger intraspecific groups should accumulate far fewer amino acid replacements than older species, and all the more so should genera.

In order to visualize how the different indices of similarity or identity are calculated, it suffices to examine the characteristics of interspecific differentiation in certain homologous polymorphic proteins (Fig. 6.10). Indeed, the difference between a species and a variety, figuratively speaking, is not so much in essence as a matter of degree: independent species differ in principle, just as populations differ within a species, that is, by the frequency of the same alleles, or by particular alleles, specific only for them. Naturally, insofar as species are genetically closed, the latter type of differentiation will be encountered more often at an inter- and not an intraspecific level, corresponding to a greater level of differences.

Let us now find out the nature of the differences among species in genetically monomorphic proteins. With this in mind, we shall turn to the results of electrophoretic analysis of the hemoglobins of Pacific salmons, and also to work on several groups of genetically monomorphic proteins in other species of animals (Figs. 6.11 and 6.12). In all cases investigated, qualitative differences were traced; for these traits, the species are as discrete and unique as are the different genotypes in one or another system of genetic polymorphism within a species. In other words, *any species is a separate individual when seen in the context of the genetically monomorphic part of a genome*, and hence the problem of species identification is resolved concurrently, irrespective of which species a researcher is dealing with, whether bisexual or unisexual. Furthermore, when one succeeds in comparing rare interspecific hybrids or species of hybrid origins with the parental species, an important law is revealed: the species' monomorphic traits act as integral genetic units by displaying a simple summation of parental types and forming hybrid products, or even a dominance–recessivity relationship (Fig. 6.12). Thus the monomorphic part of a genome is disclosed, which is marked by isofunctional families of multiple genes and is protected from segregation.

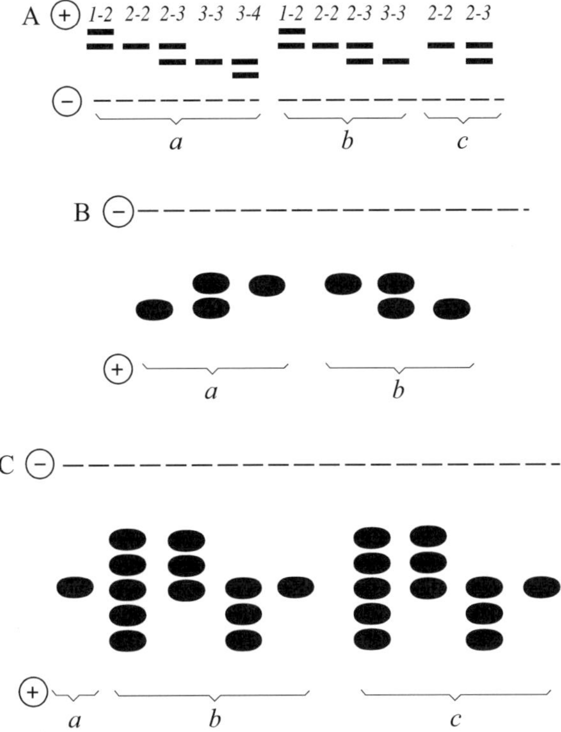

**Fig. 6.10.** Interspecific variability of genetically polymorphic proteins (from Altukhov 1974). A Genotypes at a serum esterase locus in three species of tuna: *a Thunnus maccoyii, b T. obesus,* and *c T. albacares* (electrophoresis in a starch gel); **B** genotypes of serum albumin in redfish, *Sebastes*: *a S. marinus, b S. mentella* (electrophoresis in an agar gel); and **C** genotypes of malate dehydrogenase in the muscles of three salmon species of the *Oncorhynchus* genus: *a O. nerka, b O. gorbuscha,* and *c O. keta* (electrophoresis in polyacrylamide gel)

Obviously, if we determined genetic similarity in monomorphic protein systems, encoded by multiple genes, and by qualitatively differentiating species known to be different, then we would obtain a value of "1" for any populations within a species and increasing values for differences at species or higher taxonomic levels. However, it is obvious that these evaluations would not reflect a gradual transition from a population to a species but a qualitative gap, or, in other words, a "jump". This means that interspecific genetic variation differs in essence from the intraspecific variation and does not follow from it. A similar conclusion was drawn for the case of homeobox genes in *Drosophila*: "the differences between the species do not result from sampling from intraspecific polymorphisms" (Schilthuizen 1999), i.e., species of the *melanogaster* group differ from one another in monomorphic genome regions adjacent to

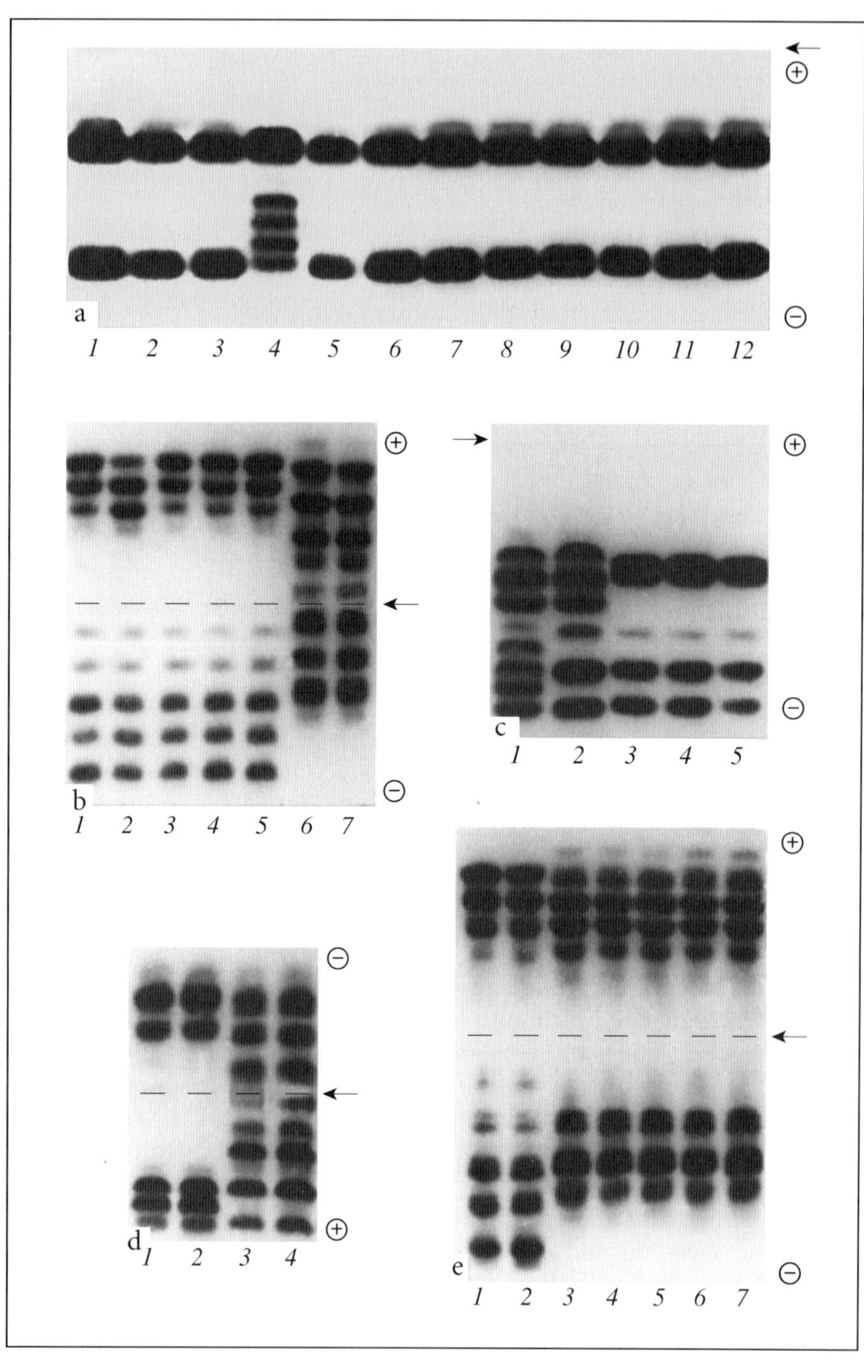

**Fig. 6.11.** Species specificity discovered in electrophoretic research of multiple hemoglobins in fish belonging to the family *Salmonidae*. **a** *1–3, 5–12 Oncorhynchus keta, 4 O. kisutch* (electrophoresis in agar gel); **b** *1–5 Salvelinus leucomaenis, 6, 7 O. gorbuscha* (electrophoresis in starch gel); **c** *1 O. gorbuscha, 2* hybrid (F$_1$) *O. gorbuscha×O. masu, 3–5 O. masu* (electrophoresis in agar gel); **d** *1, 2 O. keta, 3, 4 O. gorbuscha*; **e** *1, 2 Salvelinus leucomaenis, 3–7 O. keta* (electrophoresis in starch gel). Start position is marked with *arrows*, "+" and "–" are anode and cathode, respectively. One can see clear qualitative interspecific differences with perfect genotypic identity of individuals within species

a multigene family of regulatory loci (see Ting et al. 1998, p. 1502 for details). The results of recent *Drosophila* studies conducted with the use of interspecific complementation tests also suggest the important role of macromutation effects in transformation of some morphological structures at the species level (Gibson and Palsson 2001). Apparently, since discrete interspecific differences involve genetic traits that are invariant within the species, this phenomenon should be taken into account when developing any new concept of evolution, because putative transformational action of natural selection in this case is irrelevant (see for comparison Wu 2001; Mallet 2001). The rapid interspecific divergence of reproductive proteins is seemingly not in this scheme, but the available evidence is clearly insufficient for any definite conclusions (Swanson and Vacquer 2002).

No matter how much the idea of saltational speciation has been discredited by well-known dogmatic formulations in biology, the phenomenon of microevolution cannot be discussed objectively without taking into account all facts and approaches known to modern molecular population genetics.

Moreover, if we draw the conclusion that the genetic monomorphism of a species is as real a natural phenomenon as polymorphism, then mere recognition of this fact with allowance made for a pattern of interspecific variability in the genetically monomorphic features encoded by multiple genes, *makes it possible to treat speciation not as a gradual probabilistic process taking place at a population level, but as the result of qualitative reorganizations of the genome.* Inasmuch as multiple proteins, encoded by multiple duplicated genes, are usually the most monomorphic ones and simultaneously most charged with the characteristics of species specificity[1], the well-known mechanism of gene duplication should reflect those genetic reorganizations which, if they are "compatible" with ontogenesis, can lead to reproductive isolation in one or several consecutive steps (Lynch and Force 2000; Taylor et al. 2001).

---

[1]The surprising examples of a degree of conservativeness and species specificity also give us many ethological traits (Mayr 1968; Gilyarov 1974).

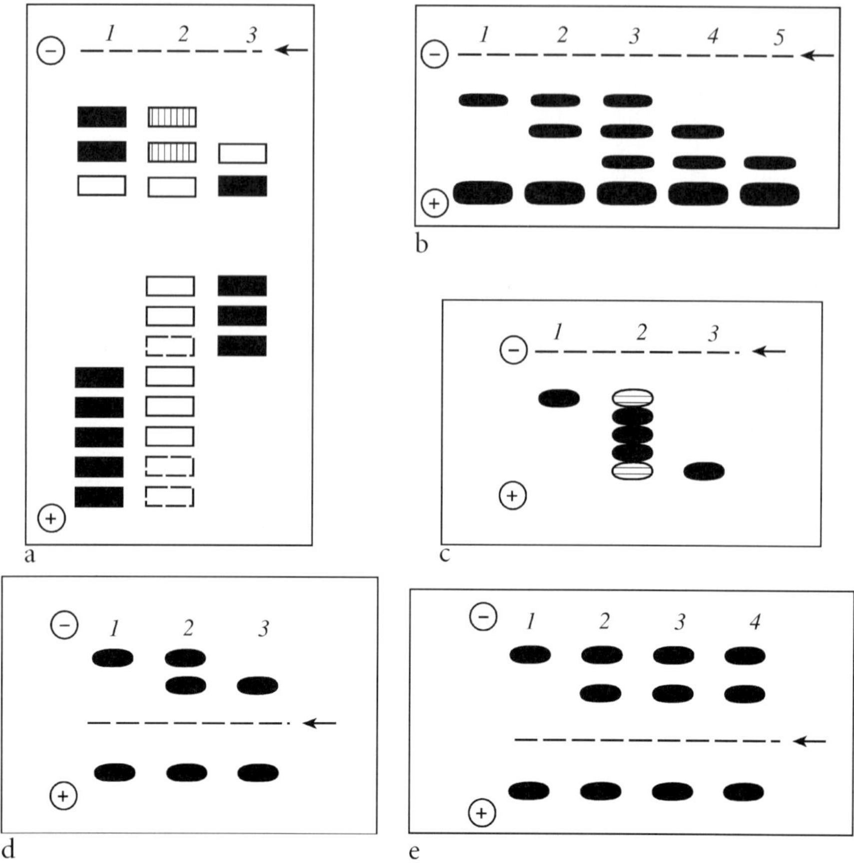

**Fig. 6.12.** Electrophoretic spectra of species-specific proteins in interspecies hybrids compared with the parental species. **a** Peroxidase of leaves: *1 Nicotiana sylvestris, 3 N. tomentosiformis, 2* interspecies hybrid $F_1$ (electrophoresis, in polyacrylamide gel; Sheen 1970). **b** Water-soluble fish muscle protein: *1 Richardsonius balteatus, 5 Mylocheilus caurinum, 3* $F_1$ hybrids, *2, 4* corresponding backcrosses (electrophoresis in starch gel) (Aspinwall and Tsuyuki 1968). **c** Lactate dehydrogenase of cardiac muscle in lizards belonging to the *Cnemidophorus* genus, *1 C. gularis,* bisexual species, *3 C. tigris,* bisexual species, *2 C. tesselatus,* parthenogenetic species of hybrid origin (electrophoresis in starch gel; Neaves and Gerald 1968). **d** Bird hemoglobins: *1 Coturnix coturnix, 3 Gallus gallus, 2* corresponding $F_1$ hybrid (electrophoresis in starch gel; Manwell et al. 1963). **e** Hemoglobin globin chains, *1* donkey, *2* mule, *3* foal, *4* horse (electrophoresis in starch gel; Trujillo et al. 1967)

Duplication of genetic material, based on local excessive self-copying of genes through polyploidy or unequal crossing over, indubitably reflects genome reorganization other than point mutations underlying hereditary protein polymorphism. Translocations, deletions, and chromosome fusions (Ayala 1976) and also interspecies hybridization (Altukhov 1974) lead to similar changes in gene sequences within a genome and correspondingly changes of gene expression. Recently, using comparisons of molecular structures of the genomes in taxonomically close species, the controlling DNA regions have been identified and shown to be very conserved (Rubin 2001).

Earlier, based on restriction analysis of repetitive nuclear DNA, the taxonoprint method has been developed, which differentiates different species practically without revealing intraspecific (geographic) variability (Mednikov et al. 1995, 2001).

In two relatively recent publications, Grechko et al. (1997) and Fedorov et al. (1999), who apparently are not acquainted with our early concept of genetic monomorphism of the species, decades later came to virtually the same conclusions: "1. All individuals from the same species have identical taxonprints. 2. Taxonprint bands can be subdivided into those specific for a single species and those specific for a group of closely related species, genera, and even families" (Grechko et al. 1997, p. 332).

This statement completely coincides with our conclusions (Altukhov 1969a, 1974; Altukhov and Rychkov 1972). Apparently, the only difference between our results and those of Grechko et al. (1997) and Fedorov et al. (1999) is the much smaller sample sizes in the latter studies, so that the intraspecific invariance of the taxonprint bands is based on the data for a few individuals. In contrast, substantiating the reality of genetic monomorphism as a natural phenomenon characterizing the species as a whole, we either examined directly or collected from literature data on numerous samples, which were relatively uniformly distributed over the species areas and consisted of hundreds and thousands of individuals (see Altukhov and Rychkov 1972; Table 6.2).

Moreover, Fedorov et al. (1999), in their classification of the monomorphic taxonprints into the so-called variable (species-specific) and invariable (genus- and family-specific), did not reach, as it seems, an important conclusion on the natural hierarchical organization of the eukaryotic genome and its correspondence to the accepted hierarchy of taxa (species, genus, family, etc.).

Other markers should be sought for higher taxonomic levels, but the phenomenology of the differences between the interspecific taxonomic ranks and higher ranks is in essence the same as that established for structural genes marked by multiple protein families: with regard to the invariant genome part, any taxon is equivalent to an individual, i.e., the individual and geographic variation are absent (Altukhov and Rychkov 1972).

The existence of genetically invariant species-specific traits undoubtedly is not only of fundamental theoretical significance, but also opens principally new perspectives for forensic medicine, food control, and establishing various phylogenies (see Grechko et al. 1997; Fedorov et al. 1999).

**Table 6.2.** Comparison of observed (in brackets) with the theoretical number of alleles obtained on the basis of Kimura and Crow's model for Drosophila and human populations. (Altukhov and Dubrova 1981)

| Frequency interval | Loci All | Loci Polymorphic | Frequency interval | Loci All | Loci Polymorphic |
|---|---|---|---|---|---|
| | *Homo sapiens* | | | *Drosophila tropicalis* | |
| < 0.01 | 16.22 (57) | 21.04 (24) | < 0.01 | 11.33 (40) | 12.21 (22) |
| 0.01–0.1 | 8.55 (8) | 10.22 (8) | 0.01–0.1 | 11.97 (30) | 13.09 (30) |
| 0.1–0.2 | 2.85 (0) | 3.32 (0) | 0.1–0.2 | 3.96 (5) | 4.25 (5) |
| 0.2–0.3 | 1.88 (3) | 2.12 (3) | 0.2–0.3 | 2.57 (1) | 2.71 (1) |
| 0.3–0.4 | 1.51 (1) | 1.64 (1) | 0.3–0.4 | 2.05 (1) | 2.12 (1) |
| 0.4–0.5 | 1.38 (2) | 1.42 (2) | 0.4–0.5 | 1.82 (2) | 1.84 (2) |
| 0.5–0.6 | 1.36 (3) | 1.35 (3) | 0.5–0.6 | 1.77 (2) | 1.72 (3) |
| 0.6–0.7 | 1.44 (1) | 1.35 (1) | 0.6–0.7 | 1.86 (1) | 1.79 (1) |
| 0.7–0.8 | 1.73 (2) | 1.45 (2) | 0.7–0.8 | 2.16 (2) | 1.89 (2) |
| 0.8–0.9 | 2.46 (0) | 1.84 (0) | 0.8–0.9 | 2.92 (3) | 2.37 (3) |
| 0.9–1.0 | 34.83 (36) | 6.21 (7) | 0.9–1.0 | 20.73 (21) | 8.27 (8) |
| Loci numbers | 42 | 13 | Loci numbers | 30 | 17 |
| Average sampling numbers | 2,648 | 3,662 | Average sampling numbers | 270 | 257 |
| Total number of alleles | 74.21 (113) | 51.96 (51) | Total number of alleles | 63.14 (108) | 52.26 (77) |
| | *Drosophila willistoni* | | | *Drosophila equinoxialis* | |
| < 0.01 | 20.05 (83) | 21.40 (59) | < 0.01 | 16.48 (54) | 17.36 (37) |
| 0.01–0.1 | 16.24 (41) | 16.96 (41) | 0.01–0.1 | 14.62 (46) | 15.10 (46) |
| 0.1–0.2 | 5.55 (8) | 5.50 (8) | 0.1–0.2 | 4.82 (3) | 4.93 (3) |
| 0.2–0.3 | 3.70 (2) | 3.50 (2) | 0.2–0.3 | 3.13 (2) | 3.16 (2) |
| 0.3–0.4 | 2.99 (1) | 2.71 (1) | 0.3–0.4 | 2.49 (1) | 2.49 (1) |
| 0.4–0.5 | 2.70 (2) | 2.25 (2) | 0.4–0.5 | 2.21 (4) | 2.16 (4) |
| 0.5–0.6 | 2.59 (2) | 2.21 (2) | 0.5–0.6 | 2.14 (1) | 2.06 (1) |
| 0.6–0.7 | 2.66 (2) | 2.19 (2) | 0.6–0.7 | 2.23 (2) | 2.07 (2) |
| 0.7–0.8 | 2.68 (0) | 2.61 (0) | 0.7–0.8 | 2.48 (1) | 2.28 (1) |
| 0.8–0.9 | 3.57 (2) | 2.88 (2) | 0.8–0.9 | 3.42 (4) | 2.94 (4) |
| 0.9–1.0 | 19.24 (23) | 9.87 (13) | 0.9–1.0 | 20.01 (21) | 11.58 (12) |
| Loci numbers | 31 | 21 | Loci numbers | 31 | 22 |
| Average sampling numbers | 568 | 559 | Average sampling numbers | 414 | 432 |
| Total number of alleles | 81.97 (166) | 72.08 (132) | Total number of alleles | 74.03 (139) | 66.14 (113) |

The loci are regarded as polymorphic if the frequencies of the rare alleles in them exceed 0.01. The observed number of alleles is in parentheses

Using PCR with random primers, Altukhov and Abramova (2000) found invariant species-specific fractions of total RNA in 13 conifer species. In the world literature, this method is referred to as RAPD (random amplified polymorphic DNA). However, our study showed the absence of intra- and interpopulation variation in a number of amplicons over wide ranges of Siberian pine species *Pinus sibirica* and *P. pumila*. The choice of conifers was not accidental: these species can be analyzed at the haploid level and thus avoid errors related to the dominant inheritance of polymorphic variants of corresponding DNA (Sect. 5.6).

Thus, analysis of the genetic material as such demonstrates its dual structural organization: in addition to polymorphic loci, another type of locus lacking both individual and geographic variation, i.e., being monomorphic, is revealed. In my view, this DNA should be distinguished from polymorphic DNA and referred to as RAMD (random amplified monomorphic DNA).

Characteristically, differences between species and higher taxa are most markedly shown precisely by monomorphic fractions (Fig. 6.13). The aim of the further studies is analysis of nucleotide sequences of the monomorphic amplicons. However, even now it should be noted that an invariant 1200-bp fragment common for *P. flexilis*, *P. monticola*, *P. pumila*, and *P. koraiensis* is detected by Southern blotting, when the same labeled DNA fragment of *P. pumila* is used as a probe. This means that the primary structure of the corresponding DNA fraction is identical in these species.

Mukha et al. (1999) showed clearly saltatory changes at the species level in the structure of ribosomal gene cluster of *Blatella* cockroaches. These authors compared the nucleotide sequences of inner transcribed spacer (ITS1) in *Blatella germanica*, *B. lituricollis*, and *B. vaga*. For *B. germanica*, they have shown that in three geographically distant populations, the ITS1 sequences virtually lacked variation within the species. These sequences are all but identical, differing only in single nucleotide substitutions in a few individuals that occur at the rate of neutral evolution (about $10^9$ per site/year; Fig. 6.14). Thus, the situation observed at the level of primary DNA structure is fully compatible with the conceptual definition of monomorphism characterizing the invariant nature of the species genome as a total entity.

By contrast, a comparison of ITS1s in three closely related cockroach species revealed marked and essentially other differences: long insertions and deletions as well as numerous point substitutions were detected (Fig. 6.15). The values of ITS1 similarity in pairs *B. germanica-B. lituricollis* and *Blatella germanica-B. vaga* are only 67.2 and 30%, respectively. The authors think that selective magnification of rRNA genes and polar gene conversion are the main mechanisms providing a possibility for a structural variant to become a major member of a multigene family. It may well be that isogenization does not occur with equal probability over the whole

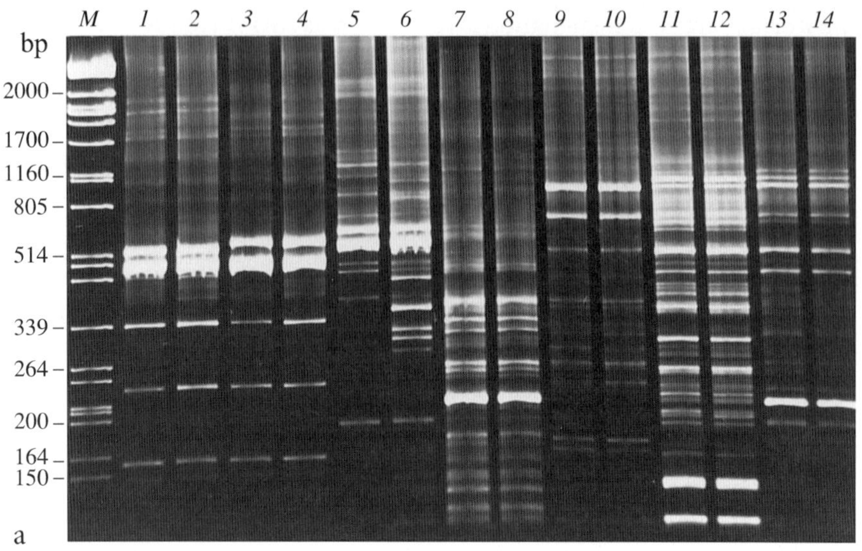

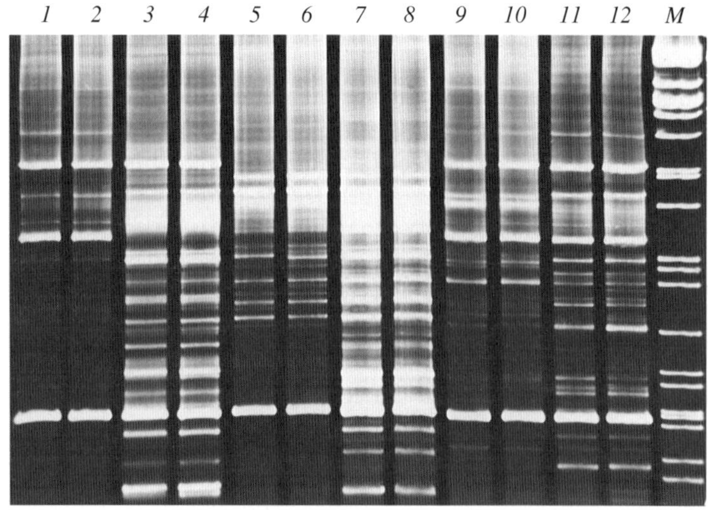

**Fig. 6.13.** Species-specific spectra of conifer DNA revealed in PCR with *OPA 09* primer (Altukhov and Abramova 2000). **a** *Abies egui-trojani (1, 2)*, *A. cilicia (3, 4)*, *Cedrus libani (5, 6)*, *Picea abies (7, 8)*, *Pinus morrisonicola (9, 10)*, *P. griffithii (11, 12)*, *P. parviflora (13, 14)*. **b** *Pinus koraiensis (1, 2)*, *P. pumila (3, 4)*, *P. sibirica (5, 6)*, *P. cembra (7, 8)*, *P. monticola (9, 10)*, *P. flexilis (11, 12)*

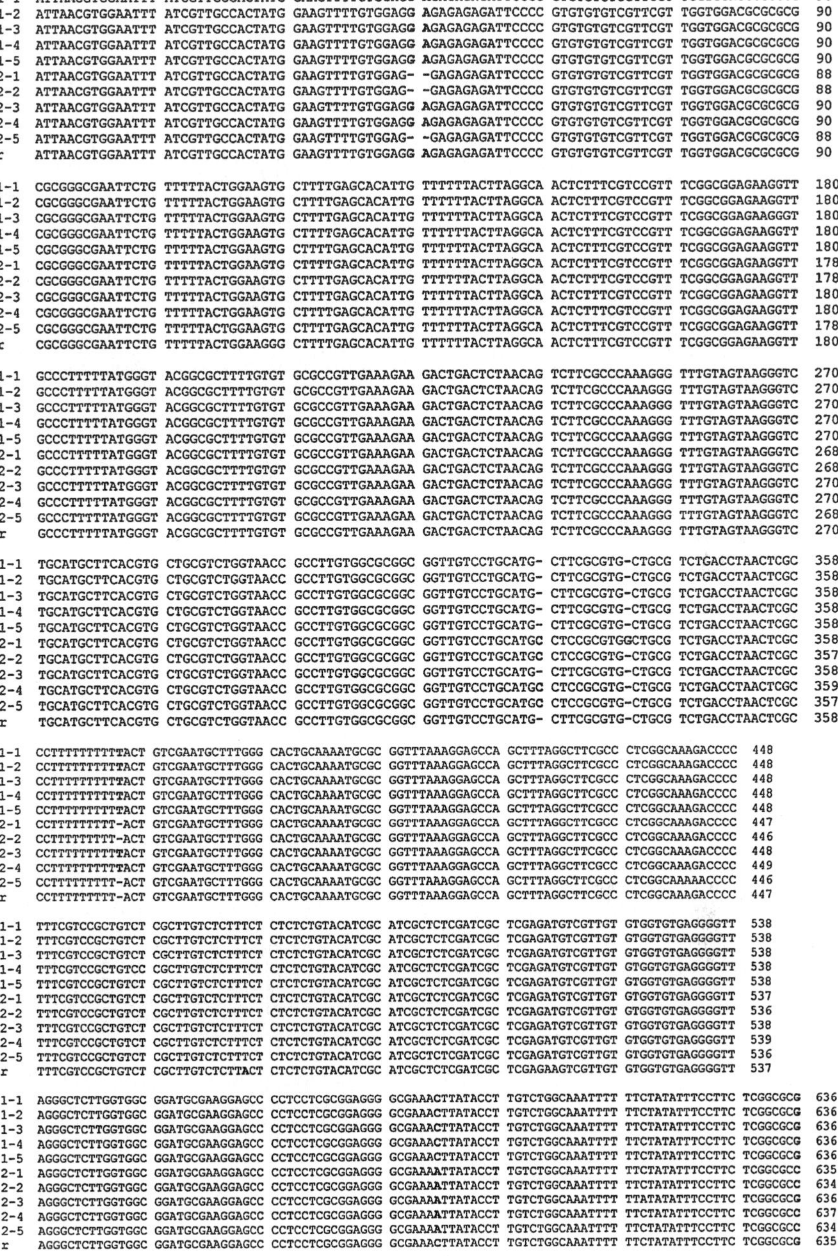

```
1-1  ATTAACGTGGAATTT ATCGTTGCCACTATG GAAGTTTTGTGGAGG AGAGAGAGATTCCCC GTGTGTGTCGTTCGT TGGTGGACGCGCGCG  90
1-2  ATTAACGTGGAATTT ATCGTTGCCACTATG GAAGTTTTGTGGAGG AGAGAGAGATTCCCC GTGTGTGTCGTTCGT TGGTGGACGCGCGCG  90
1-3  ATTAACGTGGAATTT ATCGTTGCCACTATG GAAGTTTTGTGGAGG AGAGAGAGATTCCCC GTGTGTGTCGTTCGT TGGTGGACGCGCGCG  90
1-4  ATTAACGTGGAATTT ATCGTTGCCACTATG GAAGTTTTGTGGAGG AGAGAGAGATTCCCC GTGTGTGTCGTTCGT TGGTGGACGCGCGCG  90
1-5  ATTAACGTGGAATTT ATCGTTGCCACTATG GAAGTTTTGTGGAGG AGAGAGAGATTCCCC GTGTGTGTCGTTCGT TGGTGGACGCGCGCG  90
2-1  ATTAACGTGGAATTT ATCGTTGCCACTATG GAAGTTTTGTGGAG- -GAGAGAGATTCCCC GTGTGTGTCGTTCGT TGGTGGACGCGCGCG  88
2-2  ATTAACGTGGAATTT ATCGTTGCCACTATG GAAGTTTTGTGGAG- -GAGAGAGATTCCCC GTGTGTGTCGTTCGT TGGTGGACGCGCGCG  88
2-3  ATTAACGTGGAATTT ATCGTTGCCACTATG GAAGTTTTGTGGAGG AGAGAGAGATTCCCC GTGTGTGTCGTTCGT TGGTGGACGCGCGCG  90
2-4  ATTAACGTGGAATTT ATCGTTGCCACTATG GAAGTTTTGTGGAGG AGAGAGAGATTCCCC GTGTGTGTCGTTCGT TGGTGGACGCGCGCG  90
2-5  ATTAACGTGGAATTT ATCGTTGCCACTATG GAAGTTTTGTGGAG- -GAGAGAGATTCCCC GTGTGTGTCGTTCGT TGGTGGACGCGCGCG  88
r    ATTAACGTGGAATTT ATCGTTGCCACTATG GAAGTTTTGTGGAGG AGAGAGAGATTCCCC GTGTGTGTCGTTCGT TGGTGGACGCGCGCG  90

1-1  CGCGGGCGAATTCTG TTTTTACTGGAAGTG CTTTTGAGCACATTG TTTTTTACTTAGGCA ACTCTTTCGTCCGTT TCGGCGGAGAAGGTT  180
1-2  CGCGGGCGAATTCTG TTTTTACTGGAAGTG CTTTTGAGCACATTG TTTTTTACTTAGGCA ACTCTTTCGTCCGTT TCGGCGGAGAAGGTT  180
1-3  CGCGGGCGAATTCTG TTTTTACTGGAAGTG CTTTTGAGCACATTG TTTTTTACTTAGGCA ACTCTTTCGTCCGTT TCGGCGGAGAAGGGT  180
1-4  CGCGGGCGAATTCTG TTTTTACTGGAAGTG CTTTTGAGCACATTG TTTTTTACTTAGGCA ACTCTTTCGTCCGTT TCGGCGGAGAAGGTT  180
1-5  CGCGGGCGAATTCTG TTTTTACTGGAAGTG CTTTTGAGCACATTG TTTTTTACTTAGGCA ACTCTTTCGTCCGTT TCGGCGGAGAAGGTT  180
2-1  CGCGGGCGAATTCTG TTTTTACTGGAAGTG CTTTTGAGCACATTG TTTTTTACTTAGGCA ACTCTTTCGTCCGTT TCGGCGGAGAAGGTT  178
2-2  CGCGGGCGAATTCTG TTTTTACTGGAAGTG CTTTTGAGCACATTG TTTTTTACTTAGGCA ACTCTTTCGTCCGTT TCGGCGGAGAAGGTT  178
2-3  CGCGGGCGAATTCTG TTTTTACTGGAAGTG CTTTTGAGCACATTG TTTTTTACTTAGGCA ACTCTTTCGTCCGTT TCGGCGGAGAAGGTT  180
2-4  CGCGGGCGAATTCTG TTTTTACTGGAAGTG CTTTTGAGCACATTG TTTTTTACTTAGGCA ACTCTTTCGTCCGTT TCGGCGGAGAAGGTT  180
2-5  CGCGGGCGAATTCTG TTTTTACTGGAAGTG CTTTTGAGCACATTG TTTTTTACTTAGGCA ACTCTTTCGTCCGTT TCGGCGGAGAAGGTT  178
r    CGCGGGCGAATTCTG TTTTTACTGGAAGGG CTTTTGAGCACATTG TTTTTTACTTAGGCA ACTCTTTCGTCCGTT TCGGCGGAGAAGGTT  180

1-1  GCCCTTTTTATGGGT ACGGCGCTTTTGTGT GCGCCGTTGAAAGAA GACTGACTCTAACAG TCTTCGCCCAAAGGG TTTGTAGTAAGGGTC  270
1-2  GCCCTTTTTATGGGT ACGGCGCTTTTGTGT GCGCCGTTGAAAGAA GACTGACTCTAACAG TCTTCGCCCAAAGGG TTTGTAGTAAGGGTC  270
1-3  GCCCTTTTTATGGGT ACGGCGCTTTTGTGT GCGCCGTTGAAAGAA GACTGACTCTAACAG TCTTCGCCCAAAGGG TTTGTAGTAAGGGTC  270
1-4  GCCCTTTTTATGGGT ACGGCGCTTTTGTGT GCGCCGTTGAAAGAA GACTGACTCTAACAG TCTTCGCCCAAAGGG TTTGTAGTAAGGGTC  270
1-5  GCCCTTTTTATGGGT ACGGCGCTTTTGTGT GCGCCGTTGAAAGAA GACTGACTCTAACAG TCTTCGCCCAAAGGG TTTGTAGTAAGGGTC  270
2-1  GCCCTTTTTATGGGT ACGGCGCTTTTGTGT GCGCCGTTGAAAGAA GACTGACTCTAACAG TCTTCGCCCAAAGGG TTTGTAGTAAGGGTC  268
2-2  GCCCTTTTTATGGGT ACGGCGCTTTTGTGT GCGCCGTTGAAAGAA GACTGACTCTAACAG TCTTCGCCCAAAGGG TTTGTAGTAAGGGTC  268
2-3  GCCCTTTTTATGGGT ACGGCGCTTTTGTGT GCGCCGTTGAAAGAA GACTGACTCTAACAG TCTTCGCCCAAAGGG TTTGTAGTAAGGGTC  270
2-4  GCCCTTTTTATGGGT ACGGCGCTTTTGTGT GCGCCGTTGAAAGAA GACTGACTCTAACAG TCTTCGCCCAAAGGG TTTGTAGTAAGGGTC  270
2-5  GCCCTTTTTATGGGT ACGGCGCTTTTGTGT GCGCCGTTGAAAGAA GACTGACTCTAACAG TCTTCGCCCAAAGGG TTTGTAGTAAGGGTC  268
r    GCCCTTTTTATGGGT ACGGCGCTTTTGTGT GCGCCGTTGAAAGAA GACTGACTCTAACAG TCTTCGCCCAAAGGG TTTGTAGTAAGGGTC  270

1-1  TGCATGCTTCACGTG CTGCGTCTGGTAACC GCCTTGTGGCGCGGC GGTTGTCCTGCATG- CTTCGCGTG-CTGCG TCTGACCTAACTCGC  358
1-2  TGCATGCTTCACGTG CTGCGTCTGGTAACC GCCTTGTGGCGCGGC GGTTGTCCTGCATG- CTTCGCGTG-CTGCG TCTGACCTAACTCGC  358
1-3  TGCATGCTTCACGTG CTGCGTCTGGTAACC GCCTTGTGGCGCGGC GGTTGTCCTGCATG- CTTCGCGTG-CTGCG TCTGACCTAACTCGC  358
1-4  TGCATGCTTCACGTG CTGCGTCTGGTAACC GCCTTGTGGCGCGGC GGTTGTCCTGCATG- CTTCGCGTG-CTGCG TCTGACCTAACTCGC  358
1-5  TGCATGCTTCACGTG CTGCGTCTGGTAACC GCCTTGTGGCGCGGC GGTTGTCCTGCATG- CTTCGCGTG-CTGCG TCTGACCTAACTCGC  358
2-1  TGCATGCTTCACGTG CTGCGTCTGGTAACC GCCTTGTGGCGCGGC GGTTGTCCTGCATGC CTCCGCGTGGCTGCG TCTGACCTAACTCGC  358
2-2  TGCATGCTTCACGTG CTGCGTCTGGTAACC GCCTTGTGGCGCGGC GGTTGTCCTGCATGC CTCCGCGTG-CTGCG TCTGACCTAACTCGC  357
2-3  TGCATGCTTCACGTG CTGCGTCTGGTAACC GCCTTGTGGCGCGGC GGTTGTCCTGCATG- CTTCGCGTG-CTGCG TCTGACCTAACTCGC  358
2-4  TGCATGCTTCACGTG CTGCGTCTGGTAACC GCCTTGTGGCGCGGC GGTTGTCCTGCATGC CTCCGCGTG-CTGCG TCTGACCTAACTCGC  359
2-5  TGCATGCTTCACGTG CTGCGTCTGGTAACC GCCTTGTGGCGCGGC GGTTGTCCTGCATGC CTCCGCGTG-CTGCG TCTGACCTAACTCGC  357
r    TGCATGCTTCACGTG CTGCGTCTGGTAACC GCCTTGTGGCGCGGC GGTTGTCCTGCATG- CTTCGCGTG-CTGCG TCTGACCTAACTCGC  358

1-1  CCTTTTTTTTTTACT GTCGAATGCTTTGGG CACTGCAAAATGCGC GGTTTAAAGGAGCCA GCTTTAGGCTTCGCC CTCGGCAAAGACCCC  448
1-2  CCTTTTTTTTTTACT GTCGAATGCTTTGGG CACTGCAAAATGCGC GGTTTAAAGGAGCCA GCTTTAGGCTTCGCC CTCGGCAAAGACCCC  448
1-3  CCTTTTTTTTTTACT GTCGAATGCTTTGGG CACTGCAAAATGCGC GGTTTAAAGGAGCCA GCTTTAGGCTTCGCC CTCGGCAAAGACCCC  448
1-4  CCTTTTTTTTTTACT GTCGAATGCTTTGGG CACTGCAAAATGCGC GGTTTAAAGGAGCCA GCTTTAGGCTTCGCC CTCGGCAAAGACCCC  448
1-5  CCTTTTTTTTTTACT GTCGAATGCTTTGGG CACTGCAAAATGCGC GGTTTAAAGGAGCCA GCTTTAGGCTTCGCC CTCGGCAAAGACCCC  448
2-1  CCTTTTTTTTTT-ACT GTCGAATGCTTTGGG CACTGCAAAATGCGC GGTTTAAAGGAGCCA GCTTTAGGCTTCGCC CTCGGCAAAGACCCC  447
2-2  CCTTTTTTTTT-ACT GTCGAATGCTTTGGG CACTGCAAAATGCGC GGTTTAAAGGAGCCA GCTTTAGGCTTCGCC CTCGGCAAAGACCCC  446
2-3  CCTTTTTTTTTTACT GTCGAATGCTTTGGG CACTGCAAAATGCGC GGTTTAAAGGAGCCA GCTTTAGGCTTCGCC CTCGGCAAAGACCCC  448
2-4  CCTTTTTTTTTTACT GTCGAATGCTTTGGG CACTGCAAAATGCGC GGTTTAAAGGAGCCA GCTTTAGGCTTCGCC CTCGGCAAAGACCCC  449
2-5  CCTTTTTTTTTT-ACT GTCGAATGCTTTGGG CACTGCAAAATGCGC GGTTTAAAGGAGCCA GCTTTAGGCTTCGCC CTCGGCAAAACCCC  446
r    CCTTTTTTTTT-ACT GTCGAATGCTTTGGG CACTGCAAAATGCGC GGTTTAAAGGAGCCA GCTTTAGGCTTCGCC CTCGGCAAAGACCCC  447

1-1  TTTCGTCCGCTGTCT CGCTTGTCTCTTTCT CTCTCTGTACATCGC ATCGCTCTCGATCGC TCGAGATGTCGTTGT GTGGTGTGAGGGGTT  538
1-2  TTTCGTCCGCTGTCT CGCTTGTCTCTTTCT CTCTCTGTACATCGC ATCGCTCTCGATCGC TCGAGATGTCGTTGT GTGGTGTGAGGGGTT  538
1-3  TTTCGTCCGCTGTCT CGCTTGTCTCTTTCT CTCTCTGTACATCGC ATCGCTCTCGATCGC TCGAGATGTCGTTGT GTGGTGTGAGGGGTT  538
1-4  TTTCGTCCGCTGTCC CGCTTGTCTCTTTCT CTCTCTGTACATCGC ATCGCTCTCGATCGC TCGAGATGTCGTTGT GTGGTGTGAGGGGTT  538
1-5  TTTCGTCCGCTGTCT CGCTTGTCTCTTTCT CTCTCTGTACATCGC ATCGCTCTCGATCGC TCGAGATGTCGTTGT GTGGTGTGAGGGGTT  538
2-1  TTTCGTCCGCTGTCT CGCTTGTCTCTTTCT CTCTCTGTACATCGC ATCGCTCTCGATCGC TCGAGATGTCGTTGT GTGGTGTGAGGGGTT  537
2-2  TTTCGTCCGCTGTCT CGCTTGTCTCTTTCT CTCTCTGTACATCGC ATCGCTCTCGATCGC TCGAGATGTCGTTGT GTGGTGTGAGGGGTT  536
2-3  TTTCGTCCGCTGTCT CGCTTGTCTCTTTCT CTCTCTGTACATCGC ATCGCTCTCGATCGC TCGAGATGTCGTTGT GTGGTGTGAGGGGTT  538
2-4  TTTCGTCCGCTGTCT CGCTTGTCTCTTTCT CTCTCTGTACATCGC ATCGCTCTCGATCGC TCGAGATGTCGTTGT GTGGTGTGAGGGGTT  539
2-5  TTTCGTCCGCTGTCT CGCTTGTCTCTTTCT CTCTCTGTACATCGC ATCGCTCTCGATCGC TCGAGATGTCGTTGT GTGGTGTGAGGGGTT  536
r    TTTCGTCCGCTGTCT CGCTTGTCTCTTACT CTCTCTGTACATCGC ATCGCTCTCGATCGC TCGAGAAGTCGTTGT GTGGTGTGAGGGGTT  537

1-1  AGGGCTCTTGGTGGC GGATGCGAAGGAGCC CCTCCTCGCGGAGGG GCGAAACTTATACCT TGTCTGGCAAATTTT TTCTATATTTCCTTC TCGGCGCG  636
1-2  AGGGCTCTTGGTGGC GGATGCGAAGGAGCC CCTCCTCGCGGAGGG GCGAAACTTATACCT TGTCTGGCAAATTTT TTCTATATTTCCTTC TCGGCGCG  636
1-3  AGGGCTCTTGGTGGC GGATGCGAAGGAGCC CCTCCTCGCGGAGGG GCGAAACTTATACCT TGTCTGGCAAATTTT TTCTATATTTCCTTC TCGGCGCG  636
1-4  AGGGCTCTTGGTGGC GGATGCGAAGGAGCC CCTCCTCGCGGAGGG GCGAAACTTATACCT TGTCTGGCAAATTTT TTCTATATTTCCTTC TCGGCGCG  636
1-5  AGGGCTCTTGGTGGC GGATGCGAAGGAGCC CCTCCTCGCGGAGGG GCGAAACTTATACCT TGTCTGGCAAATTTT TTCTATATTTCCTTC TCGGCGCG  636
2-1  AGGGCTCTTGGTGGC GGATGCGAAGGAGCC CCTCCTCGCGGAGGG GCGAAACTTATACCT TGTCTGGCAAATTTT TTCTATATTTCCTTC TCGGCGCC  635
2-2  AGGGCTCTTGGTGGC GGATGCGAAGGAGCC CCTCCTCGCGGAGGG GCGAAAATTATACCT TGTCTGGCAAATTTT TTCTATATTTCCTTC TCGGCGCC  634
2-3  AGGGCTCTTGGTGGC GGATGCGAAGGAGCC CCTCCTCGCGGAGGG GCGAAAATTATACCT TGTCTGGCAAATTTT TTATATATTTCCTTC TCGGCGCG  636
2-4  AGGGCTCTTGGTGGC GGATGCGAAGGAGCC CCTCCTCGCGGAGGG GCGAAAATTATACCT TGTCTGGCAAATTTT TTCTATATTTCCTTC TCGGCGCC  637
2-5  AGGGCTCTTGGTGGC GGATGCGAAGGAGCC CCTCCTCGCGGAGGG GCGAAAATTATACCT TGTCTGGCAAATTTT TTCTATATTTCCTTC TCGGCGCC  634
r    AGGGCTCTTGGTGGC GGATGCGAAGGAGCC CCTCCTCGCGGAGGG GCGAAACTTATACCT TGTCTGGCAAATTTT TTCTATATTTCCTTC TCGGCGCG  635
```

**Fig. 6.14.** Nucleotide sequence comparison of ITS1 fragments of *Blatella germanica* individuals from three different populations (Mukha et al. 1999). *1–1* to *1–5* Prestage (USA); *2–1* to *2–5* Jacksonville (USA); *r* Moscow (Russia). Variable nucleotides are in *bold*

```
1 GGAA-TTTATCGTTGCCACTATGGAAGTTTTGTGGAGGAGAGAGAGATTCCCCGTGTGTG  59
  :::: ::::: :::::::::::::  ::::  :           :::::::::::::::: :::
2 GGAACTTTATTGTTGCCACTATGAGAGTTGGG---------GAGAGATTCCCCGTG-GTG  50

1 TCGTTCGTTGGTGGACGCGCGCGCGGGCGAATTCT----GTTTTTACTGGAAGGGCTT  115
  :::::::::::::::::::: ::::       ::::::::::     :::::::: ::::
2 TCGTTCGTTGGTGGACGCACGCG----GGCGAATTCTTTTCGTTTTCACTGGAAGTGCTT  106

1 TTGAGCACATTGTTTTTTACTTAGGCAACTCTTTCGT-CCGTTTCGGCGGAGAAGGTTGC  174
  ::  ::   ::::: :::::::::::::::::::::: : :::::::::::::::::::::
2 TTTAGAGCATTGATTTTTACTTAGGCAACTCTTTCATTCCGTTTCGGCGGAGAAGGTTGC  166

1 CCTTTTTATGGGT-ACGGCGCTTTTGTGTGCGCCGTTGAAAGAAGACTGACTCTAACAGT  233
  ::: : :::::::: :::::::::: :    ::::::::::::: ::      :::: : :: ::
2 CCTCTCTATGGGTTACGGCGCTTCT---TGCGCCGTTGAAATAAA---GACT-TTACCGT  219

1 CTTCGCCCAAAGGGTTTGTAGTAAGGGTCTGCATGCTTCACGTGCTGCGTCTGGTAACCG  293
  ::::::::::::::::  : :: ::::::::::::::::::::::::::::::::::: : :::
2 CTTCGCCCAAAGGGTTCGCAGGAAGGGTCTGCATGCTTCACGTGCTGCGTCTGGCA-CCG  278

1 CCTTGTGGCGCGGCGGTTGTCCTGCATGCTTCGCGTGCTGCGTCTGACCTAACTCGCCCT  353
  :::::::::::::::::::::::: ::::::::::: ::::::::::::::: :::::::::
2 CCTTGTGGCGCGGCGGTTGTCATGCATGCTTCGCTTGCTGCGTCTGACCTACCTCGCCCT  338

1 TTTTTTT-----------------------------------------------------  360
  : : :::
2 TCTCTTTATGGGTACGGCGCTTCTTGCGACGTTGAAAGAAGACTTTCTCTCGTCTTCGCC  399

1 -----------------------------------------------------------
2 CAAAGGGTTTGCAGCAGGAAGGGTCTGCATGCTTCACGTGCTGTGCGTCTGGCACCGCCT  458

1 -----------------------------------------TAC---------------  363
                                            : ::
2 TGTGGCGGGTGTCCTGCAATGCCTCGCGTGCTGCGTCTGACTACATCTCGCCCTTCTCTC  518

1 --TGTCGAATGCTTTGGGCACTGCAAAATGCGCGGTTTAAAGGAGCCAGCTTTAGGCTTC  421
    ::::::: ::::::::::::::::::::::::::::::::::::::::::::::::: :::
2 TATGTCGAAGGCTTTGGGCACTGCAAAATGCGCGGTTTAAAGGAGCCAGCTTTAGGTTTC  578

1 GCCCTCGGCAAAGACCCCTTTCGTCCGCTGTCTCGCT-TGTCTCTTACTCTCTCT-----  475
  ::::::::::::::::::::::::::::::: :::::::: : :::::::: :::::::
2 GCCCTCGGCAAAGACCCCTTTCGTCCGCCGTCTCGCTGTCTCTCTCTCTCTCTCCATC  638

1 -GTACATCGCATCGCTCTCGATCGCTCGAGAAGTCGTTG-------TGTGGTGTG-AGGG  526
  ::  :::::::::::::::::::::::::::::: : :::: ::::::: :: : ::::
2 TGTCCATCGCATCGCTCTCGATCGCTCGAGCTGCCGTTTGCATGGCTGTGTTGAGGAGGG  698

1 GTTAGGGCTCTTGGTGGCGGATGCGAAGGAGCCCCTCC--TCGCGGAGGGGCGAAACT--  582
  ::  ::: :  :::::  :::::::::::::::::::::: :::::  ::::::::: ::::::::
2 GTAAGGACGCTTG-TGGCGGATGCGAAGGAGCTCCTCCGCTCGCGGAGGAGCGAAACTAT  757

1 ---TATACCTTGTCTGGCAAATTTTTTCTATATTTCCTTCTCGGCGCCTTGCGCGTC-GT  638
  :::::::::::::::::: :    ::::::::::::::::::: :::::::::: ::
2 TTTTATACCTTGTCTGGCAAATTTCCTTACAATTTCCTTCTCGGCGCATTGCGCGTCCGT  817

1 TTTGGATTTATATGAACTATCA-T  661
  ::::::  :: :::::::::::   :
2 TTTGGAATTGTATGAACTATCTGT  841
```

a

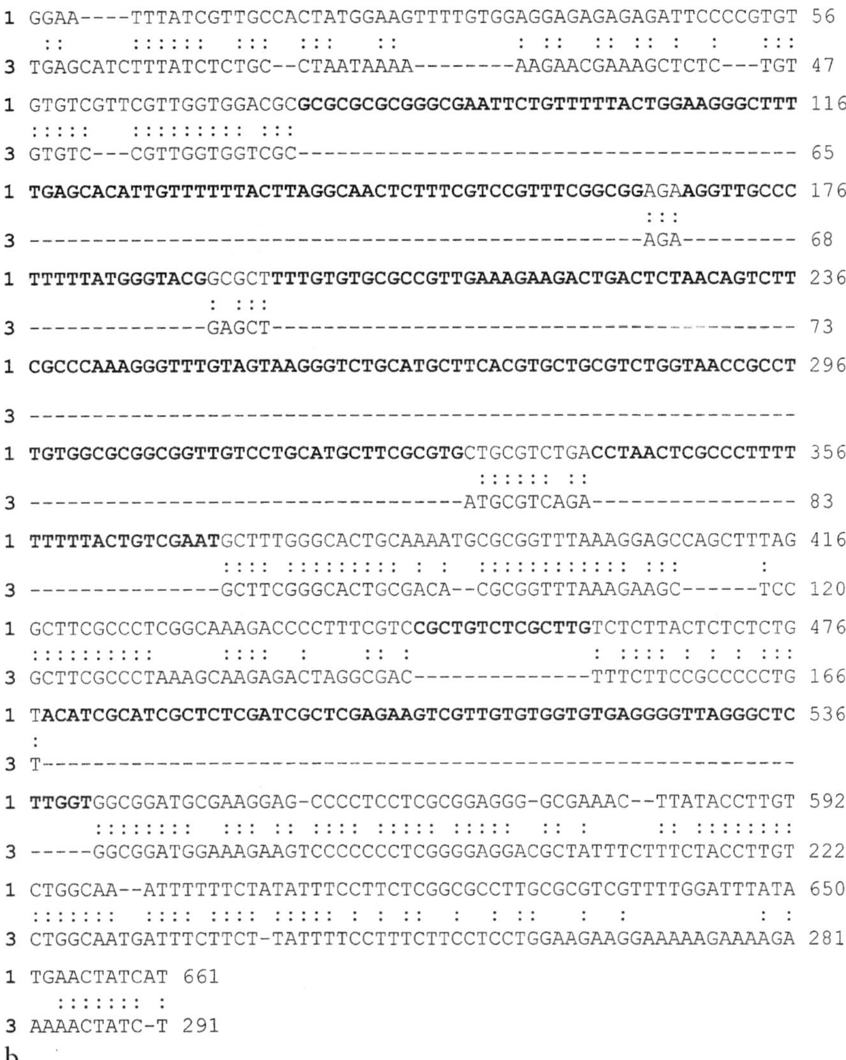

Fig. 6.15. Pairwise interspecific comparison of ITS1 nucleotide sequences (Mukha et al. 1999). **a** *B. germanica* (*1*), *B. lituricollis* (*2*); **b** *B. germanica* (1), *B. vaga* (3). Long insertions (deletions) are in *bold*

multigene family, but involve flank genes of the cluster to a lesser extent. Precisely the flanking, functionally less significant sequences of the ribosomal gene cluster are likely to provide material for forming species-specific structural variants.

Many taxonomic groups have been shown today to perform an important role of this kind in reorganizing genetic material, including those that have recently been assigned a special place in underlining qualitative differences between evolutionary processes in the animal and plant kingdoms. The most important biological significance of these rearrangements is that they transfer in a leap all genes in a genome or their certain part into a fixed heterozygous state; consequently they provide the individuals with the advantage of a qualitatively different adaptation level, freeing a future population from segregational load. At the same time this signifies increased reliability in preserving and transmitting information to generations about the vitally important functions that reflect the uniqueness of a new species.

There is a drawback in this scheme of microevolution transacted by large genome reorganizations. We stressed earlier that this kind of variability is under the constant control of purifying selection. How, then, can speciation take place?

To solve this task a special state of the environment must be postulated in which macromutations may be neutral or acquire certain selective advantages.

We suggest that these rare genomic reorganizations occur in different species all the time but can only survive when environmental shifts such as the demolition of historically formed interpopulation (and interspecies) barriers greatly raise the levels of population genetic polymorphism and the heterozygosity of the individuals that compose them. "Polymorphism produces more polymorphism" (Ohno 1970b). In this case, there may be such gene combinations, which would be considered incredible in a previous phase of stable existence of the species. This speculation made about three decades ago (Altukhov 1974) now seems to find support in the discovery of hybrid zones, which finding was made possible by the advances in DNA technologies, and in the demonstration of the fate of phylogenetic lineages during the processes of post-glacial colonization (phylogeography; Avise et al. 1987; see also Hewitt 2001). Although these zones of interspecific contacts are often regarded as "evolutionarily promising" (Hewitt 2001), many of them in fact provide secondary, recent contacts generated by extreme anthropogenic pressure on natural habitats. Under these conditions the volume of the genetic load increases markedly and a species either dies or is transformed into a new species. Therewith the usual variability is "extinguished" as a result of the systems of genetic polymorphism changing into a monomorphic state. To what extent these reorganizations are random or can be regarded as conditioned by the preceding history of a taxon's development is an open question.

In his article on "The genetics of speciation at the diploid level", Hampton Carson (1975, 1982) develops views on speciation that are close to ours. According to him, two systems of genetic variability, "open" and "closed", should be distinguished in bisexual species. The first consists of freely segregating loci that do not substantially affect viability (polymorphism of different types, subspecies or clinal variation); the second one is made up of internally balanced and co-adapted gene blocks (supergenes) that influence fitness so powerfully that any reorganization of them will eliminate natural selection (Fig. 6.16). Such blocks, protected from crossing over by inversion or simply by strong epistatic gene interactions, vary among species but never within a species. Speciation is implemented by reorganization of a genetically closed system, and new species originate from one or several founder individuals.

The parallelism of Carson's views and our own is self-evident, they are only different in terminology, research methods, the material selected, and in certain aspects connected with interpreting the state of the environment during speciation (Fig. 6.17). One should only add that at a time when some biologists were casting doubt on these ideas in one form or another, Powell (1978) carried out an extensive experiment on modeling this scheme of speciation. For the first time in experimental population genetics, and in only 50–60 generations, he achieved stable reproductive isolation among initially freely crossing populations (see Powell 1997). Nevertheless, Powell's earlier data have raised some doubts (Rice and Hostert 1993).

In recent years, many new data favoring unorthodox concepts of speciation and evolution have been obtained. The studies by Susumu Ohno and his colleagues (Ohno et al. 1968; Ohno 1970a,b) occupy a special position in these works by introducing abundant cytological and biochemical findings to support the theory that the tetraploidy of genomes has played an important integrating role in the evolution of vertebrates. Notwithstanding the skepticism of Mayr (1968; see Sect. 6.1), amphidiploid origin has been rather conclusively shown for at least two of three extant polyploid groups of animal species – three families of the herring and two genera of carp fishes (Bender and Ohno 1968; Wilkins 1970; Altukhov et al. 1972; Bushuyev 1973; Altukhov 1974; Omel'chenko and Gerasimenko 1981; Tsigenopoulos

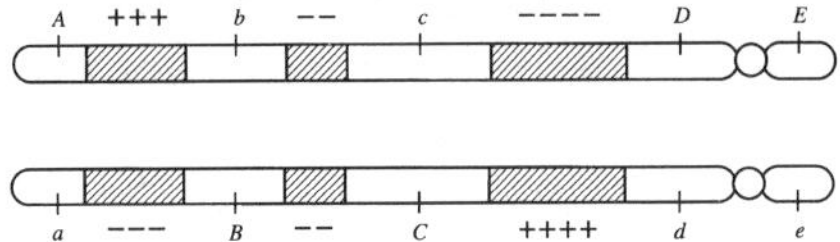

**Fig. 6.16.** Model of "open" and "closed" genetic systems. A pair of homologues that are heterozygous at 12 out of 14 loci (Carson 1975). The result of crossing over within an open system (*unhatched*) sometimes produces zygotes with relatively high fitness. Crossovers in a closed system (*hatched*) are non-adaptive in the majority of cases. The *letters* designate different genes or polygenes; *pluses* and *minuses* represent internal balanced complexes in which any "plus–minus" combinations are weakly adapted within the compass of any one block. Genetic balance preserves the majority of blocks from fixation in a homozygous state (Reproduced with permission from the University of Chicago Press)

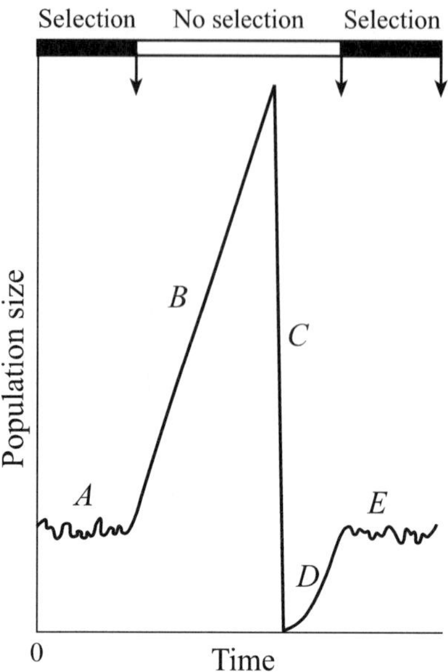


**Fig. 6.17.** Diagram of speciation based on the reorganization of a "closed" genetic system (Carson 1975). The population after the selective phase (*A*) passes through stages of flush (*B*) and crush (*C*). After surviving a catastrophe, the founder individuals (*D*) produce a new population, again stabilized by natural selection (*E*). Reorganizations of the closed genetic system of a species which are "compatible" with ontogenesis are only possible in the phase when selection weakens, when one or several "discordant" individuals may survive. (Reproduced with permission from the University of Chicago Press)

et al. 2002) – whereas tetra- and octoploid South American frogs from the family Ceratophrididae are autopolyploids (Beçak et al. 1966, 1967). A discovery of tetraploid green toads *Bufo viridis* in Mongolia suggests that polyploidy takes greater part in speciation in amphibians than has been previously thought.

Hybrid nature was demonstrated for parthenogenetic lizards of the genera *Cnemidophorus* (Neaves and Gerals 1968; Neaves 1969), *Lacerta*, and *Lepidodactylus* (Darevsky and Kulikova 1964; Darevsky and Danielyan 1969; Darevsky 1993; Kupriyanova 1997), and for several gynogenetic fishes of the family Poeciliidae (Schultz 1961, 1966, 1967, 1969; Rash et al. 1965; Rash and Prehn 1969; Volobuev et al. 1993). Hybridization, parthenogenesis (gynogenesis) and polyploidy were shown to be related in animal evolution (Astaurov 1969; Schultz 1969, 1973; Kupriyanova 1997; Alves et al. 2001).

In recent years, polyploid species series have been discovered at least in six orders and several families of anadromous and fresh-water fishes (for review, see Vasilév 1985; Altukhov 1989a; Tsigenopoulos et al. 2002): Acipenseriformes, Salmoniformes, several families from the order Cyprini-

formes (Cyprinidae, Catostomidae, Cobitidae), Siluriformes, Poeciliifor-
mes, and Atheriniformes.

The family Cyprinidae includes over 2,000 species widely spread in Europe, Asia, Africa, and North America. Just in Africa there are more than
300 species of the genus *Barbus*, represented by three ploidy levels: diploids,
tetraploids, and hexaploids. However, a recent analysis of the primary structure of the mitochondrial cytochrome b gene and a set of mtDNA haplotypes has shown that most species groups are erroneously assigned to the
genus *Barbus*. For instance, South African tetraploids are a monophyletic
group, which is different from the tetraploids from the Euro-Mediterranean
region (*Barbus. sensu stricto*). African hexaploid species are also a monophyletic (and young) group, whereas African diploid species are para-
phyletic and do not belong to *Barbus, sensu stricto*. Thus, Tsigenopoulos et
al. have discovered that polyploidy in African Cyprinidae is of a multiple
and independent origin; in the view of these authors, the species in question
are likely to be allotetraploids, having originated according to the reticular
speciation model.

Another striking example of this ("closed") speciation model is presented by the complex of diploid and polyploid forms in the Iberian min-
now *Leuciscus alburnoides*, which exhibits a reproductive mode similar to
those recorded earlier in Poeciliopsis. The parental genomes are cyclically
lost, substituted, or augmented. Note that hybrid *L. alburnoides* males are
fertile and can initiate the process of tetraploidization when a diploid clonal
sperm fertilizes a diploid egg. This amazing scheme of combinatorial speciation, based on the results of experimental crosses, is presented in Fig. 6.18
(Alves et al. 2001).

A survey of hybridogenic ("reticulate") speciation in vertebrates (Borkin
and Darevsky 1980; Kupriyanova 1997) describes the most stringently
verified cases of this kind and examines the corresponding evolutionary
scheme. It includes the following stages:

1. A diploid form develops in interspecific hybridization, which passes over
   into unisexual reproduction through gynogenesis (fishes), "hybridogenesis"[1] (fishes and excaudate amphibians), or parthenogenesis (caudate
   amphibians and reptiles).

---

[1] Borkin and Darevsky (1980) proposed the term "creditogenesis" instead of "hybridogenesis" (Schultz 1969) to denote a unique type of reproduction discovered in diploidal unisexual
forms of *Poeciliopsis* fish. As with gynogenesis, participation of males of another species
is necessary for reproduction, but in the case of hybridogenesis the genomes of the sperm
and egg cell combine and a form develops having characteristics of both parental forms – *P.
monacha-lucida*. These hybrid individuals are female without exception. During ovogene-
sis, paternal chromosomes from *P. lucida* are selectively eliminated, with the result that only
the maternal *monacha* genome remains in the mature egg cell at the moment of pairing.
Further crossing of *P. monacha-lucida* × *P. lucida* again produces hybrid individuals and so
on for every new generation. Thus the paternal genome is "lent" to one generation.

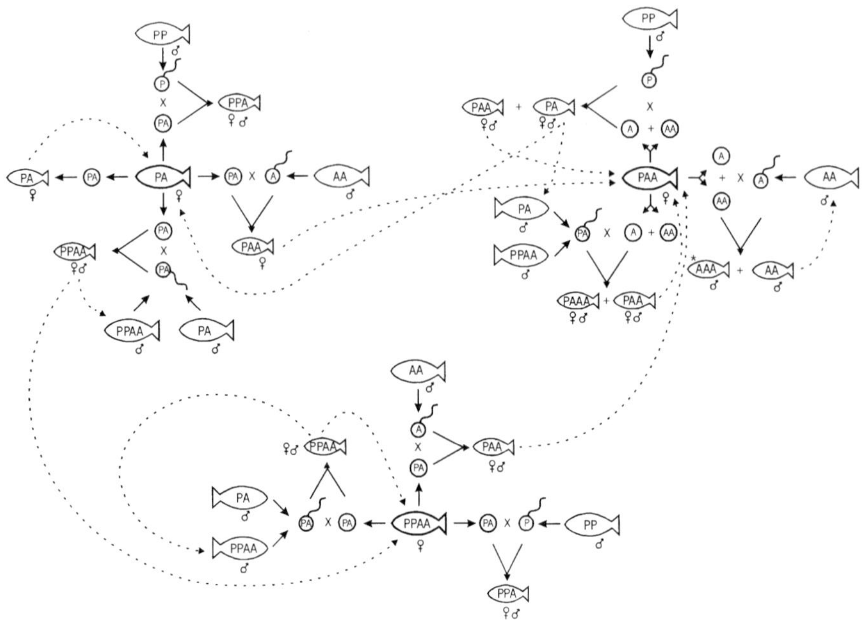

**Fig. 6.18.** Presumable relationships in the *Leuciscus alburnoides* complex inferred from the results of experimental crosses. The form obtained in the experiment but never observed in natural populations is marked by an *asterisk*. A, C and P are genomes, in two of which (*C* and *P*) the origin is known, but an ancestor with genome *A* is not known (Alves et al. 2001, p. 382)

2. On crossing these hybrid forms with one of the parental (or closely related bisexual) species, a unisexual triploidal form develops which is reproduced gyno- or parthenogenetically.

3. In its turn, the crossing of a triploid with one of the closely related bisexual diploidal species leads to the formation of a tetraploid which is capable of reverting to a bisexual mode of reproduction (Fig. 6.19; see also Vasilyev 1985).

Of course, allotetraploidal species can arise, which also avoid triploidy when hybrid forms produce non-reduced gametes, and the crossing of allotetraploidal males having an unbalanced set of chromosomes ($X_1X_2X_2Y_2$) with females $X_1X_2$ results in the progeny $X_1X_1X_2X_2$ (females) and $X_1X_1X_2Y_2$ (males); evidently, crossing within this progeny should lead to the formation of a stable allotetraploidal species (Viktorovsky 1969; Borkin and Darevsky 1980). In all these and other investigations that led to the discovery of hybrid parthenogenetic species, a decisive role is played by electrophoretic study of proteins which, being monomorphic in several cases, reliably differentiate parental forms. A second, no less important factor is

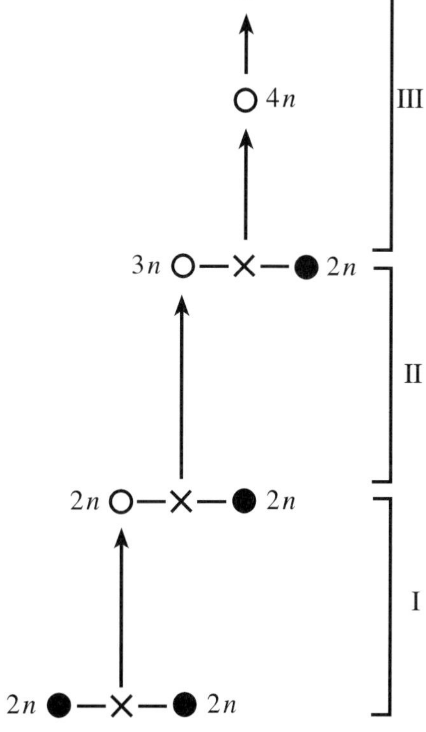



**Fig. 6.19.** Diagram of hybridogenic speciation in vertebrates (from Borkin and Darevsky 1980). *I–III* Stages of consecutive hybridization leading to an increased degree of ploidy of a hybrid form. *Dark circles* Bisexual species; *light circles* unisexual hybrid species

the exceptionally high level of heterozygosity (at many or even all of the loci studied) discovered in parthenogenetic species and forms.

Taken all together, the discoveries made during the past 20–30 years in the field of biochemical population genetics and the comparative genetics of species compel one to admit that speciation, and even more so, the macro-evolutionary process cannot be reduced merely to the substitution of alleles at gene loci that already exist, and that reproductive isolation is not a byproduct of the lengthy process of population divergence. On the contrary, reproductive isolation is the most important prerequisite for the emergence of a new species, as has long been clear from the well-argued role of certain chromosomal aberrations that, although not leading to intraspecific polymorphism, nevertheless mark the level of a species (White 1954). Of major importance in this connection is the fact that *new species are homozygotes for such macromutation, which powerfully reduces the fertility of its carriers when in a heterozygous state* (see also Chadov 2001).

The fixation potentiality of these large chromosomal aberrations is particularly great when a species is subdivided and when the $N_e$ value of the populations which constitute it is restricted (Wright 1940, 1941). It is precisely in a population system of this kind that the probability is greatly increased of two individuals, heterozygous upon mutation, meeting in one of the peripheral demes and producing normally viable homozygotes – a new species, which becomes virtually isolated reproductively from its ancestral species for only two steps[1].

Because this scheme for the origin of new species may now be regarded as demonstrable fact, there is no need for further examples. It is only necessary to stress the two following important points.

First, this "quantum" speciation, as is now becoming clear, is quite widely distributed in nature (White 1968, 1973, 1978; Fredga 1977; Bush et al. 1977; Viktorovsky and Glubokovsky 1977; Viktorovsky 1979; Lande 1979; Stegnii 1979, 1980a,b, 1984, 2001; Vorontsov 1980, 1999; Vorontsov and Lyapunova 1984; Dutrillaux 1986; Korochkin 1999).

Second, a situation involving finely subdivided population structure is indeed optimal in evolution, not in the sense of its transformation through shifts of allelic frequencies, but because of the presence of strongly isolated demes in which there is a high degree of probability that two individuals will meet that are heterozygous in rare chromosomal mutations of the pericentric inversion type, reciprocal translocations or Robertsonian fusion. Despite their reduced fertility, all these individuals are capable of crossing with each other and bequeathing perfectly viable, fertile progeny.

Thus, what occurs at an above-species level of evolution is not only or, rather, not so much a process of the appearance of new genes with new functions, as a rapid reorganization of genetic material with the subsequent development of new systems of gene interaction (and regulation) at post-transcription and post-translation levels (Korochkin 1983, 1984, 1985, 2001; see also Akam 1998; Brakefield and Zwaan 1999; Schilthuizen 1999; Gibson and Palsson 2001).

A diagram appropriated from Britten and Davidson (1971) and summarizing copious findings in literature, graphically represents the variability scale of the genome size in the *Eucariota* (Fig. 6.20). There is no need to show that these kinds of interspecific differences result, not from allelic gene substitutions, but from tandem duplications, hybridization, and polyploidy. Insofar as this diagram only embodies data up to 1970, it should be

---

[1]Discussing this circumstance in connection with the indisputable role of pericentric inversions in primate speciation (Bush et al. 1977) and the discovery of homozygosity resulting from corresponding mutations in the progeny of incest pairs in a modern human population (Betz et al. 1974), Vogel and Motulsky (1997, p. 519) remark in their capital work: "Do all human beings have in common an ancestral couple? The myth of Adam and Eve as the ancestral couple of mankind thus may even have a scientific basis."

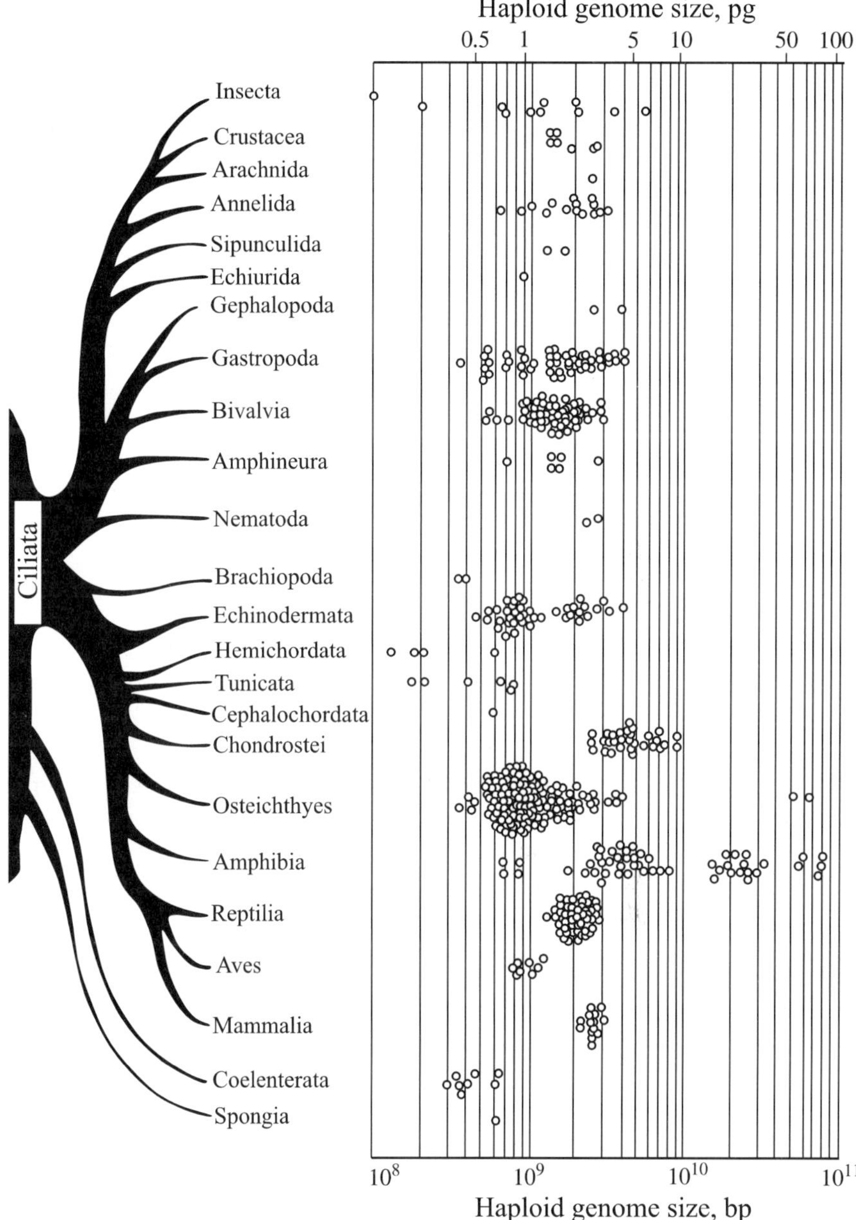


**Fig. 6.20.** Variability of genome size in different animal groups (Britten and Davidson 1971). (Reproduced with permission from the University of Chicago Press)

supplemented by more recent material (Hinegardner 1976; Ginatullin 1984; see also Petrov 2001). Analysis of the later works shows that the variability of the genome size in different taxonomic groups increases.

All these new facts mirror the considerable diversity of the animal world and undoubtedly point to evolutionary paths present in zoological species regarded until quite recently as being highly problematical or even simply impossible (Vorontsov 1966; Mayr 1968; Dobzhansky 1970), despite their dissemination in the plant kingdom (see Takhtadzhan 1983). Developing and improved techniques in systematics justify one envisaging, in the near future, even wider evidence of the part played by these genetic reorganizations in the evolution of zoological species. Clearly, even the new facts available cannot be accommodated easily by the synthetic theory of evolution (see, for example, Mayr 1968, 1974; Dobzhansky et al. 1977; Vorontsov 1980, 1984, 1999) whose main postulates, as is well known, are:

1. A population is the main unit of the evolutionary process.

2. Genetic polymorphism within populations is evidence of an ongoing, continuously operating evolutionary process.

3. The difference between a species and variety is in degree, not in essence; or, expressed in other words, all the traits used to demarcate a species are subject to geographical variability (Mayr 1968, p. 270).

Every concept consists of a system of interrelated views. A contradiction of facts in only one of the branches of a logical chain may provide the prerequisite for constructing other hypotheses. The facts at our disposal have enabled us to investigate these by the criteria of all three principal postulates and to identify the difficulties that stand in the way of their acceptance. The deduction of the structural–functional duality in the organization of a eukaryotic genome has special significance. Whereas by "preserving" the identity of an individual species, families of monomorphic genes betoken cardinal functions whose changes accompany speciation, polymorphic genes, because of their broad variability, determine only the secondary adaptive characteristics of a species.

Interpreted in this way, population polymorphism is not evidence of continuously ongoing evolution but a universal strategy by nature to ensure the preservation of species, based on the constant interaction of genetic variation, random drift, and natural selection in a normally fluctuating environment. Evolution and speciation, out of phase with the long term stability of a species, involve large shifts in the natural environment and corresponding genome reorganizations.

This conclusion, which we reached over 30 years ago (Altukhov and Rychkov 1972), recalls the concept of "punctuated equilibria" that appeared at about the same time (Eldredge and Gould 1972), and it has been widely discussed since (see, for instance, Grant 1981; Maynard Smith 1982). We wrote then that "taking into account the simultaneous existence of the phenomena of polymorphism and monomorphism, it is possible to reconstruct the evolutionary process as periods of the transformation of species by qualitative genome reorganization alternating with periods of lengthy stability. This stability is ensured by the organizational characteristics of vitally important monomorphic properties and successful adaptations, formed by natural selection on the basis of genetic polymorphism in populations" (Altukhov and Rychkov 1972 p. 228).

I shall point out, in concluding this chapter, that these views of speciation resemble the well-known concepts of G. de Vries (1904, 1912), R. Goldschmidt (1948, 1952), and B. Ushakov (1958, 1959a,b, 1964). All these earlier works discuss the problem of duality in the structural–functional organization of genes or an organism's traits and, with differing degrees of logical completeness, best expressed by Goldschmidt, postulate a qualitative difference between the actual evolutionary process and adaptive intraspecific differentiation. However, I consider that my treatment of the subject bears only an external similarity to these concepts, and represents a natural construction based on many years' work experience by our group on the biochemical genetics of populations and species. This experience convinces us that one of the most important characteristics of a eukaryote genome – the duality of its structural–functional organization – is directly reflected by the coexistence of the two real phenomena of polymorphism and monomorphism. In my opinion it is impossible to understand the mechanisms of the evolutionary transformation and adaptive stabilization of species if these phenomena are examined separately.

The findings that have been examined above would also appear to answer criticism of the speciation model set forth by us (Starobogatov 1975; Malinovsky and Mina 1976; Mina 1977; Severtsov 1981). Neither the "diversified directional population changes" within a species nor the difficulties in interpreting the "saltational" origin of a new species are ignored by our model, but are incorporated in it, taking stock of new data, both on the qualitatively different level of genetic stability of population systems in nature and the character of the difference among species in the functionally most loaded (monomorphic) part of the genome. I also hope that the material that has been presented above will help to remove several criticisms by Coulthart and his co-authors (Coulthart et al. 1984) by supplementing my previous reply to them (Altukhov 1982b, 1985b). It should be particularly emphasized that the decisive role performed in speciation by large chromosomal reorganizations, encountered in extremely low frequency in the marginal areas of a species, has also been fully accepted by Sewall Wright, one of the creators of the theory of gradual evolution described in terms of the dynamics of gene frequencies.

"It is necessary to conclude from the extreme rarity of translocations within species, but great frequency of such differences between closely related ones, that a large proportion of species have their origins in the partial reproductive isolation provided by the fixation of a translocation in a very small colony.

Multiple selective peak shifts are specially likely to have been established in the portion of the range of the parent species within which a translocation becomes fixed and also during the early history of the new species while still broken up into small transient colonies...."

And further: "This situation gives rise to a curious similarity of the course of evolution from the shifting balance process [i.e., Wright's theory] to that supposed to appear under de Vries' mutation theory. In both cases, there would be an association between the 'chromosome repatternings' of Goldschmidt that provide the basis for reproductive isolation and for the major phenotypic steps" (Wright 1977, p. 473; see also Wright 1980).

Although I completely agree with Wright's conclusion about the optimal nature of finely subdivided population structure for macromutation fixation, I cannot accept his attempt to regard this type of speciation as in no way contradicting the theory of gradual evolution. If we aim at avoiding logical error, we should admit that this kind of speciation is a saltatory event that has nothing in common with gradualism, which links the origination of a new species with the slow accumulation of adaptive allelic replacements.

In conclusion, I would like to stress that I do not consider these views to be in any way deliberately directed against the synthetic theory of evolution (see, for instance, Dubinin 1986b). I am profoundly convinced that to this day the synthetic theory of evolution is the most developed and substantiated evolutionary concept (see a recent book by Tatarinov 1987; Vorontsov 1999). At the same time, however, because it is a probabilistic concept, it cannot incorporate the phenomenon of species invariance, and the study of genetic processes at a population level, as we have seen, does not always lead to evolutionary conclusions.

Taking into account the exceptional role played by theories of evolution in the human activity of transforming nature, in the face of new problems, we should admit that we are still far from achieving a quantitative account of the evolutionary process as a whole and understanding its main driving forces. Lewontin (2002) has recently come to the conclusion that the causes and bases of the evolutionary process will forever remain a mystery. Nevertheless, the new data that have been examined from the sphere of population biochemical genetics and the speciation model constructed upon its basis have a certain practical as well as cognitive significance. This applies first and foremost to interpreting the character of the genetic processes in modern human populations and analyzing our potential role in the evolutionary process under conditions of a drastically changing environment. If there are monomorphic or, in Carson's terminology, closed genetic systems of a universal character, then clearly the existence of any species as a whole is connected with the stability of precisely these systems, which in the norm should be closed to the destructive effect of mutations or the recombinations of genes. It is evident that under the conditions of increased mutation rates or growth in the intensity of any other influences leading to the destruction of these systems, extremely negative and far-reaching consequences may be exhibited, ultimately capable of leading to the elimination of a species or its transformation into a new species. This kind of conclusion is particularly relevant to our times, when mankind is experiencing a qualitatively new phase of its history, shown by the dramatic changes of migration patterns, destruction of population structure, and possibly increased rates of recombination and mutation of genes.

In the commonly held view, every evolutionary transformation should be regarded as a progressive occurrence. However, I do not believe that it is possible for the sake of such progress to abandon genetic and social continuity with generations of our remote and close ancestors.

On the other hand, we should understand what is happening to population gene pools in the usual Mendelian genes which encode the structure of proteins or erythrocytic antigens. After all, if this polymorphism is only relatively neutral and has some degree of adaptive significance at the level of individual genes or their communities, then one can also understand those negative effects that we may encounter in our practical activity, ignoring that huge hereditary diversity, whether it be in agriculture, the industrial exploitation of natural species, or formulating the scientific bases of preventive medicine. All this constitutes the circle of those questions that merge directly within the context today of the problem, "Man and the biosphere", and which indubitably are clearly population-genetically oriented. The principal aim of the next chapter will be to examine this aspect.

# 7 Population Genetic Aspects of the Problem "Man and the Biosphere"

If we take it as evident that in the last 10,000 years the biosphere has not undergone sharp changes but only accumulated local deformations, then we must admit that today we are witnessing a turning point in the evolution of nature and the state of the environment.

Advanced civilization has brought the human race to the brink of ecological catastrophe. Such adverse processes as depletion of the ozone layer, the greenhouse effect, and acid rain are no more than the most obvious evidence of the imminent hazard. Another sign of the times is the increasing reduction of species diversity. During this century, human activity has led to the extinction of nearly 25,000 higher plant species and over 1,000 vertebrate species. Hundreds of unique breeds of domestic animals may share the same fate (World Conservation Strategy 1980; Pavlov et al. 2001). According to estimates by well-known specialists, the biosphere may lose up to 10–15% of its constituent species before we are very deep into the 21st century (Wilson 1988). The rate of extinction caused by anthropogenic influence exceeds everything that is known from paleontological records. The only possibility of preventing (more precisely, moderate) the ecological catastrophe occurring before our eyes is to revise and radically change the whole strategy of human–nature interactions. To make efforts in this direction, however, it is necessary to realize that adverse factors responsible for the increasing reduction of biological diversity are not limited to environmental pollution and destruction of habitat. Another less obvious factor is *the incompetent approach to commercial exploitation of natural resources, disregarding the genetic subdivision of species and the structure of intraspecific hereditary variation.* Negative effects accompany the artificial reproduction of biological resources, as well as their commercial use. A typical example is the situation in fishing and agriculture (Altukhov 1995, 2001). These phenomena can be revealed only by *monitoring* intraspecific genetic diversity. That is why I focus attention on the problem of monitoring.

Although the English term "monitoring" has been widely used in international genetic literature, principles of corresponding studies have not yet been developed adequately. This applies, in particular, to judicious choice of *the reference point* necessary for evaluating the changes observed and understanding their significance. The purpose of this chapter is to fill the gap in our knowledge about these principles. I will specify the key problem of genetic monitoring, review theoretical approaches to its solution, and give examples of monitoring for populations of two types: native and those exposed to various anthropogenic influences. A forecast concerning the negative consequences of anthropogenic pressure on the biosphere and practical measures for preventing such consequences will be described in the final section.

## 7.1
## The Problem of Genetic Monitoring
## and a Theoretical Approach to Its Solution

Genetic monitoring is *the long-term surveillance of the state of population gene pools intended to evaluate and predict their temporal and spatial dynamics and to determine the limit of permissible changes in their state.* However, any prediction should necessarily be based on the concept of the *norm* (*normal state or normal process*). Only this approach gives us an essential reference point and allows us to understand mechanisms underlying the negative human impact on populations, species, and entire ecosystems.

According to the theory of population genetics, there are four basic processes that, either separately or in interaction, determine evolutionary changes in populations: (1) random genetic drift, (2) gene migration, (3) gene mutations, and (4) natural selection.

*Random genetic drift* comprises the stochastic changes in gene frequencies that occur in a series of consecutive generations because the size of any real population is finite. An especially important fact is that the *genetically effective population size* ($N_e$) is virtually always much lower than the total ($N_t$) or reproductive size ($N_r$; see Chap. 1).

Genetic drift leads to a decrease in heterozygosity and, as a result, to inbreeding ($F$). The latter increases in a series of generations in the proportion $(1/2N_e)^t$, where $t$ is the number of generations from the initial time $t_0$ corresponding to heterozygosity $H_0 : F_t = 1 - (1 - 1/2N_e)^t$. Correspondingly, $H_t = H_0(1 - 1/2N_e)^t$.

These simple formulas show that the intensity of genetic drift, leading to the reduction of gene diversity and, most often, to degradation of populations, is inversely proportional to $N_e$: the lower the $N_e$ value, the more intense the genetic drift, and vice versa. This circumstance is critically important for animal and plant breeding, and specialists organizing the process of reproduction take special measures to prevent negative consequences of consanguineous crossings (Falconer 1960). In natural (native) populations, the effects of random drift are compensated by gene *migration* of a certain intensity ($m$). In other words, natural populations are usually nonhomogeneous and represent historically formed *systems of subpopulations* simultaneously affected by both random drift and migration of genes, processes that counterbalance one another (Chaps. 3, 6). Thus, under normal conditions, the negative effects of inbreeding are eliminated, and $N_e m$, an important parameter of population structure, is a measure of the absolute intensity of gene migration per generation.

According to Wright (1943a,b, 1951), this balance for the island model of a subdivided population is estimated from the level of structural inbreeding $S$ relative to the entire subdivided population $T$, so that $F_{ST} = 1/(4N_e m + 1)$ or, more precisely, $F_{ST} = (1 - m)^2/\{2N_e - [(2N_e - 1)(1 - m)^2]\}$ (Chap. 1).

As it was shown in Chap. 5, the pressure of natural selection, acting simultaneously on a set of independent gene loci, plays an important role in the microevolution of a population gene pool. If the species' structure is not disturbed, the genetic process in a population system corresponds on average to a selectively neutral one. This was estimated by $F_{ST}$ statistics by S. Wright (or $G_{ST}$ statistics by M. Nei; Chap. 1) and the $F_{ST_e}$ value, expected from the parameters of equilibrium population structure, turns out to be identical to $F_{ST_0}$, obtained empirically through averaging allele frequencies for a set of various genes. Along with this, the average coefficient of spatial genetic differentiation is formed by a certain varying in time–space proportion of gene loci groups that: (1) are selectively neutral; (2) have experienced the pressure of stabilizing selection; (3) have experienced the pressure of diversifying selection.

Further, we will see that every appreciable anthropogenic influence that disturbs population structure quickly changes the proportion of different groups in the gene pool, and the empirical $F_{ST_0}$ begins to differ from the one expected for a selectively neutral process.

Wright's $F_{ST}$ expression, as the criterion of genetic subdivision in the population and, simultaneously, of the level of inbreeding in the subpopulation, has high biological significance. It reflects a certain balance between differentiation and integration of gene pools and, importantly, become *self-regulated* when the population approaches the stationary state. Under such conditions, $N_e$ and $m$ are coupled by negative feedback: when the effective size of subpopulations comprising the population system decreases, the intensity of gene immigration increases, and vice versa.

This nonrandom structure of migrations was revealed for the first time in studies of the genetic processes in an experimentally subdivided *Drosophila melanogaster* population conforming to Wright's island model. It was found that the smaller the size of "island" subpopulations, the more intense the gene immigration from the "continent", and vice versa (for details, see Chaps. 4, 6). The same phenomenon was subsequently observed for natural populations of other species (Altukhov and Salmenkova 1994). Such self-regulation is equivalent to the maintenance of a stable ratio between homo- and heterozygous genotypes, i.e., of the balance between inbreeding and outbreeding. However, anthropogenic influences leading to significant changes in the intensity or direction of gene migrations or in the $N_e$ value can upset this evolutionarily established balance, which is characteristic for the native self-reproducing population systems. For example, damage inflicted to breeding grounds results in the reduction and fragmentation of the reproductive range, usually followed by a drastic decrease in the effective population size and the rate of gene exchange. In such cases, an increase in interpopulation gene diversity ($G_{ST}$) and a decrease in intrapopulation gene diversity ($\overline{H}_S$) should be expected. By contrast, excessive mixing and interaction of previously isolated gene pools may result in manifestations of outbreeding, i.e., in decreased viability of hybrid combinations. Such effects are particularly marked in respect to gene complexes that form the basis of the adaptive genetic structure of species and are linked with adaptive morphophysiological traits and properties (Altukhov 1984; Templeton 1986). Here, natural selection, which is usually regarded as the most systematic and important factor in evolution of populations, again becomes prominent. Viability of populations depends directly on the adaptive values (fitness) of genotypes. If the fitness of heterozygotes is increased (overdominance), each population has to "pay" for its adaptation to the particular environment by segregation of less viable homozygotes (the so-called segregational genetic load; Chap. 1). However, the cost of adaptation normally proves to be "acceptable" because the ratio of homo- and heterozygotes is autoregulated via the structural parameter $N_e m$ (see above) and maintained at a stable level. Moreover, this level is *optimal in the sense that both decreased and excessively increased heterozygosity is unfavorable for normal functioning of a population* (Chap. 5).

The concept of optimal gene diversity as a prerequisite for successful existence of populations in a normally varying environment is particularly important in view of their systemic organization. Knowing *the ratio between intra- and interpopulation components of heritable variation* under the conditions of normal reproduction or immediately before a certain anthropogenic influence, we have the unique possibility of organizing genetic monitoring of a population system, taking into account its gene pool structure. To obtain reliable information, however, the following requirements should be met. First, the population should be well-defined in terms of natural history and geography. Second, it is necessary to describe the distribution of the subpopulation structure of the system *in time* and *space*, because these two factors are closely interrelated in the stationary genetic process. Third, the material should be sampled in such a way as to comprehensively characterize the subpopulation structure by a complex of traits selected in advance.

Such traits should be selected taking into account the possibility of combining the methods and approaches of *population* and *quantitative* genetics, i.e., of analyzing both mono- and polygenic traits (Altukhov 1983; Zhivotovsky 1984). *In a suboptimal environment or under sufficiently strong artificial selection, nonrandom correlations between individual heterozygosity for a number of loci and the values of adaptive quantitative traits are observed* (Altukhov 1989a,b; Chap. 5).

Thus, in addition to properly organized sampling (even distribution of samples in space and time), the analysis should also involve: (1) data on the demographic structure of each population, including sex, age, sex ratio, size, migrations, etc.; (2) data on body weight, size, and proportions of individuals under study; and (3) evaluation of individual genotypes for the maximum possible number of polymorphic gene loci, including various protein systems and, if necessary, systems of DNA polymorphism (both nuclear and non-nuclear). This information makes it possible to analyze the distribution of poly- and monogenic traits and their joint variability, to estimate the ratios among components of gene diversity and understand the state of the genetic process in a certain population, and to estimate the contribution of random genetic drift, gene migration, and selection, either separately or in combination (for details, see Chaps. 3–5).

I omitted only one factor – *mutation pressure*. As the probability of mutation for individual genes is negligible ($10^{-5}$–$10^{-6}$ per gene per generation), the effect of this factor may be disregarded under normal conditions. At the same time, pollution of the environment with radionuclides and chemical mutagens makes *monitoring of the mutation load* an important problem, especially as concerns human populations (Altukhov 1981b, 1983, 1989a,b; Altukhov and Kurbatova 1984; Altukhov et al. 1985).

This problem as well as possible consequences of the change of population structure of *Homo sapiens* will be considered later in this chapter.

We now consider the application of the principles of genetic monitoring to natural subdivided populations (= population systems) of animals and plants, primarily on the basis of our long-term studies and ensuing views on normal reproduction of species' gene pools (Chap. 5).

## 7.2
# Genetic Monitoring of Natural Populations

Cases of long-term monitoring of genetic parameters of individual natural populations have been reported in literature (see, e.g., Mayr 1968). The genetic dynamics in these populations has been found to be typically complex, but in most cases temporal trends were revealed. As a rule, these trends were interpreted as resulting from the continuous microevolutionary (and frequently adaptive) process. However, more careful analysis of these situations[1] shows that the observed dynamics could be stochastic, reflecting the fact that the authors ignored the subpopulation structure of the population in question and thus did not carry out the appropriate sampling covering the whole system. For this reason, in the earlier years, only our series of long-term studies on commercial fish species, including Pacific salmonids, closely approached the primary aim of genetic monitoring and the principles of its implementation presented above (Altukhov 1974, 1983, 1989a,b; Altukhov et al. 1997, 2000). Later, other authors used a similar approach in studies of various salmonids and other fish species (Allendorf and Phelps 1980; Ryman and Ståhl 1980; Ståhl 1983, 1987; Cross and King 1983; Krieg and Guyomard 1985; Ryman and Utter 1987; Campton and Utter 1987; Verspoor 1988; Hindar et al. 1991; Youngson et al. 1989; reviews: Carvalho 1993, 1994). The main finding of these monitoring programs has been a decrease in gene diversity (detected using biochemical markers) in artificial populations as compared to natural populations. However, of no less importance is a discovery of the negative effects of enhanced intrapopulation gene diversity caused by selective fishing and gene pool transplantations to other parts of the range. Let us examine these studies in more detail.

## 7.2.1
## Fishing

A typical strategy of sea fishing includes two main stages: searching for fish with a special vessel and, after discovering sufficiently dense fish accumulations (stocks), catching them with a fishing fleet. The process can be dynamically represented with a series of pictures taken at certain intervals (Fig. 7.1). These pictures again show the chain of genetically different redfish populations revealed in Chap. 3 (Fig. 3.4). It is clear that such a method of fishing, the fleet always working in zones where stocks of fish are densest, leads to over-fishing in some subpopulations, whereas others are underexploited. Eventually, the natural channels of migrational communication among elements of the system are disturbed, and the genetic structure of the population deteriorates. To avoid this situation, one should *utilize the whole fish shoal uniformly, taking into account its spatial subpopulation structure.*

---

[1] Except evident shifts in the cases of industrial melanism in *Biston betularia* and of adaptation of agricultural pests to insecticides.

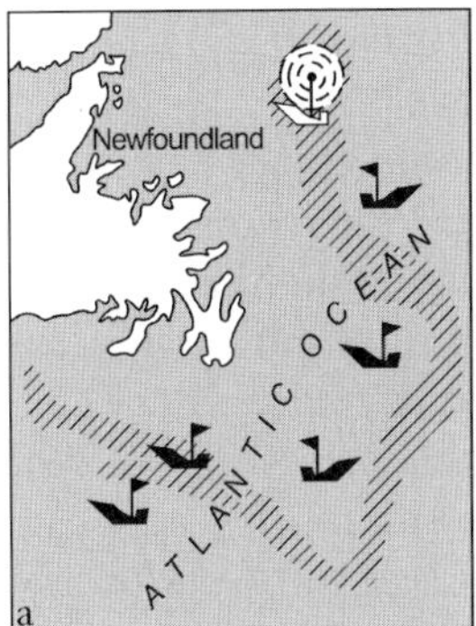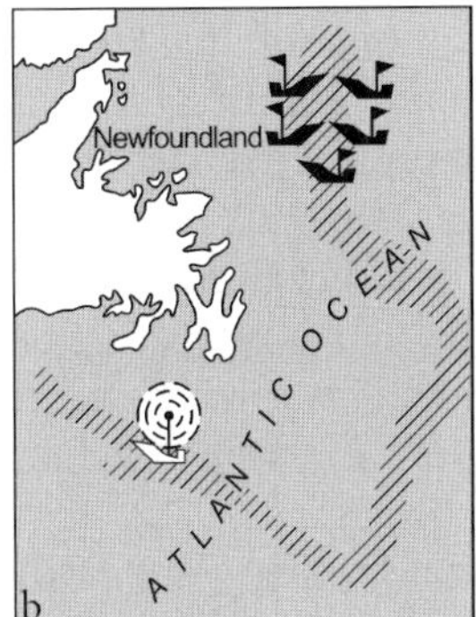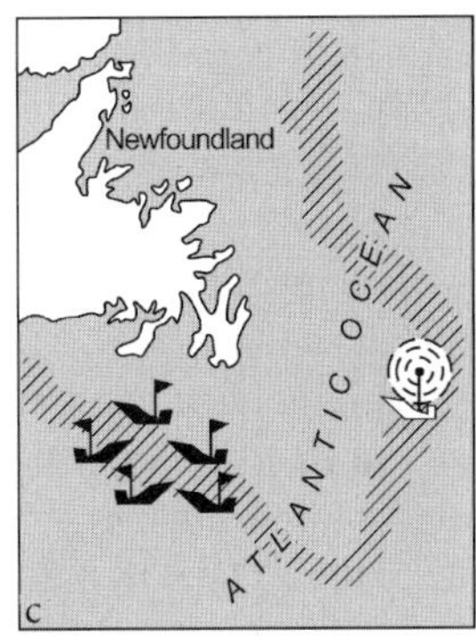


**Fig. 7.1.** Approximate diagram of the ocean fishing industry during successive changes of positions (**a–c**) by a flotilla of vessels in fishing areas of high density of redfish discovered by a search ship

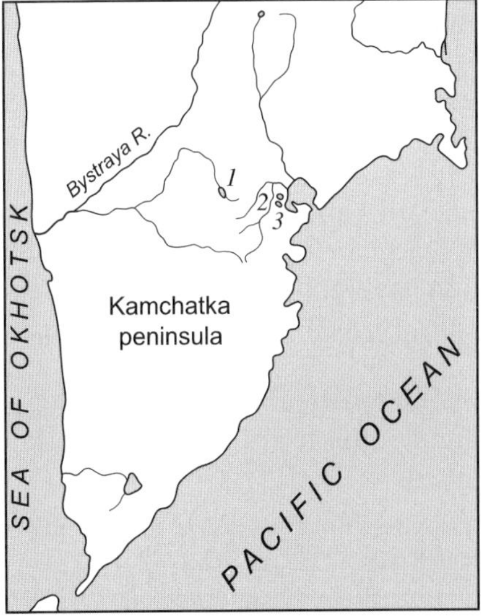


**Fig. 7.2.** The geographical location of sockeye salmon populations: *1* Lake Nachikinskoye; *2* Lake Blizhneye; and *3* Lake Dalneye. In Lake Nachikinskoye the genetic structure at the *Ldh* and *Pgm* loci of 16 subpopulations was studied, 5 in Lake Blizhneye, and 3 in Lake Dalneye

Such a seemingly abstract model was directly confirmed in a recent study of the consequences of commercial fishing for subdivided populations of sockeye salmon reproducing in three Kamchatka lakes (Altukhov and Varnavskaya 1983; Fig. 7.2). The choice of these populations was not arbitrary, along with the accessibility factor being determined by the fact that these shoals form a natural series reflecting the degree of their subjection to marine fishing in the order of its increase in lakes: Nachikinskoe (lowest fishing impact), Blizhnee, and Dalneye (highest). In additions, these populations, especially the sockeye salmon of Lake Dal'nee, were the object of systematic observations for decades (Krogius 1979), providing the researcher with unique biological information. Let us examine these data more closely.

Lakes Blizhnee and Dalneye are small neighboring reservoirs connected by canals with the Paratunka River, which flows into the Avachinsky Gulf of the Pacific Ocean (Fig. 7.2). In the 1930s, the population numbers were close to 100,000 spawners (Krogius 1979). Sockeye salmon fry spent up to 3 or 4 years in Lake Dalneye and 3 years in the sea; thus, the shoal's reproductive portion was constituted by a large "set" of up to 15 age groups. The available materials (Krogius 1979) show that after 1947 the state of the migratory portion of the sockeye shoal in Lake Dalneye changed. Over the last 35 years, the sex ratio in the reproductive portion of the population became uneven (with an excess of females), the production of small males ("jacks") increased, total population numbers decreased catastrophically, and the shoal became "younger". At the same time, relative numbers of the so-called "residual" dwarf form increased to 90% of the males so that this form has become dominant (Table 7.1).

Considerable changes occurred in the sex–age structure of the sockeye salmon in Lake Blizhnee. While large males (2 and 3 years in the sea) predominated in the population in the 1930s, now their portion is considerably reduced, whereas the proportion of jacks has significantly increased. Population numbers also dropped to, typically, 10,000–12,000 spawners during the years of the research.

Lake Nachikinskoe differs from lakes Blizhnee and Dalneye in its larger size, and its morphology and hydrography; the fry of the Nachikinskoe

**Table 7.1.** Average values of several biological characteristics of a sockeye salmon shoal in Lake Dalneye for periods from 1935 to 1976 (from Krogius 1979)

| Characteristics | Periods | | | |
|---|---|---|---|---|
| | 1935–1946 | 1947–1956 | 1957–1965 | 1966–1976 |
| Numbers of breeding fish entering the lake, in thousands | 62.6 | 10.0 | 5.7 | 1.6 |
| Proportion of females in the shoal's migrating section (%) | 52 | 54 | 59 | 68 |
| Proportion of "jacks" in the migrating males (%) | 0.2 | 0.6 | 4.3 | 37.5 |
| Numbers of "dwarfs", in the lake, in thousands | 14.9 | 7.6 | 5.8 | 5.3 |
| Proportion of "dwarfs" in all the males (%) | 26.0 | 49.4 | 74.2 | 88.8 |

sockeye salmon moves along the Plotnikova and Bolshaya rivers to the Sea of Okhotsk. This population is a typical subdivided population of many spawning subpopulations, forming the structure of the spring and summer races. An analysis of the spawners' body length distributions with respect to their sex shows that noticeable changes in the last decades affected mainly the spring race, first of all due to increased frequency of small males. The numerical size declined, but did not undergo such a significant fall as in Lake Dalneye's population (Varnavskaya 1984).

Genotypes of fishes were identified for *Ldh* and *Pgm* loci (Fig. 3.8). The objective of the study was to compare the genetic structure of three sex–age (size) groups, spawning simultaneously: large and small males, and females. The expediency of this comparison follows from the data of Sect. 5.4, where it was shown for Azabachye lake sockeye salmon that the heterozygosity level is minimal in large males and maximal in small ones, while it is intermediate in females.

A similar picture, but with better definition, was found in the same three spawning groups of Nachikinskoe, Blizhnee, and Dalneye lakes (Table 7.2).

1. Total distributions of the *LDH* genotypes for the Lake Nachikinskoe sockeye salmon are characterized by a significant deficit of heterozygotes. We ascribe this situation to the Wahlund effect at a relatively neutral locus under conditions of genetic subdivision of the shoal. The distributions of the *PGM* genotypes do not differ significantly from those expected according to Hardy–Weinberg, or else there is an excess of heterozygotes. This picture is a result of selection in favor of heterozygotes, analyzed in detail with the example of Lake Azabachye sockeye salmon (Chaps. 3–5).

2. Division of the sample into groups distinguished by sex and age reveals fine genetic differentiation: a statistically highly significant excess of heterozygotes is observed at the *PGM* locus in small males, with a clearly expressed deficit of these genotypes in large males, and a genotypic equilibrium in females. Such differentiation is not evident at the *LDH* locus, which is more selectively neutral, but there is evidence of a deficit of heterozygotes in both females and large males, with Hardy–Weinberg equilibrium in small males. The frequency of heterozygous genotypes is maximal for the groups of small males, reaching 55% at the *PGM* locus.

3. The same differences are typical for the distributions of the *PGM* genotypes in Lake Blizhnee sockeye salmon, providing evidence of heterozygotes' excess in small males, a deficit in large males, and an intermediate level of heterozygosity in females.

4. A particularly significant excess of heterozygotes over the expected value at the *PGM* locus is found in Lake Dalneye's dwarf male. The observed frequency of heterozygous genotypes (62%) is highest for this group.

**Table 7.2.** Population genetic parameters for the *Ldh* and *Pgm* loci in different sexually mature sockeye salmon groups in three lakes on the Kamchatka peninsula

| Sockeye salmon lake populations and sexually mature groups | Sample size | *Ldh* | | | | *Pgm* | | | |
|---|---|---|---|---|---|---|---|---|---|
| | | $q$ | $H$ | $D$ | $\chi^2$ | $q$ | $H$ | $D$ | $\chi^2$ |
| Lake Nachikinskoye | | | | | | | | | |
| Small males (43–56 cm) | 202 | 0.653 | 0.43 | −0.06 | 0.72 | 0.648 | 0.55 | +0.19 | 7.61** |
| Large males (57–73 cm) | 296 | 0.667 | 0.39 | −0.11 | 3.57 | 0.687 | 0.35 | −0.19 | 10.68*** |
| Females (50–69 cm) | 579 | 0.683 | 0.38 | −0.12 | 8.11** | 0.657 | 0.46 | +0.02 | 0.30 |
| Total males and females | 1,077 | 0.673 | 0.39 | −0.11 | 11.95*** | 0.664 | 0.45 | 0.00 | 0.00 |
| Lake Blizhneye | | | | | | | | | |
| Small males (31–39 cm) | 248 | 0.893 | 0.19 | 0.00 | 0.01 | 0.631 | 0.56 | +0.20 | 10.30*** |
| Large males (40–63 cm) | 104 | 0.986 | 0.03 | +0.04 | 0.03 | 0.711 | 0.36 | −0.11 | 1.25 |
| Females (43–59 cm) | 191 | 0.937 | 0.11 | −0.11 | 0.38 | 0.654 | 0.47 | +0.04 | 0.33 |
| Total males and females | 543 | 0.926 | 0.13 | −0.06 | 1.64 | 0.655 | 0.49 | +0.08 | 3.42 |
| Lake Dalneye | | | | | | | | | |
| Anadromous form, spring race[a] | 28, 30[b] | 0.946 | 0.11 | +0.05 | 0.09 | 0.783 | 0.30 | −0.12 | 0.41 |
| Anadromous form, summer race[a] | 38, 26[b] | 0.895 | 0.21 | +0.12 | 0.52 | 0.769 | 0.31 | −0.13 | 0.46 |
| Small "residuals" | 158 | 0.905 | 0.16 | −0.04 | 0.28 | 0.646 | 0.62 | +0.35 | 19.98*** |
| Total | 224, 214 | 0.908 | 0.17 | −0.01 | 0.00 | 0.680 | 0.54 | +0.23 | 11.77*** |

$q$ Frequencies of the most frequent allele; $H$ observed heterozygosity; $D$ relative deviation of the observed heterozygosity from that expected; $\chi^2$ test for the correspondence of genotypic distributions to the Hardy–Weinberg equilibrium. ** $P < 0.01$; *** $P < 0.001$.

[a] From Kirpichnikov and Ivanova's data (1977)

[b] Number of fish analyzed at locus *Pgm*

The observed differentiation is an evidence of optimal heterozygosity in females, and simultaneously of its minimal and maximal levels in males, according to their natural subdivision into two groups distinguished by age structure (Fig. 5.8). It should be stressed once again that male and female sockeye salmon represent two adaptive systems, with maximal variance of fitness in males. At the same time, the fact that small males are the most heterozygous contradicts the previous conclusion concerning heterosis associated with the locus *PGM*: at first glance, it seems illogical to discuss heterosis if higher heterozygosity is observed in small individuals. This inference, however, is premature, at least for as long as the corresponding sex and age groups remain insufficiently studied with respect to specific features of their growth and maturation. As it has been shown (Krogius 1960, 1972, 1975), small male fish are distinguished from large males by accelerated growth rate and earlier sexual maturation. Since the *Pgm* locus is selectively more significant, it contributes most to the detected differences, and the excess and empirical level of heterozygosity are especially high in fast-growing and early-maturing small males. The three compared populations of Kamchatka's sockeye salmon can be arranged in a series according to the degree of increase of small and dwarf males' frequency: Nachikinskoe, Blizhnee, and Dalneye (18.7, 45.7, and 73.8%, respectively). A similar situation is also observed for empirical heterozygosity at the *PGM-2** locus: 45, 49, and 54%. As noted above, these three populations are also arranged in the same sequence according to marine fishing impact.

This result needs special attention at least in two aspects. First, it reconfirms an earlier conclusion about heterotic selection at the *PGM* locus, and in part at the *LDH* locus. Second, it enables us to comprehend many unclear cause–effect links that form the basis of the reaction of local salmon populations to fishing impact. In terms of population genetics, the detected differentiation in sex–age groups of sockeye salmon reflects nothing but the differences in the intensity (and direction) of selection. This selection favors small, more heterozygous males in the reproductive period, and, hence, is directed against large and more homozygous males. Occupying an intermediate position between these two extremes, females do not experience such strong differentiation pressure. This regularity can easily be explained by the fact that Pacific salmonids, reproducing in various river systems at both coasts of the North Pacific, have been an object of intense fishing since the midninetinth century. The impact of fishing on the biological structure of populations is proven: in many cases; early migrating stocks, especially the largest fish with a high frequency of males, are the primary subjects of heavy exploitation (Altukhov 1974, 1981a; Krogius 1979; Ricker 1981). This concentrated fishing led to a decrease in populations' numerical size with delayed spawning migration periods, decline in body size, and rejuvenation of spawning shoals (Vaughan 1947; Krogius 1979; Ricker 1981).

Some authors associate phenomena of this kind with overfishing. However, with regard to the population genetic data described above, another conclusion can be drawn: the cause is not overfishing, but rather the distortion of the historically formed genetic structure of the shoals and their systemic organization, or their subdivision into semi-isolated subpopulations, whose biological parameters (sex ratio, age structure, rates of growth and development, spawning migration periods, etc.) are derived from their historically formed gene pools (Altukhov 1974). In particular, ordered behavior, repeated from generation to generation, is observed in sockeye salmon during the spring migration period: large fish are predominant in the beginning of the spawning migration of both races, and the level of heterozygosity changes in time, increasing toward the end of the spawning run (Altukhov and Varnavskaya 1983; Varnavskaya 1984).

Obviously, if the fishing is selective, i.e., some subpopulations are harvested more than others, this will inevitably lead to changes in genetic structure of a population as a whole.

Our findings, discussed here and in Sect. 5.4, reveal that Pacific sockeye salmon has an adaptive genetic system, marked by allozyme genes, which is linked with intrapopulation differentiation for important polygenic traits such as sex, age, growth and maturation rates, life span, and spawning migration periods. The reorganizations of this structure under fishing pressure also become clear.

The following picture emerges. Native sockeye salmon populations are characterized by historically formed genetic differentiation, which is correlated, with differences in heterozygosity levels for an aggregate of Mendelian genes, in such a way that its maximal values are typical for small heterotic males ("jacks", dwarfs), its minimal values for large males, and its average, optimal levels for females. As already noted, large, more homozygous males have an adaptive advantage due to selective mating. The females are practically unaffected by directional and disruptive selection, and their genetic structure remains relatively unchanged over generations. This is the "adaptive norm" of a population. The small males are a reserve of genetic diversity; however, their portion in the native system of populations is small. Genetic structure of the male population as a whole corresponds to the structure of female population. The genetic process in this system is stationary – selectively neutral on the average. The most important biological characteristics of a population system are highly stable in a normally fluctuating environment, and vary only within certain limits; they do not cross the bounds of a historically formed biological optimum, specific for each separate population. The sockeye populations of Lakes Nachikinskoe and, partly, Azabachye are an example of such a system.

Fishing influence causes changes in the vector (and intensity) of selection. The portion of large, more homozygous fish in a population declined, while the portion of small, more heterozygous fish increased. The optimal 1:1 sex ratio is disturbed, which leads to reduction of the effective population size. Maturation rate is accelerated because of heterosis, average life-span is lower, and, consequently, the rate of generation succession is increased. This reaction to fishing pressure is possible because of the reserve of genetic variability carried by small males. However, the biomass of a population is reduced due to a decline in fertility. A gap grows between heterozygosity levels of a population as a whole and the population of females, genetic structure of which acts as a kind of "memory" about the population's past state before it became an object of fishing. The sockeye salmon populations in Lake Blizhnee, and particularly Lake Dalneye, are examples of such populations. Lake Dalneye's heterozygosity excess at the *PGM* locus is especially high, reaching 23%. When reproductive size and capacity of a population are reduced even by fishing of constant intensity, perfectly compatible with a population's initial production potential, a numerical decline results. In this case, the excess of heterozygotes found from the relation of expected and empirical genotype numbers, and serving as an evidence of genotypic equilibrium distortion, acts as a marker of this process.

Indeed if to use the estimate of mean population fitness for sockeye salmon of Lake Dalneye at *PGM-2** locus ($\overline{w} = 0.77$), it may be shown that theoretically expected and observed numbers are almost the same in four 3-year time periods (Table 7.3).

It is clear that this unfavorable process can lead to at least partial replacement of anadromous forms by residual sockeye (if not to complete degradation), and hence, to the loss of a commercially valuable resource. However, as selective fishing is associated with increasing heterozygosity

**Table 7.3.** Theoretically expected and observed numbers of sockeye salmon in Lake Dalneye using the estimate of mean population fitness at the *PGM-2** locus

| Generations (years) | | Average annual observed numbers (thousands) | Average annual expected numbers (thousands) |
|---|---|---|---|
| 1 | 1947–1950 | 9.93 | – |
| 2 | 1951–1954 | 8.82 | 8.01 |
| 3 | 1955–1958 | 7.30 | 6.47 |
| 4 | 1959–1962 | 5.03 | 5.13 |
| 5 | 1963–1966 | 4.84 | 4.21 |

Data on the observed numbers in sockeye salmon of Lake Dalneye are taken from the comprehensive work of Krogius et al. (1969). The expected number of the second generation is estimated by multiplying the observed number of the first generation by the $\overline{W}$ value, and in the same way for the next expected numbers

of stocks, the current situation can still be improved. To this end, it is necessary to (1) revise the dates of fishing, making the fishing pressure more uniform in time; and (2) solve the problem of establishing the practice of fishing with nets of the optimal mesh size, which should vary throughout the period of spawning run.

Genetic processes described above are characteristic of not only Pacific salmon populations but also of other commercially important fish species (Altukhov 2001). In all the cases studied in detail, monitoring revealed the same pattern of changes in commercial fish shoals: a decrease in body size and individual age, an increase in the proportion of early maturing small males, and the reduction of average life span. As the vector of selection remains unchanged (in favor of heterozygotes), the intrapopulation component of gene diversity increases, whereas the interpopulation component decreases, leading to the reduction of local genetic differentiation. Under such conditions, empirical $F_{ST}$ ($G_{ST}$) values become significantly lower than expected. Thus, in the group of sockeye subpopulations from Lake Azabachye (Kamchatka), which were affected by selective fishing to the greatest extent, the value of local genetic differentiation ($F_{ST}$) was estimated at 0.008 vs. 0.059 expected in a selectively neutral process (Altukhov et al. 2000; Chap. 5). Exposure to such strong artificial selection leads to the establishment of nonrandom relationships between polygenic traits (body length, growth rate, and maturation rate) and monogenic traits (allozymes; Altukhov and Varnavskaya 1983; Altukhov 1991).

But the same undesirable effects associated with misjudging the population genetic structure of a species are not confined to the commercial fishery process. They can also be found where other quite reasonable aims are pursued, such as that of artificially propagating biological resources, which has happened in the same human food supply context of fisheries.

Hydraulic engineering, environmental changes, and other causes have already deprived the populations of several migratory fish species of their spawning grounds, and these are now being reproduced artificially by incubating fish eggs in hatcheries, raising them and releasing the young into the sea. Such artificial reproduction of salmon has reached an industrial peak. But despite increased measures to improve this activity, in many cases the effectiveness has been minimal.

We have succeeded in elucidating (Altukhov 1974, 1981a; Altukhov et al. 1980a) that once again it is a question of the failure to take into account a species' population genetic structure. On the one hand, it is caused by intensive commercial fishing of ocean salmon, on the other, by acclimatization measures – that is, by transferring artificially fertilized eggs from one river to another.

## 7.2.2
## Acclimatization

Transplantations of artificially fertilized eggs among different salmon hatcheries are conducted in connection with the Soviet practice: complete the plan of egg collection for incubation at any price. These measures have been also practiced because of the reduced number of "native" shoals, resulting in hatcheries being compelled to use "foreign" shoals. The intensity of these measures has been quite high with respect to the chum salmon *Oncorhynchus keta* (Walb.; Table 7.4).

Because genetic differences (in all the loci studied) were found among these chum salmon populations (Sect. 3.1), it is possible to differentiate populations in mixed communities (Fig. 7.3). We were able to follow the fate of transplantation measures; they proved to be unsuccessful. Transferring millions of Sakhalin fish eggs into the Amur yielded no results and after transplantation within the Sakhalin region only small returns were observed in the first generation: the coefficients of return as an integral parameter of a population's fitness[1] were much lower than in their own river

---

[1] The ratio of the number of spawners that returned to the river and number of fry released by a hatchery.

**Table 7.4.** Acclimatization transfers of chum salmon eggs in the Sakhalin-Kuril-Amur region. (Altukhov et al. 1980a)

| Year | Number of transferred eggs, in millions | Year | Number of transferred eggs, in millions |
|---|---|---|---|
| | Naiba[a] – Kalininka | | Kalininka – Naiba |
| 1956 | 4.500 | 1959 | 4.830 |
| 1960 | 2.638 | 1964 | 70.610 |
| | Naiba – Amur | 1965 | 70.000 |
| 1964 | 22.000 | 1966 | 54.123 |
| 1967 | 9.000 | 1967 | 25.665 |
| 1968 | 98.000 | 1968 | 26.310 |
| 1969 | 62.000 | 1969 | 35.563 |
| 1970 | 17.600 | 1970 | 18.820 |
| | | 1971 | 56.000 |
| | Naiba – Tym | | Kurilka – Naiba |
| 1967 | 9.487 | 1972 | 30.000 |
| | | 1973 | 3.500 |
| | Naiba – Buyuklinka | | Rivers of Iturup I. – Kalininka |
| 1968 | 5.390 | 1976 | 30.000 |
| 1976 | 3.000 | | Tym – Naiba |
| | | 1966 | 10.208 |

[a] The name of the "donor" river is followed by the name of the "recipient" river

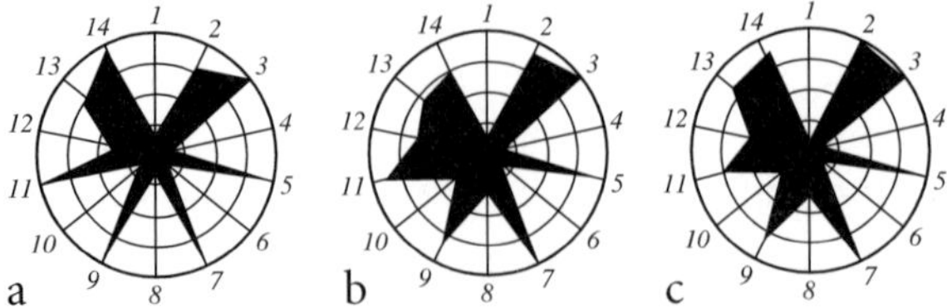

**Fig. 7.3.** Schematic representation of phenotypic and allelic frequencies of allozyme loci (plotted in radii) characteristic of chum salmon populations in the Kalininka (**a**) and Kurilka (**c**) Rivers and for samples containing an admixture from the Kurilka chum salmon in the River Kalininka (**b**). 0 frequency at the *circumference*, and 1 at the *center* of the *circle*. At radii *1*, *2*, and *3* phenotype frequencies *Mdh-A*, *Mdh-C*, and *Mdh-B*; at *4* and *5* phenotype frequencies *Aat-AA′* and *Aat-AA*; further allelic frequencies *6* and *7 Ldh-1-A′* and *Ldh-1-A*; *8* and *9 Me-2-S* and *Me-2-F*; *10* and *11 Idh-3-F* and *Idh-3-S*; *12*, *13*, and *14 Idh-2-F*, *Idh-2-S*, and *Idh-2-S′*

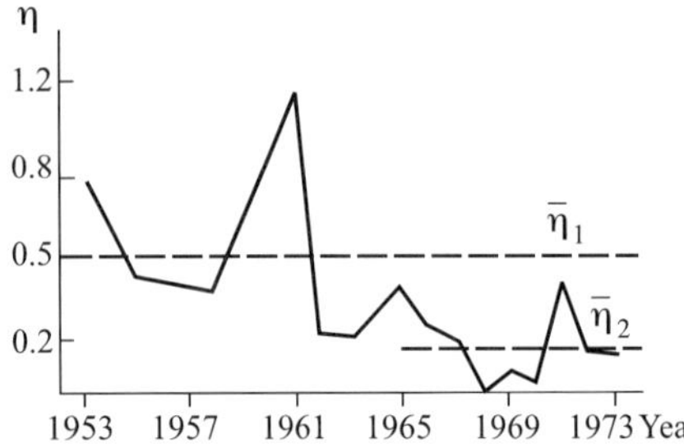

**Fig. 7.4.** The dynamics of the return rate ($\eta$) of a Naiba chum salmon population (Altukhov et al. 1980a). $\bar{\eta}_1$ and $\bar{\eta}_2$: respective average values before and after transfers of eggs from the Kalininka population (see Fig. 3.5). *x-axis* Generations; *y-axis* return coefficients

and subsequently decreased. Furthermore, there was a reduced return of the Kalininka and Naiba populations to their rivers in those generations from which a considerable portion of the gene pool had been removed for acclimatization in another section of the area. This particularly affected the Naiba population whose numbers fell sharply from 1971 onward (Fig. 7.4).

From the viewpoint of population genetics, acclimatization is adaptation to a new environment. Its efficiency can be assessed only after the formation of a self-reproducing population with a stable, integrated gene pool capable of existing indefinitely through many generations. Unfortunately, there is far too little evidence to date of these kinds of phenomena, whereas there is plenty of evidence to the contrary. Thus, Ricker (1972) summarized voluminous material about acclimatization transfers, mainly carried out with North American salmon populations, and showed that the return of first generation fish to a "foreign" river frequently occurs, but considerably less often than to a "native" one. In subsequent generations the return falls sharply or does not occur at all. The Pacific pink salmon, acclimatized to the rivers of the Kola peninsula by means of almost annual transfers of eggs from (mainly) Sakhalin, from 1956, as a rule has resulted in only some returns in odd-year strains (Dyagilev and Markevich 1979). Evidently, however, this new population almost failed to reproduce in the natural way.

Only in the last few years, when the transplantation of Pacific pink salmon eggs was carried out from the Ola river (the northern Okhotsk sea coast), are processes being observed that show probable adaptation of introduced fish to new ecological medium, presumably due to global climate warming (Gordeeva 2002; Gordeeva et al. 2003).

The negative impact of massive transfers, i.e., transfers of population gene pools from one part of a species' range into others, have been discussed in detail (Ricker 1972; 1974, 1981a, 1989a,b; Altukhov et al. 1980a; Altukhov and Salmenkova 1987a, 1990, 1991; Diagilev and Markevich 1979; Withler 1982; Hindar et al. 1991; Hansen et al. 1995). Along with decreased fitness of the hybrids between native and introduced fish (Hinder et al. 1991) there was another important finding: in a new environment, the introduced population returns for reproduction at the same time as to the native river, and the genetic structure of the introduced fish remains practically unchanged in spite of a very low return rate (Altukhov and Salmenkova 1987a).

All these and many other facts point to the uniqueness and conservativeness of local adaptations formed by natural selection over thousands of generations in a specific environment with which all the natural history of one or another population is related.

When we disrupt the links between the elements of ecosystems that have been formed over the ages, uproot a population from its "own" historically formed environment and transplant it into new surroundings, the reserves of genetic resistance are usually unequal to the task. The fact that under the new environment a transferred population returns for reproduction at the same periods of time as it would to a native river, and moreover, that the genetic structure of the transferred populations remains virtually unchanged despite an extremely low return coefficient (Altukhov et al. 1980a), means only one thing: selection under the new conditions plays a catastrophic role representing non-selective elimination of the genotypes examined.

It should be stressed, however, that successful acclimatizations are also generally undesirable because of their ecological consequences, since they result in populations of local species being ousted either on the basis of competition for food or, as more frequently happens, by the dissemination of diseases to which local forms have not developed immunity. To illustrate this one can recall the almost forgotten attempt, several decades ago, to introduce Caspian star sturgeon *Acipenser stellatus* into the Aral Sea. The prerequisite for this measure was the fact that the conditions in the Aral

Sea were apparently right for boosting sturgeon numbers as one of its own species, spiny sturgeon *Acipenser nudiventris*, had lived in the Aral Sea since ancient times. However the attempt was a total failure: not only did the star sturgeon not adapt itself to the new conditions, but it actually caused the downfall of the spiny sturgeon population by contaminating it with the parasite – the nitzschia, *Nitzchia sturionis*. This parasite was not a danger to the star sturgeon because during its lengthy evolutionary coexistence with it the star sturgeon had been able to develop an immunity, whereas the spiny sturgeon had never before encountered the nitzschia[1].

The history of salmon aquaculture in Norway provides another impressive example of such a situation (Heggberget et al. 1993). From the mid-1970s *S. salar* populations in more than 30 Norwegian rivers have been perishing from a disease caused by the ectoparasite monogenean *Gyrodactilus salaris*. The latter was supposedly brought to Norway from Swedish hatcheries together with transplanted salmon eggs or juveniles. Subsequent studies showed that Baltic *S. salar* populations are far less sensitive to this parasite than Norwegian populations. Populations of this species in Norway were also seriously damaged by furunculosis acquired from rainbow trout transferred from Denmark.

The number of these examples could be greatly increased (see, for instance, Timofeeff–Ressovsky et al. 1977; Chunikhin 1979; Harlan 1981), but they are scarcely mentioned now. Meanwhile, many acclimatization measures that in practice do not succeed are being widely publicized by the press to create a false illusion of success when this is not justified by facts.

It must be admitted that despite important results in the genetic aspect of acclimatization and the clear position adopted by the World Conservation Union (World Conservation Strategy 1980; see also Bannikov 1979), which regards acclimatization as a most important factor in the "biological" pollution of the environment, scientific circles are not united as they should be on the issue; the most valuable, unique genetic capital of natural populations continues to be destroyed.

Does this mean that we must renounce all human activity in nature to solve the scientific and practical tasks facing mankind? Obviously not, as the example of our agricultural practice shows. Clearly, modern mankind could not exist without a developed agriculture. However, there is one fundamental difference between the living conditions of agricultural and natural populations. In nature species live in a varied, heterogeneous environment independent of humans, whereas animal breeds and plant varieties exist in conditions that we control to a considerable extent. The stability of the environment, multiplied by the productive hereditary qualities of

---

[1]This dismal incident has lost none of its instructiveness when confronted today by an imminent ecological catastrophe: the Aral Sea, as we know, is being destroyed.

agricultural populations, also makes possible the controlled, predictable type of farming that may be called *intensive*.

Taking into account the successes won in the genetics of natural populations, we can introduce into our practical activities methods for the industrial utilization, artificial reproduction, and acclimatization that will make it possible to change to the intensive type of farming. But this is only practicable if one makes allowance for the special features of the internal structure of populations, preserving their historically formed genetic heterogeneity, and if the internal autoregulatory mechanisms that ensure their effective adaptation under conditions of a normally fluctuating environment are maintained. Only in this way can an optimal managing policy that envisages not just the extraction of economic advantage but also the natural preservation of the populations given to us for an unlimited time be realized in practice.

To organize a controlled biologically based industry we must have clear views about the population structure characteristics of a species under scrutiny. We must know the degree of isolation of its populations, the rate of their reproduction and the dynamics of their numbers and, finally, taking their structure into account, know how to give reasoned recommendations to ensure that the intensity of industrial exploitation will not exceed the rates at which they are naturally (or artificially) reproduced. Only in this way will it really be possible to implement rational economic activity that allows us to preserve our biological resources for a long time.

Subpopulation structure should be considered in exactly the same way for artificially reproducing populations. In doing so it should not be forgotten that each of them has its own evolutionarily formed biological optimum, determined by previous evolution and the present state of a population in the ecosystem.

The limits to the numbers of this optimum are set by their minimal and maximal levels whose stable values can only be obtained by averaging the results of systematic observations made over many years. It is clear that without a knowledge of this area of a system's stability it is extremely difficult to be successful in reproducing it artificially, or controlling it.

### 7.2.3
### Artificial Reproduction

Salmon shoals provide a graphic example for analyzing the genetic consequences of artificial reproduction. As shown above, these shoals are population systems with a complex structure consisting of multiple discrete subpopulations. To reproduce such systems artificially at hatcheries, we must collect eggs and sperm throughout the spawning period rather than limiting ourselves to using only a part of the differentiated gene pool (Altukhov 1974). The more distinct the subpopulation structure of a system, the less the chance of reproducing the whole from a part. Unfortunately, this fact is often ignored at hatcheries, and, consequently, the genetic diversity of artificially reproduced populations gradually decreases. This was demonstrated for *Salmo clarkii* (Allendorf and Phelps 1980) and *S. salar* (Ståhl 1983, 1987).

For the last few years, we have monitored three adjacent pink salmon (*Oncorhynchus gorbuscha*) populations of South Sakhalin: two native, from Firsovka and Bakhura Rivers, and one artificially reproduced in the Naiba River (Fig. 7.5). The latter is maintained by a hatchery that had specialized in reproducing the Naiba stock of the chum (*O. keta*) but, after a sharp decrease in its size in 1972, began to work with pink salmon. Judging from yields, artificial fish breeding resulted in a several fold increase in the size of the local population. In recent years, however, the biological structure of the hatchery stock has changed: the fish became considerably larger, the proportion of males increased, and the numerical size of the stock began to decrease.

To reveal the mechanisms responsible for these changes, work was organized as follows (Altukhov et al. 1989):

1. We compared the genetic characteristics of male spawners selected for breeding with those of rejected males, using a set of electrophoretically identified allozyme loci.

2. We analyzed alterations of sex ratio and body length in a series of generations of the artificially reproduced Naiba population.

3. We compared genetic and biological parameters of this population with those of two native shoals reproducing in the neighboring rivers.

Also, we were interested in determining the relationship between individual heterozygosity and biologically important parameters in males. To this end, we analyzed them for the frequency of abnormal gill rakers (bent, split, or grown together), using this frequency as a parameter of developmental stability. We distinguished weak (only one gill arch affected) and severe abnormalities (two or more gill arches affected), and only the latter were taken into account during comparative analysis.

The dynamic of body length and sex ratio in the Naiba population was estimated for the whole period of systematic artificial reproduction of pink salmon from records kept at the hatchery since 1973. Corresponding data on the self-reproducing Firsovka population have been collected in the course of our studies. Genotypes were identified by six allozyme loci.

At the hatchery, individual portions of eggs, each obtained from 50 females, are fertilized by sperm obtained from 20 to 30 males. Moreover,

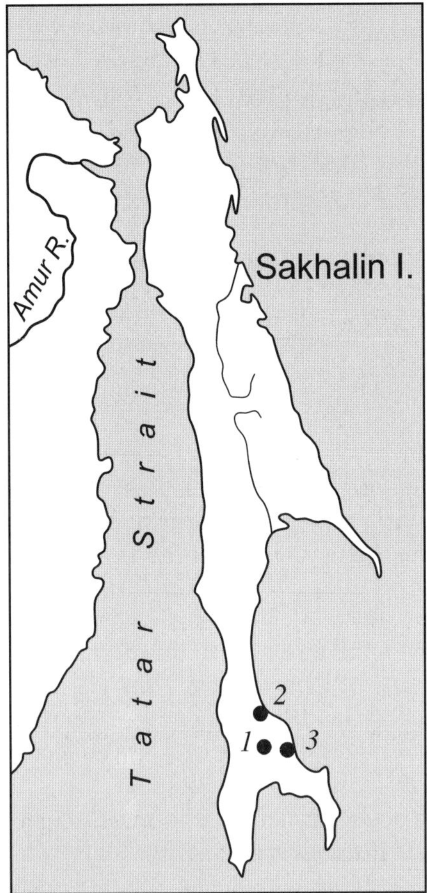



**Fig. 7.5.** Location of pink salmon populations studied: *1* Naiba River hatchery; *2* Firsovka River, *3* Bakhura River. For further explanations, see the text

gametes are usually collected at the beginning of the spawning run, when males prevail in the shoal, and in the middle of it, when the sex ratio approaches equilibrium. Comparison of males selected for breeding with rejected (control) males showed that (human) fish breeders prefer large fish (Table 7.5). No such selectivity was revealed for females. A particularly important fact is that males used at the hatchery also differed from the control males in the level of allozyme heterozygosity and in the frequency of abnormal gill rakers (Table 7.5): in large males selected for breeding, both parameters were higher than in the control group. Assuming that this selection was performed more or less regularly throughout the period of artificial reproduction of the Naiba population and that the "rearguard" part of the shoal with an excess of females was not involved in breeding, we

**Table 7.5.** Morphological and genetic differences between two groups of pink salmon males. (Altukhov et al. 1989)

| Parameter | Males used for breeding | Males not used for breeding | Significance of differences $t_d$ test |
|---|---|---|---|
| Body length, cm | 52.6±0.2 | 47.8±0.1 | 18.02*** |
| $\sigma$ | 3.87 | 2.64 | 6.33*** |
| $N$ | 300 | 293 | |
| Proportion of fish homozygous for all loci | 0.443±0.029 | 0.362±0.028 | 2.01** |
| Proportion of fish homozygous for all loci, $N$ | 300 | 293 | |
| Proportion of fish with severe abnormalities of gill rakers | 0.17±0.03 | 0.09±0.02 | 2.08** |
| $N$ | 203 | 210 | |

Loci *MDH-B1,2*, G3PDH*, PGDH*, PGM-2*, MEP-2*, PGI-A** were analyzed; $N$ is number of fish studied. ** $P < 0.05$; *** $P < 0.01$

could expect definite changes in the biological structure of this population. This applied, in particular, to a decrease in heterozygosity for allozyme loci, an increase in body length, and deviation from the optimal sex ratio toward an increase in the proportion of males, because eggs are collected from early migrating groups with a high frequency of males.

Two groups of facts confirmed this suggestion. The first was data obtained during biological monitoring of the Naiba population for several generations, and the second was the result of comparing the biological and genetic characteristics of this population with those of two native populations spawning in Firsovka and Bakhura Rivers. We found that, in consecutive generations of the Naiba population, the proportion of females decreased and body length increased in more "productive" odd years, in which artificial selection among males was more stringent (Fig. 7.6).

Knowing the difference between the values of average body length at the beginning and end of the selection cycle (the so-called selection differential $S$) and the heritability coefficient of the trait $h^2$ (for salmon, approximately 0.27; Kirpichnikov 1981), we can use Falconer's formula $R = Sh^2$ (Falconer 1960) to estimate the expected shift ($R$) in the value of the trait and to compare it with the shift actually observed in a series of generations. The results (5.8 and 5.3 cm, respectively) are so similar that any further comments seem unnecessary: fish breeding is obviously a process of the selective type. As well as under fishing pressure, a nonrandom correlation appears between variability of a polygenic trait (body length) and the integral genotype structure with respect to a complex of allozyme loci.

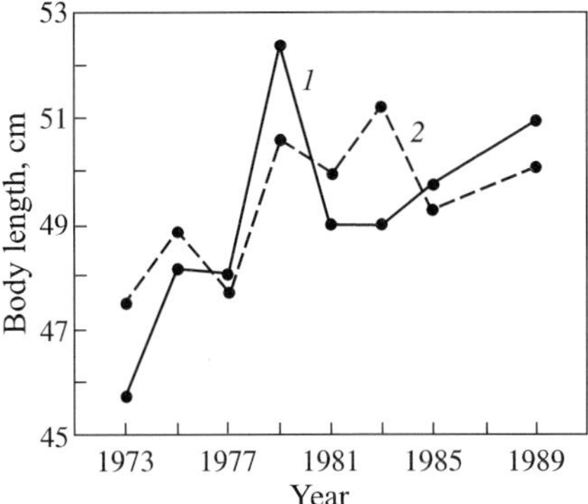



**Fig. 7.6.** Body length dynamics in *1* male and *2* female spawners from the Naiba population of pink salmon after the beginning of artificial reproduction. *Dots* on *curves* show average values calculated from data of biological analyses made at the hatchery. Standardized deviations from corresponding total means increase significantly ($r = 0.65$; $df = 14$; $P < 0.05$)

Important data are also shown in Table 7.6. These are estimates of sex ratio and average body length for five "odd" generations of pink salmon from the naturally reproducing Firsovka population. The stationary state of this population is quite obvious in comparison with the Naiba population, which undergoes marked time-dependent changes. It is also seen that the body length of both males and females in the Naiba population is significantly greater and that variance of the trait is lower than in the adjacent Firsovka and Bakhura populations (Table 7.6). Correspondingly, fish from the River Naiba have a decreased level of multilocus allozyme heterozygosity (Table 7.7).

The sharp reduction in gene diversity has been described (Allendorf and Phelps 1980) in a hatchery population of trout *Salmo clarkii* in comparison with native donor population in a study of 35 loci, coding various proteins. During the breeding of a trout population at a hatchery in Montana (USA),

**Table 7.6.** Sex ratio and average body length of pink salmon spawners from the native Firsovka river population in generations of odd years (from Altukhov et al. 1989)

| Year | No. of fish studied | Proportion of females (%) | Mean body length and variance | | | |
| --- | --- | --- | --- | --- | --- | --- |
| | | | Females | | Males | |
| | | | $\bar{x}$ | $\sigma$ | $\bar{x}$ | $\sigma$ |
| 1979 | 416 | 51.2 | 46.6 | 2.88 | 47.3 | 3.88 |
| 1981 | 349 | 48.3 | 48.0 | 2.20 | 47.9 | 2.47 |
| 1985 | 250 | 50.0 | 45.7 | 2.51 | 47.3 | 4.10 |
| 1987 | 100 | 61.0 | 47.1 | 2.18 | 47.4 | 4.54 |
| 1989 | 216 | 51.8 | 46.3 | 4.27 | 45.9 | 5.91 |

**Table 7.7.** Body length ($\bar{x}$) and average heterozygosity ($H$) for allozyme loci in three spawning populations of pink salmon. (Altukhov et al. 1989)

| River (population) | | | | Parameter | | | | | | | | | |
| --- | --- | --- | --- | --- | --- | --- | --- | --- | --- | --- | --- | --- | --- |
| | | | | Mean body length (cm) and variance ($\sigma$) | | | | | | | Heterozygosity | | |
| | | | Females | | | | | Males | | | | | Significance of differences |
| | $N$ | $\bar{x}$ | $\sigma$ | Significance of differences | | $N$ | $\bar{x}$ | $\sigma$ | Significance of differences | | $N$ | $\overline{H} \pm m$ | Populations compared / $t_d$-test |
| | | | | Populations compared | $t_d$ test | | | | Populations compared | $t_d$ test | | | |
| Bakhura (1) | 175 | 45.3 | 2.8 | 1–2 | 1.32 | 225 | 46.4 | 4.5 | 1–2 | 1.86 | 400 | 0.821±0.019 | 1–2   2.57* |
| Firsovka (2) | 135 | 45.7 | 2.5 | 1–3 | 16.88** | 115 | 47.3 | 4.1 | 1–3 | 11.02** | 250 | 0.892±0.020 | 1–3   3.67* |
| Naiba (3) | 480 | 49.3 | 2.4 | 2–3 | 14.64** | 593 | 50.2 | 4.1 | 2–3 | 6.97** | 785 | 0.730±0.016 | 2–3   6.33** |

Heterozygosity was estimated for the same loci as in Table 7.5. $N$ Number of fish studied. * $P < 0.05$; ** $P < 0.01$

the proportion of polymorphic loci fell by 57%, the average number of alleles per locus decreased by 29%, and the value of mean individual heterozygosity decreased by 21% (see other similar examples in: Altukhov and Salmenkova 1987b; Altukhov et al. 2000).

Comparing natural and artificially maintained populations of Atlantic salmon, *Salmo salar* and *S. trutta*, reveals two opposite processes associated with redistribution of the intra- and interpopulation components of gene diversity (Table 7.8). Thus, *S. salar* bred at hatcheries demonstrates a higher level of interpopulation genetic diversity but lower intrapopulation polymorphism than under natural conditions. A directly opposite but even more prominent trend is characteristic of Spanish and French stocks of the brown trout *S. trutta*. It is apparent that, in the case of *S. salar*, artificial reproduction leads to a higher level of inbreeding, and the limited number of spawners used at hatcheries contributes to this effect (Ståhl 1987). As concerns *S. trutta*, an increased intrapopulation heterozygosity and a reduced level of interpopulation genetic differentiation are due to mixing gene pools received from maternal lines of different origin or to selection in favor of heterozygotes (Krieg and Guyomard 1985; Garcia–Marin et al. 1991). In the first case, salmon stocks suffer from inbreeding, and in the second, from outbreeding.

This approach, which has been elaborated for salmon populations kept under artificial conditions (Altukhov 1974), has already been used for several years in the Sakhalin hatcheries. As for the concept as a whole, its value to the fishing industry today may be regarded as broadly recognized (Konovalov et al. 1975; Starobogatov 1975; Dubinin 1976; Aronshtam et al. 1977; Kirpichnikov 1979; Konovalov 1980; Hynes et al. 1981; MacLean and Evans 1981; Thorpe et al. 1981; Zhirmunsky and Kuzmin 1982; Ryman and Utter 1987; Thorpe 1988; Pavlov et al. 2001).

**Table 7.8.** Genetic diversity of natural and hatchery salmon populations. (Altukhov 1995)

| Species, region | Natural | | | Hatchery | | | Reference |
|---|---|---|---|---|---|---|---|
| | $H_T$ | $H_S$ | $G_{ST}$ | $H_T$ | $H_S$ | $G_{ST}$ | |
| *Salmo salar* | | | | | | | |
| Within the Baltic sea, eastern, and western Atlantic basins, average | 0.041 | 0.038 (29 populations) | 0.064 | 0.037 | 0.030 (24 populations) | 0.196 | Ståhl (1987) |
| *Salmo trutta* | | | | | | | |
| Spain | 0.069 | 0.027 (4 populations) | 0.610 | 0.092 | 0.083 (4 populations) | 0.028 | Garcia–Marin et al. (1991) |
| France | 0.111 | 0.050 (8 populations) | 0.550 | 0.077 | 0.072 (7 populations) | 0.063 | Kreig and Guyomard (1985) |

It should be stressed, however, that the practical effectiveness of hatcheries is still gauged not by the coefficients of commercial return, but by the volume of fish eggs incubated and young released. Moreover, it is clear from our exposition that this approach is only effective when the point of equilibrium has been reached between population numbers and environmental resources, primarily by supplying food in the coastal sea areas for young fish. When the optimum point has been exceeded, "environmental resistance" grows, inevitably leading to the reduced fitness of a population and a fall in its numbers.

Indeed, it is true to say that the increased release of young salmon by a fish hatchery can mark the beginning of a reduction in the return coefficients of breeding fish caused, for instance, by inadequate provision of food in the estuaries of hatchery rivers. This relationship is shown for two Sakhalin hatcheries, one of which artificially reproduces the pink salmon, the other the chum salmon (Fig. 7.7).

To ascertain optimal values for releasing young fish, one must determine the interdependence of the coefficient of commercial return ($\eta$) and the number of the young released ($x$) by means of theory-of-numbers dynamics (Ricker 1954); this dependence may be represented approximately as $\eta = Ae^{-\lambda x}$, where $\lambda$ is the proportionality coefficient of the incidence of the natural logarithm of the return coefficient with the increased release of the young fish, and $A$ is the maximal return coefficient, which corresponds to minimal release. Then the correlation of the number of the fish that have been returned ($y$) and the release of the young ($x$) can be determined by the equation (Ricker 1954) $y = Axe^{-\lambda x}$. Maximal return is accompanied by the optimal release value $x_{\mathrm{opt}} = 1/\lambda$ equaling $y_{\mathrm{max}} = A/\lambda x$.

This method enables one to find the function's maximum, making it possible to determine the number of young produced, corresponding to the ecological optimum point that is characteristic of one population or another. Such data may be obtained only during long-term monitoring of a population.

Let us consider the results of this work using an example of a chum salmon population from the small Kalininka River (about 4 km long) on the southwestern Sakhalin coast (Fig. 3.5). This population has been kept under observation since the hatchery on this river was put into operation in 1951.

Figure 7.8 shows the dynamics of the ratio between the release of fry and the return of spawners. At the initial stage, the latter parameter increased proportionally to the former. However, when the number of released fry exceeded approximately 50 million, that of returning spawners began to decrease fairly rapidly: at 80–90 million fry released, the number of spawners in 1981, 1983, and 1984 was only 20,000–40,000, i.e., similar to that at the initial stage of artificial propagation. Meanwhile, using the Ricker's

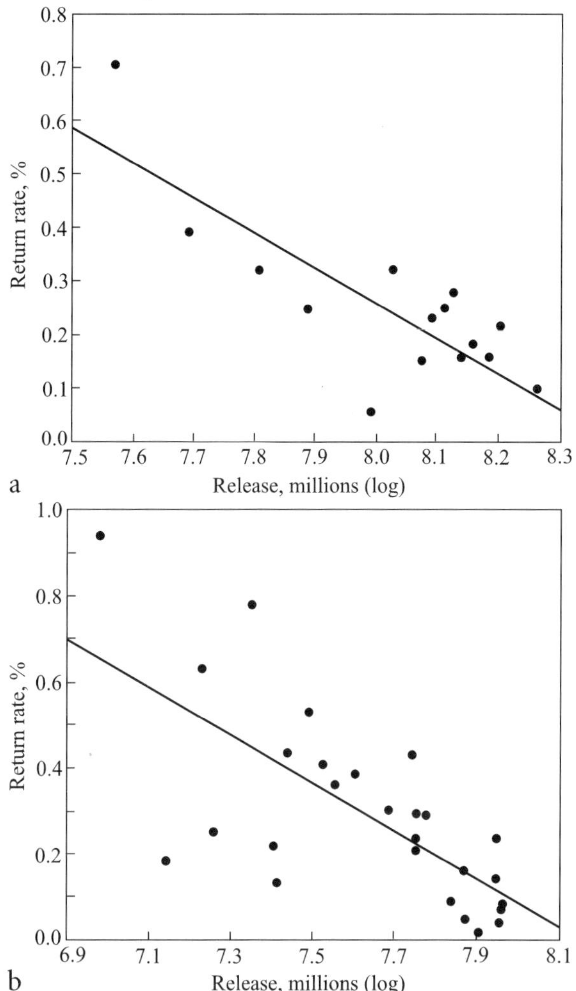

**Fig. 7.7.** Relationship between release (million fry, log) and return rate (%) at Sakhalin salmon hatcheries. **a** Naiba River pink salmon 1968–1986; $y = 5.5267-0.6587 \log x$; $r = -0.8035$; $P < 0.001$. **b** Kalininka River chum salmon 1952–1984; $y = 4.5485-0.5577 \log x$, $r = -0.7009$; $P < 0.001$

function for approximating the empirical data on the period from 1952 to 1984 (seven to eight consecutive generations, as generation length in chum salmon is four years), we determined the optimal rate of release at $x = 46$ million fry per year, which corresponds to the expected annual return of about 130,000 spawners (to the "natal" river alone). In fact, at the annual average release of 45 million fry, the annual return of spawners in the years 1985–1988 averaged 280,500.

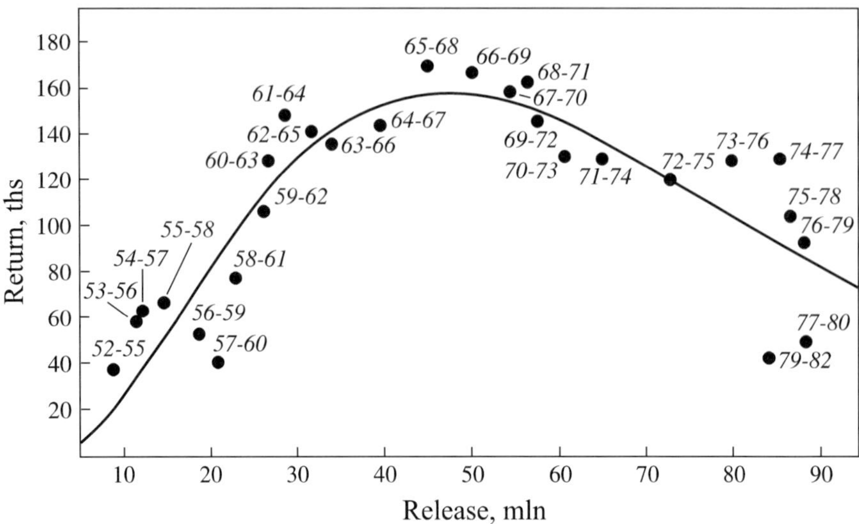


**Fig. 7.8.** Relationship between release (million fry) and return (thousand adult) in Kalininka River in 1952–1984. Approximation by Ricker's function (1954): $y = 7.4980x \times \exp[-(0.0216)x]$; $x_{max} = 46.35$; $y_{opt} = 127.86$. See details in text

It should be emphasized that analyzing the relationship between release and return in the Kalininka population of chum salmon, we excluded the data on the extremely high numbers of spawners returning to the river in 1978, 1979, and 1982, i.e., immediately after the 200-nautical mile zone was commissioned in 1977. This immediate increase in population size clearly demonstrates that chum salmon of southwestern Sakhalin had been intensively harvested for a long time by Japanese fishermen, who removed approximately two-thirds of reproductive fish from the hatchery population created and maintained by Russian specialists. Thus, throughout the preceding period of relatively steady production rate at the Kalininka hatchery (1964–1971), the annual return of spawners averaged 158,400 fish, whereas in the years 1978–1982 this value increased to 303,000 fish. The number of spawners decreased in subsequent years, but this was accounted for by overproduction of fry at the hatchery up to 1986.

Thus, monitoring natural populations of commercially important fish species reveals a rather uniform pattern of changes caused by anthropogenic pressure. In virtually all cases, these influences lead to *unfavorable genetic processes, i.e., to the type of reproduction of species gene pools that leads to deviation from the optimal ratio between intra- and interpopulation components of gene diversity and to the loss of "memory" about the previous state of the population.* This situation is the result of the failure of those involved in the industrial development of natural resources to pay attention to the historically developed subpopulation structure of affected populations. Even fish breeding, supposedly serving the cause of maintenance and artificial reproduction of biological resources, can lead to negative consequences (Makoedov 1999). As predicted in the first section, the latter are associated with redistribution of genetic diversity in such a way that its intrapopulation component $H_S$ decreases, whereas the interpopulation component $G_{ST}$ increases. The situation is typical for salmon hatcheries, where breeders either use an insufficient number of spawners, thus provoking inbreeding, or perform unintentional selection in favor of homozygotes (with the same effect in practice). This process is *inadaptive* and can lead to irreversible degradation of populations even after the influence is terminated.

Redistribution of the components of gene diversity via an increase in the level of intrapopulation heterozygosity was revealed during monitoring self-reproducing populations subjected to commercial fishing and also in Atlantic salmon populations artificially reproduced at hatcheries and grown at fish farms (Altukhov and Salmenkova 1994). Although this process is *adaptive*, it also leads to degradation of populations, because the cost of adaptation proves to be excessive. For example, highly productive anadromous populations are substituted by commercially useless resident forms, or mortality at early developmental stages increases (Ryman and Utter 1987; Altukhov 1990; Altukhov et al. 1991). Still, the opportune termination of an adverse external influence provides the possibility of the damaged subpopulation structure being restored and the genetic processes in the system returning to the norm.

These conclusions were derived from long-term observations based on the principles of monitoring discussed previously. Apparently, they are applicable to agricultural populations as well as to natural ones.

Unfortunately, this kind of approach to rationalize the salmon industry has not been applied so far, although there is no doubt about its exceptional importance. The existing criteria for gauging the effectiveness of fishing in natural stocks must also be changed, and the commercial exploitation of shoals must be combined with their artificial reproduction within the framework of a single, properly regulated system (Altukhov 1974; Altukhov et al. 1997, 2000; Altukhov and Evsyukov 2001).

The principles of population genetics are the same for all biological species. We have paid attention here to fish populations only because they in particular, on account of their practical importance, have in the last 15–20 years, using the methods of biochemical population genetics, become the subject of intense research. In turn, these works have made possible the disclosure of the systemic organization of isolated populations of other species known to have a different ecology and provide new data, as discussed in this chapter and others.

## 7.3
## Genetic Monitoring of Agricultural Populations

Reliable information obtained to date concerns varieties of barley, *Hordeum vulgare*, grown in East Siberia (Pomortsev et al. 1994) and various chicken breeds (*Gallus gallus*; Moiseeva et al. 1993).

In recent decades, about 60 new barley varieties have been produced, become commercial, and accepted for industrial use. Because virtually all these varieties are individual populations adapted to different agroecological zones, it is important to consider the problem of the direction of selection and its effect on the genetic diversity of such populations.

Table 7.9 shows data on changes in allelic composition of three hordein loci. These data were obtained by comparing the old spring barley varieties from East Siberia (Khabarovskii krai, Primorskii krai, Irkutskaya, and Chitinskaya oblast') with the new varieties adapted to these regions during the last 60 years. Alleles are designated according to Pomortsev et al. (Sozinov 1985).

Analysis of these data reveals significant changes in the heterogeneity of populations: initially, they were represented by a mixture of different genotypes (old varieties), but at present, the linear (new) varieties are predominant. The level of genetic variation in old varieties is significantly higher than in new: the proportion of forms with three or four alleles of the *Hrd B* locus reaches 30%, whereas the majority of new varieties carry only one or, at most, two alleles of hordein loci (Table 7.9). It is clear that, during the last 60 years, the genotypic composition of spring barley varieties cultivated in East Siberia has undergone significant changes manifested in the loss of genetic diversity. These changes were caused primarily by traditional breeding practice: new varieties are the progeny of a few plants or even only one plant.

The same tendency toward the loss of genetic diversity with time is also revealed by monitoring chicken populations. Among factors leading to the loss of genetic variability in poultry farming, we should note the drastically reduced number of breeds used for commercial purposes. Currently, commercial crossings involve only four to seven breeds out of 603 listed in Somes' catalogue (Somes 1985). In Russia, about 30 of 80 old breeds have not been preserved (or were not found). This corresponds to a 37.5% reduction of genetic resources (with respect to breed composition) during the last 50 years. Several other breeds are also on the brink of extinction (Moiseeva et al. 1993).

Data on the dynamics of genetic variability in poultry breeding and its more accurate quantitative estimates confirm the above facts. We analyzed the results of our own and other experimental studies on biochemical polymorphism of 48 chicken populations bred in Russia and abroad (Mediter-

**Table 7.9.** Heterogeneity of old and new barley varieties for *Hrd A*, *Hrd B*, and *Hrd F*. (Pomortsev et al. 1994)

Hrd A

| Local varieties | | | New varieties | | |
|---|---|---|---|---|---|
| Number of | | Proportion of | Number of | | Proportion of |
| Alleles | Varieties | varieties (%) | Alleles | Varieties | varieties (%) |
| 1 | 9 | 34.62 | 1 | 16 | 61.54 |
| 2 | 12 | 46.15 | 2 | 9 | 34.62 |
| 3 | 5 | 19.23 | 3 | 1 | 3.85 |

$$\chi^2 = 5.05; df = 2; P < 0.10$$

Hrd B

| Local varieties | | | New varieties | | |
|---|---|---|---|---|---|
| Number of | | Proportion of | Number of | | Proportion of |
| Alleles | Varieties | varieties (%) | Alleles | Varieties | varieties (%) |
| 1 | 9 | 34.62 | 1 | 20 | 76.92 |
| 2 | 9 | 34.62 | 2 | 6 | 23.08 |
| 3 | 6 | 23.08 | 3 | 0 | 0 |
| 4 | 2 | 7.69 | 4 | 0 | 0 |

$$\chi^2 = 9.22; df = 2; P < 0.01$$

Hrd F

| Local varieties | | | New varieties | | |
|---|---|---|---|---|---|
| Number of | | Proportion of | Number of | | Proportion of |
| Alleles | Varieties | varieties (%) | Alleles | Varieties | varieties (%) |
| 1 | 9 | 34.62 | 1 | 22 | 84.62 |
| 2 | 12 | 46.15 | 2 | 4 | 15.38 |
| 3 | 5 | 19.23 | 3 | 0 | 0 |

$$\chi^2 = 13.49; df = 2; P < 0.001$$

$$\sum \chi^2 = 27.77; df = 5; P < 0.001$$

ranean and Asian breeds), including the red jungle fowl, wild ancestor of the chicken (subspecies *Gallus gallus gallus*). The analysis was based on 16 loci encoding blood and egg proteins, including six polymorphic (*Ov, G-3, G-2, Tf, Alb, and Es-1*) and ten monomorphic loci (*AMY-3, Es-2, PGM, PHI, TO, MDH, LDH, Es-D, Hbl, and Hb2*). Each population was characterized by the allele frequencies of six loci. An expected genetic structure of a hypothetical ancestral population (prapopulation) was reconstructed by averaging the frequencies of alleles characteristic for 47 breeds.

The number of alleles per locus (Table 7.10) was usually lower in the groups of commercial and Mediterranean breeds. Relatively high estimates were obtained for the wild form, the hypothetical prapopulation, and the Asian breeds. With respect to the average level of heterozygosity, the groups

**Table 7.10.** Genetic diversity in chicken for 16 loci encoding blood and egg proteins and 18 loci controlling morphophysiological traits. (Moiseeva et al. 1993)

| Group of breeds | Number of breeds | Protein-encoding loci | | | | Number and proportion (%) of lost alleles controlling morphophysiological traits** |
| --- | --- | --- | --- | --- | --- | --- |
| | | Number of alleles per locus[a] | Percentage of polymorphic loci* | Average heterozygosity $H\pm$(SE) | Number and proportion (%) of lost alleles[b] | |
| Red Jungle Fowl | 1 | 1.44 | 31.25 | 0.091 (0.042) | 9 (39.1) | 19 (51.4) |
| Hypothetical prapopulation | 47 | 1.62 | 37.50 | 0.090 (0.042) | 7 (30.4) | 0 (0) |
| Russian and Ukrainian | 14 | 1.44 | 37.50 | 0.085 (0.041) | 9 (39.1) | 9 (24.3) |
| Mediterranean | 6 | 1.38 | 31.25 | 0.071 (0.035) | 11 (47.8) | 11 (29.7) |
| Asian | 11 | 1.62 | 37.50 | 0.097 (0.045) | 7 (30.4) | 2 (5.4) |
| Commercial | 6 | 1.38 | 31.25 | 0.093 (0.043) | 9 (39.1) | 13 (35.1) |

[a] Locus was considered polymorphic if the frequency of the rarest allele exceeded 0.01

[b] Twenty-three alleles of protein-encoding loci and 37 alleles of loci controlling morphophysiological traits were analyzed. The loss of alleles was estimated in relation to the world gene pool

of breeds were ranked in a similar way. Figure 7.9 shows the genetic profiles of the hypothetical prapopulation and several local breeds. It is seen that certain breeds have a unique structure, whereas others resemble the prapopulation because of their synthetic origin. The first group includes five populations: Orlovskaya, Pervomaiskaya, Russkaya Belaya, Leningradskaya Belaya, and Moskovskaya, and the second group includes nine other populations shown in Fig. 7.9.

The following fact is also important: local breeds similar to the prapopulation demonstrate a higher level of intrapopulation heterozygosity ($H_S = 0.213$) and lower interpopulation genetic diversity ($G_{ST} = 0.0975$) than the breeds genetically most distant from the prapopulation ($H_S = 0.183$, $G_{ST} = 0.2311$).

From the statistical standpoint, this should be expected. However, it is important that the close genetic relation to the prapopulation is linked with a lesser specialization of the breed: virtually all of the corresponding nine breeds are meat-and-egg-producing, whereas the distant five breeds

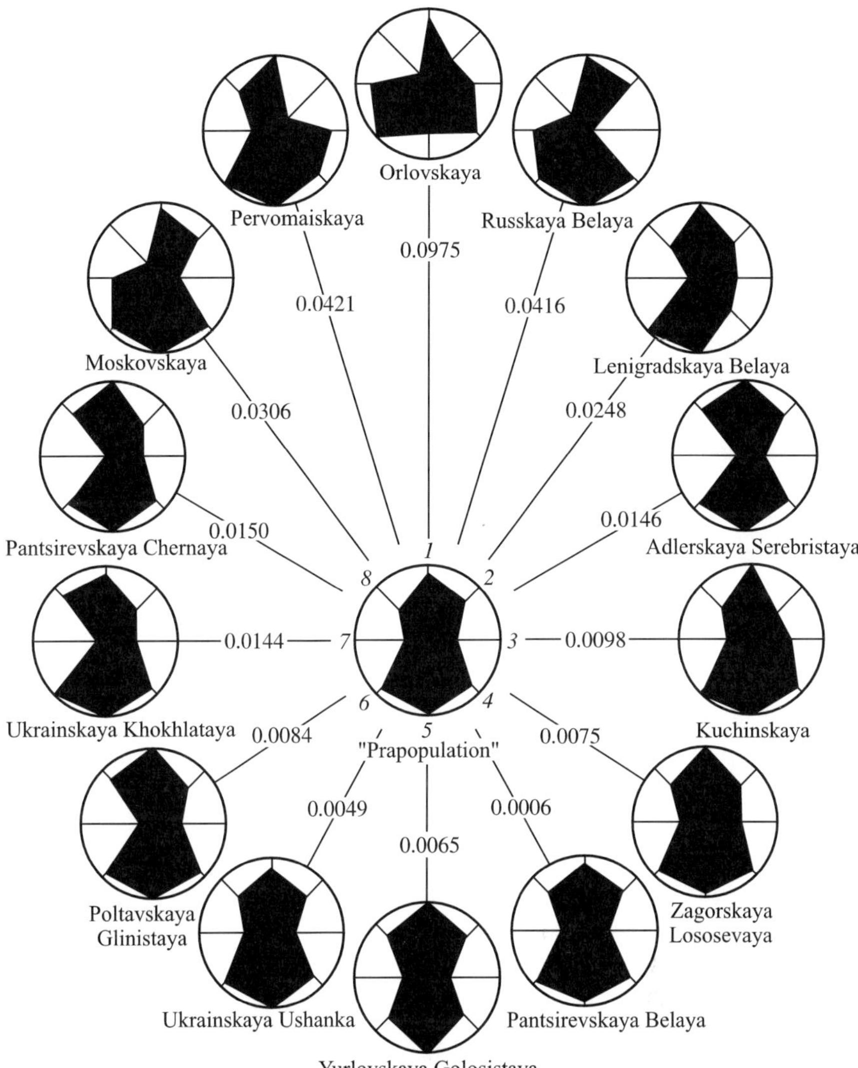



**Fig. 7.9.** Genetic profiles of several chicken breeds maintained in Russia and of their ancestral population. The *radii* show the frequencies of the following alleles: *Ov A (1)*; *G3-A (2)*; *G3-B (3)*; *G2-B (4)*; *Tf B(5)*; *Alb B (6)*; *Es-1A (7)*; *Es-1B (8)*. Interval of gene frequencies is 0–1 (in the *center* and on the *perimeter*, respectively). The *lines* connecting the breeds with the ancestral population show estimates of genetic distances according to Nei

are more specialized as either egg-producing (e.g., Russkaya Belaya or Moskovskaya) or meat-producing (e.g., Leningradskaya Belaya), and the Orlovskaya breed, the most distant from the ancestral population, was derived from fighting fowls.

A considerable interpopulation genetic differentiation, accompanied by a decreased level of heterozygosity in the specialized group of breeds, indicates that the process of selection, in this case, resulted in the loss of intrapopulation gene diversity. Breeds of another group demonstrate an opposite trend, i.e., an increase in the level of heterozygosity and the loss of breed specificity. Apparently, these results are the same as the detrimental genetic processes resulting from anthropogenic pressure that were described above for the natural populations. Evidently, in order to overcome the detrimental processes in populations, special approaches to genetic structure stabilization should be developed.

## 7.4
## The Stabilization Principles of the Genetic Structure of Agricultural Populations

In a previous section we emphasized the difference between the conditions in which populations of agricultural animals and plants are maintained and the environment inhabited by natural populations. However, this contrast practically disappears now. The wide dissemination of monocultures and industrial methods of farming, and the creation of gigantic stock-breeding complexes all lead ultimately to the kind of population-genetics problems that we encounter in the industrial utilization and artificial reproduction of different biological species in nature: the habitat of agricultural populations becomes increasingly strained, while their genetic diversity is even now being reduced by the very process of selection. As never before, the task of preserving the world's genetic resources, first mooted by N.I. Vavilov (1926, 1927), calls for urgent resolution (see also Ehrlich 1983).

To illustrate what has been said, it is sufficient to recall the events that accompanied the "green revolution". Now, years later, one can see only too well how the initial triumphs of certain wheat varieties, based on the selection of so-called minus variants, soon gave way to a more somber response as an increasing number of examples appeared of their weak genetic stability under specific environmental conditions (Chang 1979). The highly specialized varieties required the mass doses of mineral and fertilizers, and toxic substances battling their diseases became the cause of chemical pollution of environment. The "green revolution" disrupted ecological equilibrium and produced a number of other problems (Medvedev 1998). By taking formulations in the sphere of population genetics into consideration, these poor results could have been foreseen because specialization and improvement in the process of directional selection of some traits are inevitably linked with decrease and deterioration of others, accompanied by the usual "cost of selection" as a consequence of the existence of negative correlations within the system of integral ontogenesis (Schmalhauzen 1938; Belyayev and Trut 1964a,b; Belyayev 1974, 1980). But this kind of destabilization and disintegration of population genetic systems also occurs, for the same reason, in the selection of "plus variants", although even heterozygosity is unable to eradicate the negative effects – the segregation of less adapted genotypes in each generation is inevitable (albeit, of course, a population's genetic diversity can be maintained for longer periods of time).

Nevertheless, the natural question arises: should the strategy of selection always be aimed at the preferential increase of highly specialized genotypes? After all, the solution to the problem of the optimization of the selection process by improving and creating animal breeds and plant varieties maximally adapted to modern agricultural conditions requires that these populations consist of individuals that are reasonably similar in scale, form, and rate of development, that are highly productive, have broadly based nonspecific resistance, including immunity to diseases, yield high quality products, and so on. What can be done to ensure that a breed or a variety meets these demands, sometimes contradictory?

From the foregoing, it is clear that directional selection cannot solve this task.

However, there is plenty of evidence to confirm that individuals close to the population mean for quantitative characters frequently have maximum resistance to diversified fluctuations of the external and internal environment (see Chap. 5).

As shown in Chap. 5, the co-adaptive gene complexes formed by stabilizing selection are associated with an optimal "average" phenotype (adaptive norm) that ensures its "buffer state" and broad nonspecific stability, whereas "extreme" phenotypes that deviate from the optimum have other genetic characteristics, which are associated with reduced fitness. The development of methods of immunological and biochemical genetics, as already stressed, provides the opportunity, not previously available, of connecting the differentiation of population individuals by their quantitative adaptively significant traits, with differentiation based on a community of Mendelian genes.

In formulating this approach, we have developed a program in recent years for analyzing the principles of stabilizing the genetic structure of agricultural populations. Despite the fact that our studies related to such disparate species as the cotton plant and the Karakul sheep, the results are very similar. We have succeeded in showing that it is precisely the morphologically broadly "average" types which answer the needs of selection in the highest degree – in the first case, in creating varieties with increased wilt-resistance and suitability for mechanized processing and harvesting; and in the second case, in solving the problem of selecting pairs of parents for progeny with the desirable wool type (characterized by well-defined symmetry and smoothness in the wool coat) (Altukhov et al. 1976, 1978, 1980b; Altukhov and Sarsenbayev 1980; Sarsenbayev 1980).

How the idea works is simple. Because an organism's development depends on the interacting effects of many genes within a reasonably heterozygous polygenic genetic system, which determines "channeling" and the "buffered state" of ontogenesis as a whole, the following link is postulated: the more resistant ontogenesis is to diverse changes in external and internal environments, the closer such resistant individuals should approximate the average population characteristics in the aggregate of morpho-anatomical traits. By selecting these phenotypes (or obtaining them through crossings), we hoped to obtain the individuals that interested us. This necessitated analysis of the variability of several polygenic characters of constitution, concurrently with consideration of indicators of productiveness. It was also thought that the genetic heterogeneity of the initial material was fairly high.

Naturally, it is not easy to identify the average phenotype by a complex of traits. Especially if independent traits are involved, then, as their numbers increase, the numbers of "average" phenotypes in a research sample will become progressively smaller, since in multidimensional space an average type is merely an abstract. Hence our approach to the traits selected for classification was not arbitrary, the following factors being considered as paramount (Altukhov et al. 1980b):

1. Preference was given to correlated, normally distributed traits connected with the most important morphofunctional systems (growth, weight, body size, and proportions, etc.).

2. There was a priori evidence of high additive genetic variance of the initial population.

3. The heritability of the traits chosen for classification was quite high.

We designated the group of "average" individuals $M^0$, and groups having larger and smaller values with the traits $M^+$ and $M^-$, respectively. In addition, we singled out a group of so-called disproportional phenotypes – $M^\delta$ – that is, individuals with discordant combinations of indicators (for instance, "low weight", "high growth", etc.); the values of some of the traits are below the population average in some of these individuals, in others above (Zhivotovsky and Altukhov 1980).

We chose the mean (or modal) value of the $\bar{x}$ traits as our calculation point and then used the Euclidean distance to estimate the deviation of individual $x$ from the $\bar{x}$ distribution "center" by the formula:

$$d_0 = \left[ \frac{1}{p} \left(x - \bar{x}\right)^T D^{-1} \left(x - \bar{x}\right) \right]^{1/2},$$

where $p$ is the number of quantitative traits, $T$ is the transposition symbol, and $D$ the covariance matrix:

$$D = \frac{1}{N-1} \left( \sum_{i=1}^{N} x_i^T x_i - N \bar{x}^T \bar{x} \right) ,$$

with $N$ representing the sample size and $\bar{x}$ the vector of mean values.

Of course, individuals for whom $d_0 = 0$, or close to zero, hardly exist, in exactly the same way as there are very few individuals that deviate markedly from the mean in the values of all the quantitative traits. This is why the $pd_0$ distance distribution has a pronounced left-sided asymmetry that recalls the $\chi^2$ distribution when the number of degrees of freedom is more than 2. In ideal normality of the distributions of traits, the $d_0$ distance distribution coincides almost completely with the $\chi^2$ distribution (Zhivotovsky 1984).

If the distributions of traits differ from the normal (for example, in asymmetry and positive or negative excess), then the number of $M^0$ individuals may vary in one direction or another of that expected for an $\chi^2$ distribution, and this circumstance must be taken into account in selecting the traits.

For practical purposes, the $M^0$ group is identified by ranging all individuals at the $d_0$ value, the subsequent classification within certain limits. Naturally, all the other individuals come within the group of "extreme" phenotypes as having maximal $d_0$ distance values (Fig. 7.10). As a community of these individuals is heterogeneous in composition, being formed by the phenotypes $M^+$, $M^-$, and $M^\delta$, it must be further subdivided on the basis of the method of the principal components. Proceeding from the correlation matrix, the first component is found from

$$u = a_1 \frac{x_1}{S_1} + a_2 \frac{x_2}{S_2} + \ldots + a_p \frac{x_p}{S_p} ,$$

where $a_1, a_2, \ldots, a_p$ are the coordinates of the first eigenvector of the correlation matrix of the traits; $x_1, x_2, \ldots, x_p$ are the values of the traits, and $S_1, S_2, \ldots, S_p$ the estimates of their standard deviations.

The corresponding value is calculated for all "extreme" phenotypes which are then arranged in order, and among them groups $M^+$ (with maximal $u$ values), $M^-$ (with minimal $u$ values) and $M^\delta$ (with intermediate $u$ values). Individuals which compose group $M^\delta$ are characterized by the largest index of disproportionality (Zhivotovsky 1984):

$$\delta = \sqrt{\frac{\left(u - \bar{u}\right)^2}{p\lambda} - d_0^2} ,$$

where $\lambda$ is the maximal eigenvector and $\bar{u}$ the average of the sample.

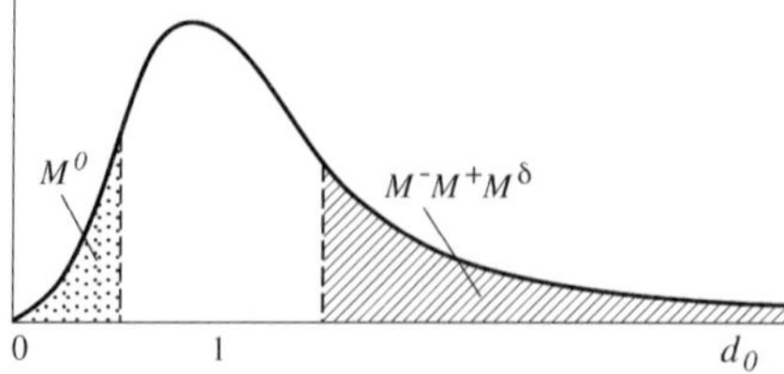

**Fig. 7.10.** The form of the distribution of $d_0$ distances, and the division of the group of "average" ($M^0$), "extreme" ($M^-$, $M^{0+}$) and disproportional phenotypes. (Zhivotovsky and Altukhov 1980)

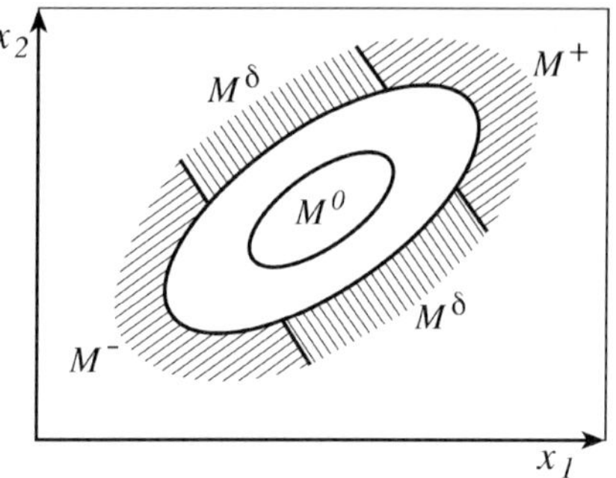

**Fig. 7.11.** The disposition of the groups $M^0$, $M^+$, $M^-$ and $M^\delta$ in space of the traits. (Zhivotovsky and Altukhov 1980)

A diagram of the location of four phenotypic groups in space of the traits is shown in Fig. 7.11.

Of course, the proposed method has defects (see Zhivotovsky and Altukhov 1980), but it is perfectly satisfactory for all practical purposes when the problem of classification is approached in an informal way, and when, in addition to statistics, intuition and research experience are taken into consideration in selecting traits. We have obtained uniform results that attest to the model's effectiveness when there is stabilization of the genetic structure of populations. Let us examine the results obtained in more detail, focusing first on experiments with *Gossypium hirsutum*, the cotton plant.

## 7.4.1
### The Effects of Modal Selection of the Cotton *Gossypium hirsutum*

Modal selection has been used once before on cotton (Manning 1955, 1956), but was subsequently declared to be ineffective (Arnold 1972). One must emphasize, however, that the theoretical premises and aims of our research and of the earlier works are different, and to be discussed at the end of this Section. Our work was conducted in the experimental station of the

Uzbek Academy's Institute for the Experimental Biology of Plants, near Tashkent. The following persons participated in the project: B. Abdullayev, L.P. Filatova, B.A. Kalabushkin, E.Ya. Tetushkin and V.D. Prokhorovskaya. Varieties of the "Tashkent" cotton plant, which have been widely used in production, were selected as experimental material. The history of creation of these varieties (Mirakhmedov and Yuldashev 1971; Mirakhmedov 1974; Simongulyan 1975) indicates their considerable genetic heterogeneity – an essential prerequisite for any work on artificial selection.

The scheme of analysis included the following stages.

**1. Collecting the Initial Material.**   The material was gathered at the end of the growing period in 1972 on allotments containing approximately 5,000 plants of the Tashkent-1, Tashkent-2, and Tashkent-3 varieties. The allotments were split up arbitrarily into squares comprising four rows of 40 – 50 plants each. One plant was selected from the center of a row, making a total of 100 plants of the Tashkent-1 variety and 40 plants of the two other varieties. This material represents the parental generation ($P$).

**2. The Morphological Standardization of Plants.**   Each plant was characterized by 17 traits connected with morphophysiological features of the vegetative organs such as height, number of fruit branches, total number of pods, percentage of shed fruit organs, etc. The distributions of several traits were transformed to the lognormal one.

Each trait was assigned three spheres of variability: $M^0$, $M^+$, and $M^-$ (Fig. 7.12). "Disproportional" plants were not included in the analysis as they rank below the $M^0$ and $M^+$ plant groups in average fitness (Tishkin and Glotov 1983).

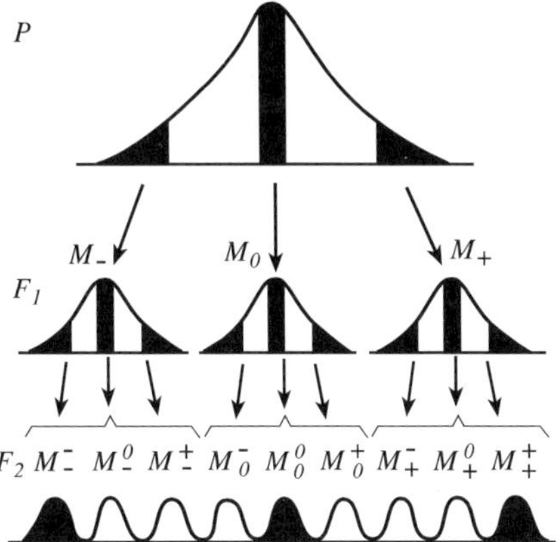


**Fig. 7.12.** Diagram of modal and directional selection based on the complex of traits for *Gossypium hirsutum*, the cotton plant. (Altukhov et al. 1976)

The variability of the generative organs linked with crop yield components was examined simultaneously (pod weight, fiber growth, fiber length, etc. – seven characters altogether).

To test the plants' progeny, seeds were sown in the spring of 1973 in the same area where they had been raised in the previous generation; in other words, the experiment's conditions were standardized for the whole of the test period. Control seeds of production crop plants were sown in the areas between the groups being compared. Sowings during the vegetation period received identical treatment (intercrop processing, watering, fertilization, etc.)

The progeny of each of the three groups of plants ($F_1$) were again classified on the same principles, so that the $F_2$ was represented by nine subgroups (Fig. 7.12). In the spring of 1974 the seeds from these subgroups and the control material were sown again under the same conditions. Wilt vulnerability of the plants was determined at the end of the growing period.

All the groups and subgroups were compared with each other and with the control population obtained from the production-sowing seed with respect to all the characters being used for the studies.

Calculations were made for the mean, dispersion, and variation coefficients in each of the subgroups and for the varieties overall, as well as for correlation coefficients between all the traits. The degree of genetic heterogeneity of the selected plant groups was estimated from the data for their progeny by means of the $R_W$ coefficient of intraclass correlation (Ginsburg 1968; see also Altukhov et al. 1976, 1978).

Additionally, the characteristics of the morphobiological differentiation of varieties were investigated with the object of deciding whether selection should be made only within each variety or, if there were no substantial differences among the varieties, whether it would be possible to utilize the entire initial plant group as a whole. In this connection the cotton plant varieties were compared with each other for all the research traits (taking correlated links into account) three times – in 1972, 1973, and 1974. It proved to be the case that there were no fundamental differences among the Tashkent-1, -2, and -3 varieties in the majority of traits and hence, for the purposes of our experiment the plants should not be contrasted with each other as varieties, but must be regarded as an overall parental community selected from a heterogeneous Tashkent cotton plant population that was, nevertheless, single in its origin. However, subsequent research concentrated chiefly on the Tashkent-1 variety as being the most representative of the material studied (Table 7.11).

Table 7.12 provides an evaluation of the population's genetic variation with regard to different traits of the vegetative organs, from which it follows that the most variable traits are nos. 1, 2, 3, 5, and others. The genetic determination of variability traits nos. 12, 15, 16, and others is somewhat

**Table 7.11.** Morphological features of the "Tashkent-1" variety of the cotton-plant (initial material 1972)

| Trait number | Trait heading | Statistical parameters | | |
|---|---|---|---|---|
| | | $X \pm m$ | $\sigma$ | C. V. |
| 1 | Height of plants, cm | 99.5±1.9 | 19.4 | 19.4 |
| 2 | Number of fruit branches | 17.1±0.2 | 2.5 | 14.5 |
| 3 | Number of internodes on the main stalk | 16.6±0.2 | 2.6 | 15.4 |
| 4 | Total length of internodes, cm | 81.5±1.9 | 19.5 | 23.9 |
| 5 | Average length of internodes, cm | 5.3±0.3 | 3.5 | 66.3 |
| 6 | Number of nodes on fruit branches | 35.0±1.4 | 14.0 | 40.0 |
| 7 | Average length of internodes on fruit branches, cm | 6.3±0.2 | 1.8 | 28.7 |
| 8 | Total length of all the fruit branches, cm | 222.1±11.3 | 113.4 | 51.1 |
| 9 | Height of plant to first fruit branches, cm | 6.1±0.1 | 1.0 | 15.9 |
| 10 | Total number of pods per shrub | 17.1±0.9 | 8.6 | 50.6 |
| 11 | Number of opened pods ($\approx 15.X$) | 9.9±0.6 | 6.1 | 61.9 |
| 12 | Percentage of open pods (15.X) | 60.1±2.5 | 24.7 | 41.2 |
| 13 | Number of shed fruit-bearing organs | 19.2±1.6 | 16.1 | 83.8 |
| 14 | Percentage of shed fruit-bearing organs | 51.4±1.0 | 10.5 | 20.4 |
| 15 | Branching type | 1.9±0.1 | 0.9 | 49.7 |
| 16 | Planting level of low pods, cm | 18.0±0.9 | 8.7 | 48.1 |
| 17 | Width of shrub, cm | 53.5±1.9 | 19.6 | 36.7 |

**Table 7.12.** Coefficients of intraclass correlation ($R_w$) in a group of plants obtained during the modal selection process

| Statistical parameters | Year | Trait number | | | | | | | | |
|---|---|---|---|---|---|---|---|---|---|---|
| | | 1 | 2 | 3 | 4 | 5 | 6 | 7 | 8 | 9 |
| $R_w$ | 1973 | 0.18 | 0.22 | 0.30 | 0.14 | 0.33 | 0.25 | 0.29 | 0.21 | 0.13 |
| $F$ | | 3.86 | 4.58 | 6.35 | 3.02 | 7.35 | 5.23 | 6.14 | 4.41 | 2.89 |
| $R_w$ | 1974 | 0.26 | 0.18 | 0.10 | 0.22 | 0.24 | 0.17 | 0.33 | 0.17 | 0.12 |
| $F$ | | 9.26 | 6.20 | 3.79 | 7.72 | 8.50 | 5.88 | 12.8 | 5.96 | 4.22 |

| Statistical parameters | Year | Trait number | | | | | | | |
|---|---|---|---|---|---|---|---|---|---|
| | | 10 | 11 | 12 | 13 | 14 | 15 | 16 | 17 |
| $R_w$ | 1973 | 0.21 | 0.25 | 0.09 | 0.24 | 0.15 | 0.11 | 0.12 | 0.11 |
| $F$ | | 4.44 | 5.28 | 2.29 | 4.93 | 3.22 | 2.53 | 2.52 | 2.51 |
| $R_w$ | 1974 | 0.10 | 0.08 | 0.18 | 0.18 | 0.16 | 0.07 | 0.21 | 0.07 |
| $F$ | | 4.43 | 3.76 | 3.07 | 6.45 | 5.77 | 2.73 | 7.33 | 2.82 |

All the $F$ values are significant at a level of $P \leq 0.01 - 0.001$

**Table 7.13.** The coefficients of intraclass correlation, showing measurement of genetic determination of population diversity in cotton from the characteristics of its generative organs

| Trait number | Trait heading | $x$ | $S_e^2$ | $S_{mg}^2$ | $S_a^2$ | $R_w$ | $F$ |
|---|---|---|---|---|---|---|---|
| 1 | Number of sections per pod | 4.45 | 0.59 | 2.97 | 0.12 | 0.167 | 5.02** |
| 2 | Number of seeds per pod | 28.32 | 46.49 | 99.07 | 2.64 | 0.054 | 2.13 |
| 3 | Fiber weight per pod, g | 2.44 | 0.44 | 2.46 | 0.10 | 0.187 | 5.62** |
| 4 | Seed weight per pod, g | 3.63 | 0.95 | 4.74 | 0.19 | 0.167 | 4.97 |
| 5 | Pod weight, g | 5.71 | 2.68 | 15.22 | 0.63 | 0.190 | 5.68 |
| 6 | Fiber yield, % | 36.43 | 3.24 | 13.40 | 0.50 | 0.136 | 4.14* |
| 7 | Fiber length, mm | 33.83 | 7.07 | 9.63 | 0.13 | 0.018 | 1.36 |

* Significant at the level of $P = 0.05$. ** Significant at the level of $P = 0.01$

less, but even in these cases intergroup variance reliably exceeds the intragroup.

The values of intraclass correlation coefficient $R_W$, were also determined in the control population for the characteristics of the structure and development of the generative organs (Table 7.13). On average, as one would expect, the heritability of these characters was lower when compared with the characters for the plants' vegetative organs.

**3. Variability and Expression of Economic Value Traits in Groups of Plants Subjected to Modal and Directional Selection.**   Even a glance at the graphed characteristics of variability in the compared plant groups shows the effects of selection – in the $M^0$ group several traits reveal that deviations of individual plants from mean values in the control are substantially less than in the $M^+$ and $M^-$ groups (Fig. 7.13). Modal selection led to reduced variability, already clearly exhibited in the second generation (Fig. 7.14).

The data grouped together in Table 7.14 show that the plants of the $M^0$ group are almost as good as the control population in yield, weight,

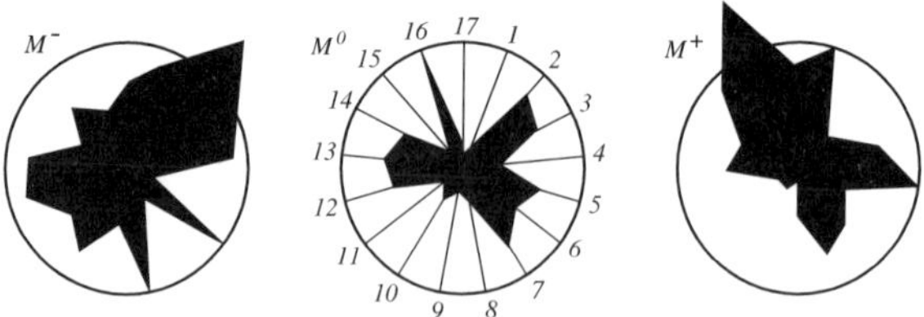


**Fig. 7.13.** Characteristics of the normalized deviations of average trait values in the $M^0$, $M^+$, and $M^-$ cotton plant groups from average control values (*circle* with a unit radius; Altukhov et al. 1976). A 0 frequency at the *perimeter* and 1 frequency at the *center* of the circle. The probability of randomness of differences for the majority of traits is $P < 0.01–0.001$

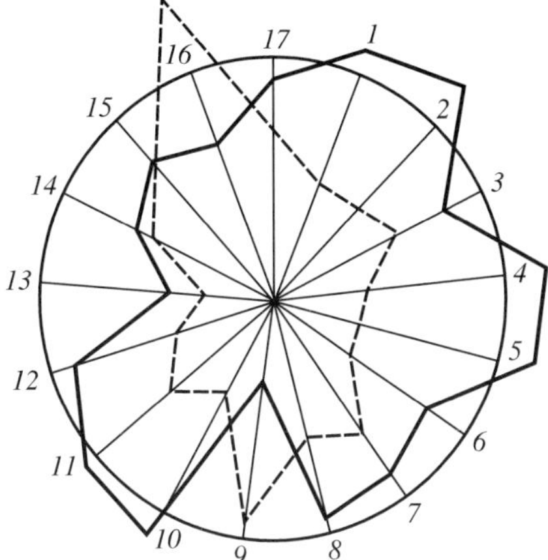


**Fig. 7.14.** The variability of structural and developmental characteristics of the vegetative organs of the selected $M^0$ plant group compared with the control (*circumference*; Altukhov et al. 1976). The *paths* plot the ratio of the dispersions in the $M^0$ group to trait dispersions in the control. *Continuous line* First generation of modal selection; *dotted line* second generation of selection. The differences between the $M^0$ group and the control in 16 traits are statistically significant, $P < 0.01–0.001$

fiber length, and especially in the rate at which the pods open (Table 7.15). This latter trait is exceedingly important when one considers that usually, because of protracted ripening, the cotton plant is harvested by machines at least twice a season.

But of no less significance is the fact that the $M^0$ plant group also undoubtedly surpasses the $M^+$ and $M^-$ as well as the control group in the degree of phenotypic uniformity. In virtually all the traits, the modal population is typified by less dispersion, so that the proportion of plants which most meet the requirement of mechanized harvesting is maximal in this group (Table 7.16).

The plants obtained in the process of modal selection also proved to be more wilt-resistant, both when compared with extreme variants and also (which is particularly important) in comparison with the plants grown from the seed of the normal field crop (Table 7.17). Similar differentiation is also found with respect to the dynamics of field seed germination (Fig. 7.15).

In discussing the question of plant genotypes having an "optimal" phenotype, it should be noted that in a cross-pollinated population[1] this may be accompanied by a high level of heterozygosity (Chap. 5).

---

[1] The cotton plant is self-pollinated, although the quantitative assessment of the potential contribution made by cross-pollination to the formation of different populations under varied conditions of the environment still has not been fully clarified. This question is particularly complex regarding the Tashkent strains and requires investigation.

**Table 7.14.** Comparison for crop yield of groups of plants, obtained by modal and directional selection

| Compared groups and subgroups of plants | Sampling volumes | Number of ripe pods at the end of the growth stage | Weight of one pod (g) | Number of analyzed pods | Fiber output (%) | Fiber length (mm) | Crop yield per plant (g) | Shedding of fruit bearing elements (%) |
|---|---|---|---|---|---|---|---|---|
| Initial material (1972) | 180 | 18.3±0.6 | 5.7±0.1 | 60 | 35.99±0.35 | 33.23±0.16 | 102.3±11.7 | 52.6±0.8 |
| Control (1973) | 174 | 22.4±0.9 | 5.9±0.2 | 60 | 36.04±0.37 | 33.44±0.22 | 126.2±12.8 | 52.9±0.7 |
| $M^-$ | 166 | 18.0±0.7 | 5.7±0.2 | 20 | 37.34±0.57 | 33.63±0.35 | 101.0±10.3 | 56.8±0.7 |
| $M^0$ | 357 | 23.0±0.6 | 6.0±0.1 | 60 | 39.81±0.28 | 33.80±0.12 | 136.0±13.8 | 51.2±0.4 |
| $M^+$ | 213 | 29.1±1.1 | 6.3±0.2 | 38 | 36.10±0.47 | 33.11±0.30 | 181.0±18.3 | 52.2±0.6 |
| Control (1974) | 152 | 18.6±0.7 | 5.8±0.2 | 60 | 35.67±0.44 | 32.75±0.23 | 106.6±10.6 | 53.2±0.7 |
| $M^-$ | 114 | 12.5±0.6 | 6.0±0.3 | 40 | 36.43±0.74 | 33.06±0.29 | 72.7±7.2 | 51.3±0.8 |
| $M^0$ | 644 | 15.9±0.3 | 6.4±0.1 | 40 | 36.20±0.33 | 33.55±0.20 | 101.2±5.4 | 47.8±0.3 |
| $M^+$ | 130 | 19.5±0.6 | 6.6±0.4 | 24 | 35.50±0.83 | 32.19±0.36 | 130.1±12.7 | 54.9±0.7 |
| Control (1976) | 300 | 13.6±0.2 | 5.7±0.2 | 150 | 35.68±0.31 | 32.75±0.19 | 75.5±12.8 | 53.0±0.2 |
| $M^0$ | 300 | 15.2±0.2 | 6.2±0.2 | 150 | 36.60±0.32 | 33.48±0.20 | 92.2±6.9 | 50.6±0.3 |

**Table 7.15.** Early ripening of different groups of cotton

| Plant groups | Sampling volumes | Number of days from source until the first pods begin to open | Percent of opened pods (about 15.$X$) | Percentage of plants with 100% opening of pods (about 15.$X$) |
|---|---|---|---|---|
| Control (1973) | 174 | 137.2±1.6 | 70.0±3.5 | 16.1±2.6 |
| $M^-$ | 166 | 133.9±1.7 | 83.3±2.9 | 37.3±3.4 |
| $M^0$ | 357 | 135.6±0.9 | 77.7±2.2 | 19.3±1.7 |
| $M^+$ | 213 | 139.6±1.3 | 47.7±3.4 | 5.2±2.1 |
| Control (1974) | 152 | 140.4±1.5 | 88.4±2.6 | 27.0±1.6 |
| $M^-$ | 114 | 132.0±2.4 | 97.8±1.3 | 78.9±1.6 |
| $M^0$ | 644 | 134.7±0.5 | 93.6±0.9 | 63.2±0.9 |
| $M^+$ | 130 | 142.9±1.7 | 87.8±2.8 | 28.5±1.9 |
| Control (1976) | 300 | 135.6±1.5 | 72.3±0.4 | 21.0±1.2 |
| $M^0$ | 300 | 131.1±1.2 | 87.0±2.5 | 41.7±1.9 |

**Table 7.16.** The suitability of the cotton groups for mechanized harvesting

| Plant groups | Sampling volumes | Proportion of optimal type plants (%) according to: | | |
|---|---|---|---|---|
| | | Height | Planting level of low pods | Branching type |
| Control (1974) | 152 | 44.7±4.0 | 50.6±4.0 | 62.5±3.9 |
| $M^-$ | 114 | 35.1±4.5 | 50.0±4.7 | 63.7±4.4 |
| $M^0$ | 644 | 59.3±1.9 | 61.8±1.9 | 73.6±1.7 |
| $M^+$ | 130 | 28.5±3.9 | 41.5±4.3 | 50.0±4.3 |
| Control (1976) | 300 | 37.7±2.6 | 53.8±2.4 | 40.0±2.6 |
| $M^0$ | 300 | 70.3±2.7 | 59.7±2.5 | 76.0±2.7 |

**Table 7.17.** The degree of wilt susceptibility of three groups of cotton, obtained by directional and modal selection

| Plant groups | Sampling volumes | Proportion of observed plants (about 15.$X$) (%) | | | |
|---|---|---|---|---|---|
| | | Slightly affected | Moderately affected | Greatly affected | Total |
| Initial material (1972) | 180 | 8.88 | 2.77 | 2.22 | 13.4±2.5 |
| Control (1973) | 174 | 7.47 | 4.02 | 2.29 | 13.8±2.6 |
| $M^-$ | 166 | 10.84 | 6.02 | – | 16.9±3.0 |
| $M^0$ | 357 | 5.32 | 1.68 | 2.24 | 9.2±1.5 |
| $M^+$ | 213 | 4.22 | 7.51 | 2.34 | 14.1±2.4 |
| Control (1974) | 152 | 3.28 | 4.60 | 2.63 | 10.9±2.4 |
| $M^-$ | 114 | 7.89 | 0.87 | – | 8.8±2.7 |
| $M^0$ | 644 | 2.01 | 1.55 | – | 3.6±0.9 |
| $M^+$ | 130 | 5.38 | 2.30 | 10.76 | 18.6±3.4 |
| Control (1976) | 900 | 2.88 | 2.88 | 2.11 | 7.9±0.6 |
| $M^0$ | 900 | 2.00 | 1.55 | 1.33 | 4.9±0.6 |

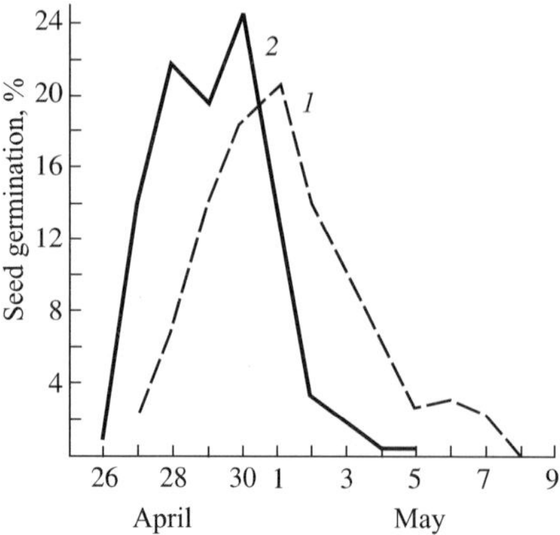

Fig. 7.15. The dynamics of field seed germination of plants obtained by modal selection.
1 Control; 2 $M^0$

If this is so, then the classical model of genetic homeostasis enables one to understand the reason for the increased stability and hence, adaptability of the group of plants resulting from the modal selection process. It is clear, too, that a population of this kind can only be maintained with adequate continuity of generation intervals through crossing the more homozygous strains, which in this case consist of groups that have been formed under the effects of directional selection.

Another genetic model based on Wright's concepts of an "intermediary optimum" can also be enlisted to explain the effects of modal selection (see, also Latter 1960); in this instance, it is not necessary for the plants to be represented as heterozygous genotypes; the selection of self-pollinators presents no obstacles as regards the problem of preserving "average" phenotypes or those close to it. Finally, a third model is possible which combines the first two and presupposes an optimal average level of heterozygosity in the $M^0$ group. In any case the identification of plant genotypes with minimum deviation from the average in the aggregate of morphobiological traits is important in the seed production for the next generation.

The further analysis of the effects of modal selection in terms of the adaptive norm concept is exceptionally important for the solution of problems faced in selection programs.

Our findings demonstrate the effectiveness of modal selection, which, as indicated, contradicts the results of work by Arnold (1972) who reached a negative conclusion. Let us try to explain the cause of this difference in findings.

1. Manning (1955, 1956) proposed modal selection as a method of obtaining a fairly stable control variety of the BP-52 plant that could serve as a standard for the crop yield of different selected strains of the Ugandan

cotton plant. *Of greatest importance is the fact that the* BP-52 *variety comes from a single plant selected from a population.* This is the first fundamental difference between Manning's approach and ours, which is based on obtaining randomized sampling from a reasonably heterogeneous population.

2. In the selection program, Manning (1955, 1956), Walker (1964), and Arnold (1972) used only three to five traits, linked to harvest yield or the generative organs: the number of pods per plant, namely, the number of seeds in a pod, fiber length, fiber output, and pod size. Thus, the morphological characteristics of plants, based on their vegetative organs, were not included in the selection scheme and, as we have seen, the heritability of the traits of the generative organs is small. This is the second way in which our approach differs fundamentally from that adopted by these researchers. It is only by utilizing the whole complex of traits and characteristics of genetically heterogeneous original material that one can achieve the desired result in modal selection.

In fact, there is no way in which the effects revealed by us can be ascribed to environmental influences (as Arnold postulates in the case of the BP-52 variety), as they are directly connected with the genetic characteristics of the populations; despite sharp population differences caused by the environment in 1973 and 1974, the difference between the control and the modal group of plants became perceptibly more accentuated in the second generation of selection.

A further task is to formulate methods of assessing the structure of selected cotton populations directly from their allozyme genes (Shapoval et al. 1992). Nevertheless the data obtained can be used even now as a requisite basis for seed-production aimed at improving not only cotton plant varieties but other valuable agricultural crops as well (Kodzoev et al. 2001). It is evident that a standardized variety *with an average but stable harvest yield* has greater economic value than a *specialized* variety *with a potentially high, but strongly fluctuating* harvest yield.

Let us now analyze the possibility of implementing our model in livestock breeding, by looking for links between the dressed Karakul lambskin pattern type and the external characteristics of Karakul sheep in the early stages of postnatal ontogenesis.

## 7.4.2
### The Pattern of Dressed Karakul Lambskin Related to Morphologically "Average" and "Extreme" Types

A very important trend in breeding Karakul sheep is the choice of parents for obtaining progeny that have a desirable type of lambskin. Although, as is well known, tastes change with time, specialists are unanimous in their belief that the jacket type of lambskin is, and has been for many years, one of the best (Vasin 1939, 1971; Shirinsky 1962, 1975; D'yachkov 1968, 1975, 1980; Koshevoy 1975). The chief merit of this kind of dressed lambskin is its symmetrical pattern and the uniformity of the chief traits that characterize the quality of a wool coat (Fig. 7.16).

Selection for the jacket type began in the 1930s (Nikolsky 1976), but the selection process of concentrating specifically on Karakul lambskin traits reached a plateau beyond which further efforts appeared to be of no avail (Shirinsky 1962; Baatar 1968). For example, even in 1977, the proportion of jacket-type lambs raised in Kazakhstan did not exceed 35.7%, while the frequency of animals having the most valuable lambskin ("jacket-1") was only 2.3%. Evidently, this kind of picture was observed because the development of the main morphological characteristics of the Karakul type also depends on the interaction of many genes within the polygenic system discussed above.

If we wish to increase the proportion of progeny with the Karakul lambskin pattern that interests us we should select the parents so that the

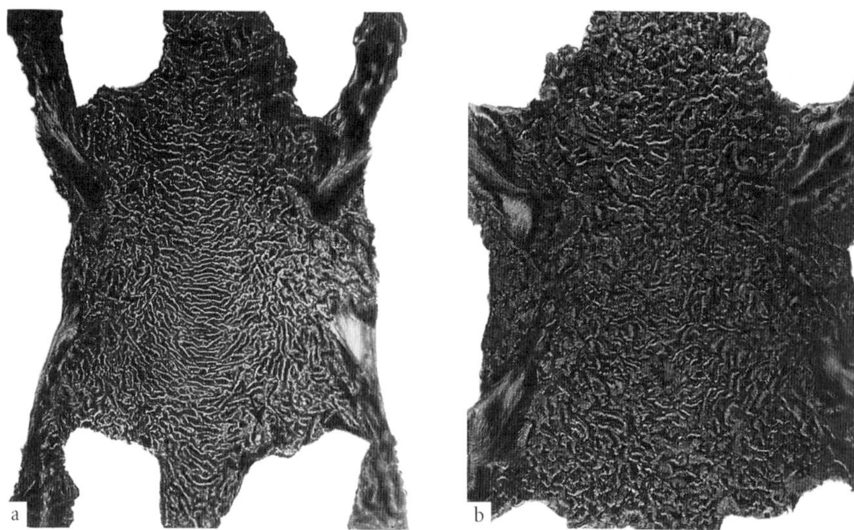

**Fig. 7.16.** Depicts "desirable" (**a**) and "undesirable" (**b**) types of dressed lambskin coat (Altukhov and Sarsenbayev 1980). **a** "Jacket" type; average curls, long semicircular rolls, clear concentric and symmetrical pattern, good luster and silkiness, reasonably fine skin; **b** "Caucasian" type; excessively large curls, short rolls, large bobs; the silkiness and luster of the hair are poor, and the pattern is indeterminate

progeny, roughly speaking, have "average" constitutional characteristics. The "Zadarinsky" State Pedigree Farm (Kazakh SSR) provided us with a base for putting this idea through its paces. From the day when it was founded in 1940, the farm has specialized in raising black Karakul sheep introduced from five Uzbekistani farms and four Turkmenian farms. This was a very important factor in our research, as it indicated the herd's considerable genetic variability. The selection program was in principle the same as for the cotton plant.

**Collecting the Initial Material.**   The material was gathered from a herd of some 3,000 sheep. Sampling was randomized in order to cover the population's gene pool reasonably comprehensively. Immediately after birth each lamb was categorized according to a series of polygenic traits reflecting constitutional characteristics (weight, snout-to-vent length, chest girth, etc.) and also lambskin quality (curl length and width, hair length, skin thickness, pattern type, etc.). In addition, electrophoresis was used to determine the genotypes in 19 pedigree rams, 101 ewes, and 256 lambs at transferrin, hemoglobin, and serum esterase loci (Altukhov et al. 1980b) with a view to studying population genetic structure.

**Morphological Standardization and the Division into "Average" and "Extreme" Types.** The division into "average" and "extreme" types was accomplished simultaneously by means of five correlated traits reflecting constitutional characteristics (Table 7.18). Subsequently we used three traits: weight at birth, oblique snout-to-vent length, and chest girth.

**Table 7.18.** The variability of newborn Karakul lambs by five quantitative-traits ($n = 618$)

| No. | Researched traits | $\bar{x} \pm m$ | $\sigma$ | Correlation coefficient, $r \pm SE$ |
|---|---|---|---|---|
| 1 | Weight at birth, kg | 4.40±0.03 | 0.69 | 0.82±0.007 |
|   |   |   |   | 0.80±0.008 |
|   |   |   |   | 0.83±0.007 |
|   |   |   |   | 0.90±0.006 |
| 2 | Height at the withers, cm | 37.68±0.09 | 2.23 | 0.77±0.009 |
|   |   |   |   | 0.73±0.011 |
|   |   |   |   | 0.82±0.007 |
| 3 | Depth of chest, cm | 14.27±0.05 | 1.17 | 0.74±0.010 |
|   |   |   |   | 0.78±0.009 |
| 4 | Oblique snout-to-vent length, cm | 32.31±0.08 | 2.00 | 0.83±0.007 |
| 5 | Chest girth, cm | 38.11±0.10 | 2.43 | |

"Average" individuals were those that deviated at each trait by not more than 1.0 $\sigma$ to the right and 0.5 to the left of the distribution center[1]. We designated this group as $M^0$ (the "adaptive norm"), representing the animals by the following trait values: (1) from 4,100 to 4,700 g; (2) from 31 to 34 cm; and (3) from 37 to 40 cm. The trait values for the $M^+$ and $M^-$ groups were: (1) > 5, 100 and < 3,600 g; (2) > 36 and < 29 cm; and (3) > 41 and < 35 cm. Lambs with "discordant" trait combinations were excluded from the analysis: here we shall only examine sheep which are "biologically" average or extreme types, using a series of three traits.

Following this standardization, the average group, the "adaptive norm", ($M^0$) consisted of 216 sheep, and the $M^-$ and $M^+$ groups comprised 101 and 97, respectively.

**Evaluating Lambskin Quality.** The following traits were used to characterize the optimal type of lambskin, based on the formulas of several researchers concerned with Karakul lambskin quality (D'yachkov 1968, 1975; Vasin 1971; Koshevoy 1975; Shirinsky 1975; and "Instructions for the commercial estimation of Karakul lambs" in Shirinsky 1975):

| | |
|---|---|
| Karakul lambskin type | Jacket |
| Pattern type | Parallel–concentric |
| Curl length, mm | 30.0 and over |
| Curl width, mm | 4.0 to 8.0 |
| Hair length, mm | Not more than 10.0 |
| Skin thickness, mm | 1.8 to 2.5 |
| Silkiness of wool coat | Strong, normal |
| Luster of wool coat | Strong, normal |

From these criteria we were able to determine the frequency of animals having the most valuable lambskin properties among the three lamb groups ($M^-$, $M^0$ and $M^+$) based only on constitutional characteristics.

**Karakul Lambskin Pattern Features and the Genetic Characteristics of Morphologically Different Karakul Groups.** Because variability of quantitative traits, such as weight and growth, largely incorporate an environmental component, we constructed our classification to take into account the variability of these characters in newborn lambs; that is, at a time when individual differences, determined by genotype, are more in evidence than at subsequent stages of ontogenesis.

Indeed, if the correlation between the exterior traits and those that characterize lambskin quality is examined during research on the same animals at different stages of ontogenesis, not only are the links found to

---

[1] This classification scheme is connected with the fact that a population's maximum fitness, estimated by the variability of weight–growth factors, is often displaced somewhat "to the right" of the distribution center (see Karn and Penrose 1951; Altukhov et al. 1979b, 1981; Zhivotovsky and Feldman 1992).

be weakened but their character also changes. So, for instance, whereas the weight at birth correlates positively with curl length and width, hair length, and skin thickness, at 1.5 years these links become unreliable, and in two cases their sign reverses (Altukhov 1983). This tendency is already found in the group of 4–5 month old lambs; moreover, the weakened correlation is observed both when constitutional characteristics are compared with the wool coat traits, and in the constitutional features themselves: whereas correlation at birth between the live weight and four other external traits is fairly strong ($\bar{r} = 0.68$), by the age of 18 months this association diminishes sharply in strength ($\bar{r} = 0.13$).

There is nothing unexpected about these results, they only reconfirm, for the several analyzed traits, what had already been observed for individual traits (Glembotsky and Bogolyubova 1940). However, at the same time, this means that when one attempts to establish constitutional traits and lambskin quality one must not base the work on the constitutional types of the adult sheep, but take the earliest stage of postnatal ontogenesis into account.

Indeed, in this way, we found a clear frequency difference in the type of lambskin that interests us among these groups of lambs that differ in their phenotypes: in the $M^0$ group the frequency of the jacket-type lambs with a parallel-concentric lambskin pattern is 43.05%, compared to 18.8% and 22.7%, respectively, in the $M^-$ and $M^+$ groups (Table 7.19). Of the $M^0$ group, 28.2% had a "jacket-1" type of lambskin. It is clear even from the lambskin quality traits that the average Karakuls are also close to the optimum with respect both to the magnitude of the traits (Table 7.20) and the scale of their variation – the maximum uniformity of traits such as hair length and curl length and width are characteristic of the $M^0$ group animals (Fig. 7.17).

Hence, our findings establish a clear connection between lambskin type and external animal characteristics at early stages of ontogenesis. Con-

**Table 7.19.** The frequency (as %) of different Karakul lambskin types in three groups of lambs, classified by constitutional characteristics

| Lambskin type | $M^-$ ($n = 101$) | $M^0$ ($n = 216$) | $M^+$ ($n = 97$) | Statistical estimate of differences with frequency of the jacket type ($F$-test) |
|---|---|---|---|---|
| Jacket | 18.80 | 43.05 | 22.70 | |
| Ribbed | 5.90 | 2.80 | 8.20 | $M^-, M^+ -2.18$ |
| Flat | 1.98 | 3.24 | 4.12 | $M^-, M^0 -18.7^{***}$ |
| Caucasian | 73.32 | 50.91 | 64.98 | $M^+, M^0 -6.5^{**}$ |
| Total | 100 | 100 | 100 | |

The average population frequency of lambs having parallel concentric pattern configurations is 28.48±1.82%

**Table 7.20.** Lambskin quality traits in three groups of Karakul lambs differing in constitutional characteristics

| Compound groups | Sampling volume | Fleece length (mm) ($\bar{x} \pm m$) | Fleece width (mm) ($\bar{x} \pm m$) | Hair length (mm) ($\bar{x} \pm m$) | Skin thickness (mm) ($\bar{x} \pm m$) |
|---|---|---|---|---|---|
| $M^-$ | 101 | 39.33±1.69 | 4.88±0.09 | 9.30±0.12 | 1.69±0.03 |
| $M^0$ | 216 | 47.70±1.43 | 5.19±0.05 | 9.54±0.07 | 2.10±0.02 |
| $M^+$ | 97 | 48.13±2.64 | 5.48±0.07 | 10.15±0.13 | 2.69±0.05 |

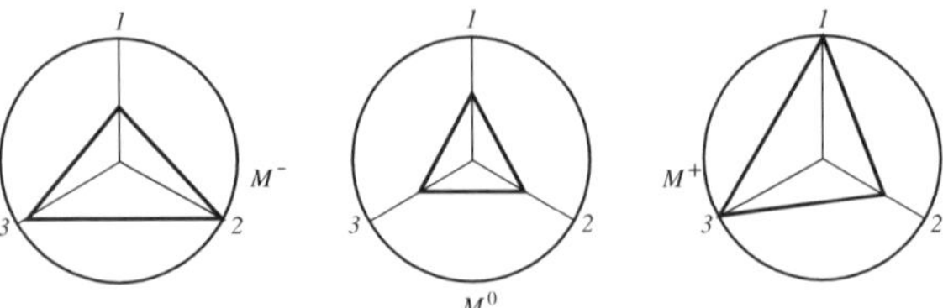


**Fig. 7.17.** Comparison of the coefficient of variation of three quantitative traits that characterize the skin in Karakul lamb groups $M^0$, $M^-$, and $M^+$. The maximal variability value of each trait in the group is taken as 100% and corresponds to the *radius* of the circle. *1* Curl length; *2* curl width; and *3* hair length

sequently, the way has been paved for the more purposeful selection of parents in order to obtain more progeny having the desirable type of lambskin. It is evident that the crossing of extreme morphotypes, or those close to them is of special interest; that it is important to analyze different variants of crossings in the process; and that a wider range of qualitative traits that characterize a lamb's morphological features could be used to identify early the most promising individuals. Unfortunately, it is not so easy to carry out this work, as the characteristics of stud rams on the "Zadarinsky" Pedigree Farm conform mainly to type $M^+$. Also there is the further consideration that the possibility of error in pedigree records cannot be ruled out (Sarsenbayev and Afanas'ev 1984).

Nevertheless, 136 eighteen-month-old experimental sheep were artificially inseminated by six stud rams and were classified at birth in accordance with the scheme outlined above. Table 7.21 shows the results. As expected, the maximum frequency of morphologically average lambs was found in the progeny of mothers belonging to group $M^-$.

Although the size of the groups compared is not large, the trends revealed are quite evident. If, in order to increase the sample size, we combine the data for the $M^0$ and $M^+$ animal types, the difference in the frequency of progeny having the desired type of lambskin, between this mixed group, on one hand, and the $M^-$ groups of mothers, on the other, is significant – 40% compared to 62% ($F_\varphi = 4.07$; $P < 0.05$). Repeated experiments have given similar results (Altukhov et al. 1987b).

**Table 7.21.** Frequency of Karakul lambs of morphologically "average" ($M^0$) and "extreme" ($M^+$ and $M^-$) types in the progeny of the corresponding group of mothers

| Mothers | $M^0$ | $M^-$ | $M^+$ |
|---|---|---|---|
| $M^-$ | 0.61±0.10 | 0.04±0.04 | 0.35±0.10 |
| $M^0$ | 0.38±0.07 | 0.10±0.04 | 0.52±0.08 |
| $M^+$ | 0.43±0.09 | 0.20±0.07 | 0.37 ±0.09 |

Let us examine the genetic features of the three Karakul lamb groups for the blood hemoglobin, transferrin, and esterase loci. The main results follow (for details, see Altukhov et al. 1980b).

1. The heterozygosity level at the *Hb* locus was higher for the $M^0$ and $M^+$ groups, genetically identical in this trait, than for the $M^-$ group. A similar finding was seen for the transferrin locus, although the $M^+$ was more heterozygous than the $M^0$ group. Groups $M^-$ and $M^+$ were somewhat more heterozygous than group $M^0$ at the esterase locus.

2. When comparing the morphologically average and extreme groups of lambs at the three protein loci simultaneously, it transpired that minimal and maximal levels of total heterozygosity were characteristic of the $M^-$ and $M^+$ groups, respectively, whereas the heterozygosity level of group $M^0$ was average. However, one should point out that these kinds of differences should only be regarded as a tendency, because a reliable evaluation of heterozygosity requires a larger number of loci to be examined.

3. Rare genotypes of individual loci were found most often in lambs belonging to the extreme morphological types of $M^+$ and $M^-$. This was particularly noticeable in the case of the transferrin (*DD*, *JS*, and *MS*) and esterase (*FS* and *SS*) genotypes. The *Tf S* rare allele is encountered only in the $M^+$ and $M^-$ groups.

Reduced viability is characteristic of rare genotypes, and the effects of negative selection are more considerable in the extreme $M^-$ group, where the frequency of rare genotype combinations is substantially higher, than in the morphologically "average" sheep (Table 7.22). Indeed, in the older age group, the frequency of the heterozygotes at the transferrin locus increased as the proportion of *MM* homozygotes (0.0083 for the stud rams, against 0.0125 in newborn lambs) and *JJ* homozygotes (0.05 against 0.09) fell. A similar picture was observed at the esterase locus. At the same time genotypes with average frequency predominated in the average $M^0$ group.

Hence, it can now be accepted that a correlation exists not only between lambskin pattern type and constitutional characteristics at birth, but also between these two characteristics of an individual and its genotype. It will be realized that these connections being statistical in nature only register

**Table 7.22.** Frequencies of different genotypic combinations

| Researched groups | Sample size | Genotypic combinations | | | $F_\varphi$ |
| --- | --- | --- | --- | --- | --- |
| | | "Frequent" (>0.12) | "Average" (0.04–0.12) | "Rare" (<0.04) | |
| $M^0$ | 67 | 0.1642 | 0.3721 | 0.4627 | $M^0, M^- -5.00{**}$ |
| $M^-$ | 64 | 0.1406 | 0.2031 | 0.6563 | $M^0, M^+ -0.35$ |
| $M^+$ | 74 | 0.1486 | 0.3378 | 0.5135 | $M^-, M^+ -2.92$ |

$**P < 0.01$

the "average" at a population level. It is also clear that further interpretation of this kind of interrelationship between genotype and phenotype calls for an extended range of biochemical markers. Nevertheless, at least one conclusion can be drawn – that the lamb group with low weight–growth traits at birth has a lower level of heterozygosity.

Regardless of whether a morphologically average type would be subsequently associated with the "maximum" or "optimum" total heterozygosity (as the set of the analyzing genetic markers is being extended), the following clearly emerges: the generally accepted selection scheme includes only a selection of breeding rams of the $M^0 \times M^+$ type and so, because of complex segregation cannot lead to a marked increase in Karakuls with the most valuable lambskin parallel-concentric configuration.

Nevertheless, this scheme still preserves genetic diversity of population, because if the opposite were true it would be difficult to implement our model on the basis of an available gene pool, especially considering the widespread use of artificial insemination on farms.

The discovery of the link between the constitutional characteristics of the Karakul lambs and the lambskin pattern type is not something new – an analysis of literature on Karakul farming shows that this kind of correlation has already been noticed in one form or another (Yudin 1955; Koshevoy 1975). However, the point is that no attempts have been made before this to approach the analysis of this question from the viewpoint of population genetics, especially taking into account the clear connection between a genotypic structure and the stability of ontogenesis as a whole. With this in mind, it has been found that the morphologically average type is characterized not only by increased viability, but also by the broad nonspecific stability, a "buffered state", in early ontogenetic stages. Apparently the appropriate gene complex controls the optimal, harmonic development of any morphofunctional system in which the qualities of symmetry and "standardization" occupy leading positions in the expression of the traits.

The realization of this model in practice will make it necessary to revise several outdated selection principles at present widely used in Karakul breeding, as well as in other branches of animal husbandry. In these traditional systems, selection is aimed against the $M^-$ group of individuals, and partly against those in the $M^0$ group, hence it will inevitably lead to the loss of rare genotypes, a reduction in population genetic diversity, and diminished population fitness. The importance of our approach, which aims to surmount these negative signs by compromising between directional and stabilizing selection, is confirmed by a whole series of studies that have been conducted in recent years (Bell 1978; Gorin et al. 1978, 1980; Bondarenko et al. 1979; Kovalenko et al. 1980; Podstreshny 1981; Andriyasheva and Lokshina 1985, 1987; Kodzoev et al. 2001).

All these works have shown that individuals of a modal class are characterized by maximal fitness and are economically more profitable with respect to the strategy of attaining long-term, lasting effects. It is curious that one publication 20 years ago, devoted to the practical tasks of poultry farming, was actually entitled "Compromise essential to success in breeding" (Moultrie 1984).

On a theoretical level, the further formulation of the question may play a decisive role in bridging the gap existing between population genetics expressed, on the one hand, in terms of the dynamics of genotype and gene frequencies, and on the other, in terms of the variability of polygenic morphological traits. This gives greater understanding of the functional organization characteristics of eukaryotic genes and also of the mechanisms of population genetic stability (= adaptation) as determined by the joint effect of many genes. There is no doubt that this stability, which we have called *nonspecific* (Altukhov et al. 1979b), is more widespread than the *specific* stability controlled by individual Mendelian genes.

Despite the large numbers of Mendelian loci known to exist in complex organisms, I do not anticipate any great technical difficulties in realizing this program, which postulates the simultaneous study of an aggregate of polygenic, adaptively significant traits and polymorphic loci that encode the synthesis of different proteins. In fact, although work is only beginning in this direction, already the result can clearly be seen in a general way – a reduced level of heterozygosity in individuals, associated with low values of constitutional traits. In the cases of the better-adapted "average" phenotypes and those that deviate in the aggregate of traits such as "to the right" of the distribution center, these are evidently the effects, respectively, of selection for optimum and for maximum gene diversity, as discussed in Chap. 5. The crux of the matter is that we may encounter manifold effects of individual genes or combinations of them, on fitness. Apart from overdominance, the simple adaptive effects of independent loci cannot be excluded, nor the stable co-adapted gene blocks that are non-randomly distributed in the field of the normal variability of separate quantitative traits or their aggregates. Evidence in support of this possibility has been obtained experimentally (Marshall and Allard 1970a,b; Allard et al. 1972; Allard and Kahler 1973; Sozinov et al. 1973; Weir et al. 1974; Sozinov and Poperelya 1979; Sozinov 1985, 1985) and in the theoretical formulation of the problem (Lewontin 1978a; Zhivotovsky 1981a, 1982; Altukhov et al. 2000).

In any case, we would like to emphasize that optimizing selection (or the corresponding choice of pairs to obtain the "average" phenotypes) on constitutional traits should be coupled with moderate directional selection for commercial traits. This is of principal importance for preservation of valuable properties of the populations, stabilization of their genetic structure, and enhancement of nonspecific resistance to various fluctuations of both the external and internal environment (Altukhov et al. 1987b).

Let us now consider the significance of factors and conditions of genetic stability of populations in view of the state of the environment and specific genetic processes in populations of us humans.

## 7.5
## Genetic Processes in Modern Human Populations: the Environment and the Problem of Genetic Load

The same micro-evolutionary processes that we encounter in studying different animal and plant populations are also significant for the human species. There is only one difference: by transforming the world that surrounds him, man is able to change the environment in which he lives and his own population organization in a way that no other living things on the earth are able to do. Therefore, while speaking about the unlimited scope of man's social progress we should not forget the need to preserve the genetic status of *Homo sapiens* as a biological species formed in the process of a long evolutionary development. This problem of preserving the human gene pool in a drastically changing environment is of universal interest today. In the concluding act for European security and cooperation, signed in Helsinki on 1 August 1975, the participating states declared that "the protection and improvement of the environment, as well as the preservation of nature and the rational utilization of its resources in the interests of present and future generations is a task of great importance for the prosperity of the nations and the economic development of all countries ..." The most important spheres of cooperation indicated include "... predicting potential genetic changes in the plant and animal worlds produced by contamination of the environment and ... gauging the effects of levels of contamination and degradation of the environment on man's health."

Indeed, these tasks are crucial. In addition to radiation, today humans are affected by a plethora of diverse chemical substances widely used in industry, agriculture, medicine, and everyday life. For instance, in the mid-1970s, over 500,000 chemicals were in use in the world with about 10,000 of them produced in the quantity of 0.5 to 1.0 million kg each (The International Registry of Potentially Toxic Chemicals, UNEP Sp. Iss., Sept. 1975). Tens of thousands of various chemicals are in everyday use. The most reliable data on this issue are presented by the Chemical Abstract Services (CAS) registry. According to its official site, as of 14 January 2002, 19,149,111 organic and inorganic chemicals have been registered, of which 3,302,383 are commercially available. Documents regulating their use exist for 226,372 of the inventoried chemicals (http://www.cas.org/cgi-bin/regreport.pl).

## 7.5.1
## Test Systems and Mutagenesis

Chemical mutagens were discovered in the 1940s (Öehlkers 1943; Rapoport 1946; Auerbach and Robson 1946) and abundant research has been carried out on them since. However, it has only become apparent quite recently that some chemical compounds are being converted from an experimental tool into a permanent element of the biosphere in our times.

The danger of environment pollution by mutagens lies in the fact that the vast majority of the newly emerging mutations, which have not been ground on evolution's "polishing stone", have a negative effect on all organisms, including man. Where damage of the genetic apparatus of gametic cells is concerned, the negative consequences are accompanied by increased frequency in the population of mutant genes and chromosomes; in other words, by the growth of the mutational load. Such processes in somatic cells can increase the incidence of cancers. As to the mutational load dynamics in human populations, this problem is still unclear. However, UNGAR, based on the results of monitoring the populations of Hungary and British Columbia and on the advances in understanding the nature of multifactorial human disorders, has made a principal revision of the earlier estimates of the mutation load, which is currently represented by the striking figure 738,000/1,000,000 (including multigenic pathologies and extrapolating for the total life span up to 70 years of age) (Table 7.23). If these estimates are correct, they require a revision of a number of medical genetic and demographic forecasts made by UNCAR in the 1970s and 1980s (Shevchenko 2001).

Because genetic adaptation to such mutagenic factors cannot have evolved, the need for immediate measures to prevent their entry into the environment is patent. Another no less important task is to assess the extent to which the changes that have already taken place in the biosphere can affect mutation rates and the mutational load in human populations.

**Table 7.23.** UNGAR Estimation of the mutational load per 1 million of newborns in human populations (Shevchenko 2001)

| Classification of diseases | 1982 | 1986 | 1993 | 2001 |
|---|---|---|---|---|
| Autosomal dominant and X-linked | 10,000 | 10,000 | 10,000 | 16,500 |
| Autosomal recessive | 2,500 | 2,500 | 2,500 | 7,500 |
| Chromosomal | 3,400 | 3,800 | 4,000 | 4,000 |
| Chronic multifactorial | 90,000 | 600,000 | 650,000 | 650,000 |
| Congenital disorders | 90,000 | 60,000 | 60,000 | 60,000 |
| Total | 105,900 | 676,300 | 726,500 | 738,000 |

There are two main ways of solving these problems that face modern genetics: (1) by creating sensitive experimental *test systems* for detecting mutagenic environmental factors and (2) by formulating *genetic monitoring* principles for long-term observation of the dynamics of the hereditary burden of human populations.

In recent years, various laboratories around the world have devised over 200 diversified test systems. Some of them were already known from studies of the mutagenic effects of radiation, while others have been developed to allow for the specifics of chemical mutagenesis. Nonetheless, practical application for the estimate of genetic safety is found in no more that 20 test systems (Abilev and Poroshenko 1986; Poroshenko and Abilev 1988; Instruction 1989; Shigaeva et al. 1994; Hughes et al. 1998; Keshava and Ong 1999; Snyder and Green 2001).

The main theoretical and experimental requirements of a test system are that it should:

1. reveal all types of genetic damage – genomic, chromosomal and genic mutations;
2. be sensitive enough to pinpoint the effects of even small doses of mutagens;
3. be explicit and cheap, providing reliable and reproducible results;
4. detect compounds which, harmless in standard use, can become mutagens as a result of intracellular metabolic activation when they enter the human body;
5. allow quantitative evaluation of genetic risk to man to be extrapolated from the results.

Clearly, a single, universal test system that would meet all these requirements is hardly possible (Bochkov et al. 1975). Nevertheless, by utilizing a series (or combinations) of different tests, specialists are quite successful in establishing the mutagenic effects of many chemical compounds, in reaching practical conclusions, and making constructive recommendations. Thus the discovery of a highly reliable correlation between the mutagenic and carcinogenic effects of the same chemical compounds – a connection found in at least 60–90% of cases (McCan and Ames 1975, 1976; Rosenkranz et al. 1999) – must rank as one of the most significant results in applying test systems.

An extensive literature has been devoted to the many aspects of formulating and implementing test systems (see, for instance, Hollander 1971, 1973, 1976, 1978; Kilbey 1977; Dubinin and Pashin 1978; Hollstein et al. 1978; Altukhov 1980b; International commission 1982; Regulatory approaches 1983; Ehling et al. 1985; Favor 1986a,b; Dubinin 1986a; Ehling 1986, 1988; Bochkov 2001). At present, efforts are focused on developing standard international guidelines for testing chemicals for mutagenic activity. This is related to the global integration drive urging unification of the national testing guidelines. In 1995, an expert group of the International Program on Chemical Safety (IPCS) advanced the project "Harmonization of approaches to evaluation chemical impact risk." In Europe, the corresponding work has been conducted under the auspices of the Organization on Economic Cooperation and Development (OECD) (Kikuchi et al. 1999; Albertini et al. 2000).

New variants of test systems based on the advances of immunological and molecular genetics have been devised (Waldren 1983; Waldren et al. 1986; Dubrova et al. 1993; Shelby and Tindall 1997; Kassie et al. 2000; Kohara et al. 2001). Attempts at computer-aided prediction of mutagenic activity of chemical substances have been made, based on their structures (Lyubimova et al. 2001).

In several countries growing numbers of synthesized chemical compounds have reached the legislative stage of testing for mutagenicity and carcinogenicity (International Commission 1982; Regulatory Approaches 1983). More than 3,000 of these compounds with mutagenic effects have been found over the years; some of them have been eliminated from the environment, while previous estimates of the "safe" doses of others have been revised.

Nevertheless, it should be clearly recognized that test systems designed to discover the mutagenic activity of different environmental pollutants do not enable one to conceptualize the mutational dynamics process directly in human populations. As we have pointed out, this requires genetic monitoring. In this approach the populations themselves ought to become indicators of the state of the environment. When unfavorable processes are revealed, the underlying principles of their causes and effects should be identified and a system of measures specified to avert detrimental tendencies. Methods for analyzing the principles of genetically monitoring human populations are being formulated in different countries (Vogel 1970; Crow 1971; Neel 1972, 1974, 1980; Altukhov et al. 1976; Dubinin 1976, 1986a; Bochkov 1977,

1980; Altukhov 1980b; Shram 1981; Shram and Kuleshov 1984; etc.). This research can be run on specially selected population groups (for instance, high risk occupational groups) as well as on children in large populations by cytological, biochemical, or molecular genetic mass screening.

## 7.5.2
## Genetic Monitoring

However, the difficulties created by polluting the biosphere with mutagens is only one genetic aspect of the problem of "Man and his environment." An area even less well studied is the potentially undesirable effects of radical changes in the population genetic organization of our own species, constituted by a multiplicity of isolated subpopulations for most of its natural history, the latter characterized by a certain inbreeding level equilibrated by small gene migrations.

This type of structure is now to be found only on the perimeter of the ecumene, in ancient isolates of "primitive" peoples (Neel 1970; Neil et al. 1964; Rychkov 1968), while the overwhelming mass of present day population is typified by something completely different: the direction and intensity of migration are changing so much that from a state of subdivision, populations are transiting to a broad panmixia phase. This is particularly well illustrated by urban populations of economically developed countries. Although almost no studies have been made of the consequences of this process, it is clear that a modern multimillion city population has an enormous genetic effective size. It represents a giant cauldron, a center of panmixia, intensively brewing together genetic information from a large number of neighboring and remote populations whose gene pools have been differentiated both by random drift and by adaptation to differing environmental conditions (Kurbatova 1975; Rychkov 1979; Altukhov and Kurbatova 1984).

We are faced with the far from simple task of quantitatively evaluating the contribution made by the main micro-evolutionary factors to the genetic dynamics of human populations at the modern stage of their development. Random genetic drift, having lost its former importance, is being upstaged by the mutation process, by migration and selection, as we have stressed. Regarding the latter, many authors believe that it continues to lose its previous role because of successes in medicine. Whether in fact this is so, we shall consider somewhat later; but now, in order not to break the sequence of our exposition, let us return to the problem of gene migration.

What is the intensity of gene migration in a modern urban population?

The following unique information was obtained by O.L. Kurbatova (1975, 1977) during her genetic demographic research projects. According to her findings, the present day coefficient of gene migration to a large Moscow population may be in excess of 50% per generation; this is in stark contrast to what is known about the ancient systems of isolates, whose migration intensity did not exceed a few percent per generation (Rychkov and Sheremet'yeva 1977). Such a powerful gene flow can lead to an urban population's original gene pool being almost entirely replaced in three to five generations. The large urban populations are often not self-reproducing and have no real analogs in known population theoretical models.

The radius of centripetal migration is being continuously increased in time thanks to the development of communication and transportation, leading to increased population mobility. This leads to a considerable enlargement in the "marriage circle" and the extension of exogamous marriages. For example, whereas the range of the marriage contacts of a rural population in the European part of Russia at the end of the last century was mostly limited to 10 km (Zhomova 1965), even before the end of the Soviet era the radius for those living in its capital covered almost the entire country (Kurbatova 1975, 1977; Rychkov 1979; Kurbatova et al. 1984).

Noting this world population trend, Vogel and Motulsky (1997) suggested that it may merge into one huge panmictic population in the not-too-distant future. However, recently it became clear that even in a large urban population there exists positive assortative mating for ethnic, demographic, and anthropometric characters, which counteracts the effects of panmixia. From the viewpoint of genetics, the processes of migration and amalgamation of different ethnic components (*outbreeding*) are usually regarded as the antithesis of *inbreeding* that is, as being biologically favorable and leading (because of the growth of heterozygosity in a population) to the "enrichment" of heredity (heterosis), the concealment of harmful recessive genes, and so forth (Efroimson 1968; Dubinin and Shevchenko 1976). However, there is an opposite view, formulated most clearly by N.P. Bochkov (1978, pp. 172–173): "Mankind has never before been confronted by outbreeding on such a scale; there is great need to know what its consequences for a population's health are.... The results of the few studies undertaken in an effort to resolve this matter are contradictory. Conjectures about the advantages to man of heterozygous states that lead to reduced incidence of recessive diseases are not proved and contrary data can be adduced (such as the destruction of gene complexes and increased spontaneous mutation rate in heterozygotization, etc.). Research on inbreeding can and should be supplemented by research into the effects of outbreeding in human populations."

In fact, if the heterosis is not constant, each population is compelled to pay for it by the segregation of less adapted genotypes, and the cost of adaptation, as we know, can be extremely high (Chap. 2). And so we come to the problem of evaluating the intensity and directions of natural selection in modern human populations. Many social measures and the successes of medicine are believed to have considerably reduced the effects of selection, especially in the industrially developed countries. Nevertheless, as shown by L. Penrose's (1955) estimates for the European population, 15% of human embryos die in the early stages of development (spontaneous abortions), 3% are stillborn, 2% die neonatally, 3% die before reaching the age of reproduction, 20% of individuals do not marry, and 10% of marriages are infertile. These figures have not changed much over the last 50 years, so one concludes that *no less than half a primary* gene pool fails to reproduce itself in a *following generation.* Even if one admits that the genetic component of all these phenomena is no more than 20–30%, the selection intensity is not lower than 10–15% per generation!

Whether the character of selection has changed is another matter, as the total selection index developed by Crow (1958) shows. An estimate can be made from demographic statistical data by singling out components of fitness such as differential mortality ($I_m$) and differential fertility ($I_f$) : $I_m = P_d/P_s$, $I_f = V_k/\bar{k}^2$, where $P_d$ is the proportion of individuals that have died before reaching reproductive age; $P_s$ is the proportion of persons surviving to reproductive age; $\bar{k}$ is the mean number of offspring per person surviving to reproductive age; and $V_k$ is the variance in the number of children.

As defined by Crow (1958), the total selection index $I = I_m + 1/P_s I_f$ regards all differences of fitness as genetically conditioned; therefore, it represents a higher potential estimate of the intensity of selective pressure.

Turning to populations at different stages of development, it is easy to see that the $I_m$ component is maximal for a population with a primitive agrarian economy and minimal for industrially developed countries in which the $I_f$ component that is linked with differential fertility, conversely, reaches maximal magnitude (Table 7.24). For the populations of the industrialized countries, this implies a dramatic and prolonged reduction of mortality at the late stages of postnatal ontogeny, which has led to an unprecedented rise of life expectancy and aging of the population owing to the beneficial sociocultural environment (Table 7.25).

Naturally, these differences should not all be ascribed to genetic causes. Although the estimation of the genotypic component of individual and group fitness differences in humans seems problematic, this does not mean that no efforts should be made in this direction. Quite the contrary, this is the only way to develop the approaches toward quantitative estimation of mutational and segregational components of hereditary burden in the modern human populations and thus to prediction of adverse genetic processes. To do this, we should go back to the notions of the adaptive norm and genetic load, which, albeit already formulated in the general sense in modern genetics (see Sect. 1.3), are not sufficiently developed for human populations.

**Table 7.24.** Index of total selection ($I$) and its components, linked with differential death-rate ($I_m$) and differential fertility ($I_f$) in different human populations. (Neel and Chagnon 1968)

| Parameter | Indians of Xavanta | Uganda, Bantu, Basoga | Uganda, Bantu, Benoro | E. Pakistan, rural population | USA, white population | USA, non-white population |
|---|---|---|---|---|---|---|
| $P_d$ | 0.33 | 0.64 | 0.63 | 0.31 | 0.03 | 0.05 |
| $I_m$ | 0.49 | 1.78 | 1.70 | 0.45 | 0.03 | 0.05 |
| $\bar{k}$ | 4.7 | 7.0 | 3.7 | 7.0 | 2.2 | 2.7 |
| $\sigma_k^2$ | 6.1 | 7.2 | 7.4 | 10.9 | 4.3 | 9.8 |
| $I_f/P_S$ | 0.41 | 0.42 | 1.46 | 0.32 | 0.92 | 1.41 |
| $I$ | 0.90 | 2.20 | 3.16 | 0.77 | 0.95 | 1.46 |

**Table 7.25.** Life expectancy (years) (Demographic Yearbook of Russia 2001)

| Country | Males | Females | Δ | Country | Males | Females | Δ |
|---|---|---|---|---|---|---|---|
| Great Britain | 74.7 | 80.0 | 5.3 | USA | 73.5 | 80.2 | 6.7 |
| France | 74.4 | 82.1 | 7.7 | Japan | 76.9 | 83.0 | 6.1 |
| Germany | 74.1 | 80.3 | 6.2 | Average | 74.7 | 81.1 | 6.4 |

## Segregational Load

It must be remembered that views about the adaptive norm were formed on the strength of work carried out in the 1940s and 1950s on Drosophila and chickens that led to the discovery of co-adapted polygenic systems in populations, destructible both during inbreeding and outbreeding. In Lerner's theory of genetic homeostasis, the best-adapted genotype is associated with heterozygosity, while the nonadapted phenotype – the "phenodeviant" – is connected with failure to reach a certain threshold of heterozygosity. On the basis of these views, Wallace and Madden (1953) defined the adaptive norm as a community of individuals whose viability is within the limits of two standard deviations higher and lower than the average viability of heterozygotes, and hence they assigned genetic load to individuals whose fitness is more than two standard deviations lower than the average fitness of the population. However, this approach is beset with difficulties, primarily because genotype fitness can change under different conditions of the environment and, figuratively speaking, what may be all very well here and now can be bad there and then. This factor also has a bearing on the concept of the genetic load that provoked heated discussion as soon as Crow (see Sect. 1.3) defined it in 1958 as the relative reduction of a population's fitness compared with the optimal genotype. As is well known, attempts to apply what was considered the optimal genotype (phenotype) as a denominator for man proved unsuccessful.

Furthermore, the concept of the optimum (norm) is of fundamental importance to biology and medicine, especially when it involves distinguishing an unfavorable state or process. These cannot be assessed and prognosticated without a base reference point.

I believe that the increased fitness of a morphologically "average" phenotype may provide the solution to this difficult situation – the relevant data for different animal and plant populations have already been examined in previous chapters. As far as man is concerned, the clearest analogy is to be found in the well-known article of Karn and Penrose (1951); they used copious material to study distribution of mortality in newborn children in relation to body weight (Fig. 7.18). The association they showed, confirmed by later work (Sansing and Chinicci 1976; Terrenato et al. 1981a,b; Ulizzi et al. 1981; Promboon et al. 1983; Terrenato 1983), has turned into a typical example for demonstrating the action of stabilizing selection and "canalization" of development (in Waddington's sense) at early stages of ontogenesis in the given case. But evidently *this approach can also be used for monitoring the genetic load in human populations*, or, which means the same, *for tracing a population's fitness in time*. However, before proceeding to a more detailed analysis of this model, the two following items must be discussed.

First, it is expedient if the term *segregational load* is defined more widely to signify the presence in a population of the genotypes of one or more polymorphic (segregating) loci that have a negative effect on the fitness of their carriers; this provides independence from a model with overdominance.

Second, one must recall formulated views about *normal and unfavorable genetic processes in populations* (Altukhov 1978, 1984). By *normal process* we mean the type of transmission of hereditary information in

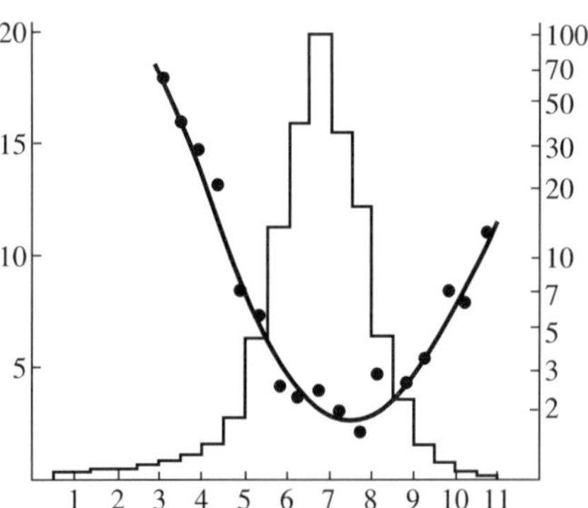

**Fig. 7.18.** The correlation between the body weight of newborn children (in pounds, histogram) and their mortality (*curve*), *y*-axis, *left* the proportion of children with corresponding body mass (%); *y*-axis, *right* mortality (%, logarithmic scale). Based on Karn and Penrose (1951); from Cavalli–Sforza and Bodmer (1971). (Reproduced with permission from W.H. Freeman and Company/Worth Publishers)

generations where the genetic load remains constant and corresponds to the historically fixed point of ecological equilibrium. In these conditions the gene pool of a population is preserved unchanged and is faithfully reproduced in subsequent generations, thereby maintaining equilibrium within a stable environment. By an *unfavorable process* we mean the type of genetic dynamics connected to an increased load of hereditary burden in the generations of a population compelled to adapt itself to a deteriorating environment. Under these conditions, the level of genetic variability in the population will increase, leading to its reduced fitness.

These comments make it clear that if an unfavorable genetic process occurs in a population, then in terms of the adaptive-norm concept, expressed by the optimal "average" phenotype, this means nothing other than the reduction in time of the frequency of the most adapted, morphologically average individuals, and the increase in the proportion of individuals with extreme and, especially, small values of quantitative traits. In a statistical sense, the increased frequency should result in a significant deviation of distribution of polygenic anthropometric traits from the initial distribution taken as the base reference point (Fig. 7.19).

Despite the simplicity of the proposed approach, one should point out the difficulties connected with its implementation.

First, Karn and Penrose showed the maximum fitness of an average phenotype with respect to such extreme characteristics as child mortality at birth. But what about other, less dramatic indicators of viability?

Second, they took only one trait, though also polygenic – body weight. But it is obvious that every phenotype is a regulated community of traits in multidimensional space.

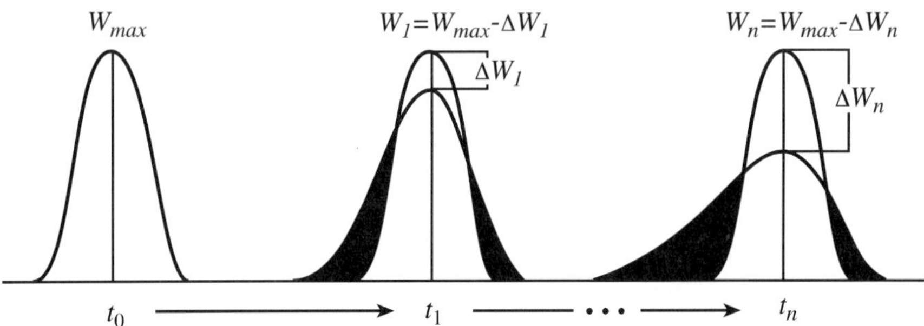


**Fig. 7.19.** Method of tracing population fitness in time from the variability of polygenic traits. It is assumed that a stable population at a certain moment has maximal fitness ($W_{max}$), which may be reduced in time ($t_1...t_n$) by a value of $\Delta W_1...\Delta W_n$ because of increased genetic load. As a result, the proportion of "average" phenotypes (based, for instance, on weight and body length at birth) falls, whereas the proportion of phenotypes with extreme value traits ("phenodeviants") rises

Third, although the heritability of the polygenic traits is fairly high, one should not forget the considerable environmental component that influences their external expression.

Fourth, however soon the task is set of exploring the effects of gene mutation and recombination on fitness, one cannot dispense with several demographic and genealogical findings that are of fundamental importance.

And fifth, it is essential to have evidence of stable associations between genetic markers and the differentiation of individuals, based on polygenic traits (see Chap. 5).

With these remarks in mind, we worked to analyze this proposed monitoring model. We established that the effects first shown by Karn and Penrose also remain unchanged with respect to any other parameters of fitness, and not merely mortality. This is illustrated by data obtained from examining 2-year-old children. These deviated minimally at birth from the population average in a complex of anthropometric traits (group $M^0$), and were least affected by most pathologies, including acute respiratory diseases, pneumonia, and even leukemia (Altukhov et al. 1979b, 1981; Botvinyev et al. 1980; Kurbatova et al. 1982; Fig. 7.20). However, the most important result of our investigations is that behind the "average" and "extreme" morphotypes there stand different genotypes of the polymorphic blood group loci, and rare genotypes (or combinations of genotypes) are concentrated on the edges of multidimensional distributions of polygenic quantitative

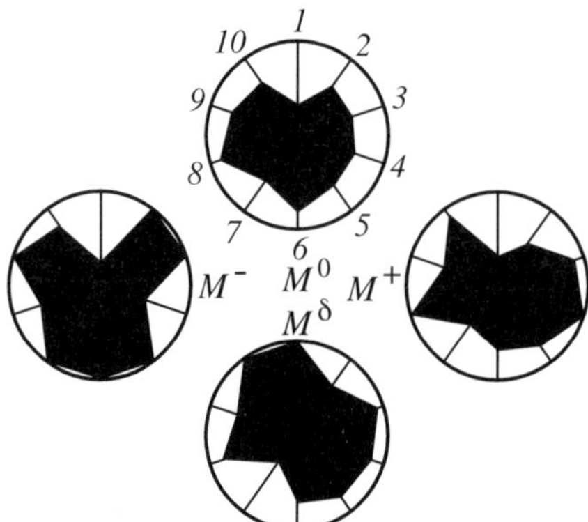


**Fig. 7.20.** The relative incidence of several illnesses in four groups of children, assigned on the basis of weight and body length at birth. The relative incidence (as %) of illness in a given group is plotted at each radius: 0 is the *center* of a *circle* and its *circumference* is 100%. The maximum incidence of a given illness in the groups is taken as 100%. *1* Genetic illnesses; *2* sepsis; *3* pneumonia; *4* suppurative skin diseases; *5* posthypoxic encephalopathy; *6* hypotrophy; *7* anemia; *8* exudation diathesis; *9* jaundice; and *10* other illnesses

traits – that is, in groups $M^+$, $M^\delta$ and, particularly, $M^-$ (Altukhov 1984; Altukhov and Kurbatova 1984, 1990). Curiously, the same empiric correlation between the frequency of a phenotypic combination in the blood groups and fitness is also observed at later ontogenetic stages. For instance, among women who have reached reproductive age, fitness is highest in all cases in complex geno(pheno)types, represented in a population at intermediate frequencies (Kurbatova 1977). This means that the morphological and genetic "adaptive norms" coincide and that the average environmental component of observed variability does not overlap its genotypic component. *Rare genotypes* have particularly low fitness, and their increased frequency can result, first, from reduced intensity of selection (due to successes in medicine) and second, from changes in the historically evolved structure of gene migration. This matter merits special consideration as it is exceptionally important for evaluating the state of the genetic process that is taking place in modern human populations (Kurbatova 2001).

We made a special analysis of the genealogy of children of the extreme phenotypic classes ($M^+$ and $M^-$) and children within the range of the adaptive norm ($M^0$). The first category mostly comprised marriages between people from geographically remote or close populations, whereas nearly half the marriages in the group of parents with morphologically average children fell within the limits of the historically evolved Russian population system in the central part of the USSR's European territory (Table 7.26). This result is also confirmed by demographic findings, which establish a more significant increase in the proportion of urban population and, hence, of non-local marriages in the parental generation of the less fit $M^-$ and $M^+$ groups (Table 7.26). Thus, negative effects of outbreeding are revealed that mark an unequal distribution of genetic information in groups of individuals differing from the population mean in their morphological traits.

These results may be interpreted as evidence of the adaptive significance of the differentiation discovered and, simultaneously, as indicators of the need for more careful studies of the question of co-adapted gene complexes in man, their existence having hitherto been doubted (see, for example, Dobzhansky et al. 1977; Sutton 1980; Vogel and Motulsky 1997). However, this kind of conclusion seems to be associated less with strict scientific evidence than with the perfectly understandable response of humanist scientists to well-known reactionary doctrines predicting mankind's downfall if different races intermingle. But modern genetics do not justify this prediction, since most of the considerable genetic variation characteristic of *Homo sapiens* is connected not with interracial but with intrapopulation differences (Cavalli–Sforza 1966; Nei 1975; Rychkov et al. 2000). It must be emphasized that our data also do not relate to interracial marriages, as certain authors mistakenly contend (see, for instance, Dubinin 1986a,b,

**Table 7.26.** Characteristics of three groups of newborn children from the frequencies of antigen combinations and certain demographic parameters

| Researched groups | Percentage of urban population in the two preceding generations I and II | | Percentage of marriages between individuals from | | | Antigenic combinations | | |
|---|---|---|---|---|---|---|---|---|
| | I | II | Geographically remote populations | Adjacent populations | Same population | Frequent | Average | Rare |
| $M^0$ | 37.84 | 55.41 | 29.27 | 46.34 | 24.39 | 0.3699 | 0.6164 | 0.0137 |
| $M^-$ | 37.50 | 69.23 | 32.14 | 25.00 | 42.96 | 0.4559 | 0.4853 | 0.0588 |
| $M^+$ | 37.93 | 70.69 | 41.38 | 31.03 | 27.59 | 0.4921 | 0.4603 | 0.0476 |

The significance of the differences of frequencies of genotypic combinations, $F$-test for $M^0$ and $M^-$: 5.79; for $M^-$ and $M^+$: 0.33; and for $M^0$ and $M^+$: 6.82, of which the first and third values of $F_\varphi$ are significant at $P < 0.01$. The frequent combinations range from $P = 0.03$ to 0.13, average combinations from $P = 0.01$ to 0.03, and rare combinations are $P < 0.01$

pp. 204–205), but exclusively to inter-local marriages within a single ethnic group (see Dubinin et al. 1976; Dubinin and Altukhov 1977). The new facts examined in Sect. 2.2 suggest a need for more careful study of the interconnections between the increased heterozygosity of modern panmictic populations and a possible increase in the frequency of intragenic recombination producing the alleles that can negatively influence the fitness of their carriers when environmental conditions change.

This approach to the problem of the segregational load suggests that the specific features of the genetic process in a modern urban population are not obligatorily linked with gene co-adaptation, which is apparently formed only in the lengthy process of evolution. One can theorize, for instance, that the genotype combinations of the loci studied by us have been neutral (or almost neutral) during most of the evolution of *Homo sapiens* and only now attain some selective significance in a changing environment. However, even now the possibility is open for a stronger genetic explanation of a key event in human biology, *acceleration*, i.e., an increase in body size and rate of sexual maturation in the populations of virtually all industrially developed countries (Harrison et al. 1964; Volkova 1988; Dubrova and Kurbatova 1995), if regarded as resulting from intrapopulation heterosis caused by intense mixing of genetically differentiated subethnic groups. It is clear that resolution of this issue requires studying the distribution of individual biochemical heterozygosity at a number of codominant genes "along" the variability curve of polymorphic morphophysiological characters, as done in Sect. 5.4 for other organisms. Such a study has been conducted by A.V.

Shurkhal (1982), who examined blood samples of 1,035 male ethnically Russian donors at 22 biochemical loci, using electrophoresis of blood proteins in polyacrylamide and starch gels. He found that body size and mean heterozygosity of the groups increased with the level of exogamy, although this relationship was probably nonlinear. This result is directly related to the issue of heterosis in humans and thus required support.

The possibility of testing this hypothesis directly arose only recently, with the finding of the universal polygenic system connecting genome heterozygosity with a number of demographic parameters.

Comparative biometric analysis of 77 animal and 30 plant species showed that high allozyme heterozygosity increases rates of growth and sexual maturation while decreasing post-reproductive growth and life span (Altukhov 1996a, 1998, 1999; Altukhov et al. 2000b; Fig. 7.21). Even a cursory examination of the diagram is sufficient to understand that association 1–4 (between heterozygosity and body size) is not related to the acceleration consequences, namely, to the fact of an increase in body size. This effect can only result from *positive somatic heterosis*, whereas the relationship found in animals and plants corresponds to *negative somatic heterosis*.

How can this result obtained for a wide variety of biological species be related to the strictly opposite pattern characteristic of acceleration in humans? To answer this question, we (Altukhov et al. 2000b) compared mean heterozygosities and mean body sizes in a number of native ethnic groups from North Eurasia. The genetic information was retrieved from the

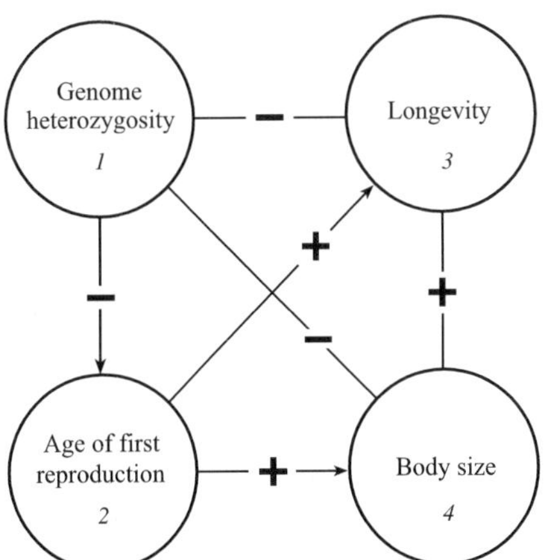


**Fig. 7.21.** Functional polygenic system, determining correlations between *1* genome heterozygosity, *2* age of first reproduction, *3* longevity, and *4* body size of biological species. Cause-and-effect relations are indicated by *arrows*. *Plus* and *minus* are the signs of the correspondent correlations

database of the Laboratory of Human Genetics, Vavilov Institute of General Genetics, Russian Academy of Sciences, which had been developed by Yu.G. Rychkov and his associates (Rychkov et al. 2000). The anthropometric data were taken from more than 100 sources including expedition archives and works of prominent Russian anthropologists (V.V. Bunak, G.F. Debets, Ya.Ya. Roginskii, M.G. Levin, V.P. Alekseev, T.I. Alekseev, N.N. Miklashevskaya, M.Ya. Gremyatskii, and many others).

Since heterosis, or at least its form known as *euheterosis* (the term coined by Th. Dobzhansky), is an intrapopulational phenomenon, our scheme requires special comments that take into account the structure of intraspecific gene diversity. Here they are:

The average estimates of intrapopulation heterozygosity and body size in members of different, greatly diverged branches of the *Homo sapiens* phylogenetic tree (e.g., Mongoloids and Caucasoids) cannot be correlated. Heterozygosity and body size in the former are smaller than in the latter, and, consequently, the sought correlation will be evidently positive. However, it by no means reflects a causal relationship. Different, not closely related population systems have their own levels of heterozygosity. These heterozygosities reflect the ultimate result of the long process of evolution, and thus all random and systematic microevolutionary factors related to this process, which, interacting with one another, have formed the current gene–geographic landscape and the proportions of intra- and intergroup components of genetic variability. To put it differently, we have absolutely no grounds for assigning a specific biological significance (in terms of fitness) to heterozygosity differences of populations belonging to different population systems. We should look for the sought associations only within sets of closely related groups, considering each ethnos (or subethnos) as a generalized, integral genotype. This approach implies finding *intragroup associations* and excluding *intergroup ones*, such as those presented in the above example on Mongoloids and Caucasoids (see the substantiation of this approach in Altukhov 1998).

Consider now the results of such analysis of 50 populations examined in the recent decades for numerous polymorphic genes controlling blood group systems (AB0, Di, Fy, Kell, and others) and proteins (Hp, Tf, PGD, PGM, and others; on average 22 polymorphic loci per population) and for body size (height) of adult men in the 1930s–1950s and partially in the 1960s–1980s (on average, 761 subjects per population). This total sample was divided into two subsamples, one of which included 39 ethnic and subethnic groups, and the other 19 ethnic groups partially overlapping with the former. The second subsample was characterized by the fact that its populations exhibited a statistically significant increment in body size (by more than 2 cm) in the 1960s–1980s (167.05 ± 0.594) as compared to the 1930s–1950 s (164.83 ± 0.601), which conclusively testifies in favor of acceleration having occurred during one or two successive generations. In this case, the body size estimates in the 1960s–1980s were exactly compared to the heterozygosity parameters, whereas in the group of 39 ethnoses, the corresponding data for the 1930s–1950s were taken into account. Clusters of most closely related ethnoses were formed within each subsample on the basis of ethnolinguistic classification, which was followed by comparing the genetic and anthropometric parameters at the population level. The correlation was estimated using the software package STATISTICA for Windows (1998, StatSoft, Tulsa, OK, USA).

In view of the necessity to exclude the intergroup variation component discussed above, we estimated relative indices of heterozygosity and body size rather than absolute ones, i.e., the deviations of the traits from the intragroup means standardized by the corresponding variances. This allowed us to eliminate the masking effect of the intergroup differences and analyze absolutely different ethnic groups as average genotypes within the population as an integral entity (see Altukhov 1998 for details).

The results of this study are presented in Figs. 7.22 and 7.23. They demonstrate crucially different proportions of the compared variables: in one case, the correlation between multilocus genome heterozygosity is negative (Fig. 7.22), i.e., the association is similar to that established earlier for diverse biological species (Altukhov 1996a,b, 1998), whereas, in the other, the correlation is of the same strength, but positive in sign, which shows positive somatic heterosis (Fig. 7.23a). We should like to emphasize that the values of the19 subethnoses' relative heterozygosity and body size, presented in Fig. 7.23a, are characterized by the same positive correlation between these variables, even when the body size estimates include those for the 1930s–1950s rather than only the 1960s–1980s. This relationship is depicted in Fig. 7.23b, clearly indicating that the acceleration caused by positive heterosis started in the corresponding ethnoses considerably earlier than in some others, but, judging by the correlation value, the association earlier was weaker than in the 1960s–1980s (Fig. 7.23a).

This result is the first direct proof of the genetic nature of acceleration in humans, demonstrated in time and space at the population level. The most striking evidence of that is positive somatic heterosis found in the ethnic groups characterized by a high level of panmixia (e.g., Ukrainians, Kazakhs, Bashkirs, and others). By contrast, this phenomenon is not observed in relatively isolated subpopulations (Fig. 7.22).

In conclusion, we would like to return to the issue addressed at the beginning of it: since rapid sexual maturation leads to significant deceleration or even suppression of growth in many biological species, what is the explanation of the positive correlation between multilocus genome heterozygosity and body size in human, despite early sexual maturation connected to acceleration?

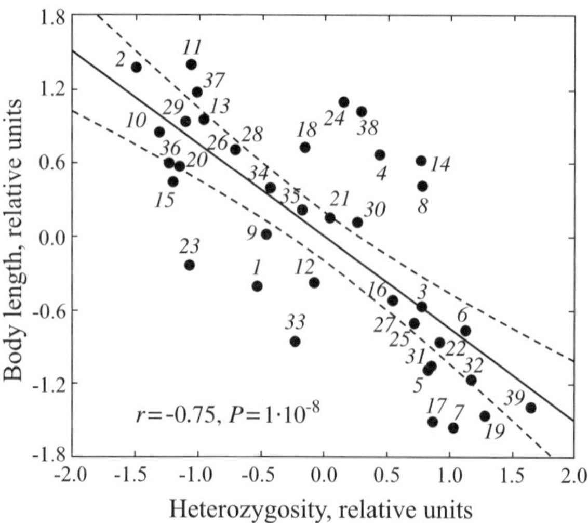



**Fig. 7.22.** The relationship between the mean intrapopulation heterozygosity and mean body size at the ethnic level (relative values). *Caucasian language family*: *Abkhaz-Adyg group*: *1* Abazins; *2* Abkhazs; *3* Adygs; *4* Kabardins; *5* Cherkes; *Nakh-Dagestan group*: *6* Rutuls; *7* Avars; *8* Dargins; *9* Laks; *10* Lezgins; *11* Ingushes; *12* Chechens. *Altaian family*: *Turkic group*: *13* Yakuts; *14* northern Altaians; *15* southern Altaians; *16* Khakassians; *17* Tofalars; *Tungus-Manzhurian group*: *18* eastern Evenks; *19* western Evenks; *20* Evens; *21* Nanais. *Group of isolated languages*: *22* Kets; *23* Yukagirs; *24* Nivkhs. *Chukotka-Kamchatka family*: *25* Koryaks; *26* Chukchi. *Eskimo-Aleutian group*: *27* Aleuts; *28* Eskimos. *Uralian family*: *Samodian group*: *29* Nganasans; *30* Forest Nenets; *31* Tundra Nenets; *Finno-Ugric group*: *32* Khants; *33* Mansi; *34* Komi; *35* Komi-Permyaks; *36* Mari; *37* Mordovians; *38* Karels; *39* Saami

The answer to this question is provided by the energy model of longevity, taking into account genotype–environment interactions that modify the metabolism structure during growth and development (Altukhov 1998, 1999). The greater the pressure of stabilizing selection and the number of overdominant loci involved in the processes of growth and development, the higher the proportion of energy costs for this period of ontogeny, and the earlier is maturity and deceleration of post-reproductive growth (Altukhov 1998). In this case, *negative somatic heterosis is coupled with positive reproductive heterosis* (for sexual maturation rate).

A completely different situation is characteristic of acceleration. The mixing of subethnic gene pools and an increase of intrapopulation heterozygosity underlying this phenomenon were accompanied, as is known, by an improvement of the sociocultural environment, including advances in medicine and health care, conditions of everyday life, better nutrition, etc. Under such conditions, stabilizing selection dramatically decreases, thus, enhancing possibilities of compensation of the energy expenditure of

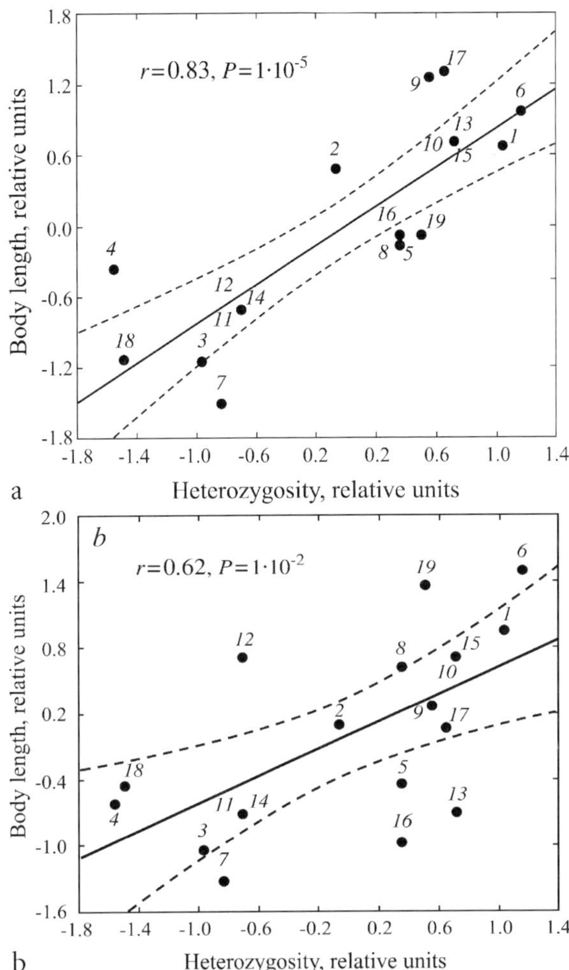


**Fig. 7.23.** The relationship between the intrapopulation heterozygosity and body size (in relative units) in successive generations of the same ethnic groups exhibiting positive somatic heterosis. **a** 1960s–1980s; **b** 1930s–1950s. *Indo-European language family*: *1* Ukrainians; *2* Belarussians; *3* Armenians. *Altaian family*: *Turkic group*: *4* Kazakhs; *5* Bashkirs; *6* Kumyks; *7* Tatars; *8* Uzbeks; *9* Azerbaijanis; *Mongolian group*: *10* Buryats; *11* Kalmyks; *Tungus-Manchurian group*: *12* Evens; *13* western Evenks. *Uralian family*: *Finno-Ugric group*: *14* Mari; *15* Mordovians. *Caucasian family*: *Nakh-Dagestan group*: *16* Avars; *17* Rutuls; *18* Lezgins; *19* Dargins

a growing organism, not only on sexual maturation, but also on somatic growth. Correlation of heterozygosity of the genome (genotype) with processes of growth and development – one of the most important adaptive connections – has changed in modern human populations. This is the main

**Table 7.27.** Life expectancy in Russia (years) (Demographic Yearbook of Russia 2001)

| Year | Males | Females | $\Delta$ | Year | Males | Females | $\Delta$ |
|------|-------|---------|------|------|-------|---------|------|
| 1960 | 63.78 | 72.38 | 8.6 | 1996 | 59.00 | 72.60 | 13.6 |
| 1970 | 63.21 | 73.55 | 10.34 | 1997 | 60.80 | 72.90 | 12.1 |
| 1987 | 64.90 | 74.60 | 9.7 | 1998 | 61.30 | 72.90 | 11.6 |
| 1994 | 57.60 | 71.30 | 13.7 | 1999 | 59.93 | 72.38 | 12.5 |
| 1995 | 58.30 | 71.70 | 13.4 | 2000 | 59.00 | 72.20 | 13.2 |

reason for distinguishing modern panmictic human populations from all other organic forms. We would like to emphasize here that, though excessive heterozygosity of such populations remains on average selectively neutral in beneficial environments, it must nevertheless be regarded as a "delayed action bomb", capable of drastically changing the demographic situation if and when triggered by deterioration of environment, particularly by stress. Since the rate of sexual maturation and longevity are strongly negatively correlated, this seems to explain the situation currently existing in Russia, namely, an unprecedented drop in male life expectancy observed in recent decade (Altukhov 1998; Table 7.27).

It seems likely that this extremely high male mortality is related to the elimination of potential long-lived persons characterized by higher homozygosity, who proved to be less adapted to the dramatic changes of the last decade, including the so-called "shocking therapy" in Russian economy in the early 1990s (Altukhov et al. 2000b). The appearance of the same adverse process in other countries that underwent acceleration is only a question of time. The global ecological balance is irreversibly disturbed, and thus the known prediction of the impending long-term aging of humankind (Laslett 1997; Schneider 1999) may require considerable correction if not a complete revision.

Summarizing the discussed facts, we would like to draw the following conclusions. First, there are no grounds for the statements on the allegedly exclusively beneficial consequences of the current stage of the *Homo sapiens* biological evolution, which implies unprecedented admixture of previously differentiated populations. As we have seen, at least in Russia this process is accompanied with the appearance of less viable genotypes that contribute to the population genetic load in the broadest sense of the word. Second, the efficiency of our approach for detecting such genotypes, which drastically differ from the population mean by a set of anthropometric traits, via selective screening of early ontogenetic stages has become evident. Let us now consider whether this approach can be employed for detecting the load of gene mutations.

## Mutational Load

James Neel (1972, 1974) was the first to propose a program for monitoring the mutation rates in man from electrophoretic findings about rare protein variants. His method presupposes *mass screening*, based on the notion of the selective neutrality of de novo mutations that change protein structure. To date, Neel and his colleagues (Satoh et al. 1984; Neel et al. 1986), have used a series of biochemical markers to study an enormous number of samples (934,370 tested genes) and have discovered five "fresh" mutations corresponding to a mutation rate of $1.07 \times 10^{-5}$ per gene, per generation (including electrophoretically "silent" substitutions).

Our approach, with its point of departure in concepts of the harmfulness of recently emerging mutations of protein loci (especially monomorphic ones) in a heterozygote, proposes *selective screening* to permit a reduction in the sample size for estimating rates of mutation at the corresponding genome loci. This method has been used in the electrophoretic examination of blood samples of premature children, children with gross multiple development defects, and also for examining spontaneous abortion material; in all cases the frequency of rare electrophoretic variants has proved significantly higher than in the control (Dubinin and Altukhov 1979; Altukhov 1980b; Altukhov and Dubrova 1982; Afans'ev et al. 1984). In the case of some of the variants found in children with congenital development defects, it would be wrong to exclude the possibility that they are basically de novo mutations and not inherited from preceding generations. A preliminary estimate of the population's mutation rate (calculated for the frequency of anomalous children, and with correction for electrophoretically "silent" alleles) gave a value of $2 \times 10^{-5}$ per gene, per generation (Dubinin and Altukhov 1979; Altukhov 1983).

Neel and Mohrenwieser (1984) also investigated a small sample of premature children and those with congenital defects, but were unable to repeat our result. Twenty years ago we made a detailed analysis of the possible causes of these divergences (Altukhov et al. 1985) and reached the conclusion that there were no grounds for revising the principal result of our research, that de novo mutations of protein loci are eliminated during the early stages of ontogenesis. Moreover, the same conclusion was confirmed experimentally in tests on Drosophila exposed to the effects of radiation and chemical mutagens (Krutovsky et al. 1983a,b; Krutovsky and Milishnikov 1983). A similar result was obtained in research on the mutation process in the Scots pine, as described in detail in Sect. 5.3; and Waldren and his co-authors (Waldren et al. 1986) have demonstrated the elimination of radiation-induced gene mutations in human cell cultures.

Also of great interest in this connection is the statistical analysis of the anomalous human $\alpha$ and $\beta$ globins, conducted by Yu. E. Dubrova (1986).

It is well known that a lower frequency of $\alpha$-globin variants is observed in human populations than $\beta$-chain variants, which enabled Fitch (1974) to hypothesize their prenatal elimination. We pointed out in Chap. 2 that $\alpha$- and $\beta$-globin chains form the main hemoglobin tetramer $A(\alpha_2\beta_2)$ in an adult's erythrocytes. However, whereas the $\beta$ locus begins functioning only 2 months after birth, the $\alpha$-locus is actively expressed at 2-months' intrauterine development; it is then that the $\alpha$ subunit forms tetramers with the products of the $\varepsilon$ and $\gamma$ genes. Dubrova (1986) summarized data from the literature for the primary structure of 123 anomalous $\alpha$ and 211 anomalous $\beta$-human globins. He showed that the average number of amino acid replacements per site was 1.66 times greater for the $\beta$- than for the $\alpha$-globins. Comparative analysis of the variability of functionally different parts of $\alpha$- and $\beta$-chains, their conformational stability, and the effect of amino acid replacements in them resulted in the conclusion that 40% of mutant $\alpha$-globins undergo intrauterine elimination. Comparison of functionally different segments of a molecule revealed a significant reduction in the variability of the amino acid radical groups responsible for forming hydrophobic contacts between the spiral areas within a protein globule and between subunits. By taking into consideration the fact that a hemoglobin tetramer is typified by a degree of cooperation among interacting subunits, without which it would be unable to function normally, an opportunity is provided for explaining the dominance effects of newly emerging mutations which, as we have seen, are eliminated in a heterozygotic state in the very first generation. Evidently it requires only one molecule with impaired capacity to contact its tetramer neighbor for the latter's activity to be seriously disrupted. Amino acid substitutions in the areas responsible for forming the hydrophobic core of a molecule may result in an even greater suppression of the capacity to form intermolecular contacts (Dubrova 1986; see also Kimura 1983; Vogel and Motulsky 1997).

In the light of these results, the recent publications by N.P. Dubinin (1986b, 1988), rejecting the thesis of selection directed against recently arising mutations in protein loci, cannot be credited. Without having recourse to any experimental facts, Dubinin introduces nothing new into the problem under discussion; at the same tune the author commits one substantial error that contradicts his conclusion. By taking the data on one group of the rare protein variants discovered by us for new mutations (27.3% of those found in family analyses), Dubinin refutes his own conclusion about their selective neutrality. Indeed, whereas Neel, Satoh and their colleagues found five fresh mutations at 930,000 tested "loci" (while studying not newborn children, as Dubinin assumed, but children of the parents who have survived the Hiroshima and Nagasaki atomic bombings), we discovered in our research at least three new mutations in 29,200 tests; which means a 17-fold difference in the sampling volumes needed to reveal a comparable number of new mutations! In other words, far fewer new mutations occur in the reproductive part of a population than those that arose originally; but this also provides *evidence of the effects of strong negative selection.* This is not the same as the possibility of the selection's fitness coefficients being somewhat lower in man than we had first thought – but the matter has been discussed thoroughly in our 1985 work. Hence, it is not surprising that electrophoretic screening of even large population groups, especially in the late stages of ontogenesis, enables one to observe de novo mutations of enzyme loci only when very large samples are analyzed. This conclusion is of fundamental importance for organizing the genetic monitoring of human populations, since if one fails to account for the adverse effects of selection upon newly emerging mutations, a general evaluation of the mutation rate may prove to be materially reduced. And although this problem requires further study, our accumulated experience shows that the proposed *genetic monitoring model,* constructed on the concepts of *normal and unfavorable* genetic processes in populations, is effective for discovering both the segregational load and the load of newly arising mutations. Moreover, there is no need for mass analysis of blood samples by biochemical gene markers – all that is required is the selective screening of non-random samples of newborn children (or material from spontaneous abortions) sharply deviating from the norm, with accurate evaluations of their incidence in the whole of a controlled population. By way of illustration, I reproduce a scheme for estimating the mutation process rate based on findings of the frequency of rare electrophoretic protein variants in newborn children suffering from gross and multiple developmental defects (Altukhov et al. 1985):

| | |
|---|---|
| Number of "loci" tested | 29,204 |
| Number of rare protein variants (heterozygotes) | 21 |
| Average frequency of rare alleles | $21/29{,}204 = 0.72 \times 10^{-3}$ |
| Proportion of non-inherited variants | $3/11 = 0.273$ |
| Frequency of mutation in the sampling of children with congenital development anomalies | $0.00072 \times 0.273 = 1.97 \times 10^{-4}$ |
| With correction for electrophoretically silent alleles | $2 \times 1.97 \times 10^{-4} = 0.394 \times 10^{-3}$ |
| Frequency of anomalous children in the total population of newborn children | $1{,}057/50{,}950 = 0.02$ |
| Population mutation rate | $0.000394 \times 0.02 = 0.78 \times 10^{-5}$ |

Of course, the method developed by us needs further improvements, particularly in obtaining an adequately stable evaluation of the ratio between newly arising and inherited alleles of the protein loci in the group of newborn children studied. The above mode of calculating is not intended as a definitive evaluation of the gene mutation rate in man, but merely as an illustration of the possibility offered by such an evaluation, based on selective screening.

The fact that an increased frequency of rare protein variants in man is associated with these forms of congenital pathology that have been least studied with regard to their causes (multiple defects of the central nervous system; for details, see Altukhov et al. 1985), is of considerable interest for deciphering the etiology of the corresponding pathological forms. Of course, it is not possible in every individual case to speak of a direct link between this or that pathology and the carrying of a rare protein variant. As we obtained our result in the course of research into population genetics, we are dealing with a legitimacy that has been established statistically. At the same time it is important that the following question be discussed: how should the negative effects on the viability of rare electrophoretic alleles be explained from a physiological–biochemical standpoint?

The negative effect on the fitness of a null-allele group, mutations for instance, leading to total suppression of polypeptide synthesis or its loss of enzymatic activity, has been well demonstrated in research projects on *Drosophila melanogaster* (O'Brien and MacIntyre 1972; Gvozdev et al. 1977) and in a study of metabolic disturbances connected with the activity of enzymes in man (Harris 1977). All these sorts of defects belong to the category of so-called inborn errors of metabolism and they act as if autosomal recessive. The main pathologic mechanism, prominently expressed in homozygotes, is an enzyme deficit, usually structurally anomalous or unstable. Heterozygotes usually have 50% of the activity of the corresponding protein and are most often clinically normal, at least they are in humans. In our works we have encountered autosomal-dominant lethal or semilethal mutations of protein loci eliminated in the early stages of ontogenesis, including even the gametic phase. What are the possible reasons for this elimination?

Above, using hemoglobins as an example, we discussed the mechanisms that interfere with multimer protein function even when only one mutant subunit appears. This mechanism very likely also has fairly universal significance for any other vitally important multimer proteins, enzymes and non-enzymes. Deserving special attention, too, are the proteins of cellular membranes, receptors, or those responsible for the normal functioning of the cytoskeleton. One could also name several other mechanisms of the molecular dominance of newly emerging mutations, but there is no need as this question has received detailed examination in a recently published monograph (Vogel and Motulsky 1997 pp. 345–353).

Seen from the viewpoint of a population geneticist, the elimination of rare mutant heterozygotes of protein loci in the early stages of ontogenesis may also be associated with threshold effects, when viability depends on the number of these mutations impinging upon a genome simultaneously. If their frequency in a group of children with severe development defects comprises in fact $10^{-3} - 10^{-4}$, and the total number of genes functioning in an individual lies between $5 \times 10^4$ and $10^5$, then the probability of carrying two or more rare genes of mutational origin simultaneously is not so slim. But since rare heterozygotes at electrophoretically distinct alleles are nevertheless present in the reproductive portion of a human population, clearly the coefficients of selection directed against these heterozygotes should be less than unity in several cases, depending on the specifics of a gene locus, the type of mutation damage (functionally more or less important parts of protein molecule that are implicated) and, possibly, the genotypic integral structure determining the "buffered state" and nonspecific resistance of ontogenesis to varied disturbance factors (see Altukhov et al. 1979b, 1996). Moreover, if in its early stages selection conflicts with multiple heterozygosity of polymorphic loci, the probability of this elimination being intensified may be greatly accentuated by de novo mutation at any monomorphic gene, thus bringing us again to the threshold model.

I believe that all these questions are of key importance for the problem of unraveling the genetic causes of infertility, spontaneous abortion, and the heterogeneous group of congenital developmental defects, which are the least-studied aspects of modern human genetics. They should all be subject to more intensive medico-genetic research. But as far as population and evolution genetics are concerned, the following is of fundamental importance: if the conclusion about the strong effects of purifying selection on de novo mutations is correct, all the evaluations of the spontaneous gene mutation rate in man made up till now must be regarded as underestimates. This means that one must also revise the calculations of the mutational load to which, as is now well known, such mutation-like events as intragenic recombination make a considerable contribution (see Chap. 2). By acting on preexisting genetic variability, this process, combined with genic conversion and postmeiotic segregation, can change allelic frequencies to produce new alleles (Leslie and Watt 1986). Seen in the context of the increasing heterozygosity of modern human populations, the importance of this kind of genetic dynamics can only grow, which calls for constant global or regional monitoring.

At the beginning of this section, we discussed two possible ways of monitoring genetic load in human populations. On the one hand, we proposed the *mass screening* of blood samples from representative randomized samplings of newborn children and, on the other, the *selective screening* of specially selected groups, such as those with occupational injuries or examples given by the author. Both methods have their merits and failings, but selective screening seems preferable since it permits a considerable reduction of the sampling size necessary for detection of newly emerging mutations. It is true that this requires additional research to estimate the average population frequency of a specially selected group of children with developmental anomalies among all those born (in our studies we had to analyze 50,950 cases of the development of newborn children), but the cost of this work was incomparably less than mass electrophoretic analysis of blood samples of many thousands of individuals.

Nevertheless, we should like to stress that both these approaches, *selective* and *mass screening*, need further improvement, using more powerful technical tools, such as two-dimensional protein electrophoresis and DNA analysis (Jeffreys et al. 1985; Dubrova et al. 1993).

Only relatively recently, Yu. A. Dubrova in collaboration with the Department of Genetics of Lester University (United Kingdom) has succeeded in developing a new approach to this problem, using hypervariable minisatellite human DNA. These studies for the first time have shown experimentally that the mutation rate in human germ-line cells can increase after exposure to ionizing radiation. Employing this method, Dubrova et al. found a twofold higher rate of gene mutations (more precisely, mutation-like events) in children from the regions polluted by radiation after the Chernobyl meltdown (Fig. 7.24; Dubrova et al. 1997, 1998) and nuclear ground tests on the Semipalatinsk testing site in 1949 1951, 1953, and 1956 (Dubrova et al. 2002).

However, the attempts at obtaining evidence of a direct link between the appearance of a de novo mutation and the external factor impairing the parental gamete are most problematic. As we have seen, there is a multitude of such factors, and their effects can be indirect. We believe that the great complexity and heterogeneity of the current environment of humans makes science all but helpless in face of this challenge (except maybe for some extreme situations). It all shows that today humankind has encountered a global problem of unprecedented scale and possible social consequences.

Any possible solution requires careful investigation of specific genetic processes in modern human populations. In addition to evaluations of gene mutations and recombination rates in varied environments, the problem of acceleration and heterosis demands greater attention. Indeed, as has been pointed out, subsequent generations must "pay" for heterosis in the first generation by the segregation of less fit genotypes, which supplant a population's segregational load in subsequent generations. However, what is accepted as ordinary in all other biological species acquires quite special significance for man, since the increased volume of genetic load in a population of us is fraught with consequences incompatible with commonly held principles of morals and ethics – with everything that bears the name of humanism. Clearly a broad study of the genetic processes in human populations is of great importance for the development of preventive medicine. However dubious the well-known doctrine, "cure the patient,

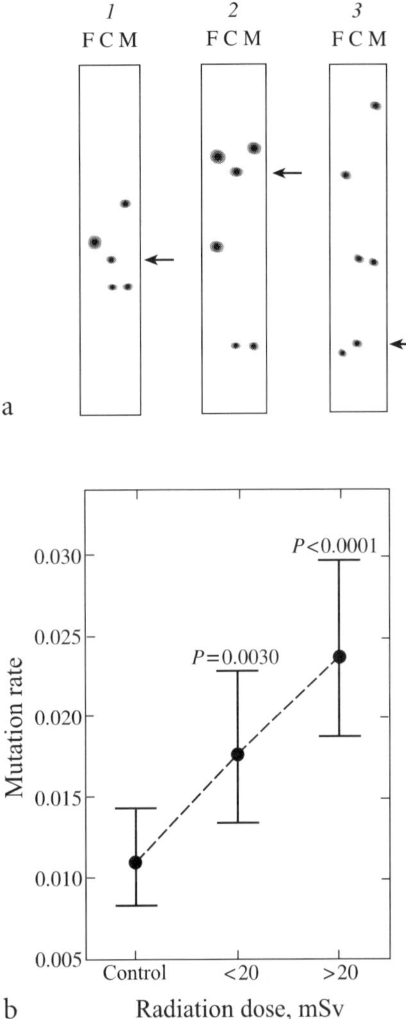

Fig. 7.24. Radiation dose and frequency of de novo mutations in human (data of Yu.E. Dubrova and the team of A. Jeffreys). **a** Autoradiograms of de novo mutations (shown by *arrow*) of minisatellite DNA appearing in germ-line human cells. *F* Father; *M* mother; *C* child; *1–3* different families

not the disease," proper attention is not being given to the fact that every ill person is an integral component of one population or another, which has its own gene pool and exists under the conditions of a certain environment. Efforts to eliminate the external causes of illnesses may have little effect if nothing is known about the specifics of the genetic process in a population and the state of the environment. Further study of all these questions based on a broad multidisciplinary approach is very necessary, with a key role being played by specialists in population, demographic, and clinical genetics.

Summarizing the data considered in this chapter, it is probably necessary to emphasize once more that nature and society are a united dynamic system, and humans, as both the creators and subjects of history, affecting the environment by their activities, induce in plant and animal species the same processes as those currently observed in human populations: the population system evolves to its extreme states, i.e., to panmixia or to maximal subdivision, when interpopulation differences disappear or become excessive. The former is characteristic for the European human population (excluding the ex-USSR), in which smoothing of interpopulation differences is accompanied by an increase in intrapopulation polymorphism ($\overline{F}_{ST_0}$ = 0.0279). The latter process is observed in the indigenous Mongoloid populations of northern Asia and America ($\overline{F}_{ST_0}$ = 0.081 and 0.082, respectively) that have already approached the limit of spatial differentiation (Rychkov et al. 1982; Rychkov and Yashchuk 1983). Obviously, the two processes are the evolutionary costs paid for "civilization" and the "primitive way of life." Which of the two states is the more favorable for human society should be discussed separately. However, as for wild species, the answer is unambiguous: *both processes lead to decreased fitness or even degradation of population systems, bringing them to the limits of maintaining their integrity.*

Indeed, the results obtained during genetic monitoring of commercially important species are discouraging. Gene diversity at various evolutionarily established levels is disturbed by commercial activities (e.g., fishing) and, moreover, by actions undertaken with good intentions, such as breeding and improvement of agricultural plants and animals (e.g., barley and chicken) or artificial reproduction of salmon at hatcheries. In all cases, disproportionate removal of certain genotypes and insufficient utilization or nonuniform reproduction of other genotypes give rise to unfavorable processes that lead to a decrease in the fitness of populations. The mechanism underlying these phenomena involves both a decrease and an increase in genetic diversity, relative to the optimum established in the course of natural history.

Moreover, this is due to both mutation pressure and some other influences. Such influences change systemic relations, historically formed, and distort a balance between the processes of gene-pool differentiation and integration and result in unfavorable impacts. When an intrapopulation component of gene diversity increases, the consequence can be equivalent to outbreeding with increasing segregation load; and in growing in an interpopulation component, to inbreeding. In the first case, the genetic process is adaptive, and timely termination of unfavorable pressure leaves a chance for the restoration of genetic structure and normalization of reproduction in population gene pools. In the second case, the genetic process is inadaptive, and probably results in population degradation.

As to genetic processes in modern human populations related to increasing outbreeding and the phenomenon of acceleration, this area should attract particular attention of geneticists, anthropologists, demographers, and medical specialists. In any case, the results discussed above give no indication that the predicted fusion of humankind into a huge panmictic population is a process beneficial for the health of forthcoming generations.

# Conclusion

In this book, we have focused attention chiefly on the factors and conditions of genetic stability at the population level, endeavoring to examine new evidence favoring a previous conclusion about qualitative differences between adaptation and processes of evolution. It has become increasingly obvious that evolutionary consequences do not necessarily emerge from the results of studying natural and experimental populations with respect to their systemic organization.

On the contrary, it has been reconfirmed that the genetic variability of the simplest populations, traditionally regarded as the elementary units of the evolutionary process, represents nothing less than the mechanism for stabilizing the hierarchical, historically formed structure of a species.

In other words, the theoretically possible variant of polymorphism serving as a transient phase in gene evolution – *the so-called transient polymorphism* – does not occur in native population system even in the case of mild anthropogenic influence. The intraspecific genetic differentiation turns out not to be a Markov chain, in which the population dynamics cannot be predicted further than one generation ahead, but a process having memory whose structure and volume can be reconstructed. As to the depth of this memory, in a number of cases, it was shown to be comparable to the duration of the history of an isolated population system (Chaps. 4 and 6).

According to the population system concept, subdivision of a population cannot promote its evolution. On the other hand, under natural conditions, intraspecific genetic differentiation involves increasing addition of new levels of population structure that *slow down* rather than *accelerate* the evolution of its genetic structure.

The phenomenon of genetic stability of population systems and the existence of the adaptive optimum of gene diversity of these systems allow one to formulate a statement that is different in principle from the commonly accepted view on a population as the elementary unit of the evolutionary process: by shifting from its historically formed optimum, a polymorphic population (panmictic or subdivided) is subject to degradation (or, at best, narrow specialization) without acquiring any evolutionary novelty properties. Furthermore, if we accept the existence of the monomorphic part of the genome, whose reorganization leads to interspecific differences, we will have accepted a refutation of Darwinian theory as a theory of evolution (although not of adaptation! – *Yu.A.*). Nevertheless, genetic monomorphism has been overlooked by the authors in the field of evolution beyond Russia. Because of this, the branch of population genetics developed by our Western colleagues gives principal attention to equating population dynamics with the process of evolution.

The exceptional part that this view plays in shaping our present concepts of man's place and role in the surrounding world is common knowledge. However, to support my statements with evidence, I shall quote from two well-known reports on evolution: "Nothing in biology makes sense except in the light of evolution" (Dobzhansky et al. 1977). And: "Life is a dynamic system of populations in a state of constant change. The earth's environment oscillates and shifts as a result of natural causes and the activities of man.... Some evolutionary processes lead to the formation of new species and some species become extinct. The biosphere continues to evolve" (Ayala and Valentine 1979b).

It is an indisputable fact that when one turns to biological phenomena and processes one cannot ignore the variability and changes that are found at this level. However, it is no less indisputable that permanence and dynamic stability have been essential features of these processes from the very beginning of time.

The emergence of a scientific evolutionary theory has given wings to the progress of our knowledge and its successful utilization in a practice whose entire main thrust has been inspired by a belief in man's unlimited transforming potential. However, today we are well aware that these kinds of effects upon the biosphere inexorably lead in many cases to situations that cannot be interpreted as other than crises. Consequently, it is impossible to equate the variability of natural, internally inherent life with the variability introduced into it by man's transforming activities.

Indeed, when we examine the reproduction of the gene pool of a native population system, we find that this process is self-regulated via non-random gene migrations in such a way that the proportion of the intra- and interpopulation components of the gene diversity is optimal and consistently preserved in time and space (Chaps. 3–5).

Meanwhile the problem of biological optimum should be considered at two levels of research, individual and population. At the individual level, this problem intimately involves the concept of the adaptive norm, and is of fundamental importance. For instance, medicine is based on distinguishing norm and pathology, and any therapy is aimed at returning the patient to the normal (i.e., healthy) state. However, studying human populations, we were the first to demonstrate that the norm and the optimum of genetic diversity represent, in fact, the same phenomenon. When intrapopulation heterozygosity deviates from the optimum, disturbances ensue in various functional systems – reproductive, immune, respiratory, cardio-vascular, etc. – and, in extreme cases, results in an elevated frequency of rare genotypes and developmental abnormalities. The same relationship was observed in populations of animal species. For instance, in the Pacific salmon *Oncorhynchus gorbuscha* Walb., high individual heterozygosity is associated with high growth rate. However, in suboptimal environments, the progeny of highly heterozygous individuals was shown to have damaged gill structure and high mortality at the early ontogenetic stages. For plant species, the same relationship was reported: in a number of conifer species, the level of individual allozyme heterozygosity is positively correlated with growth rate. At the same time, in Norway spruce and Siberian dwarf pine, both extreme (high and low) levels of heterozygosity of a maternal tree were shown to be associated with an increase in mortality in seeds of these plants. In short, high individual heterozygosity is found to be beneficial for an individual but harmful for its progeny, and thus for the popula-tion as a whole, because numerous segregated genotypes are not adaptive in suboptimal environments. In other words, the optimum heterozygos-ity corresponds to an equilibrium between inbreeding (homozygosity, an increase in frequency of genes identical by descent) and outbreeding (an increase in heterozygosity and a decrease in frequency of genes identical by descent). This phenomenon, which was shown to be universal for a wide range of organisms, corresponds to an analogous rule of the information theory: normal function of a system is disturbed by both an excess and a deficit of information. This analogy is not superficial but reflects the very essence of the genetic process, because heterozygosity of an organism re-flects its information capacity, and can be quantitatively evaluated using Shannon's information index. Thus, according to the genetic optimum con-cept, adaptation is regarded as a compromise between the individual and population components of it. The following universal mechanisms main-taining genetic optimum in natural populations were demonstrated: (1) restriction of free genetic recombination; (2) various types of natural se-lection varying in direction and intensity at different ontogenetic stages or in different sexes; and (3) systemic organization of population determining non-random mating structure (Chaps. 4–6).

Let us again turn to the analysis of the ratio of intra- and interpopulation components of genetic diversity. In view of this, another important prop-erty of Wright's $F_{ST}$ statistic should be emphasized. This estimate of genetic diversity measures both the degree of genetic subdivision of a population and structural inbreeding. $F_{ST}$ is of primary biological significance as it reflects the proportion of homo- and heterozygous genotypes in popula-tion systems and, consequently, the equilibrium between inbreeding and outbreeding, which was discussed above in relation to the simple unstruc-tured population. In other words, at the level of native population systems, a stable ratio of intra- and interpopulation components of genetic diversity reflects an equilibrium between integration and differentiation of the species gene pool.

Monitoring of genetic processes in natural and agricultural populations shows that deviations from the optimum value associated with redistribution of intra- and interpopulation components within the total genetic variation are unfavorable for population fitness. Thus, like genetic diversity itself, genetic process generating it can have normal or adverse population effects. *A normal genetic process is defined as the mode of reproduction of the gene pool that maintains the ratio of intra- and interpopulation components of genetic variation at the optimal species-specific level.*

This process reflects natural variability inherent to life. If we use such information as a reference point measuring the norm, we shall see that many natural and agricultural populations under anthropogenic influence are subject to *non-rational management* that does not take into account the systemic organization of the species and the structure of intraspecific hereditary variation discussed above.

The optimal proportions of the intra- and interpopulation components of gene diversity are disturbed not only through fishing and hunting but also through artificial reproduction and breeding of agricultural plants and animals, as well as artificial reproduction of salmon stocks in hatcheries or fish farms. In all cases, disproportional elimination of some genotypes and underuse or irregular reproduction of other genotypes create negative consequences resulting in low population survival and reduction in productivity. This is caused by *unfavorable genetic processes, i.e., a mode of reproduction of the species gene pools that disturbs the optimal ratios of the intra- and interpopulation components of gene diversity* and erase the memory of the original state of the species.

Being concerned with the fate of the biosphere, scientists usually emphasize aspects of reduction in its genetic diversity. However, our results indicate that the problem has a wider scope. Excessive artificial increase in hereditary diversity over its historical optimum is also hazardous for populations. A dramatic example of this is the increase of mutation rate in radioactively or chemically polluted environments.

Still another aspect of the problem is biological pollution. This term, aptly coined by the International Union of Nature Conservation, is applied to often-practiced acclimation measures associated with the transfer of population gene pools to other parts of the species range. This issue was thoroughly investigated in Pacific salmon species and presented in numerous publications (Chap. 7). In addition to low fitness of hybrids between local and transferred fishes, we demonstrated that in novel environments, the transplanted population returns for reproduction at the same time as it did in its place of origin. And in spite of extremely low levels of the return coefficient, genetic structure of this population remains virtually constant. These facts suggest a non-directional character of selection in new environments, which leads to catastrophic results. When centuries-old associations among components of an ecosystem are disrupted and a population is deprived of its proper historically formed environment and transferred to a new one, genetic plasticity generally is not sufficient for adaptation.

Our results and literature data on transplantation of Pacific salmon demonstrate that attempts to create new anadromous stocks within the natural range were unsuccessful. This shows the uniqueness and conservative nature of local adaptations that have been created by selection for hundreds and thousands of generations in the local environments related to the natural history of populations.

Note that even successful acclimation typically leads to adverse ecological consequences, which are caused by substitution of local populations due either to competition for food resources or, more often, to spread of infections for which local forms have no immunity.

We can state that, in many cases, or at least in those where intrapopulation polymorphism decreases and nonadaptive spatial differentiation increases, permissive genetic changes exceed their limit.

This conclusion constitutes the core of our long-term prognostication of genetic consequences of man's activities on natural and agricultural populations. This forecast is obviously unfavorable. Hence, in contrast to the existing views on the ongoing evolution of the biosphere, I am inclined to draw a different conclusion: the biosphere is undergoing degradation before our very eyes. To prevent this outcome, the strategy of interaction between Man and Nature must be revised, as discussed in the beginning of Chapt. 7. Concluding this book, we return to the same problem but with one specification: if an earlier conclusion on a deplorable state of the biosphere might seem to a certain extent unfounded, now it should be considered in light of all the facts and observations discussed above. These facts suggest that interaction of the mankind and nature must follow the lines that will not destroy the systemic organization of populations, while the intra- and interpopulation gene diversity is sustained at the optimal level.

This approach implies: (1) preserving genetic diversity of still-existing population systems depleted by hunting, fishing, and artificial reproduction (non-exhaustive management of natural resources); (2) restoring systems whose structures have already been damaged; (3) creating new population systems in regions with appropriate natural, historical, and economic conditions.

These principles of preserving evolutionarily formed optimal diversity lie at the basis of a new scientific field, *conservation genetics*. Clearly, such rules pertain to all levels of biological integration, including ecosystems. On the other hand, although anthropogenic pressure may destroy the biosphere as a whole, only *populations presenting historically evolved self-reproducing intraspecies groups of individuals* are an object of direct external exposure (consider at least hunting, forestry, and fishing). Except for anthropogenic pressures of such an extent that they can be equated to natural catastrophes (for instance, the destruction of the Aral Sea with its flora and fauna), all other types of human activity, and particularly commercial exploitation and artificial reproduction, are realized only through local populations, gradually involving species, ecosystems, and finally the biosphere.

Hence, no matter whether a biologist is a "globalist", "systemic ecologist", or a "populationist", and no matter to what extent he is concerned with the problem of life conservation, on the level of the whole biosphere, individual ecosystem, or its constituent species, he cannot get away from populations.

In other words, *it is necessary to conserve population systems characterized by several levels of subpopulation hierarchy with their own unique sex and age structure, which is ordered in time and space and imbued with size dynamics, productivity, tempo of succession of generations, resistance to diseases, etc.*

There are no other populations in nature. There are none other in agriculture, either, because all populations of agricultural species of animals and plants appeared not suddenly but were introduced gradually, and thus they are related to some extent, which reflects *the process of successive differentiation of ancestral gene pools.*

In the beginning of the book, we emphasized that it is the specificity of a gene pool that determines the most important biological and demographic features of populations and species.

Professor A.S. Serebrovsky (1928), the author of the term "gene pool", gave the following definition of it in 1925: "The totality of all genes of a given species of animals ... has been called by me 'gene pool' in order to stress the idea that in the gene pool we have the same national wealth as in the oil, gold and coal reserves hidden in our entrails." I would make only one addition to this figurative, now commonly accepted definition: the minerals belong to the category of non-renewable resources that will sooner or later become exhausted. The population gene pool, on the contrary, is a renewable resource and therefore theoretically inexhaustible. This means, however, that successful renewal, reproduction of gene pools and populations containing them are critical for conservation biology – which is just intended to work out a proper strategy.

Ecosystems and population systems within a species have the same prerequisite for stability: autoregulation via interaction of relatively independent structural components that exchange information on their own state and the state of the environment. Both continuous existence of a protected or a newly created community and the ability of this community to respond to external factors lying within the limits of the historical optimum are possible only on the basis of conservation, restoration, and imitation of directions and intensity of these information fluxes.

To preserve genetic diversity in agricultural animals and plants, we have designed a method of breeding and seed production combining moderate directional selection for productivity traits and simultaneous stabilizing selection for adaptive characters. This approach changes the traditional breeding strategy. Routine breeding techniques are aimed at an increase of productivity of breeds and cultivars and disregard the important parameter of genetic stability, the ability to resist adverse environmental effects and various diseases. However, it is known that, along with an increase (improvement) in selected traits and properties, directional selection inevitably leads to a reduction and weakening of other traits. This typical correlated response, termed selection cost, is explained by negative correlations in the ontogenetic system. For instance, in cattle breeding, an increase in milk production is associated with a reduction in protein and fat content in meat; in chicken breeding, directional selection results in loss of the original hereditary diversity of valuable lines and crosses and in their susceptibility to diseases. An enthusiasm for sunflower cultivars having seeds with high oil content proved catastrophic: in breeding these cultivars, genes of high oil content were selected together with genes of late maturity. As a result, the cultivars created ripened in autumn, i.e., in the rainy season, when environmental conditions are beneficial for root molds. The highly productive cultivars virtually perished.

The most illustrative example of the one-sided narrowly specialized breeding is the well-known "green revolution". The triumphal advent of some highly productive wheat cultivars, based on short-stem genotypes, soon gave way to degradation of these cultivars in specific environments, primarily due to their low disease resistance. Narrowly specialized cultivars required massive doses of fertilizers and pesticides, which resulted in chemical pollution of the environment. Other problematic issues have arisen. These errors can be avoided by taking into account regularities of intraspecific hereditary diversity reflected in our method of breeding and seed production. This method allows one to create populations that resist adverse environments and diseases and, at the same time, have productivity sufficient for commercial purposes.

By means of this method, employed since 1975, A.I. Gundaev (Vavilov Institute of General Genetics) restored and improved the early-maturing sunflower cultivar Enisei. This cultivar is technological, resistant to weather changes and diseases, has stable productivity, and, for decades, ranked among the cultivars with the largest sowing areas in Russia. This example convincingly demonstrates that a long-lived cultivar having a moderate but genetically stable productivity is more commercially valuable over the long haul than a specialized cultivar with a potentially high but fluctuating productivity.

Realization of the methods of stabilization of genetic structure of agricultural population and principles of management of hereditary variation of natural populations will promote a stable existence of the human–biosphere system for generations rather than extensive growth and associated destruction of biosphere gene pools.

A similar conclusion based on the genetic stability of population systems was made over three decades ago (Altukhov and Rychkov 1970) but remained almost overlooked. In later years, I have many times and in different circumstances advocated this strategy of optimizing relationships between the mankind and the nature. However, only relatively recently, at a UNO Conference in Rio de Janeiro, did this approach seem to prevail, generating a "Convention on Biological Diversity" that called for the creation of a strategy of *sustained development*. The equivalent of this expression is *socially ecological optimum model* (Altukhov 1983). Maybe a possibility is now open to full awareness of the significance of population genetics for conservation of biological diversity and its control in the process of rational management.

Clearly, the strategy of conservation of gene pools of the species that are involved in active management and artificial reproduction should be supplemented by traditional approaches consisting in organizing biosphere reserves, sanctuaries, and national parks – that is, in creating in a dramatically changing environment stable reserves of life which themselves change little, or, ideally, not at all (Sokolov 1979, 1981; Wilson 1988; Pavlov et al. 2001).

Let us hope that the implementation of all these measures aimed at conservation of the genetic diversity of populations will in fact reorganize the strategy of our interaction with nature. The irreversible destruction of genetic resources of the biosphere and the heredity of man himself must be prevented. We should come into awareness of the significance of the population structure in which differentiation and integration processes balance each other. This population organization (in case of its reconstruction) might be based on an intermediate variant involving both subdivision and panmixia. However, here I only mention this problem since it requires special careful consideration.

Today, particular attention must be focused on implementing at regional and global levels a program of estimation and prediction of genetic load dynamics in the human population, which is a task almost completely neglected for the past 10–15 years. We have no reasons to believe that the unprecedented ecological crisis that is being played out right in front of us is in the least subsiding, and that the genetic processes in modern civilized mankind could be regarded as beneficial.

# References


Aagaard JE, Krutovsky KV, Strauss SH (1998) RAPDs and allozymes exhibit similar levels of diversity and differentiation among populations and races of Douglas-fir. Heredity 81:69–78

Abbot CL, Double MC (2003) Genetic structure, conservation genetics and evidence of speciation by range expansion in sky and white-capped albatrosses. Mol Ecol 12:2953–2962

Abilev SK, Poroshenko GG (1986) Rapid methods of predicting mutagenic and blastogenic properties of chemical substances. Itogi Nauki i Tekhniki Ser Toxikologiya. VINITI 14, 171 pp

Abramoff P, Darnell RM, Balsano JS (1968) Electrophoretic demonstration of the hybrid origin of the gynogenetic teleost *Poecilia formosa*. Am Nat 102:555–558

Adams WT (1983) Application of isozymes in tree breeding. Isozymes in plant genetics and breeding. Elsevier, Amsterdam, pp 381–400

Afanas'ev KI, Suskov II, Altukhov YP (1984) Frequency of rare protein variants in congenitally malformed newborns. Tsitol Genet 2:129-134

Akam M (1998) Hox genes: from master genes to micromanagers. Curr Biol 8:76–78

Albertini RJ, Anderson D, Douglas GR, Hagmar L, Hemminki K, Natarajan AT, Norppa H, Shuker DE.G, Tice R, Waters MD, Aitio A (2000) IPCS guidelines for monitoring of genotoxic effects of carcinogens in humans. Mutat Res 463:11–172

Alexandrov DA (1985) Population systems and speciation: theory, nature, experiment. Priroda. 3:122–125

Allard RW, Kahler AL (1973) Multilocus genetic organization and morphogenesis. Brookhaven Symp Biol 25:329–343

Allard RW, Kahler AL, Weir BS (1972) The effect of selection on esterase allozymes in barley population. Genetics (US) 72(N4):489–503

Allendorf FW, Leary RF (1986) Heterozygosity and fitness in natural populations of animals. In: Soule ME (ed) Conservation biology: the science of scarcity and diversity. Sinauer, Sunderlend, MA, pp 57–76

Allendorf FW, Phelps SR (1980) Loss of genetic variation in a hatchery stock of cutthroat trout. Trans Am Fish Soc 109:537–543

Allendorf FW, Seeb LW (2000) Concordance of genetic divergence among sockeye salmon populations at allozyme, nuclear DNA and mitochondrial DNA markers. Evolution. 54:640–651

Allendorf FW, Thorgaard GH (1984) Tetraploidy and the evolution of Salmonid fishes. In: Turner BJ (ed) Evolutionary genetics of fishes. Plenum Press, New York, pp 1–53

Allendorf FW, Utter FM (1979) Population genetics. Fish physiology, vol 8. Academic Press, New York, pp 407–454

Altukhov YP (1969a) Recent physiological, biochemical and immunological studies on the problem of intraspecific differentiation in marine fish. Rapp et Proc-Verb Reun Cons Perm Int Explor Mer 161:103–108

Altukhov YP (1969b) Immunogenetical approach to the problem of intraspecific differentiation in fish. Usp Sovrem Gonet, vol 2. Genet Nauka, Moscow, pp 161–195

Altukhov YP (1969c) On the relation between monomorphism and polymorphism of haemoglobins in fish microevolution. Dokl Akad Nauk SSSR 189:1115–1117

Altukhov YP (1970) Biochemical polymorphism in fishes and the problem of intraspecific differentiation. XII Int Conf Anim Blood Groups and Biochem Polymorphisms. Budapest

Altukhov YP (1971) Population genetics of fish. Priroda 3:44–57

Altukhov YP (1973a) Rational management of marine biological resources in the light of population genetics. Propagation marine resources Pacific Ocean. Tokai Univ, Tokyo, pp 31–39

Altukhov YP (1973b) Local fish shoals as genetically stable population systems. Biokhimicheskaya genetika ryb (Fish biochemical genetics). Nauka, Leningrad, pp 43–53

Altukhov YP (1974) Populatsionnaya genetika ryb (Fish population genetics). Pischevaya Promyschlennost, Moscow, 245 pp

Altukhov YP (1975) Genetics of natural populations and biosphere resources. Vestn Akad Nauk SSSR 10:37–45

Altukhov YP (1977) Problems of populational-genetic organization of species in fish. Zh Obshch Biol 38:893–906

Altukhov YP (1978) Genetic monitoring of populations in relation to changing environment. Tesisy Dokladov XIV Mezhdunarodnogo Geneticheskogo Kongressa. Nauka, Moscow, pp 16–17

Altukhov YP (1980a) Environmental conditions and genetic monitoring of populations. Well-being of mankind and genetics. Proc XIV Int Congr Genet Moscow. Mir 1b(1): 238–256

Altukhov YP (1980b) Genetic consequences of environmental pollution. Ekologicheskaya kooperatsya (Ecological cooperation). COMECON Bull Bratislava 1:53–66

Altukhov YP (1981a) The stock concept from the viewpoint of population genetics. Can J Fish Aquat Sci 38:1523–1538

Altukhov YP (1981b) Genetic monitoring of populations as related to environmental conditions. In: Vartanyan ME (ed) Genetika i blagosostoyanie chelovechestva (Genetics and human welfare). Nauka, Moscow, pp 205–220

Altukhov YP (1982a) Biochemical population genetics and evolution. Molekularniye Mekhanizmy Geneticheskikh Protsessov (Molecular mechanisms of genetic processes). Nauka, Moscow, pp 89–112

Altukhov YP (1982b) Biochemical population genetics and speciation. Evolution 36:1168–1181

Altukhov YP (1983) Geneticheskye protsessi v populatsiakh (Genetic processes in populations), 1st edn Nauka, Moscow, 279 pp

Altukhov YP (1984) Adaptive norm concept and the problem of outbreeding. Vestn Akad Med Nauk SSSR 7:16–21

Altukhov YP (1985) Two kinds of genetic variability and the problem of speciation. Evolution 39:223–226

Altukhov YP (1985) Molecular evolution of populations. Molekularnye mekhanizmy geneticheskikh protsessov (Molecular mechanisms of genetic processes). Nauka, Moscow, pp 100–131

Altukhov YP (1989a) Geneticheskiye protsessi v populatsiakh (Genetic processes in populations), 2nd edn Nauka, Moscow, 328 pp

Altukhov YP (1989b) Balancing selection as the factor of maintenance of allozyme polymorphism. Usp Sovrem Biol 107:323–340

Altukhov YP (1990) Population genetics: diversity and stability. Harwood, London, 367 pp

Altukhov YP (1991) The role of balancing selection and over-dominance in maintaining allozyme polymorphism. Genetica (Neth) 85:79–90

Altukhov YP (1992) Effects of fishing on the genetics resources of aquatic organisms. FAO Expert Consultation, Rome, 9–14 Nov 1992, background paper, pp 1–30

Altukhov YP (1994a) Principles of population monitoring for conservation genetics. Loeschcke V, Tomiuk L, Jain SK (eds) Conservation genetics. Birkhauser Verlag, Basel, pp 337–350

Altukhov YP (1994b) Genetic consequences of selective fishing. Genetika 30(1):5–21

Altukhov YP (1995) Intrapopulation genetic diversity: monitoring and conservation. Genetika 31(10):1331–1357

Altukhov YP (1996a) Allozyme heterozygosity, sexual maturation rate and longevity. Genetika 34:908–919

Altukhov YP (1996b) Genome heterozygosity, sexual maturation rate and longevity. Dokl Ros Acad Nauk 348:842–845

Altukhov YP (1998) Allozyme heterozygosity, sexual maturation rate, and longevity. Rus J Genet 34(7):908–919

Altukhov YP (1999) Conservation genetics. Ecology in Russia at the beginning of the 21st Century. Nauchnyi Mir, Moscow, pp 9–26

Altukhov YP (2001) Genetic consequences of selective fishery and hatchery. Vopr Rybolovstva 2(4):562–603

Altukhov YP, Abramova AB (2000) Monomorphic species-specific DNA detected in polymerase chain reaction with random primers. Rus J Genet 36(12):1674–1681

Altukhov YP, Bernashevskaya AG (1978) Experimental modelling gene frequency dynamics in the system of partially isolated populations. Dokl Akad Nauk SSSR 238(3): 712–714

Altukhov YP, Bernashevskaya AG (1981) Experimental modelling of genetical processes in population system of *Drosophila melanogaster*, corresponding to the circular stepping stone model. Genetika 17(6):1052–1059

Altukhov YP, Blank ML (1991) Computer modeling of genetic processes in subdivided populations. Dokl Akad Nauk SSSR 219(6):1467–1472

Altukhov YP, Blank ML (1992) Genetic dynamics of population systems with variable parameters of structure and selection. Dokl Akad Nauk SSSR 326(6):1068–1072

Altukhov YP, Dubrova YE (1981) Biochemical polymorphism of populations and its biological significance. Usp Sovrem Biol 91(3):467–480

Altukhov YP, Dubrova YE (1982) Frequencies of rare protein variants in human spontaneous abortions. Dokl Akad Nauk SSSR 262(4):982–985

Altukhov YuP, Evsyukov AN (2001) Overproduction of juveniles by hatcheries as a reason of degradation of Volga's Russian sturgeon stock. Dokl Rus Acad Nauk 380:273–275

Altukhov YP, Kalabushkin BA (1974) Stable polymorphism in modern and ancient populations of molluscs, *Littorina squalida*. Dokl Akad Nauk SSSR 215(6):1477–1480

Altukhov YP, Klimenko VV (1978) Positive correlation between level of individual heterozygosity and ability for complete thermal parthenogenesis in silkworm, Bombyx mori L. Dokl Akad Nauk SSSR 239(2):460–462

Altukhov YP, Kurbatova OL (1984) Human heredity and environment. In: Altukhov YP (ed) Nasledstvennost' cheloveka i okruzhayushchaya sreda (Human heredity and environment). Nauka, Moscow, pp 7–34

Altukhov YP, Kurbatova OL (1990) The problem of the adaptive norm in human populations. Genetika 26:583–598

Altukhov YP, Livshits GM (1978) Factors of differentiation and integration of gene pools in isolated population of mollusc, Chondrus bidens. Dokl Akad Nauk SSSR 238(24): 955–958

Altukhov YP, Pobedonostseva EY (1978) Experimental modelling of genetical processes in subdivided populations. Dokl Akad Nauk SSSR 238(2):466–469

Altukhov YP, Pobedonostseva EY (1979a) Biological dynamics in experimental population system of *Drosophila melanogaster*. Zh Obshch Biol 40(3):368–376

Altukhov YP, Pobedonostseva EY (1979b) Genetical dynamics in experimental population system of *Drosophila melanogaster*. Zh Obshch Biol 40(6):916–923

Altukhov YP, Rychkov YG (1970) Population systems and their structural components: genetic stability and variability. Zh Obshch Biol 31(5):507–526

Altukhov YP, Rychkov YG (1971) Genetic variability on the levels of population systems and their structural components. Nauchn Soobshch Inst Biol Morya. Nauka, Vladivostok, pp 16–20

Altukhov YP, Rychkov YG (1972) Genetic monomorphism of species and its biological significance. Zh Obshch Biol 33(3):281–300

Altukhov YP, Salmenkova EA (1981) Applications of the stock concept to fish populations in the USSR. Can J Fish Aquat Sci 38:1591–1600

Altukhov YP, Salmenkova EA (1987a) Stock transfer relative to natural organization, management, and conservation of fish populations. In: Ryman N, Utter F (eds) Population genetics and fishery management. Univ Washington Press, Seattle, pp 333–343

Altukhov YP, Salmenkova EA (1987b) Population genetics of cold water fish. Proceedings of the world symposium of selection, hybridization, and genetic engineering in aquaculture, vol 1. Heenemann, West Berlin, pp 3–29

Altukhov YP, Salmenkova EA (1990) Introduction of distinct stocks of chum salmon, *Oncorhynchus keta* (Walbaum), into natural populations of the species. J Fish Biol 37 [Suppl A]:25–33

Altukhov YP, Salmenkova EA (1991) The genetic structure of salmon populations. Aquaculture 98:11–40

Altukhov YP, Salmenkova EA (1994) Straying intensity and genetic differentiation in salmon populations. Aquacult Fish Manage 25 [Suppl 2]:99–120

Altukhov YP, Sarsenbayev NA (1980) Correlation between the karakul patterns and variability of polygenic constitutional traits of lambs. Dokl Akad Nauk SSSR 253(6):1469–1473

Altukhov YP, Varnavskaya NV (1983) Adaptive genetic structure and its connection with intrapopulational differentiation for sex, age and growth rare in sockeye salmon, *Oncorhynchus nerka* (Walb.). Genetika 19(3):796–807

Altukhov YP, Apekin VS, Limansky VV (1964) The basic principles of serological investigations of intra- and inter-specific differentiation in fishes. Voprosi fisiologii ryb Chernogo i Asovskogo morei (Questions of the fish physiology of the Black Sea and the Sea of Azov). Pischevaya Promyshlennost, Moscow, pp 53–71

Altukhov YP, Altukhova EP, Ivankov VN, Salmenkova EA, Sachko GD (1968a) Frequencies of lactate dehydrogenase genes in the chum and pink salmon populations in different rivers of Sakhalin Island. Ref Nauchn Rabat Inst Biol Morya Nauka Vladivostok, pp 11–14

Altukhov YP, Altukhova EP, Salmenkova EA, Sachko GD (1968b) Electrophoretic study of intrapopulational variability proteins in chum salmon, *Oncorhynchus keta* (Walb.). Ref Nauchn Rabat Inst Biol Morya Nauka, Vladivostok, pp 19–20

Altukhov YP, Limansky VV, Payusova AN, Truveller KA (1969a) Immunogenetical analysis of intraspecific differentiation of the European anchovy from the Black and Azov Seas. 1. The blood groups and possible mechanism of gene control. The diversity of the Azov race. Genetika 5(4):50–64

Altukhov YP, Limansky VV, Payusova AN, Truveller KA (1969b) Immunogenetical analysis of intraspecific differentiation of the European anchovy from the Black and Azov Seas. 2. Elementary populations of anchovy and its place in population-genetical structure of species. Genetika 5(5):81–94

Altukhov YP, Salmenkova EA, Sachko GD (1970) Duplication and polymorphism of lactate dehydrogenase genes of the Pacific salmon. Dokl Akad Nauk SSSR 195(3):711–714

Altukhov Yu, Salmenkova EA, Omelchenko VT, Sachko GD, Slynko VI (1972) Number of mono- and polimorphous loci in the population of tetraploid salmon species *Oncorhynchus keta* Walb. Genetika 8(2):251–259

Altukhov YP, Salmenkova EA, Konovalov SM, Pudovkin AI (1975a) Stationary distributions of the frequencies of lactate dehydrogenase and phosphoglucomutase genes in the system of subpopulations of a local fish shoal of *Oncorhynchus nerka* (Walb.). 1. Stability of the shoal over generations with simultaneous variability of the component subpopulation. Genetika 11(4):44–53

Altukhov YP, Pudovkin AI, Salmenkova EA, Konovalov SM (1975b) Stationary distributions of the frequencies of lactate dehydrogenase and phosphoglucomutase genes in the system of subpopulations of a local fish shoal of *Oncorhynchus nerka* (Walb.). 2. Random genetic drift, migration and selection as factors of stability. Genetika 11(4):54–62

Altukhov YP, Khilchevskaya RI, Dubinin NP (1976a) Genetic monitoring of human populations. Abstr V intern congr hum genet Mexico, pp 176–177

Altukhov YP, Zhivotovsky LA, Sadykov SS et al (1976b) Effects of modal and directional selection for a set of quantitative traits in cotton *Gossipium hirsutum*. Dokl Akad Nauk SSSR 227(1):212–215

Altukhov YP, Abdullaev BA, Sadykov SS (1978) The possibility of using the principle of modal selection for stabilizing and improving cotton cultivars. Genetics and selection of quantitative traits in cotton. Fan, Tashkent, pp 19–32

Altukhov YP, Bernashevskaya AG, Milishnikov AN, Novikova TA (1979a) Experimental modelling of genetic processes in the populational system of *Drosophila melanogaster*, corresponding to the circular stepping-stone model. 1. The approach to the problem and local differentiation of *6-Gpd* and *Est-6* allele frequencies. Genetika 15(4):646–655

Altukhov YP, Botviniyev OK, Kurbatova OL (1979b) Population-genetical approach to the problem of non-specific biological resistance of human organism. 1. The problem of definition and principles of the approach. Parameters of distributions of anthropometric traits of healthy and diseased newborns. Genetika 15(2):352–360

Altukhov YP, Salmenkova EA, Ryabova GD, Kulikova NI (1980a) Genetic differentiation of chum salmon *Oncorhynchus keta* (Walb.) populations and effectiveness of some acclimatization measures. Biol Morya 3:23–38

Altukhov YP, Sarsenbaev NA, Afanas'ev KI, Malinina TV, Milishnikov AN, Pobedonostseva EY (1980b) Peculiarities of the karakul pattern and genetic structure of karakul sheep groups belonging to morphologically "average" and "extreme" types. Genetika 16(10):1871–1883

Altukhov YP, Kurbatova OL, Botvinyev OK, Afanas'ev KI, Malinina TV, Kholod ON et al (1981) Gene markers and diseases: genetic, anthropometric and clinical characteristics of children with acute pneumonia. Genetika 17(5):920–931

Altukhov YP, Salmenkova EA, Omelchenko VT, Efanov VN (1983) Genetic differentiation and population structure in the pink salmon of Sakhalin-Kuril region. Biol Morya 2:46–51

Altukhov YP, Novoselskaya AY, Salmenkova EA, Ryabova GD, Goncharova AA, Konovalov SM et al (1983a) Factors of differentiation and integration of genetical structure of a native system of *Oncorhynchus nerka* (Walb.) subpopulations in Lake Azabachye (Kamchatka). Zh Obshch Biol 44(3):316–331

Altukhov YP, Pobedonostseva EY, Bernashevskaya AG (1983b) Experimental modelling of the genetical processes in subdivided populations. Biologicheskie osnovy rybovodstva: genetika i selektsia (The biological bases of fishery: genetics and selection). Nauka, Leningrad, pp 51–70

Altukhov YP, Naidich VA, Zhivotovsky LA (1984) Computer simulation of the interaction between random genetic drift and gene migration in a system of partially isolated populations. Genetika 20(4):605–609

Altukhov YP, Suskov II, Afanas'ev KI, Malinina TV, Shurkal AV, Rakitskaya TA. et al (1985) Frequency of rare protein variants in normal and congenially malformed infants. Genetika 21(12):2031–2043

Altukhov YP, Krutovsky KV, Gafarov NI, Dukharev VA, Morozov GP (1986a) Allozyme polymorphism in a natural population of Norway spruce, *Picea abies* (L.). 1. Polymorphic systems and mechanisms of genic control. Genetika 22(8):2135–2147

Altukhov YP, Gafarov NI, Krutovsky KV, Dukharev VA (1986b) Allozyme polymorphism in a natural population of Norway spruce, *Picea abies* (L.). 2. Correlation between the level of individual heterozygosity and relative amount of non-viable seeds. Genetika 22(12):2825–2830

Altukhov YP, Salmenkova EA, Omelchenko VT, Rubtzova GA, Dubrova YE (1987a) Balancing selection as a possible factor maintaining uniformity of allele frequencies of enzyme loci in populations of pink salmon, *Oncorhynchus gorbuscha* Walb. Genetika 23(10):1884–1896

Altukhov YP, Sarsenbaev NA, Shirinskii MA et al (1987b) Modal selection for morphological traits in the karakul sheep breed. Intensive Technology of Karakul Sheep Breeding. Kainar, Alma-Ata, pp 63–70

Altukhov YP, Mezhzherin SV, Salmenkova EA, Omelchenko VT (1989) Effect of selective breeding on adaptive genetic and biological structure of Oncorhynchus gorbuscha (Walb.) population. Genetika 25(10):1843–1853

Altukhov YP, Salmenkova EA, Kartavtzev YF (1991) Relationship of allozyme heterozygosity with viability and growth rate of pink salmon. Tsitologiya i Genetika 25(1):47–51

Altukhov YP, Zhivotovsky LA, Gundaev AI (1994) A method of selection and seed production. Patent no 1445645. Rospatent, pp 1–4

Altukhov YP, Korochkin LI, Rychkov YG (1996) Hereditary biochemical diversity in the evolution and development. Genetika 32(6):1450–1473

Altukhov YP, Salmenkova EA, Omelchenko VT (1997) Population genetics of salmonid fishes. Nauka, Moscow, 288 pp

Altukhov YP, Salmenkova EA, Omelchenko VT (2000a) Salmonid fishes: population biology, genetics and management. Blackwell Scientific, Oxford, 368 pp

Altukhov YP, Sheremetyeva VA, Rychkov YG (2000b) Heterosis as a Cause of Acceleration in Humans. Dokl Ros Acad Nauk 370(1):130–133

Alves MJ, Coelho MM, Collares-Pereira MJ (2001) Evolution in action through hybridization and polyploidy in an Iberian freshwater fish: a genetic review. Genetica 111:375–385

Anderson PK (1970) Ecological structure and gene flow in small mammals. Symp Zool Soc Lond 26:299–325

Andriyasheva MA, Lokshina AB (1985) Method for increasing heterozygosity level of Coregonid fish populations in the process of artificial reproduction. Molekularniye mekhanismy geneticheskikh protsessov (Molecular mechanisms of genetic processes). Nauka, Moscow, pp 212–218

Andriyasheva MA, Lokshina AB (1987) Populational-genetic approach to the breeding of peled. Genetika 23(12):2085–2097

Anxolabéhère D (1976) Modèle de sélection an locus sepia chez Drosophila melanogaster. Bull Soc Zool Fr 101:1035

Arber W (1979) Promotion and limitation of genetic exchange. Science 205:361–365

Armour JAL, Alegre SA, Miles S et al (1999) Minisatellites and mutation processes in tandemly repetitive DNA. In: Goldstein DB, Schlotterer C (eds) Microsatellites. Evolution and applications. Oxford Univ Press, Oxford, pp 24–33

Arnold MH (1972) Modal selection in BP-52. Cotton Grow Rev 49:107–125

Aronshtam AA, Borkin LY, Pudovkin AI (1977) Isozymes in population and evolutionary genetics. Genetika isofermentov (The genetics of isoenzymes). Nauka, Moscow, pp 199–249

Aspinwall N (1974) Genetic analysis of North American populations of the pink salmon, *Oncorhynhcus gorbuscha*. Possible evidence for the neutral mutation-random drift hypothesis. Evolution 28:295–305

Aspinwall N, Tsuyuki H (1968) Inheritance of muscle proteins in hybrids between the redside shiner (*Richardsonius balteatus*) and the peamouth chub (*Mylocheilus caurinum*). J Fish Res Board Canada 25:1317–1322

Astaurov BL (1969) Experimental polyploidy and origin of natural polyploids. Genetika 5(7):129–148

Auerbach C, Robson TM (1946) Chemical production of mutations. Nature 157:302–305

Avise JC (1994) Molecular markers, natural history and evolution. Chapman and Hall, New York, 511 pp

Avise JC (2000) Phylogeography: the history and formation of species. Harvard Univ Press, Cambridge, MA, 447 pp

Avise JC, Arnold J, Ball RM et al (1987) Intraspecific phylogeography: the mitochondrial DNA bridge between population genetics and systematics. Annu Rev Ecol Syst 18: 489–522

Awadalla P, Eyre-Walker A, Smith JM (1999) Linkage disequilibrium and recombination in hominid mitochondrial DNA. Science 286:2524–2525

Ayala FJ (1972) Darwinian versus non-Darwinian evolution in natural populations of Drosophila. Proc VI Berkeley Symp Math Stat Probl Berkeley, vol 5, pp 211–236

Ayala FJ (1975) Genetic differentiation during the speciation process. Evol Biol 8:1–78

Ayala FJ (1976) Molecular evolution. Sinauer, Sunderland, MA, 277 pp

Ayala FJ (1978) The mechanism of evolution. Sci. Amer. 239:48–61

Ayala FJ (1983) Genetic polymorphism: from electrophoresis to DNA sequences. Experientia 39:813–823

Ayala FJ, Gilpin ME (1974) Gene frequency comparisons between taxa: support for the natural selection of protein polymorphism. Proc Natl Acad Sci USA 71:4847–4849

Ayala FJ, Powell JR (1972) Enzyme variability in the *Drosophila willistoni* group. Biochem Genet 7:331–345

Ayala FJ, Tracey ML (1974) Genetic differentiation within and between species of the Drosophila willistoni group. Proc Natl Acad Sci USA 71:999–1002

Ayala FJ, Valentine LW (1979a) Genetic variability in pelagic environment: a paradox? Ecology 60:24–29

Ayala FJ, Valentine JW (1979b) Evolving: the theory and processes of organic evolution. Benjamin Cumings, Menlo Park, CA, 452 pp

Ayala FJ, Powell JR, Dobzhansky T (1971) Polymorphisms in continental and island populations of *Drosophila willistoni*. Proc Natl Acad Sci USA 68:2480–2483

Ayala FJ, Tracey ML, Barr LG et al (1974a) Genetic variation in natural populations of five *Drosophila* species and the hypothesis of the selective neutrality of protein polymorphisms. Genetics (US) 77:343–384

Ayala FJ, Tracey ML, Hedgecock D et al (1974b) Genetic differentiation during the speciation process in *Drosophila*. Evolution 28:576–592

Ayala FJ, Valentine JW, Barr LG et al (1974c) Genetic variability in a temperate intertidal phoronid, *Phoronopsis viridis*. Biochem Genet 11:413–427

Baatar D (1968) Analysis of the breeding "plateau" development in the process karakul lamb selection. D Sci Thesis Inst Gen Genetics, USSR Acad Sci Moscow

Bailey DW (1966) Heritable histocompatibility changes: lysogeny in mice? Transplantation 4:482–488

Bailey GS, Cocks CT, Wilson AC (1969) Gene duplication in fishes: malate dehydrogenases of salmon and trout. Biochem Biophys Res Com 34:605–612.

Bailey GS, Wilson AC, Halver JE et al (1970) Multiple forms of supernatant malate dehydrogenase in salmonid fishes: biochemical, immunological and genetic studies. J Biol Chem 245:5927–5940

Balanovskaya EV, Rychkov YG (1990) Ethnogenetics: adaptive structure of the mankind gene pool according to data on human polymorphic genetic markers. Genetika 26(4):739–748

Balloux F, Lugon-Moulin N (2002) The estimation of population differentiation with microsatellite markers. Mol Biol 11:155–165

Banks MA, Eichert W (2000) WHICHRUN (version 3.2): a computer program for population assignment of individuals based on multilocus genotype data. J Hered 91:87–89

Banks MA, Rashbrook VK, Calavetta MJ et al (2000) Analysis of microsatellite DNA resolves genetic structure and diversity of Chinook salmon (*Oncorhynchus tshawytscha*) in California's Central Valley. Can J Fish Aquat Sci 57:915–927

Bannikov AG (1979) "World strategy of conservation" and protection of animal diversity. Priroda 5:24–28

Barauskene VK, Lekyavichus EK, Altukhov YP (1985) Experimental analysis of interrelation between the type of reproduction in *Daphnia* and the level of individual allozymic heterozygosity. Genetika 21(4):591–600

Barlow R (1981) Experimental evidence for interaction between heterosis and environment in animals. Anim Breed Abstr 49:715–737

Bartley D, Bendy B, Brodziak J et al (1992) Geographic variation in population genetic structure of Chinook salmon from California and Oregon. Fishery Bull US 90: 77–100

Barton NM, Hewitt GM (1985) Analysis of hybrid zones. Annu Rev Ecol Syst 16:113–148

Basset P, Bernard Y, Garel MC et al (1978) Isoelectric focusing of human hemoglobin: its application to screening, to the characterization of 70 variants, and to the study of modified fractions of normal hemoglobins. Blood 51:971–982

Beacham TD, Withler RE, Gould AP (1985) Biochemical genetic stock identification of pink salmon (*Oncorhynchus gorbuscha*) in southern British Columbia and Puget Sound. Canad J Fish Aquat Sci 42:1474–1488

Beacham TD, Withler RE, Murray CB, Barner LW (1988) Variation in body size, morpholgy, egg size, and biochemical genetics of pink salmon in British Columbia. Trans Am Fish Soc 117:109–126

Beardmore JA, Dobzhansky Th, Pavlovsky O (1960) An attempt to compare the fitness of polymorphic and monomorphic experimental populations of *Drosophila pseudoobscura*. Heredity 14:19–33

Beardmore JA, Shami SA (1979) Heterozygosity and the optimal phenotype under stabilising selection. Aquilo Ser Zool 20:100–110

Beattie AJ, Culver DC (1979) Neighbourhood size in Viola. Evolution 33:1226–1229

Beaumout MA, Nichols RA (1996) Evaluating loci for use in the genetic analysis of population structure. Proc R Soc Lond 263:1619–1626

Beçak ML, Beçak W, Rabello MN (1966) Cytological evidence of constant tetraploidy in the bisexual South American frog *Odontophrynus americanus*. Chromosoma 19:188–193

Beçak ML, Schwantes AR, Schwantes ML (1968) B. Polymorphism of albumin-like proteins in the South American tetraploid frog *Odontophyrnus americanus* (Salientia: Ceratophrydidae). J Exp Zool 168:473–476

Beçak W, Beçak ML, Lavalle P, Schreiber GS (1967) Further studies on polyploid amphibians (Ceratophrydidae). Chromosoma. 23:14–23

Beet EA (1949) The genetics of sickle cell trait in a Bantu tribe. Ann Eugenics 14:279–286

Begon M (1977) The effective size of a natural *Drosophila subobscura* populations. Heredity 38:13–18

Bekkevold D, Hansen MM, Loeschcke V (2002) Male reproductive competition in spawning aggregations of cod (*Gadus morhua*, L.). Mol Ecol 11:91–102

Bell D (1978) Use body weight to improve your income. Poultry Tribune 84:18–22

Belyaev DK (1974) Some aspects of stabilizing and destabilizing selection. Istoriya i teoriya evolutsionnogo ucheniya (The history and theory of the concept of evolution). Nauka, Leningrad, pp 76–84

Belyaev DK (1980) Destabilizing selection as a factor of domestication. Well-being of mankind and genetics. Proc XIV Int Congr Genet, Mir, Moscow 1(1):64–80

Belyaev DK, Trut LN (1964a) Behaviour and reproductive function of animals. 1. Correlation of behaviour traits, the time of reproduction and fecundity. Byull Mosk Obshch Ispyt Prir Otd Biol 69(3):5–18

Belyaev DK, Trut LN (1964b) Behaviour and reproductive function of animals. 2. Correlation changes observed in the course of domestication. Byull Mosk Obshch Ispyt Prir Otd Biol 69(4):5–14

Bender K, Ohno S (1968) Duplication of the autosomally inherited 6-phosphogluconate dehydrogenase gene locus in tetraploid species of cyprinid fish. Biochem Genet 2: 101–107

Bentzen P, Harris AS, Wright JM (1991) Cloning of hypervariable minisatellite and simple sequence microsatellite repeats for DNA fingerprinting of important aquacultural species of salmonids and tilapia. In: Bruke T, Dolf G, Jeffreys AJ, Wolf R (eds) DNA fingerprinting approaches and applications. Birkhauser, Basel, pp 243–262

Bentzen P, Wright JM, Bryden LT, Sargent M, Zwanenburg KCT (1998) Tandem repeat polymorphism in the mitochondrial control region of redfishes (Sebastes: Scorpaenidae). J Hered 89:1–7

Berger EM (1971) A temporal survey of allelic variation in natural and laboratory populations of Drosophila melanogaster. Genetics (US) 67:121–136

Berger EM, Weber L (1974) The ribosomes of Drosophila. 2. Studies on intraspecific variation. Genetics (US) 78:1173–1183

Bernatchez L, Wilson CC (1998) Comparative phylogeography of nearctic and palearctic fishes. Mol Ecol 7:431–452

Bernatchez L, Dempson JB, Martin S (1998) Microsatellite gene diversity analysis in anadromous arctic char, *Salvelinus alpinus*, from Labrador, Canada. Can J Fish Aquat Sci 55:1264–1272

Betz A, Turleau L, de Grouchy J (1974) Hétérozygotie et homozygotie pour une inversion péricentrique du 3 humain. Ann Genet 17:77

Bielawski JP, Gold JR (1996) Unequal synonymous substitution rate within and between two protein-coding mitochondrial genes. Mol Biol Evol 13:889–892

Billington N, Hebert PDH (1991) Mitochondrial DNA diversity of fishes and its implications for introductions. Can J Fish Aquat Sci 48 [Suppl 1]:80–94

Birley AJ, Beardmore JA (1972) Manifold large selective effects in an enzyme polymorphism. 5th Eur Mar Biol Symp Piccin, Padova, pp 81–100

Black DL (2000) Protein diversity from alternative splicing: a challenge for bioinformatics and post-genome biology. Cell 103:367–370

Blair WF (1960) The rusty lizard: a population study. Univ Texas Press, Austin

Blank ML, Altukhov YP (1992) Genetic dynamics in populations with complex migration structure. Dokl Akad Nauk SSSR 324(6):1309–1313

Blank ML, Altukhov YP (1995) Disequilibrium genetic dynamics of population systems. Dokl Akad Nauk 344(1):136–139

Blier PU, Dufresne F, Burton RS (2001) Natural selection and the evolution of mtDNA-encoded peptides: evidence for intergenomic co-adaptation. Trends Genet 17:400-406

Bochkov NP (1974) Mutational process in man. Lectsii po meditsinskoi genetike (Lectures on medical genetics). Meditcina, Moscow, pp 70-88

Bochkov NP (1977) Genetical monitoring of human populations in connection with environmental pollution. Tsitol Genet 2(11):195-206

Bochkov NP (1978) Human genetics. Meditcina, Moscow, 382 pp

Bochkov NP (1980) Chemical mutagenesis in man and prognosis of its effects. Well-being of mankind and genetics. Proc XIV Int Congr Genet, Moscow, Mir 1b(1):215-224

Bochkov NP (2001) Clinical Genetics. GEOTAR-MED, Moscow, 290 pp

Bochkov NP, Shram RYa, Kuleshov NP, Zhurkov VS (1975) Evaluation of mutagenicity of chemicals for man: general principles, practical recommendations and further developments. Genetika 11(10):156-169

Bodmer WF, Cavalli-Sforza LL (1968) A migration matrix model for the study of random genetic drift. Genetics (US) 59:565-592

Bogart JP (1980) Polyploidy in evolution of amphibians and reptiles. In: Lewis HL (ed) Polyploidy: biological relevance. Plenum Press, New York, pp 341-369

Bondarenko YV, Kovalenko VP, Kutnyuk PI (1979) Efficacy of modal selection in poultry populations. Nauchn-Tekh Byull Ukr Nauchn-Issled Inst Ptitsevod 7:3-7

Borkin LY, Darevsky IS (1980) Reticulate (hybridogenic) speciation in vertebrates. Zh Obshch Biol 41(4):485-506

Borkin LY, Terbish H, Tsaune IA (1986) Tetraploid and diploid populations of *Bufo viridis* from Mongolia. Dokl Akad Nauk SSSR 287(3):760-764

Botvinyev OK, Kurbatova OL, Altukhov YP (1980) Populational-genetic approach to the problem of non-specific biological resistance of human organism. 2. Clinical characteristics, congenital anomalies, and genetic structure of diseased children with respect to their body weight and length at birth. Genetika 16(10):1884-1894

Bowcock AM, Bucci C, Hebert JM et al (1987) Study of 47 DNA markers in five populations from four continents. Gene Geogr 1:47-64

Bowen BW, Meylan AB, Ross JP, Limpus CJ, Balazs GH, Avise JC (1992) Global population structure and natural history of the green turtle (*Chelonia mydas*) in terms of matriarchal phylogeny. Evolution 46:865-881

Braitseva OA (1968) Stratigrafiya Chetvertichnikh Otlozhenii i Oledeneniya Kamchaiki (The stratigraphy of quaternary deposits and glaciation in Kamchatka). Nauka, Moscow, 226 pp

Brakefield PM, Zwaan B (1999) Towards a new synthesis. Trends Ecol Evol 14:84-85

Brannon EL (1972) Mechanisms controlling migration of sockeye salmon fry. Int Pacif Salmon Fish Commiss 21:1-86

Brererton J (1962) Evolved regulatory mechanisms of population control. In: Leeper GW (ed) Evolution of living organisms. Melbourne University Press, Cambridge University Press, Parkville, Victoria, Australia, pp 81-93

Brett D, Pospisil H, Valcarcel J, Reich J, Bork J (2002) Alternative splicing and genome complexity. Nature Genet 30:29-30

Britten RJ, Davidson EH (1969) Gene regulation for higher cells: a theory. Science 165:349-357

Britten RL, Davidson EH (1971) Repetitive and non-repetitive DNA sequences and a speculation on the origin of evolutionary novelty. Q Rev Biol 46:111-138

Britten RJ, Kohne DE (1968) Repeated sequences in DNA. Science 161:529-540

Brown AJL, Langley CH (1979) Reevaluation of level of genic heterozygosity in natural populations of *Drosophila melanogaster* by two-dimensional electrophoresis. Proc Natl Acad Sci USA 76:2381-2384

Brown WM, George M, Wilson AC (1979) Rapid evolution of animal mitochondrial DNA. Proc Natl Acad Sci USA 76:1967–1971

Brunner PC, Douglas MR, Bernatchez L (1998) Microsatellite and mitochondrial DNA assessment of population structure and stocking effects in Arctic charr *Salvelinus alpinus* (Teliostei, Salmonidae) from Central Alpine lakes. Mol Ecol 7:209–223

Brykov VA, Polyakova N, Skurikhina LA, Kukhlevsky AD (1996) Geographical and temporal mitochondrial DNA variability in populations of pink salmon. J Fish Biol 48:899–909

Brykov VA, Polyakova NE, Skurikhina LA, Kukhlevsky AD, Malinina TV, Minakhin LS, Altukhov YP (1999) Population-genetic structure of pink salmon, *Oncorhynchus gorbuscha* (Walbaum), by the results of restrictase analysis of mitochondrial DNA: dynamics of diversity in generations. Genetika 35:666–673

Bumpus HC (1899) The elimination of the unfit as illustrated by the introduced sparrow, *Passer domesticus*. Marine biological laboratory, Woods Hall, pp 209–226

Burger CV, Scribner KT, Spearman WJ et al (2000) Genetic contribution of three introduced life history of sockeye salmon to colonization of Frazer lake, Alaska. Can J Fish Aquat Sci 57:2096–2111

Buri P (1956) Gene frequency in small populations of mutant Drosophila. Evolution 10:367–402

Bush GL (1975) Modes of animal speciation. Annu Rev Ecol Syst 6:339–364

Bush GL, Case SM, Wilson AC et al (1977) Rapid speciation and chromosomal evolution in mammals. Proc Natl Acad Sci USA 74:3942–3946

Bushuyev VP (1973) The two-component haemoglobins in salmon are associated with its allotetraploid origin. In: Troshin AS (ed) Biokhimicheskaya genetika ryb (Fish biochemical genetics). Nauka, Leningrad, pp 62–66

Bushuyev VP, Omelchenko VT, Salmenkova EA (1975) Species-specificity and intraspecific constancy of electrophoregrams and thermal resistance of haemoglobins in some fishes (Clupeiformes). Zh Obshch Biol 36(4):569–578

Caetano-Anolles G, Gresshoff PM (eds) (1997) DNA markers: protocols, applications and overviews. Wiley-VCH, New York, 364 pp

Caizergues A, Rätti O, Helle P, Rotelli L, Ellison L, Rasplus J-Y (2003) Population genetic structure of male black grouse (*Tetro tetrix* L) in fragmented vs. continuous landscapes. Mol Ecol 12:2297–2305

Campton DE, Utter FM (1987) Genetic structure of anadromous cutthroat trout (*Salmo clarki clarki*) populations in the Puget Sound area: evidence for restricted gene flow. Can J Fish Aquat Sci 44:573–582

Cann RL, Stoneking M, Wilson AC (1987) Mitochondrial DNA and human evolution. Nature 325:31–36

Cargill M, Altshuler D, Ireland J et al (1999) Characterization of single-nucleotide polymorphism in coding regions of human genes. Nat Genet 22:231–238

Carson HL (1958) Increase in fitness in experimental populations resulting from heterosis. Nat Genet 44:1136–1141

Carson HL (1975) The genetics of speciation at the diploid level. Am Nat 109:83–92

Carson HL (1982) Speciation as a major reorganization of polygenic balances. In: Barigozzi C (ed) Mechanisms of speciation. Liss, New York, pp 411–433

Carr A, Carr MH, Meylan AB (1978) The ecology and migration of sea turtles. 7. The West Caribbean green turtle colony. Bull Am Mus Nat Hist 162:1–46

Carvalho GR (1993) Evolutionary aspects of fish distribution: genetic variability and adaptation. J Fish Biol 43 [Suppl A]:53–73

Carvalho GR (1994) Molecular genetics and the stock concept in fisheries. In: Carvalho G, Pitcher TJ (eds) Molecular genetics in fisheries. Chapman and Hall, London, pp 326–350

Cavalli-Sforza LL (1966) Population structure in human evolution. Proc R Soc Lond B 164:362–379

Cavalli-Sforza LL (1998) The DNA revolution in population genetics. Trends Genet 14(2): 60–65

Cavalli-Sforza LL, Bodmer WF (1971) The genetics of human populations. Freeman, San Francisco, 962 pp

Cavalli-Sforza LL, Barrai J, Edwards AWF (1964) Analysis of human evolution under random genetic drift. Cold Spring Harbor Symp Quant Biol 29:9–20

Chadov BV (2001) Mutations capable of inducing speciation. Evolutionary biology, vol 1. Tomsk State Univ, Tomsk, pp 138–162

Chakraborty R (1977) Distribution of nucleotide differences between two randomly chosen cistrons in a population of variable size. Theor Pop Biol 11:11–22

Chakraborty R (1987) Biochemical heterozygosity and phenotypic variability of polygenic traits. Heredity 59:19–28

Chakraborty R, Fuerst PA, Nei M (1978) Statistical studies on protein polymorphism in natural populations. 2. Gene differentiation between populations. Genetics (US) 88: 367–390

Chakraborty R, Fuerst PA, Nei M (1980) Statistical studies on protein polymorphism in natural populations. 3. Distribution of allele frequencies and the number of alleles per locus. Genetics (US) 94:1039–1063

Chakravarti A (1999) Population genetics-making sense out of sequence. Nat Genet 21:56–60

Chang Te-Tzu (1979) Genetics and evolution of the green revolution. Replies from biological research. Conselo Superior de Investigationes Cientificas, Madrid, pp 187–209

Chesser RK (1983) Genetic variability within and among populations of the black-tailed prairie dog. Evolution 37:320–331

Chetverikov SS (1926) Some aspects of evolutionary process from the viewpoint of modern genetics. Zh Eksp Biol 1:3–54

Chinnery PP, Turnbull DM (2000) Mitochondrial DNA mutations in the pathogenesis of human diseases. Mol Med Today 6:425–432

Chunikhin SP (1979) Ecological stresses. Khimiya i Zhizn' 2:32–35

Churikov D, Matsuoka M, Luan X et al (2001) Assessment of concordance among genealogical reconstructions from various mtDNA segments in three species of Pacific salmon (genus *Oncorhynchus*). Mol Ecol 19:2329–2339

Clegg MT, Epperson BK (1985) Recent developments in population genetics. Advances in genetics, vol 23. Academic Press, London, pp 235–265

Cohen PTW, Omenn GS, Motulsky AG et al (1973) Restricted variation in glycolitic enzymes of human brain and erythrocytes. Nat New Biol 241:229–233

Comings DE (1972) Evidence for ancient tetraploidy and conservation of linkage groups in mammalian chromosomes. Nature 238:455–457

Cooper DN, Schmidke J (1984) DNA restrictions fragment length polymorphism and heterozygosity in human genome. Hum Genet 66:1–16

Cope OB (1957) Races of cutthroat in Yellowstone lake. US Fish and Wildl Serv Spec Rep Fish 208:74–84

Coulthart MB, Rhomberg LR, Singh JRS (1984) The nature of genetic variation for species formation. Evolution 38:689–692

Coyne A (1992) Genetics and speciation. Nature 355:511–515

Coyne JA, Barton NH, Turelli M (2000) Is Wright's shifting balance process important in evolution? Evolution 54:306–317

Crick F (1979) Split genes and RNA splicing. Science 204:264–271

Cross TF, King J (1983) Genetic effects of hatchery rearing in Atlantic salmon. Aquaculture 33:33–40

Crow JF (1954) Breeding structure of populations. 2. Effective population number. Statistics and mathematics in biology. Iowa State College Press, Ames (Iowa), pp 543–556

Crow JF (1958) Some possibilities for measuring selection intensities in man. Hum Biol 30:1–13

Crow JF (1971) Human population monitoring. Chemical mutagens, vol 2. Plenum Press, New York, pp 591–605

Crow JW, Morton NE (1955) Measurement of gene frequency drift in small populations. Evolution 9:202–214

Curtsinger JW (1990) Frequency-dependent selection in *Drosophila*: estimation of net fitness in pseudohaploid populations. Evolution 44:857–869

Danieli GA, Costa R (1977) Transient equilibrium at the Est-6 locus in wild populations of *Drosophila melanogaster*. Genetics (US) 47:37–41

Danley PD, Kocher TD (2001) Speciation in rapidly diverging systems: lessons from Lake Malawi. Mol Ecol 10:1075–1086

Danzmann RG, Ferguson MM, Allendorf FW (1988) Heterozygosity and components of fitness in strains of rainbow trout. Biol J Linnean Soc 33:285–304

Darevsky IS (1993) Evolution and ecology of parthenogenetic reproduction in reptiles. Current problems of the evolution theory. Nauka, Moscow, pp 80–109

Darevsky IS, Danielyan FD (1969) Diploid and triploid specimens in progeny of parthenogenetic females of rocky lizards as a consequence of their natural crossing with males of closely related bisexual species. Dokl Akad Nauk SSSR 184(3):727–730

Darevsky IS, Kulikova VN (1964) Natural triploidy in polyploid group of Caucasian rocky lizards, *Lacerta saxicola* Eversmann, as a consequence of hybridization between bisexual and parthenogenetic forms of this species. Dokl Akad Nauk SSSR 158(1): 202–204

Datta U, Datta P, Mandal RK (1988) Cloning and characterization of highly repetitive fish nucleotide sequence. Gene 62:331–336

Davidson EH, Britten RJ (1973) Organization, transcription and regulation in the animal genome. Quart Rev Biol 48:565–613

Davidson EH, Galau GA, Angerer RC, Britten RJ (1975a) Comparative aspects of DNA organization in metazoa. Chromosoma 51:253–259

Davidson EH, Hough BR, Klein WH et al (1975b) Structural genes adjacent to interspersed repetitive DNA sequences. Cell 4:217–238

De Jong G, de Ruiter JR, Harring R (1994) Genetic structure of a population with social structure and migration. In: Conservations genetics. Löeschke V et al (eds) Birkhauser, Basel, pp 337–35

Demographic yearbook of Russia (Statistical Handbook) (2001) Goskomstat of Russia, Moscow, pp 397–398

De Vries H (1901) Die Mutationstheorie. Bd I. Veit, Leipzig

De Vries H (1906) Species and varietes, their origin by mutation. In: Macdougal T (ed) Lectures delivered at University of California. 2nd edn Univ of Chicago Press, Chicago

DeWoody JA, Avise JC (2000) Microsatellite variation in marine, freshwater and anadromous fishes compared with other animals. J Fish Biol 56:461–473

De-Xing Z, Hewitt GM (2003) Nuclear DNA analyses in genetic studies of populations: practice, problems and prospects. Mol Ecol 12:563–584

Diagilev SE, Markevich NB (1979) Different maturation time of even- and odd-years' pink salmon as main factor accounting for different results of its acclimatization on the north of Europe of the USSR. Vopr Ichtyol 19: 230–245

Diver C (1929) Fossil records in Mendelian mutants. Nature 124:183

Dobzhansky T (1943) Genetics of natural populations. 9. Temporal changes in the composition of populations of *Drosophila pseudoobscura*. Genetics (US) 28:162–186

Dobzhansky T (1951) Genetics and the origin of species, 3rd edn Columbia Univ Press, New York, 364 pp

Dobzhansky T (1955a) A review of some fundamental concepts and problems of population genetics. Cold Spring Harbor Symp Quant Biol 20:1–15

Dobzhansky T (1955b) Evolution, genetics and man. Wiley, New York, 380 pp

Dobzhansky T (1970) Genetics of the evolutionary process. Columbia Univ Press, New York, 505 pp

Dobzhansky T, Pavlovsky O (1953) Indeterminate outcome of certain experiments on *Drosophila* populations. Evolution 7:198–210

Dobzhansky T, Anderson WW, Pavlovsky O et al (1964a) Genetics of natural populations. 25. A progress report on genetic changes in populations of Drosophila pseudoobscura in the American south-west. Evolution 18:164–176

Dobzhansky T, Lewontin RC, Pavlovsky O (1964b) The capacity for increase in chromosomally polymorphic and monomorphic populations of Drosophila pseudoobscura. Heredity 19:597–614

Dobzhansky T, Ayala FJ, Stebbins GL et al (1977) Evolution. Freeman, San Francisco, 572 pp

Dubinin NP (1931) Genetical automatic processes and their influence on the mechanisms of organic evolution. Zh Eksp Biol 7(5/6):463–479

Dubinin NP (1948) Experimental analysis of the integration of hereditary systems in the evolution of populations. Zh Obshch Biol 9(3):203–244

Dubinin NP (1966) Evolutsiya populyatsii i radiatsiya (Population evolution and radiation). Atomizdat, Moscow, 743 pp

Dubinin NP (1976) Obshchaya genetika (General genetics). Nauka, Moscow, 590 pp

Dubinin NP (1978) Mosaic organization of eukaryotic genes. Dokl Akad Nauk SSSR 243(4):1059–1061

Dubinin NP (1979) The problem of gene in view of molecular organization of eukaryotes and prokaryotes. Usp Sovrem Biol 87(3):331–344

Dubinin NP (1986a) Novoe v sovremennoi genetike (New findings in modern genetics). Nauka, Moscow, 222 pp

Dubinin NP (1986b) Monitoring of gene mutations in human populations. Dokl Akad Nauk SSSR 291(6):1496–1498

Dubinin NP (1988) Rare variants of blood proteins in human populations. Genetika 24(2):197–203

Dubinin NP, Altukhov YP (1977) Monitoring of genetic consequences of environmental pollution: general approach. Geneticheskie posledstviya zagryazneniya okruzhayushchei sredy (The genetic consequences of polluting the environment). Mysl', Moscow 2:15-45

Dubinin NP, Altukhov YP (1979) Gene mutations (de novo) found in electrophoretic studies of blood proteins of infants with anomalous development. Proc Natl Acad Sci USA 76:5226–5229

Dubinin NP, Pashin YV (1978) Mutagenez i okruzhayushchaya sreda (Mutagenesis and the environment). Nauka, Moscow, 128 pp

Dubinin NP, Romashov DD (1932) The genetic structure of a species and its evolution. Biol Zh 1(5/6):52–95

Dubinin NP, Shevchenko YG (1976) Nekotorye voprosy biosotsial'noy prirody cheloveka (Certain questions of man's biosocial nature). Nauka, Moscow, 235 pp

Dubinin NP, Tiniakov GG (1946) Inversion gradients and natural selection in ecological races of *Drosophila funebris*. Genetics (US) 31:537–545

Dubinin NP, Geptner MA, Bessmertnaya SY (1934) Experimental study of *Drosophila melanogaster* ecogenotypes. Biol Zh 3:166–216

Dubinin NP, Romashov DD, Geptner MA et al (1937) Aberrant polymorphism in *Drosophila fasciata* Meig. Biol Zh 6:311–384

Dubinin NP, Altukhov YP, Kurbatova OL, Suskov II (1976) The integral genetic characteristics of "adaptive norm" in human populations. Dokl Akad Nauk SSSR 230(4):957–960

Dubinin NP, Altukhov YP, Suskov II (1979) Experimental basis for the monitoring of gene mutations in man. Dokl Akad Nauk SSSR 243(5):1313–1316

Dubreuil P, Charcosset A (1998) Gene diversity within and among maize populations: a comparison between isozyme and nuclear RFLP loci. Theor Appl Genet 96:577–587

Dubrova YE (1980) Stabilizing selection and distribution of allele frequencies of the loci coding of protein structure. Dokl Akad Nauk SSSR 252(2):472–475

Dubrova YE (1986) Statistical analysis of abnormal $\alpha$- and $\beta$-globins of man. Genetika 22(6):1040–1046

Dubrova YE (1990) Optimum of heterozygosity in human populations. Molekulyarnye Mekhanizmy Geneticheskikh Protsessov (Molecular mechanisms of genetic processes). Thesis, All-Union Symposium, Moscow:119

Dubrova YE, Gavrilets SY (1989) Epistatic interaction of genes in progeny from distant marriages within Russian population. Dokl Akad Nauk SSSR 309(2):211–215

Dubrova YE, Kurbatova OL (1995) Secular growth trend in two generations of the Russian population. Human Biol 67:755–767

Dubrova YE, Altukhov YP, Kucher AN, Ikramov KM (1988) Heterozygosity for biochemical and immunological gene markers and variability of morphophysiological characters in man. Genetika 24(3):556–563

Dubrova YE, Karaphet TM, Sukernik RI, Goltzova TV (1990) Heterozygosity and fertility relationship in the forest nentzy and nganasans. Genetika 26:122–129

Dubrova YE, Dambueva LK, Kholod ON, Prokhorovskaya VD, Pushkina EI (1991) Influence of mother's heterozygosity on the variation of anthropometric traits in newborn. Genetika 27:2128–2176

Dubrova YE, Jeffreys AJ, Malashenko AM (1993) Mouse minisatellite mutations induced by ionizing radiation. Nature Genet 5:92–94

Dubrova YE, Salmenkova EA, Altukhov YP, Kartavtsev YF, Kal'kova EV, Omelchenko VT (1995) Family heterozygosity and progeny body length in pink salmon *Oncorhynchus gorbuscha* (Walbaum). Heredity 75:281–289

Dubrova YE, Nesterov VN, Krouchinski NG et al (1996) Human minisatellite mutation rate after the Chernobyl accident. Nature 380:683–686

Dubrova YE, Nesterov VN, Kroushinski NG et al (1997) Further evidence for elevated minisatellite mutation rate in Belarus after 8 years the Chernobyl accident. Mutat Res 381:267-278

Dubrova YE, Plumb M, Brown J et al (1998) Stage specificity, dose response, and doubling dose for mouse minisatellite germ-line mutations induced by radiation. Proc Natl Acad Sci USA 95:6251–6255

Dubrova YE, Bersimbaev RI, Djansugurova LB et al (2002) Nuclear weapons tests and human germline mutation rate. Science 295:1037

Dufresne F, Bourget E, Bernatchez L (2002) Differential patterns of spatial divergence in microsatellite and allozyme alleles: further evidence for locus-specific selection in the acorn barnacle, *Semibalanus balanoides*? Mol Ecol 11:113–124

Duran S, Pascual M, Estoup A, Turon X (2004) Strong population structure in the marine sponge *Crambe crambe* (*Poeciloscleriada*) as revealed by microsatellite markers. Mol Ecol 13:511–522

Dutrillaux B (1986) The evolutionary role of chromosomes: a new interpretation. Ann Genet 29:69–75

D'yachkov IY (1968) Methods and organization of karakul lamb breeding. D Sci Thesis Alma-Ata, USSR

D'yachkov IY (1975) Proizvodstvo chemykh karakul'skikh shkurok zhaketnykh sortov (Production of black Karakul lamb skins of jacket quality). Kolos, Moscow

D'yachkov IY (1980) Plemennoye delo v karakul'skom ovtsevodstve (Bloodstock in Karakul sheep breeding). Fan, Tashkent, 132 pp

Dyagilev SE, Markevich NB (1979) Time differences in the maturation of *Oncorhynchus gorbuscha* (Walb.) generations of odd and even years as the main factor influencing their acclimatization in the north of the European part of the USSR. Vopr Ikhtiol 19:230–245

Eanes WF, Koehn RK (1977) The correlation of rare alleles with heterozygosity: determination of the correlation for neutral model. Genet Res 29:223–230

Efroimson VP (1968) Vvedeniye v mediisinskuyu genetiku (Introduction to medical genetics). Meditsina, Moscow, 395 pp

Ehling UH (1986) Estimation of genetic risk due to environmental mutagens. In: Marimuthu KM, Gopinath PM (eds) Recent trends in medical genetics. Pergamon Press, New York, pp 1–15

Ehling UH (1988) Quantification of genetic risk of environmental mutagens. Risk Anal 8:45–57

Ehling UH, Charles DJ, Favor J et al (1985) Induction of gene mutation in mice: the multiple endpoint approach. Mutat Res 150:393–401

Ehrlich PR (1965) The population biology of the butterfly, Euphydryas editha. 2. Structure of the Jasper Ridge colony. Evolution 19:327–336

Ehrlich P (1983) Strategy of nature protection, 1980–2000. Biologiya okhrany prirody (The biology of nature conservation). Mir, Moscow, pp 368-386

Ehrlich PR, Holm RW (1963) The process of evolution. McGraw-Hill, NY 280 pp

Ehrlich P, Holm R (1966) Protsess evolutsii (The process of evolution). Mir, Moscow, 330 pp

Ehrlich PR, Raven PH (1969) Differentiation of populations. Science 165:1228–1232

Ehrlich PR, Ehrlich AH, Holdren JP (1977) Ecoscience: population, resources, and environment. Freeman, San Francisco

Eldredge N, Gould SJ (1972) Punctuated equilibria: an alternative to phyletic gradualism. Models in paleobiology. Freeman, San Francisco, pp 82–115

Engel W, Opt'Hof J, Wolf U (1970) Genduplication durch polyploide Evolution: die Isoenzyme der Sorbitoldehydrogenase bei herings und lacksartigene Fishen. Hum Genet 9:157–163

Estoup A, Angers B (1998) Microsatellites and minisatellites for molecular ecology: theoretical and empirical considerations. In: Carvalho GR (ed) Advances in molecular ecology. Nato Sciences Series. IOS Press, Amsterdam pp 55–86

Estoup A, Presa P, Krieg F, Vaiman D, Guyomard R (1993) (CT)n and (GT)n microsatellites: a new class of genetic markers for *Salmo trutta* L (brown trout). Heredity 71:488–496

Evsyukov AN, Zhukova OV, Rychkov YG (1999) The geography of genetic processes in human population: gene migration in Northern Eurasia (the Central Asian Region). Genetika 35:83–94

Ewing B, Green P (2000) Analysis of expressed sequence tags indicates 35,000 human genes. Nature Genet 25:232–234

Excoffier L, Smouse PE, Quattro JM (1992) Analyses of molecular variance inferred from metric distances among DNA haplotypes: application to human mitochondrial DNA restriction data. Genetics 131:479–491

Eyre-Walker A (1999) Evolutionary genomics. Trends Ecol Evol 14:176

Eyre-Walker A (2000) Do mitochondria genome recombine in humans? Philos Trans R Soc Lond B 355:1573–1580

Falconer DS (1960) Introduction to quantitative genetics. Oliver and Boyd, Edinburgh, 365 pp

Favor J (1986a) The frequency of dominant cataract and recessive specific locus mutations in mice. Mutat Res 162:69–80

Favor J (1986b) A comparison of the mutation rates to dominant and recessive alleles in germ cells of the mouse. Genetic toxicology of environmental chemicals. Liss, New York, pp 519–526

Fay JG, Wyckhoff GJ, Chung-I Wu (2001) Positive and negative selection on the human genome. Genetics 158:1227–1234

Fedorov AN, Fedorova LV, Grechko VV, Ryabinin DM, Sheremet'eva VA, Bannikova AA, Lomov AA, Ryskhov AP, Darevsky IS (1999) Variable and invariable DNA repeat characters revealed by taxonprint approach are useful for molecular systematics. J Mol Evol 48:69–76

Felsenstein J (1993) PHYLIP (Phylogeny inference package), version 3.4. Department of Genetics, SK-50. Wash Univ, Seattle, WA

Ferguson MM (1992) Enzyme heterozygosity and growth in rainbow trout: genetic and physiological explanations. Heredity 68:115–122

Ferguson MM (1994) The role of molecular genetic markers in the management of cultured fishes. In: Carvalho G, Pitcher TJ (eds) Molecular genetics in fisheries. Chapman and Hall, London, pp 351–373

Ferguson A (1995) Molecular genetics in fisheries: current and future perspectives. In: Carvalho GR, Pitcher TG (eds) Molecular genetics in fisheries. Chapman and Hall, London, pp 111–115

Ferris SD, Berg WJ (1987) The utility of mitochondrial DNA in fish genetics and management. In: Ryman N, Utter F (eds) Population genetics and fishery management. Univ Washington Press, Seattle, pp 277–301

Ferris SD, Sage RD, Prager EM, Ritte U, Wilson AC (1983) Mitochondrial DNA evolution in mice. Genetics 105:681–721

Fisher RA (1930) The genetical theory of natural selection. Clarendon Press, Oxford, 272 pp

Fisher SE, Shaklee JB, Ferris SD et al (1980) Evolution of five multilocus isozyme systems in the chordates. Genetica (Neth) 52:73–85

Fitch WM (1974) A comparison between evolutionary substitutions and variant hemoglobins. Ann NY Acad Sci 241:439–441

Fitch WM, Margoliash E (1970) The usefulness of amino acid and nucleotide sequences in evolutionary studies. Evol Biol 4:67–109

Foerster RE (1968) The sockeye salmon. Bull Fish Res Board Can 162, 422 pp

Fontaine PM, Dodson JJ, Bernatchez L, Slettan A (1997) A genetic test of metapopulation structure in Atlantic salmon (*Salmo salar*) using microsatellites. Can J Fish Aquat Sci 54:2434–2442

Ford E (1940) Polymorphism and taxonomy. The new systematics. Clarendon Press, Oxford, pp 493–513

Ford EB (1964) Ecological genetics. Methuen, London 410 pp

Fowler DP (1965) Effects of inbreeding in Red pine, *Pinus resinosa*. Silvae Genet 14:76–81

Fowler DP, Park YS (1983) Population studies of white spruce. 1. Effects of self-pollination. Can J For Res 13:1133–1138

Frankham R (1995) Effective population size. adult population size ratios in wildlife: a review. Genet Res Cambr 66:95–107

Fraser A, Burnell D (1971) Computer models in genetics. McGraw-Hill, New York, 210 pp

Fraser DJ, Lippé C, Bernatchez L (2004) Consequences of unequal population size, asymmetric gene flow and sex-biased dispersal on population structure in brook chars (*Salvelinus fontinalis*). Mol Ecol 13:67–80

Fredga K (1977) Chromosomal changes in vertebrate evolution. Proc R Soc Biol 199:377–397

Fürst M, Nyman L (1969) Isoenzyme polymorphism in *Mysis relicta* Loven. J Inst Freshwater Res Drottningholm 49:44–48

Gagal'chii NG (1986) Biochemical polymorphism of pink salmon, *Oncorhynchus gorbuscha* (Walb.) in the Kamchatka region. 2. Genetika 22(12):2839–2846

Galau GA, Chamberlin ME, Hough BR et al (1976) Evolution of repetitive and nonrepetitive DNA. In: Ayala F (ed) Molecular evolution. Sinauer, Sunderland, MA, pp 200–224

Garcia-Marin JL, Jorde PE, Ryman N et al (1991) Management implications of genetic differentiation between native and hatchery populations of brown trout (*Salmo trutta*) in Spain. Aquaculture 85:235–249

Garton TW (1986) Relationship between multiple locus heterozygosity and physiological energetics of growth in the estuarine gastropod *Thais laemellosa*. Physiol Zool 57: 530–543

Geist V (1971) Mountain sheep, a study in behaviour and evolution. Univ Chicago Press, Chicago, 383 pp

Georgiev GP (1969) On the structural organization of operon and the regulation of RNA synthesis in animal cells. J Theor Biol 25:473–490

Georgiev GP, Varshavsky AJ, Ryskov AP et al (1974) On the structural organization of the transcriptional unit in animal chromosomes. Cold Spring Harb Symp Quant Biol 38:869–884

Gershenson SM (1941) "Mobilizing reserve" of intraspecific variation. Zh Obshch Biol 2(1):85–107

Gibson G (2002) Microarrays in ecology and evolution: a preview. Mol Ecol 11:17–24

Gibson G, Palsson A (2001) A complement for evolutionary genetics. Curr Biol 11:74–76

Gillespie J (1977) A general model of account for enzyme variation in natural populations. Evolution 31:85–90

Gillespie JH, Kojima K (1968) The degree of polymorphism in enzymes involved in energy production compared to that in nonspecific enzymes in two *Drosophila ananassae* populations. Proc Natl Acad Sci USA 61:581–585

Gillespie JH, Langley CH (1974) A general model to account for enzyme variation in natural populations. Genetics 76:837–884

Gilpin M (1991) The genetic effective size of a metapopulation. Biol J Linn Soc 42:165–175

Gilyarov MS (1974) Ecological and ethological traits in systematics and phylogenetics of insects. Zh Obshch Biol 35(1):13–33

Ginatulin AA (1984) Struktura, organizatsiya i evolutsiya genoma pozvonochnykh (The structure, organization and evolution of the genome of vertebrates). Nauka, Moscow, 293 pp

Ginsburg EH (1968) Application of dispersion analysis in biological studies. Teoreticheskiye osnovy selektsii zhivotnykh (The theoretical bases for animal selection). Kolos, Moscow, pp 205–252

Glembotsky YL, Bogolubova GV (1940) Body weight of lambs at birth and their subsequent growth. Vestn Skh Nauki 2:80–92

Gliddon C, Gondet J (1994) The genetic structure of metapopulations and conservation biology. In: Loeschcke V, Tomiuk J, Jain SK (eds) Conservation genetics. Birkhauser Verlag, Basel, pp 107–114

Glotov NV (1975) Population as a natural historical structure. Genetika i evolutsiya prirodnykh populyatsii rastenii (The genetics and evolution of natural plant populations). Trudy Dagestanskogo filiala Akad nauk SSSR, Makhachkala. 1:17-25

Glubokovsky MK, Zhivotovsky LA (1986) Population structure in pink salmon: a system of fluctuating stocks. Biol Morya 2:39–44

Glubokovsky MK, Zhivotovsky LA, Viktorovsky RM, Bronevsky AM, Afanas'ev KI, Efremov VV, Ermolenko LN, Kalabushkin BA, Kovalyev VG, Makoedov AN, Malinina TV,

Pustovoiyt SP, Rubtzova GA (1989) Population structure of pink salmon. Genetika 25(7):1275–1285

Goldschmidt RB (1948) Ecotype, ecospecies and microevolution. Experientia 4:465–472

Goldschmidt RB (1952) Evolution as viewed by one geneticist. Am Sci 40:84–98

Goldstein DB, Schlotterer C (eds) (1999) Microsatellites. Evolution and application. Oxford Univ Press, New York, 352 pp

Goldstein DB, Linares AR, Cavalli-Sforza LL (1995) An evaluation of genetic distances for use with microsatellite loci. Genetics 139:463–471

Golikov NV, Skarlato OA (1967) Molluscs from the Bay Boussyet (the Japanese Sea) and their ecology. Tr Zool Inst Akad Nauk SSSR 42:5–155

Goossens B, Chikhi L, Taberlet P, Waits LP, Allaine D (2001) Microsatellite analysis of genetic variation among and within Alpine marmot populations in the French Alps. Mol Ecol 10:41–52

Gordeeva NV (2002) Genetic changes in pink salmon transplanted into the White Sea basin. Dokl Ros Acad Nauk 384:250–253

Gordeeva NV, Salmenkova EA, Altukhov YP, Makhrov AA, Pustovoit SP (2003) Genetic changes in pink salmon *Oncorhynchus gorbuscha* Walbaum during acclimatization in the White sea basin. Genetika 39:402–412

Gordon M (1947) Speciation in fishes. Adv Genet 1:95–132

Gorin V, Kopylovskaya G, Merson S, Konovalov B (1978) Application of stabilizing selection in poultry breeding. Ptitsevodstvo 11:28–31

Gorin VG, Kopylovskaya GY, Sches' GP, Merson SL (1980) Efficacy of stabilizing selection in milk-cattle breeding. Tr Vses Skh Inst Zaochn Obuchen, Moscow, pp 3–12

Govindaraju DR, Dancik BP (1987a) Environmental stress and the relationships among allozyme heterozygosity, biomass and biomass components in jack pine (*Pinus banksiana* Lamb.). Genetica (Neth) 74:173–179

Govindaraju DR, Dancik BP (1987b) Allozyme heterozygosity and homeostasis in germinating seeds of jack pine. Heredity 59:279–283

Grant V (1980) Gene flow and homogeneity of species populations. Biol Zentralbl 99:157–169

Grant V (1981) The genetic goal of speciation. Biol Zentralbl 100:473–482

Grant WS, Milner GB, Krasnowski P, Utter FM (1980) Use of biochemical genetic variants for identification of sockeye salmon stocks in Cook Inlet, Alaska. Can J Fish Aquat Sci 37:1236–1247

Graur D (1986) The evolution of electrophoretic mobility of proteins. J Theor Biol 118:443–469

Grechko VV, Fedorova LV, Fedorov AN, Slobodyanyuk SY, Ryabinin DM, Melnikova MM, Bannikova AA, Lomov AA, Sheremet'eva VA, Gorshkov VA, Sevostyanova GA, Semenova SK, Ryskov AP, Mednikov BM, Darevsky IS (1997) Restriction endonuclease analysis of highly repetitive DNA as a phylogenetic tool. J Mol Evol 45:332–336

Greenwood JJD (1975) Effective population number in the snail, *Cepaea nemoralis*. Evolution 28:513–526

Greenwood JJD (1976) Effective population number in *Cepaea*: a modification. Evolution 30:186

Griffin AR, Lindgren D (1985) Effect of inbreeding on production of filled seeds in *Pinus radiata* – experimental results and a model of gene action. Theor Appl Genet 71:334–343

Griffing B, Zsiros E (1971) Heterosis associated with genotype-environment interaction. Genetics (US) 68:443–455

Gvozdev VA, Gerasimova TI, Kogan GZ et al (1977) Investigations on the organization of genetic loci in *Drosophila melanogaster*: lethal mutations affecting 6-phosphogluconate dehydrogenase and their suppression. Mol Gen Genet 152:121–128

Guries RP, Ledig FT (1982) Genetic diversity and population structure in pitch pine (*Pinus rigida* Mill.). Evolution. 36:387–402

Haldane JBS (1937) The effect of variation of fitness. Am Nat 71:337–349

Haldane JBS (1955) On the biochemistry of heterosis and the stabilization of polymorphism. Proc R Soc Lond B 144:143–221

Haldane JBS (1957) The cost of natural selection. J Genet 55:511–524

Haldane JBS (1960) More precise expressions for the cost of natural selection. J Genet 57:351–360

Halushka MK, Fan J-B, Bentley K et al (1999) Pattern of single-nucleotide polymorphism in candidate genes for blood-pressure homeostasis. Nature Genet 22:239–247

Hammer M (1994) A recent insertion of an Alu element on the Y-chromosome is a useful marker for human population studies. Mol Biol Evol 11:749–761

Hamrick JL, Linhart YB, Mitton JB (1979) Relationships between life history characteristics and electrophoretically detectable genetic variation in plants. Annu Rev Ecol Syst 10:173–200

Hancock JM (1999) Microsatellites and other simple sequences: genomic context and mutational mechanisms. In: Goldstein DB, Schlotterer C (eds) Microsatellites. Evolution and application. Oxford Univ Press, New York, pp 1–9

Hansen MM, Hynes RA, Loeschcke V, Rasmussen G (1995) Assessment of the stocked or wild origin of anadromous brown trout (*Salmo trutta* L) in a Danish river system, using mitochondrial DNA RFLP analysis. Mol Ecol 4:189–198

Hansen MM, Kenchington E, Nielsen E (2001) Assigning individual fish to populations using microsatellite DNA markers. Fish and Fisheries 2(2):93–112

Hanski I (1991) Single-species metapopulation dynamics. In: Gilpin M, Hanski I (eds) Metapopulation dynamics: empirical and theoretical investigation. Academic Press, London, pp 17–38

Hanski I (1999) Metapopulation ecology. Oxford Univ Press, Oxford, 313 pp

Hanski I, Gilpin ME (eds) (1997) Metapopulation biology. Academic Press, Boston, 512 pp

Hanski I, Simberloff D (1997) The metapopulation approach, its history, conceptual domain, and application to conservation. In: Hanski LA, Gilpin ME (eds) Metapopulation biology. Academic Press, Boston, pp 5–26

Hanson AJ, Smith HD (1967) Mate selection in a population of sockeye salmon of mixed age groups. J Fish Res Board Canada 24:1955–1977

Harlan JR (1981) Who's is charge here? Can J Fish Aquat Sci 38:1459–1463

Harris H (1966) Enzyme polymorphisms in man. Proc R Soc Lond B 164:298–310

Harris H (1977) The principles of human biochemical genetics. North-Holland, Amsterdam, 473 pp

Harris H, Hopkinson DA (1972) Average heterozygosity per locus in man: an estimate based on the incidence of enzyme polymorphisms. Ann Hum Genet 36:9–20

Harris H, Hopkinson DA (1978) Handbook of enzyme electrophoresis in human genetics. North-Holland, Amsterdam

Harris H, Hopkinson DA, Robson EB (1974) The incidence of rare alleles determining electrophoretic variants: data on 43 enzyme loci in man. Ann Hum Genet 37:237–253

Harris H, Hopkinson DA, Edwards YH (1977) Polymorphism and the subunit structure of enzymes: a contribution to the neutralist-selectionist controversy. Proc Natl Acad Sci USA 74:698–701

Harrison GA, Weiner JS, Tanner JM, Barnicot NA (1964) Human biology. Oxford Univ Press, New York

Harrison RG (1989) Animal mitochondrial DNA as a genetic marker in population and evolutionary biology. Trends Ecol Evol 4:6–11

Hartmann WL, Raleigh RV (1964) Tributary homing of sockeye salmon at Brooks and Karluk lakes, Alaska. J Fish Res Board Can 21:485–504

Heath DD, Iwama GK, Delvin RH (1994) DNA fingerprinting used to test for family effects on precocious sexual maturation in two populations of *Oncorhynchus tshawytscha* (Chinook salmon). Heredity 73:616–624

Hedrick PW (1999) Perspective: highly variable loci and their interpretation in evolution and conservation. Evolution 53:313–318

Heggberget TG, Jonsen BO, Hindar K et al (1993) Interaction between wild and cultured Atlantic salmon: a review of the Norwegian experience. Fish Res 18:123–146

Hewitt GM (2001) Speciation, hybrid zones and phylogeography – or seeing genes in space and time. Mol Ecol 10:537-549

Hey J (1999) The neutralist, the fly and the selectionist. Trends Ecol Evol 14:35–38

Hindar K, Ryman N, Utter F (1991) Genetic effects of cultured fish on natural fish populations. Can J Fish Aquat Sci 48:945–957

Hinegardner R (1976) Evolution of genome size. Molecular evolution. Sinauer, Sunderland, MA, pp 179–199

Hirsch L, Erlenmeyer-Kimling L (1962) Studies in experimental behaviour in genetics. 4. Chromosome analysis for geotaxis. J Comp Physiol Psychol 55:732–739

Hollander A (ed) Chemical mutagens: principles and methods for their detection. Plenum Press, New York, 1, 1971; 2, 1972; 3, 1973; 4, 1976; 5, 1978

Hollstein MH, McCan J, Angelosanto FA et al (1978) Short term tests for carcinogens and mutagens. Mutat Res 65:133–226

Hood L, Campbell JH, Elgin SCR (1975) The organization, expression, and evolution of antibody genes and other multigene families. Annu Rev Genet 9:305–353

Horai S, Hoyasaka K, Kondo R et al (1995) Recent African origin of modern humans revealed by complete sequences of hominoid mitochondrial DNAs. Proc Natl Acad Sci USA 92:532–536

Huang T, Cottingham R, Ledbetter D, Zoghbi H (1992) Genetic mapping of four dinucleotide repeat loci, DXS453, DXS458, DXS454 and DXS424 on X chromosome using multiplex polymerase chain reaction. Genomics 13:375–380

Hubby JL, Lewoniin RC (1966) A molecular approach to the study of genic heterozygosity in natural populations. 1. The number of alleles at different loci in *Drosophila pseudoobscura*. Genetics (US) 54:577–594

Hughes CR, Queller DC (1993) Detection of highly polymorphic microsatellite loci in a species with little allozyme polymorphism. Mol Ecol 2:131–137

Hughes TJ, Claxton LD, Houk VS (1998) Genotoxicity of industrial wastes and effluents. Mutat Res 410:237–243

Hunter RL, Markert CL (1957) Histochemical demonstration of enzymes separated by zone electrophoresis in starch gels. Science 125:1294–1295

Huxley JS (1942) Evolution, the modern synthesis. Alien and Unwin, London

Hynes JD, Brown EH, Helle JH et al (1981) Guidelines for the culture of fish stocks for resource management. Canad J Fish Aquat Sci 38:1867–1876

Il'in VE, Konovalov SI, Shevlyakov AG (1983) Migration coefficient and spatial structure of Pacific salmon. Biologicheskiye osnovy razvitiya lososevogo khozyaistva v vodoyomakh SSSR (Biological bases of the development of salmon farming in USSR reservations). Nauka, Moscow, pp 9-31

Ingman M, Kaessmann H, Paabo S, Gyllensten U (2000) Mitochondrial genome variation and the origin of modern humans. Nature 48:708–712

Ingram VM (1957) Gene mutations in human haemoglobin: the chemical difference between normal and sickle cell haemoglobins. Nature 180:326–329

Ingram VM (1960) Haemoglobin and its abnormalities. Pergamon Press, Oxford, 240 pp

Ingram VM (1961) Gene evolution and the haemoglobins. Nature 189:704–708

Ingram VM (1963) The haemoglobins in genetics and evolution. Columbia Univ Press, New York, 166 pp

International commission for protection against environmental mutagens and carcinogens (1982) Mutagenesis testing as an approach to carcinogenesis. Committee 2 report. Mutat Res 99:73–91

Jackson MA, Stack HF, Waters MD (1999) Genetic toxicology data in evaluation of potential human environmental carcinogens. Mutat Res 437:21–49

Jeffreys AJ, Wilson V, Lay Thein S (1985) Hypervariable "minisatellite" regions in human DNA. Nature 314:67–73

Jeffreys AJ, Royle NJ, Wilson V, Wong Z (1988) Spontaneous mutation rates to new length alleles at tandem repetitive hypervariable loci in human DNA. Nature 332:278–281

Jehle B, Arntzen JW, Burke T, Krupa AP, Hödl W (2001) The annual number of breeding adults and the effective population size of syntopic newts (*Triturus cristatus, T. marmoratus*). Mol Ecol 10:839–850

Jin L, Chakraborty R (1994) Estimation of genetic distance and coefficient of gene diversity from single-probe multilocus DNA fingerprinting data. Mol Biol Evol 11:12–127

Jobling MA, Tyler-Smith C (2000) New uses for new haplotypes: the human Y chromosome, disease and selection. Trends Genet 16(8):356–362

Johnson F, Shaffer H (1973) Isozyme variability in species of the genus *Drosophila*. 7. Genotype-environment relationships in populations of *Drosophila melanogaster* from the eastern United States. Biochem Genet 10:149–163

Johnson GB (1973) Importance of substrate variability to enzyme polymorphisms. Nat New Biol 243:151–153

Johnson GB (1974) Enzyme polymorphism and metabolism. Science 184:28–37

Johnson GB (1976) Genetic polymorphism and enzyme function. In: Ayala F (ed) Molecular evolution. Sinauer Assoc, Sunderland, MA, pp 45–59

Johnson KR, Wright JE Jr, May B (1987) Linkage relationships reflecting ancestral tetraploidy in Salmonid fish. Genetics (US) 116:579–591

Jorde PE, Ryman N (1995) Temporal allele frequency change and estimation of effective size in populations with overlapping generations. Genetics 139:1077–1090

Jukes TH (1971) Comparisons of polypeptide chains of globins. J Mol Evol 1:46–62

Kalabushkin BA (1976) Genetic variability in modern and mid-holocoenotic populations of *Littorina squalida*. Zh Obshch Biol 3(3):369–377

Kalabushkin BA, Zhivotovsky LA (1979) Spatial structure of the mollusc population *Littorina squalida* on the morphophysiological data. Proc 14th Pacific Scientific Congr, Section 11, Marine Biology. Nauka, Moscow, pp 163–164

Kalnin VO, Kalnina OV (1982) Population-genetic characteristics of European anchovy. Tr 2 Vses Konf po Morskoi Biol Vladivostok, pp 75–77

Kalnina OV, Kalnin VO (1984) Genetic differences and internal heterogeneity of the Azov and Black Sea anchovy, *Engraulis encrasicholus* (L.). Genetika 20(2):309–313

Kan YW (1986) The William Allan memorial address: Thalassemia: molecular mechanism and detection. Am J Hum Genet 38:4–12

Kan YW, Dozy AM (1978) Polymorphism of DNA sequence adjacent to the $\beta$-globin structural gene. Its relation to the sickle mutation. Proc Natl Acad Sci USA 75:5631–5635

Kaplan NO (1965) Evolution of dehydrogenases. Evolving genes and proteins. Academic Press, New York, pp 245–279

Karl SA, Avise JC (1992) Balancing selection at allozyme loci in oysters: implications from nuclear RFLPs. Science 256:100–102

Karl SK, Bowen BW, Avise JC (1992) Global population genetic structure and male-mediated gene flow in the green turtle (*Chelonia mydas*): RFLP analyses of anonymous nuclear loci. Genetics 131:163–173

Karn MN, Penrose LS (1951) Birth weight and gestation time in relation to maternal age, parity and infant survival. Ann Eugenics 16:147–164

Karp A, Edwards K (1997) DNA markers: a global overview. In: Caetano-Anolles G, Gressnoff PM (eds) DNA markers: protocols, applications and overview. Wiley-VCH, New York, pp 1–13

Kartavtzev YF, Salmenkova EA, Rubtzova GA, Afanas'ev KI (1990) The family analysis of allozyme variability and its relationship with body length and viability of progeny in pink salmon *Oncorhynchus gorbuscha* (Walb.). Genetika 26:1610–1619

Kashi Y, Soller M (1999) Functional roles of microsatellites and minisatellites. In: Goldstein DB, Schlotterer C (eds) Microsatellites. Evolution and application. Oxford Univ Press, New York, pp 10–23

Kassie F, Parzefall W, Knasmuller S (2000) Single cell gel electrophoresis assay: a new technique for human biomonitoring studies. Mutat Res 463:33–51

Kerster HW (1964) Neighbourhood size in the rusty lizard, *Sceloporus olivacens*. Evolution 18:445–457

Kerster HW, Levin DA (1968) Neighbourhood size in *Litosphermum carolinense*. Genetics (US) 60:577–583

Keshava N, Ong T-M (1999) Occupational exposure to genotoxic agents. Mutat Res 437:175–194

Kilbey BT (ed) (1977) Handbook of mutagenicity test procedures. Elsevier, Amsterdam, 385 pp

Kijima A, Fujo Y (1979) Geographical distribution of IDH and LDH isozymes in chum salmon populations. Bull Jpn Soc Sci Fish 45:287–295

Kikuchi Y, Sofuni T, Tweats D, Schechtman L, Probst G, Muller L, Shimada H (1999) ICH-Harmonized guidances on genotoxicity testing of pharmaceuticals: evolution, reasoning and impact. Mutat Res 436:195–225

Killik SR, Clemens WA (1963) The age, sex ratio and size of Eraser River sockeye salmon, 1915–1960. Int Pac Salmon Fish Commiss 14:1–140

Kimura M (1953) "Stepping stone" model of population. Annu Rep Nat Inst Genet Mishima 3:63–65

Kimura M (1960a) Optimum mutation rate and degree of dominance as determined by the principle of minimum genetic load. J Genet 57:21–34

Kimura M (1960b) Genetic load of a population and its significance in evolution. Jpn J Genet 35:7–33

Kimura M (1961) Some calculations on the mutational load. Jpn J Genet 36:179–190

Kimura M (1968a) Evolutionary rate at the molecular level. Nature 217:624–626

Kimura M (1968b) Genetic variability maintained in a finite population due to mutational production of neutral and nearly neutral isoalleles. Genet Res 11:247–269

Kimura M (1983) The neutral theory of molecular evolution. Cambridge Univ Press, New York, 367 pp

Kimura M, Crow JF (1963) The measurement of effective population number. Evolution 17:279–288

Kimura M, Crow JF (1964) The number of alleles that can be maintained in a finite population. Genetics (US) 49:725–738

Kimura M, Maruyama T (1971) Pattern of neutral polymorphism in a geographically structured population. Genet Res 18:125–131

Kimura M, Maruyama T, Crow JF (1963) The mutation load in small populations. Genetics (US) 48:1303–1312

Kimura M, Ohta T (1971) Theoretical aspects of population genetics. Princeton Univ Press, Princeton, 219 pp

Kimura M, Ohta T (1973) Mutation and evolution at the molecular level. Genetics (US) 73 [Suppl]:19–35

Kimura M, Ohta T (1975) Distribution of allelic frequencies in finite population under stepwise production of neutral alleles. Proc Natl Acad Sci USA 72:2761–2764

Kimura M, Weiss GH (1964) The stepping stone model of population structure and the decrease of genetic correlation with distance. Genetics (US) 49:561–576

King JL, Jukes TH (1969) Non-Darwinian evolution: random fixation of selectively neutral mutations. Science 164:788–798

King RC, Stansfield WD (1997) A dictionary of genetics, 5th edn Oxford Univ Press, New York

Kirpichnikov VS (1967) General theory of heterosis. Genetika 3(10):48–64

Kirpichnikov VS (1979) Geneticheskie osnovy selektsii ryb (Genetic bases of fish selection). Nauka, Leningrad, 392 pp

Kirpichnikov VS (1981) Genetic bases of fish selection. Springer, Berlin Heidelberg New York, 410 pp

Kirpichnikov VS (1987) Genetika i selektsiya ryb (Genetics and selection of fish). Nauka, Leningrad, 520 pp

Klug WS, Cummings MR (2000) Concepts of genetics. Prentice Hall, Englewood Cliffs, 816 pp

Kocher TD (2004) Adaptive evolution and explosive speciation: the cichlid fish model. Nat Genet 5:288–298

Kocher TO, Thomas WK, Meyer A et al (1989) Dynamics of mitochondrial DNA evolution in animals: amplification and sequencing with conserved primers. Proc Natl Acad Sci USA 86:6196–6200

Kodzoev M, Berezkin A, Gundaev A, Ermolenko I (2001) From scientific substantiation to new strategy in seed production of vegetable, melonfield and fodder crops. Int Agric J N4:53–57

Koehn RK, Eanes WF (1976) An analysis of allelic diversity in natural populations of Drosophila: the correlation of rare alleles with heterozygosity. Population genetics and ecology. Academic Press, New York, pp 377–421

Koehn RK, Gaffney PM (1984) Genetic heterozygosity and growth rate in *Mytilus edulis*. Mar Biol 82:1–7

Kohara A, Sizuki T, Honma M, Hirano N, Ohsawa K-I, Ohwada T, Hayashi M (2001) Mutation spectrum of 0-amonoazotoluene in the cII gene of lambda/lacZ transgenic mice. Mutat Res 491:211-220

Kojima K, Yarbrough KN (1967) Frequency-dependent selection at the esterase-6 locus in *Drosophila melanogaster*. Proc Natl Acad Sci USA 57:645–649

Kojima K, Gillespie J, Tobari YN (1970) A profile of Drosophila species enzymes assayed by electrophoresis. Biochem Genet 4:627–637

Kolmogorov AN (1935) Deviation from Hardy's formulae in the case of partial isolation. Dokl Akad Nauk SSSR 3(3):129–132

Kondrashov FA, Koonin EV (2003) Evolution of alternative splicing: deletions, insertions and origin of functional parts of proteins from intron sequences. Trends Genet 19:115–119

Kondzela CM, Guthrie CM, Hawkins SL, Russell CD, Helle JH, Gharrett AJ (1994) Genetic relationships among chum salmon populations southeast Alaska and northern British Columbia. Can J Fish Aquat Sci 51 [Suppl 1]:50–64

Konovalov SM (1971) Differentsiatsiya local'nykh stad nerki (Differentiation of local sockeye salmon shoals). Nauka, Leningrad, 229 pp

Konovalov SM (1980) Populyatsionnaya biologiya tikhookeanskikh lososei (Population biology of Pacific salmon). Nauka, Leningrad, 237 pp

Konovalov SM, Shapiro AP, Leibovich GE (1975) Utilization of biological resources in connection with spatial structure of a species. Biol Morya 6:26-35

Korochkin LI (1977) Vzaimodeistviye genov v razvitii (Gene interaction in development). Nauka, Moscow, 280 pp

Korochkin LI (1981) Gene regulation in development. Mol Biol 15(5):965–988

Korochkin LI (1983) Evolutionary meaning of mobile genetic elements: a hypothesis. Tsitologiya i Genetica 4:67–77

Korochkin LI (1984) Developmental genetics and certain molecular moments of evolution. Mol Genet Biofiz 9. Vischa Shkola, Kiev, pp 75–82

Korochkin LI (1985) Parallelisms in molecular organization of genome and problems of evolution. Molekulyarnye Mekhanizmy Geneticheskikh Protsessov (Molecular mechanism of genetic processes). Nauka, Moscow, pp 132–146

Korochkin LI (1999) Introduction to developmental genetics. Nauka, Moscow, 249 pp

Korochkin LI (2001) Genes, ontogeny and problems of evolutionary development. Evolutionary biology 1. Tomsk State Univ, Tomsk, pp 49–72

Korochkin LI, Serov OL, Pudovkin AI et al (1977) Genetika izofermentov. Nauka, Moscow, 278 pp

Koshevoy MA (1975) Selektsiya i usloviya razvedeniya karakul'skikh ovets (Selection and breeding conditions of karakul sheep). Fan, Tashkent, 248 pp

Koski V (1971) Embryonic lethals in *Picea abies* and *Pinus sylvestris*. Commun Inst Forest Fenn 75:1–13

Kovalenko VP, Kosenko NF, Podstreshny OP (1980) Characteristic of the chickens with modal body weight at the age of five months. Nauchn-Tekh Byull Ukr Nauchn-Issl Inst Ptitsevod 9:3–6

Kozdoev M, Berezkin A, Gundaev A, Ermolenko I (2001) From a scientific justification to a new strategy in seed production of legumes, melons, gourds, and fodder root vegetables. Mezhdunarodnyi Selskokhozyaistvennyi Zhurnal 4:53–57

Krieg F, Guyomard R (1985) Population genetics of French brown trout (*Salmo trutta* L): large geographical differentiation of wild populations and high similarity of domesticated stocks. Genet Select Evol 17:225–242

Krimbas CB, Tsakas S (1971) The genetics of *Dacus oleae*. V. Changes of esterase polymorphism in a natural population following insecticide control – selection or drift? Evolution 25:454–460

Krogius FV (1960) Growth rate and the age groupings of sockeye salmon, *Oncorhynchus nerka* (Walb.) in the sea. Vopr Ikhtiol 17:67–88

Krogius FV (1972) The linear growth of *Oncorhynchus nerka* fry in Lake Dal'neye. Izv TINRO 82:19–31

Krogius FV (1975) Population dynamics and growth of the sockeye salmon fry in Lake Dal'neye (Kamchatka). Vopr Ikhtiol 93:612–629

Krogius FV (1979) On the relationship between the freshwater and marine life-spans of sockeye salmon from Lake Dal'neye. Biol Morya 3:24–29

Krogius FV, Krokhin EM, Menshutkin VV (1969) Soobshchestvo pelagicheskikh ryb ozera Dal'nego (A community of pelagic fish in Lake Dalneye). Nauka, Leningrad, 85 pp

Krutovsky KV, Neale DB (2002) Forest genomics for conserving adaptive genetic diversity. Forest Genetic Resources Working Papers. Working Paper FGR (July 2001), Forest Resources Development Service, Forest Resources Division. FAO, Rome

Krutovsky KV, Milishnikov AN (1983a) Selection against induced mutations of allozyme loci in *Drosophila melanogaster*. Genetika 19(9):1469–1476

Krutovsky KV, Milishnikov AN, Altukhov YP (1983b) Frequency of induced null-mutations in three allozyme loci at different stages of *Drosophila melanogaster* ontogenesis. Drosophila Inform Serv 59:71

Krutovsky KV, Milishnikov AN, Gulemetova RM, Cholakova NI (1983c) Frequency of the induced mutations of allozyme loci in *Drosophila melanogaster*. Genetika 19(6):1013–1019

Krutovsky KV, Bergmann F (1995) Introgressive hybridization and phylogenetic relationships between Norway, *Picea abies* (L.) Karst., and Siberian, *P. obovata* Ledeb., spruce species studied by isozyme loci. Heredity 74:464–480

Krutovsky KV, Vollmer SS, Sorensen FC et al (1998) RAPD genome map of Douglas-fir. J Hered 89:197–205

Kuprina NP (1970) Stratigrafiya i istoriya osadkonakopleniya pleistotsenovykh otlozhenii tsentral'noi Kamchatki (Stratigraphy and history of deposit accumulating in the Pleistocene sedimentation of central Kamchatka). Nauka, Moscow, 186 pp

Kupriyanova LA (1997) Some cytogenetic regularities of reticular (hybridogenic) speciation in unisexual species of lizards and other vertebrates. Tsitologiya 39(12):1089–1109

Kurbatova OL (1975) A genodemographic study of large panmictic populations. Genetic structure of the two successive generations of Muscovites. Vopr Antropol 50:30–45

Kurbatova OL (1977) Genetic processes in urban population: a genodemographic study of the population of Moscow. D Sci Thesis Inst Anthropology, Moscow, 22c

Kurbatova OL (1996) Revealing adaptive value of human blood group polymorphism by means of analysis of locus set. Genetika:32:996–1006

Kurbatova OL (2001) Ecological genetic aspects of migration. Ecology of human: from past to future. In: Tatevosov PV (ed) Nauch Trudy MNEPU (Proceedings of MNEPU). Ecology, iss 1. Izdat MNEPU Moscow, pp 44–61

Kurbatova OL, Botvin'yev OK, Deschekina MF, Altukhov YP (1982) Non-specific resistance and predisposition to acute leukosis in children. Genetika 18(7):1173–1182

Kurbatova OL, Pobedonostseva EYu, Imasheva AG (1984) The role of migration processes in the formation of marriage structure of Moscow population. Genetika 20(3):501–511

Lacerda DR, Acedo MDP, Lemos Filho JP, Lovato MB (2001) Genetic diversity and structure of natural populations of *Plathymenia reticulata* (Mimosoideae), a tropical tree from the Brazilian Cerrado. Mol Ecol 10:1143–1152

Lahn BT, Page DC (1997) Functional coherence of the human Y chromosome. Science 278:675–680

Lamotte M (1951) Recherches sur la structure génétique des populations naturelles de *Cepaea nemoralis* (L.). Bull Biol Fr Belg Suppl 35:1–239

Lande R (1979) Effective deme sizes during long-term evolution estimated from rates of chromosomal rearrangement. Evolution 33:234–251

Lander ES (1996) The new genomics: global views of biology. Science. 274:536–539

Larson A, Wake DB, Yanev KP (1984) Measuring of gene flow among populations having high levels of genetic fragmentation. Genetics (US) 106:293–308

Laslett P (1997) Interpreting the demographic changes. Philos Trans R Soc Lond B 352:1805–1809

Latta RG, Mitton JB (1997) A comparison of population differentiation across four classes of gene marker in limber pine (*Pinus flexilis* James). Genetics 146:1153–1163

Latter BDH (1960) Natural selection for an intermediate optimum. Aust J Biol Sci 3:30–35

Latter BDH (1975) Enzyme polymorphism: gene frequency distributions with mutation and selection for optimal activity. Genetics (US) 79:325–331

Lawler RR, Richard AF, Riley MA (2003) Genetic population structure of the white sifaka (*Propithecus verreauxi verreauxi*) at Beza Mahafaly Special Reserve, southwest Madagascar (1992–2001). Mol Ecol 12:2307–2317

Lebedev NV (1946) Elementary populations of fish. Zool Zh 25(2):121–135

Lebedev NV (1967) Elementarnye populyatsii ryb. Pischevaya Promyshlennost', Moscow, 211 pp

Lee C-Y, Charles D, Brouson D et al (1979) Analysis of mouse and Drosophila proteins by two-dimensional gel electrophoresis. Mol Gen Genet 176:303–311

Lehman N (2001) Please release me, genetic code. Curr Biol 11:63-66

Lerner IM (1954) Genetic homeostasis. Wiley, New York, 134 pp

Leslie JF, Watt WB (1986) Some evolutionary consequences of the molecular recombination process. Trends Genet 2:288–291

Levanidov VY (1969) Reproduction of the Amur salmon and the food reserves for its fry in the Amur tributaries. Izv TINRO 67:3–242

Levin DA, Kerster HW (1968) Local gene dispersal in Phlox. Evolution 22:130–139

Levin DA, Kerster HW (1971) Neighbourhood structure in plants under diverse reproductive methods. Am Nat 105:345–354

Levins R (1968) The genetic consequences of natural selection. Teoreticheskaya i matematicheskaya biologiya (Theoretical and mathematical biology). Mir, Moscow, pp 401–419

Levins R (1970) Extinction. Some mathematical questions in biology. Providence. Am Math Soc 2:77–107

Levinson G, Gutman GA (1987) Slipped-strand mispairing: a major mechanism for DNA sequence evolution. Mol Biol Evol 4:203–221

Lewin B (1987) Genes. Mir, Moscow, 544 pp

Lewin B (1998) Genes. Oxford Univ Press, Oxford, 1260 pp

Lewontin RS (1978a) Geneticheskiye osnovy evolutsii (The genetic bases of evolution). Mir, Moscow, 351 pp

Lewontin RS (1978b) The genetic heterogeneity of electrophoretic alleles. Abstracts of 14th international congress of genetics. Nauka, Moscow, p 465

Lewontin RC (2002) Directions in evolutionary biology. Annu Rev Genet 36:1–18

Lewontin RC, Hubby JL (1966) A molecular approach to the study of genic heterozygosity in natural populations. 2. Amount of variation and degree of heterozygosity in natural populations of *Drosophila pseudoobscura*. Genetics (US) 54:595–609

Lewontin RC, Krakauer J (1973) Distribution of gene frequency as a test of the theory of the selective neutrality of polymorphisms. Genetics (US) 74:175–195

Lewontin RC, Ginzburg LR, Tuljapurkar SD (1978) Heterosis as an explanation for large amounts of genetic polymorphism. Genetics (US) 88:149–170

Li CC (1955) Population genetics. Univ Chicago Press, Chicago, 346 pp

Li CC (1967) Genetic equilibrium under selection. Biometrics 23:397–484

Li CC (1976) First course in population genetics. Boxwood Press, Pacific Grove, CA

Li C (1978) Vvedeniye v populyatsionnuyu genetiku (A first course in population genetics). Mir, Moscow, 555 pp

Lidicker WZ Jr, Patton JL (1987) Patterns of dispersal and genetic structure in populations of small rodents. In: Chepko-Sade BD, Halpin ZT (eds) Mammalian dispersal patterns. The effects of social structure on population genetics. Univ Chicago Press, London, pp 144–161

Limansky VV, Payusova AN (1969) On the immunogenetic differences between the elementary populations of anchovy. Genetika 5(6):109-118

Limborskaya SA (1981) The systems of globin genes. Itogi Nauki i Tekhniki. Molekulyarnaya Biologiya (Results of science and technology. Molecular biology, vol 19). VINITI, Moscow, pp 84–116

Limborskaya SA, Khusnutdinova EK, Balanovskaya EV (2002) Ethnogenomics and gene geography of peoples from eastern Europe. Nauka, Moscow, 262 pp

Lindsey CC, Northcote TG, Hartman GF (1959) Homing of rainbow trout to inlet and outlet spawning streams at Loon Lake, British Columbia. J Fish Res Board Can 16:695–719

Liskauskas AP, Ferguson MM (1991) Genetic variation and fitness: a test in a naturalized population of brook trout (*Salvelinus fontinalis*). Can J Fish Aquat Sci 48:2152-2162

Livshits G, Kobyliansky E (1985) Lerner's concept of developmental homeostasis and the problem of heterozygosity level in natural populations. Heredity 55:341–353

Loeschke V, Tomiuk J, Jain SK (eds) (1994) Conservation genetics. Birkhauser Verlag, Basel, 440 pp

Long M, de Souza SJ, Gilbert W (1995) Evolution of the intron–exon structure of eukariotic genes. Curr Opin Genet Dev 5:774–778

Loukas M, Vergini Y, Krimbas CB (1981) The genetics of *Drosophila subobscura* populations. 17. Further genic heterogeneity within electromorphs by urea denaturation and their effect of the increased genic variability on linkage disequilibrium studies. Genetics (US) 97:429–441

Lozano R, Ruiz Rejon C, Ruiz Rejon M (1992) A comparative analysis of NORs in diploid and triploid salmonids: implications with respect to the diploidization process occurring in this fish group. Heredity 69:450-457

Lucchini V, Galov A, Randi E (2004) Evidence of genetic distinction and long-term population decline in wolves (*Canis lupus*) in the Italian Appennines. Mol Ecol 13:523–536

Luikart G, England PR (1999) Statistical analysis of microsatellite DNA data. Trends Ecol Evol 14:253–256

Lynch M (1990) The similarity index and DNA fingerprinting. Mol Biol Evol 7:478–484

Lynch M, Force AG (2000) The origin of interspecific genomic incompatibility via gene duplication. Am Nat 156:590–605

Lynch M, Walsh B (1998) Genetics and analysis of quantitative traits. Sinauer, Sunderland, MA

Lyubimova IK, Abilev SK, Gal'berstam NM et al (2001) Computer-aided prognostication of mutagenic activity of substituted polycyclic compounds. Izvestiya Ros Akad Nauk Ser Biol 2:180–186

MacLean JA, Evans DO (1981) The stock concept, discreteness of fish stocks, and fisheries management. Canad J Fish Aquat Sci 38:1889–1898

Makhrov AA, Kuzishchin KV, Altukhov, YP (1997) Association of allozyme heterozygosity with growth rate and ecological differentiation in brown trout *Salmo trutta* L. Genetika 33:681–686

Makoedov AN (1999) Kariology, biochemical genetics and population phenetics of Salmonoidei of Siberia and Far East: comparative aspects. Psychology. UMK, Moscow 291 pp

Malécot G (1948) Les mathematiques de l'hérédite. Masson, Paris

Malécot G (1955) Decrease of relationship with distance. Cold Spring Harb Symp Quant Biol 20:52–53

Malécot G (1967) Identical loci and relationship. Proc V Berkeley symp math stat prob, vol 4. Univ California Press, Berkeley, pp 317–332

Malécot G (1975) Heterozygosity and relationship in regularly subdivided populations. Theor Popul Biol 8:212–241

Malinovsky AA, Mina MB (1976) A review of the book by YP Altukhov "Populyatsionnaya genetika ryb", Pischevaya Promyshlennost', Moscow (1974). Zh Obshch Biol 37(5):788–790

Mallet J (2001) The speciation revolution. J Evol Biol 14:887–888

Malukina GA (1969) A sense of smell and its role in the behaviour of fish. Itogi Nauki. Biologiya (Results of science biology). Nauka, Moscow, pp 32–78

Malyuchenko OP, Altukhov YP (2002) The influence of individual heterozygosity on fruitfulness traits in *Pinus pumila* (Pall) Regel. Dokl RAN 384:418–421

Manchenko GH (1994) Handbook of detection of enzymes on electrophoretic gels. CRC Press, Boca Raton, 341 pp

Manning HL (1955) Response to selection for yield in cotton. Cold Spring Harb Symp Quant Biol 20:103–110

Manning HL (1956) Yield improvement from a selection index technique with cotton. Heredity 10:303–322

Manwell C, Baker CM (1970) Molecular biology and the origin of species: heterosis, protein polymorphism and animal breeding. Washington Univ Press, New York, 310 pp

Manwell C, Backer A, Childres W (1963) The genetics of haemoglobin in hybrids. 1. A molecular basis for hybrid vigor. Comp Biochem and Physiol 10:103–129

March RE, Putt W, Hollyoake M, Ives JH, Lovegrove JU, Hopkinson DA, Edwards YH, Whitehouse DB (1993) The classical human phosphoglucomutase (PGM 1) isozyme polymorphism is generated by intragenic recombination. Proc Natl Acad Sci USA 99:10730–10733

Mariott RA (1964) Stream catalog of the Wood River Lake system, Bristol Bay, Alaska. Spec Sci Rep Fish 494:1–210

Markert CL, Moller F (1959) Multiple forms of enzymes: tissue, ontogenetic, and species specific patterns. Proc Natl Acad Sci USA 45:753–763

Marshall DR, Allard RW (1970a) Isozyme polymorphisms in natural populations of *Avena fatua* and *A. barbata*. Heredity 25:373–382

Marshall DR, Allard RW (1970b) Maintenance of isozyme polymorphisms in natural populations of *Avena barbata*. Genetics (US) 66:393–399

Maruyama T (1970a) On the rate of decrease of heterozygosity in circular stepping stone models of populations. Theor Pop Biol 1:101–119

Maruyama T (1970b) Rate of decrease of genetic variability in a subdivided population. Biometrika 57:299–311

Maruyama T (1971a) Analysis of population structure. II. Two-dimensional stepping stone models of finite length and other geographically structured populations. Ann Hum Genet 35:179–196

Maruyama T (1971b) The rate of decrease of heterozygosity in a population occupying a circular or a linear habitat. Genetics 67:437–454

Maruyama T (1972a) Rate of decrease of genetic variability in a two-dimensional continuous population of finite size. Genetics 70:639–651

Maruyama T (1972b) The rate of decay of genetic variability in a geographically structured finite population. Math Biosci 14:325–335

Maruyama T (1977) Stochastic problems in population genetics. Springer, Berlin Heidelberg New York

Maruyama T, Kimura M (1980) Genetic variability and effective population size when local extinction and recolonization of sub-populations are frequent. Proc Natl Acad Sci USA 77:6710–6714

Massaro EL, Markert CL (1968) Isozyme patterns of salmonid fishes: Evidence for multiple cistrons for lactate dehydrogenase polypeptides. J Exp Zool 168:223–238

Mather K (1943) Polygenic inheritance and natural selection. Biol Rev 18:32–64

Mathisen OA (1962) The effect of altered sex ratios on the spawning red salmon. Studies of Alaska red salmon. Washington Univ Press, Washington, DC, pp 137–245

Maurer G (1971) Disk-electroforez (Disc electrophoresis). Mir, Moscow, 247 pp

Maynard Smith J (1982) A century since Darwin. Nature 296:599–601

Mayr E (1942) Systematics and origin of species. Columbia Univ Press, New York, 334 pp

Mayr E (1963) Animals species and evolution. Harvard Univ Press, Cambridge, 659 pp

Mayr E (1968) Zoologicheskii vid i evolutsiya (Animal species and evolution). Mir, Moscow, 398 pp

Mayr E (1974) Populyatsii, vidy i evolutsiya (Populations, species and evolution). Mir, Moscow, 460 pp

Mayr E (1999) Understanding evolution. Trends Ecol Evol 14:372–373

McAtee WL (1937) Survival of the ordinary. Q Rev Biol 12:47–64

McCan J, Ames BN (1975) Detection of carcinogens in the Salmonella microsome test: assay of 300 chemicals. 1. Proc Natl Acad Sci USA 72:5135–5139

McCan J, Ames BN (1976) Detections of carcinogens in the Salmonella microsome test: assay of 300 chemicals. 2. Proc Natl Acad Sci USA 73:950–954

McCart P (1969) Digging behaviour of Oncorhynchus nerka spawning in streams at Babine Lake, British Columbia. Proc Symp Salmon and Trout in Streams. Univ Brit Columbia Press, Vancouver, pp 39–52

McConkey EH, Taylor BL, Phan D (1979) Human heterozygosity: a new estimate. Proc Natl Acad Sci USA 76:6500–6504

McElroy D, Moran P, Bermingham E, Kornfield J (1991) RFLP: the restriction enzyme analysis package, Version 4.0. Maine Univ, Orono

McIntyre RL, Wright TRF (1966) Responses of esterase-6 alleles of *Drosophila melanogaster* and *Drosophila simulans* to selection in experimental populations. Genetics (US) 53:371–387

Mednikov BM, Bannikova AA, Lomov AA, Melnikova MN, Shubina EA (1995) Restrictase analysis of repeating nuclear DNA, species criterion and mechanism of evolution. Mol Biol 29(6):1308–1319

Mednikov BM, Shubina EA, Mel'nikova MN (2001) Molecular mechanisms of genetic isolation. Priroda 5:40–47

Medvedev ZA (1998) The end of the "green revolution". Vestnik Ros Akad Nauk 68(9):817–827

Mettler LE, Gregg TG (1969) Population genetics and evolution. Prentice-Hall, Englewood Cliffs

Mettler L, Gregg T (1972) Genetika populyatsyi i evolutsiya (Population genetics and evolution). Mir, Moscow, 324 pp

Mina MV (1977) On the population structure of a species in fish: evaluating some hypotheses. Zh Obshch Biol 39(3):453–460

Mirakhmedov SM (1974) Novel wilt-resistant varieties of cotton. Vestn Skh Nauki 4:36–39

Mirakhmedov SM, Yuldashev SH (1971) Vihoustoichiviye sorta khlopchatnika "Tashkent" (Wilt-resistant varieties of the "Tashkent" cotton plant). Tashkent, Uzbekistan, 42 pp

Mitton JB (1997) Selection in natural populations. Oxford Univ Press, Oxford, 240 pp

Mitton JB, Grant MC (1984) Associations among protein heterozygosity, growth rate, and developmental homeostasis. Annu Rev Ecol Syst 15:479–499

Moiseeva I, Bannikova L, Altukhov Y (1993) The state of poultry breeding in Russia: genetic monitoring. Mezhdunarodnyi Selskokhozyaistvennyi Zhurnal 5/6:66–69

Moran PAP (1959) The theory of some genetical effects of population subdivision. Aust J Biol 12:109–118

Moran PAP (1962) Statistical processes of the evolutionary theory. Oxford Univ Press, Oxford

Moran P (1973) Statisticheskiye protsessi evolutsionnoy teorii (The statistical processes of evolutionary theory). Fizmatgiz, Moscow, 273 pp

Moritz C, Dowling TÅ, Brown WM (1987) Evolution of animal mitochondrial DNA: relevance for population biology and systematics. Annu Rev Ecol Syst 18:269–292

Morris DB, Richard KR, Wright JM (1996) Microsatellites from rainbow trout (*Oncorhynchus mykiss*) and their use for genetic study in salmonids. Can J Fish Aquat Sci 53:120–126

Morton NE (1960) The mutational load due to detrimental genes in man. Am J Hum Genet 12:348–364

Morton NE, Crow JF, Muller HJ (1966) An estimation of mutational damage in man from data on consanguineous marriages. Proc Natl Acad Sci USA 42:855–863

Moultrie F (1984) Compromise essential to success in breeding. Poultry Int 21:82–83

Mousseau TA, Ritland K, Heath DD (1998) A novel method for estimating heritability using molecular markers. Heredity 80:218–224

Movsesyan AA (1973) Some genetic aspects of modern and ancient human populations of Siberia. Vopr Antropol 45:77–84

Mukai T, Cockerham CC (1977) Spontaneous mutation rates at enzyme loci in *Drosophila melanogaster*. Proc Natl Acad Sci USA 74:2514–2517

Mukha DV, Vigmann BM, Shal K (1999) Saltational changes in the ribosomal gene cluster during the evolution of cockroaches of the genus *Blatella*. Dokl Ros Akad Nauk 364(1):134–139

Mueller UG, Wolfenbarger LL (1999) AFLP genotyping and fingerprinting. Trends Ecol Evol 14:389–394

Muller H (1925) Why polyploidy is rarer in animals than in plants. Am Nat 59:346–353

Muller HJ (1950) Our load of mutations. Am J Hum Genet 2:111–176

Mullis K, Faloona F, Scharf S et al (1986) Specific enzymatic amplification of DNA in vitro: the polymerase chain reaction. Cold Spring Harb Symp Quant Biol 51:263–273

Nagylacky T (1989) Gustave Malécot and the transition from classical to modern population genetics. Genetics 122:253–268

Narise T (1968) Migration and competition in *Drosophila*. 1. Competition between wild and vestigial strains of *Drosophila melanogaster* in a cage and migration-tube populations. Evolution 22:301–306

Narise T (1969) Migration and competition in Drosophila. 2. Effect of genetic background on migratory behaviour of *Drosophila melanogaster*. Jpn J Genet 44:297–302

Narise T (1974) Relation between dispersive behaviour and fitness. Jpn J Genet 49:131–138

Neave F (1958) The origin and speciation of *Oncorhynchus*. Trans R Soc Canada 52:25–49

Neaves WB (1969) Adenosine deaminase phenotypes among sexual and parthenogenetic lizards in the genus *Cnemidophorus* (Teiidae). J Exp Zool 171:175–183

Neaves WB, Gerald PS (1968) Lactate dehydrogenase isozymes in parthenogenetic teiid lizards. Science 160:1004–1005

Neel JV (1949) The inheritance of sickle cell anemia. Science 110:64–66

Neel JV (1970) Lessons from a "primitive" people. Science 170:815–822

Neel JV (1972) The detection of increased mutation rates in human populations. Mutagenic effects of environmental contaminants. Academic Press, New York, pp 99–119

Neel JV (1974) Developments in monitoring human populations for mutation rates. Mutat Res 26:319–328

Neel JV (1980) Some considerations pertinent to monitoring human populations for changing mutation rates. Well-being of mankind and genetics: Proc XIV Int Congr Genet. Mir, Moscow 1b(1):225–238

Neel JV (1983) Frequency of spontaneous and induced "point" mutations in higher eukaryotes. J Hered 74:2–15

Neel JV, Chagnon NA (1968) The demography of two tribes of primitive, relatively unaccultured American Indians. Proc Natl Acad Sci USA 59:680–689

Neel JV, Mohrenweiser HW (1984) Failure to demonstrate mutations affecting protein structure or function in children with congenital defects or born prematurely. Proc Natl Acad Sci USA 81:5499–5503

Neel JV, Rothman ED (1978) Indirect estimates of mutation rate in tribal American Indians. Proc Natl Acad Sci USA 75:5585–5588

Neel JV, Salzano FM (1966) Further studies on the Xavante Indians. 10. Some hypotheses-generalizations resulting from these studies. Am J Hum Genet 19:554–574

Neel JV, Schull W (1958) Nasledstvennost' cheloveka (Man's hereditary nature). Izdat Inost Lit, Moscow, 388 pp

Neel JV, Salzano FM, Junqueira PC et al (1964) Studies on the Xavante Indians of the Brazilian Mato Grosso. Hum Genet 16:52–140

Neel JV, Mohrenweiser HW, Meisler MH (1980) Rate of spontaneous mutation at human loci encoding protein structure. Proc Natl Acad Sci USA 77:6037–6041

Neel JV, Satoh Ch, Goriki K et al (1986) The rate with which spontaneous mutation alters the electrophoretic mobility of polypeptides. Proc Natl Acad Sci USA 83:389–393

Nei M (1972) Genetic distance between populations. Am Nat 106:283–292

Nei M (1975) Molecular population genetics and evolution. North-Holland, Amsterdam, 278 pp

Nei M (1976) Discussion to article of BDH Latter. Population genetics and ecology. Academic Press, New York, pp 409

Nei M (1983) Genetic polymorphism and the role of mutation in evolution. Evolution of genes and proteins. Sinauer, Sunderland, MA, pp 165–190

Nei M (1987) Molecular evolutionary genetics. Columbia Univ Press, New York, 512 pp

Nei M, Imaizumi Y (1966a) Genetic structure of human populations. 1. Local differentiation of blood group gene frequencies in Japan. Heredity 21:9–35

Nei M, Imaizumi Y (1966b) Genetic structure of human populations. 2. Differentiation of blood group gene frequencies among isolated human populations. Heredity 21:183–190

Nei M, Roychoudhury AK (1974) Genic variation within and between the three major races of man, Caucasoids, Negroids, and Mongoloids. Am J Hum Genet 26:421–443

Nei M, Tajima F 1981. DNA polymorphism detectable by restriction endonucleases. Genetics (US) 105:207–217

Nevo E (1978) Genetic variation in natural populations: patterns and theory. Theor Pop Biol 13:121–177

Nevo E, Beilis A, Ben-Shlomo R (1984) The evolutionary significance of genetic diversity: ecological, demographic and life history correlates. In: Mani GS (ed) Evolutionary dynamics of genetic diversity. Springer, Berlin Heidelberg New York, pp 13–213

Newnan RA, Squire T (2001) Microsatellite variation and fine-scale population structure in the woody frog (*Rana sylvatica*). Mol Ecol 10:1087–1100

Nielsen R (2001) Statistical tests of selective neutrality in the age genomics. Heredity 86:641–647

Nikolsky YF (1976) The first steps in investigations into karakul lamb breeding. Karakulevodstvo 5:54–58

Noguès RM (1977) Population size fluctuation in the evolution of experimental cultures of *Drosophila subobscura*. Evolution 31:200–213

Novosel'skaya AY, Novosel'sky YI, Altukhov YP (1982) Physico-chemical characteristics of the spawning sites and hereditary heterogeneity of the sockeye salmon population of Lake Azabachye. Genetika 18(6):1004–1011

O'Brien SJ, MacIntyre RJ (1972) The L-glycerophosphate cycle in Drosophila melanogaster. 2. Genetic aspects. Genetics (US) 71:127

Öehlkers F (1943) Die Auslosung von Chromosomenmutationen in der Meiosis durch Einwirkung von Chemicalen. Z Indukt Abstamm Vererbungsl 81:313–343

O'Gower AK, Nicol PJ (1968) A latitudinal cline of haemoglobins in a bivalve mollusk. Heredity 23:485–491

Ohba S (1967) Chromosomal polymorphism and capacity for increase under near optimal conditions. Heredity 22:169–189

Ohdachi S, Dokuchaev NE, Hasegava M, Masuda R (2001) Intraspecific phylogeny and geographical variation of six species of northeastern Asiatic Sorex shrews based on the mitochondrial cytochrome b sequences. Mol Ecol 10:2199–2213

Ohno S (1970a) Evolution by gene duplication. Springer, Berlin Heidelberg New York,178 pp

Ohno S (1970b) The enormous diversity in genome size of fish as a reflection of nature's extensive experiments with gene duplications. Trans Am Fish Soc 99:120–131

Ohno S (1973) Geneticheskiye mekhanizmy progressivnoy evolutsii (The genetic mechanism of progressive evolution). Mir, Moscow, 227 pp

Ohno S, Wolf U, Atkin NB (1968) Evolution from fish to mammals by gene duplications. Hereditas 59:169–187

Ohno S, Stenius C, Christian LC et al (1969) De novo mutation-like events observed at the 6-Pgd locus of the Japanese quail and the principle of polymorphism breeding more polymorphism. Biochem Genet 3:417–428

Ohta T (1975) Statistical analyses of Drosophila and human protein polymorphism. Proc Natl Acad Sci USA 72:3194–3196

Ohta T (1976) Role of very slightly deleterious mutations in molecular evolution and protein polymorphism. Theor Pop Biol 10:254–275

Ohta T (1978) Sequence variability of immunoglobulins considered from the standpoint of population genetics. Proc Natl Acad Sci USA 75:5108–5112

Ohta T (1980) Evolution and variation of multigene families. Lecture notes in biomathematics. Springer, Berlin Heidelberg New York

Ohta T (1983) On the evolution of multigene families. Theor Pop Biol 23:216–240

Ohta T (1984) Population genetics theory of concerted evolution and its application to the immunoglobulin V gene tree. J Mol Evol 20:274–280

Ohta T (1987a) A model of evolution for accumulating genetic information. J Theor Biol 124:199–211

Ohta T (1987b) Simulation evolution by gene duplication. Genetics (US) 115:207–213

Ohta T (2000) Mechanisms of molecular evolution. Philos Trans R Soc Lond B 355:1623–1626

Ohta T, Kimura M (1973) A model of mutation appropriate to estimate the number of electrophoretically detectable alleles in a finite population. Genet Res 22:201–204

Okada N (1991) SINEs: short interspersed repeated elements of the eucaryotic genome. Trends Ecol Evol 6:358–361

Okazaki T (1982) Genetic study on population structure in chum salmon (*Oncorhynchus keta*). Bull Far Seas Fish Res Lab 19:25–36

Omel'chenko VT (1973) Species-specificity and intraspecific constancy of haemoglobin electrophoregrams in some fishes of the Far East. Biokhimicheskaya genetika ryb (Fish biochemical genetics). Nauka, Leningrad, pp 67-71

Omel'chenko VT (1974) Electrophoretic analysis of haemoglobins in fish of the Far East. Genetika 10(9):35–43

Omel'chenko VT, Gerasimenko TP (1981) Molecular organization of the genomes and diploid/tetraploid ratios of kisutch and herring. Genetika 17(2):338–347

Omenn GS, Cohen PTW, Motulsky AG (1971) Genetic variation in glyc-olitic enzymes in human brain. Excerpta Med Int Congr Ser 233:135

Oota H, Settheetham-Ishida W, Tiwawech D et al (2001) Human mtDNA and Y-chromosome variation is correlated with matrilocal versus patrilocal residence. Nat Genet 29:20–21

O'Relly PT, Canino MF, Bailey KM, Bentzen P (2004) Inverse relationship between $F_{ST}$ and microsatellite polymorphism in the marine fish, walleye pollock (*Theragra chalcogramma*): implications for resolving weak population structure. Mol Ecol 13:1799–1814

Orr HA (1995) The population genetics of speciation: the evolution of hybrid incompatibilities.Genetics 139:1805–1813

Otto SP (2000) Detecting the form of selection from DNA sequence data. Trends Genet 16:526–529

Paetkau D, Calwert W, Stirling I, Strobeck C (1995) Microsatellite analysis of population structure in Canadian polar bears. Mol Ecol 4:347–354

Paglia G, Morgante M (1998) PCR-based multiplex DNA fingerprinting techniques for the analysis of conifer genomes. Mol Breed 4:173–177

Panov EN (1970) Obscheniye v mire zhivotnykh: evolutsionniye i populyatsionnye aspecty povedeniya zhivotnykh (Social life in the animal world: evolution and population aspects of animal behaviour), vol 1. Znaniye, Moscow, pp 3–48

Park LK, Moran P (1994) Developments in molecular genetic techniques in fisheries. Rev Fish Biol 4:272–299

Park LK, Brainard MA, Dingman DA, Winans GA (1993) Low levels of intraspecific variation in the mitochondrial DNA of chum salmon (*Oncorhynchus keta*). Mol Mar Biol Biotechnol 2:362–370

Pauling L, Itano HH, Singer SJ et al (1949) Sickle cell anemia: a molecular disease. Science 110:543–548

Pavlov DS, Alimov AA, Altukhov YP et al (2001) National strategy of biodiversity conservation of Russia. Russian Academy of Sciences and Russian Federation Ministry of Natural Resources, Moscow, 76 pp

Pedersen AA, Loeschke V (2001) Conservation genetics of peripheral populations of the mygolomorph spider *Atypus affinis* in northern Europe. Mol Ecol 10:1133–1142

Penrose LS (1955) Evidence of heterosis in man. Proc R Soc Lond B 144:203–213

Petras ML (1967) Studies of natural populations of Mus. 1. Biochemical polymorphism and their bearing on breeding structure. Evolution 21:259–274

Petrov DA (2001) Evolution of genome size: new approaches to an old problem. Trends Genet 17:23–28

Philipp DP, Childers WF, Whitt GS (1985) Correlations of allele frequencies with physical and environmental variables for populations of largemouth bass, *Micropterus salmoides* (Lacepede). J Fish Biol 27:347–365

Podstreshny OP (1981) Efficacy of modal selection in the lines of hens. Ptakhivnitstvo 32:23–26

Pogson GH, Mesa KA, Boutilier RG (1995) Genetic population structure and gene flow in the Atlantic cod, *Gadus morhua*: a comparison of allozyme and nuclear RFLP loci. Genetics (US) 139:375–385

Politov DV, Krutovsky KV (2001) Phylogenetics, genogeography and hybridization of five-needle pines in Russia and neighboring countries. In: Sniezko R (ed) Breeding and genetic resources of five-needle pines: growth, adaptability and pest resistance. Proc Conf IUFRO Working Party, 23–27 July 2001. Rocky Mountain Res Station. Ogden, USA

Politov DV, Krutovsky KV, Altukhov YP (1992) Characterization of the gene pools of Siberian pine populations at a set of isozyme loci. Genetika (Moscow) 28(1):93–118

Pollak E (1983) A new method for estimating the effective population size from allele frequency changes. Genetics 104:531–548

Pomortsev AA, Kalabushkin BA, Ladogina MP, Blank ML (1994) Gene geography and regularities in the distribution of allelic variants of three hordein-coding loci in spring barley in the former USSR. Rus J Genet 30(6):708–718

Pope TK (1992) The influence of dispersal patterns and mating system on genetic differentiation within and between populations of the red howler monkey (*Alouatta seniculus*). Evolution 46:1112–1128

Poroshenko GG, Abilev SK (1988) Anthropogenic mutagens and natural antimutagens. Itogi Nauki i Tekhniki Ser Obshchaya Genetika 12, VINITI, 206 pp

Powell IR (1978) The founder-flush speciation theory: an experimental approach. Evolution 32:465–474

Powell JR (1997) Progress and prospects in evolutionary biology: the Drosophila model. Oxford Univ Press. Oxford

Prakash S, Lewontin RC, Hubby TL (1969) A molecular approach to the study of genic heterozygosity in natural populations. 4. Patterns of genic variation in central, marginal and isolated populations of *Drosophila pseudoobscura*. Genetics (US) 61:841–858

Prehn LM, Rash RM (1969) Cytogenetic studies of *Poecilia* (Pisces). 1. Can J Genet Cytol 2:880–895

Primmer CR, Aho T, Piironen J et al (1999) Microsatellite analysis of hatchery stocks and natural populations of arctic charr, *Salvelinus alpinus*, from the Nordic region: implications for conservation. Hereditas 130:277–289

Prodohl PA, Taggart JB, Ferguson A (1994) Single locus minisatellite variation in brown trout, *Salmo trutta* L, populations. In: Beaumont AR (ed) Genetics and evolution of aquatic organisms. Chapman and Hall, London, pp 263–270

Promboon S, Mi MP, Chaturachinda K (1983) Birth weight and gestation time in relation to natural selection in Thailand. Ann Hum Genet 47:133–141

Przeworski M, Hudson RR, di Rienzo A (2000) Adjusting the focus on human variation. Trends Genet 16:296–282

Pustovoit SP (1994) Within-population genetic variation and among-population differentiation in Asiatic sockeye salmon, *Oncorhynchus nerka* (Walbaum). Genetika 30(1):101–106

Quatro JM, Vrijenhoek RC (1989) Fitness differences among remnant populations of the endangered Sonoran topminnow. Science 245:976–978

Quinn TP, Fresh K (1984) Homing and straying in Chinook salmon (*Oncorhynchus tshawytscha*) from Cowlitz River hatchery, Washington. Can J Fish Aquat Sci 41:1078–1082

Racine RR, Langley CH (1980) Genetic heterozygosity in a natural population of *Mus musculus* assessed using two-dimensional electrophoresis. Nature 283:855–857

Ramshaw JAM, Coyne JA, Lewontin RC (1979) The sensitivity of gel electrophoresis as a detector of genetic variation. Genetics (US) 93:1019–1037

Rand DM (1993) Endotherms, ectotherms and mitochondrial genome-size variation. J Mol E37:281–295

Rannala B, Hartigan JA (1995) Identity by descent in island-mainland populations. Genetics 139:429–437

Rapoport IA (1946) Carbonyl compounds and the chemical mechanism of mutations. Dokl Akad Nauk SSSR 54(1):65–67

Rash EM, Prehn LM (1969) Cytogenetic studies of Poecilia (Pisces). 1. J Cell Biol 43:112

Rash EM, Darnell RM, Kallman CD et al (1965) Cytophotometric evidence for triploidy in hybrids of the gynogenetic fish, *Poecilia formosa*. J Exp Zool 160:155–169

Rash EM, Prehn LM, Rash RM (1970) Cytogenetic studies of Poecilia (Pisces). 2. Chromosoma 31:18–40

Rasmussen DI (1964) Blood group polymorphism and inbreeding in natural populations of the deer mouse, *Peromyscus maniculatus*. Evolution 18:219–229

Rasmussen DI (1970) Biochemical polymorphism and genetic structure in populations of Peromyscus. Symp Zool Soc Lond 26:335–349

Ratner VA (1977) Matematicheskaya populyatsionnaya genetika (Mathematical population genetics). Nauka, Novosibirsk, 126 pp

Ratner VA, Zharkikh AA, Kolchanov NA, Rodin SN, Solov'yev BB, Shamin VV (1985) Problemy teorii molekulyarnoy evolutsii (Problems of the theory of molecular evolution). Nauka, Novosibirsk, 263 pp

Raven PH (1979) Future directions in plant population biology. Topics in plant population biology. Columbia Univ Press, New York, pp 461–481

Raybould AF, Goudet J, Mogg RJ, Gliddon CJ, Gray AJ (1996) Genetic structure of a linear population of *Beta vulgaris* ssp. maritima (see beet) revealed by isozyme and RFLP analysis. Heredity 76:111–117

Raymond S (1964) Acrilamide gel electrophoresis. Ann NY Acad Sci. 121:350–365

Reinitz GL (1977) Inheritance of muscle and liver types of supernatant NADP-dependent isocitrate dehydrogenase in rainbow trout (*Salmo gairdneri*). Biochem Genet 15:445–454

Rice WR, Hostert EE (1993) Laboratory experiments on speciation: what have we learned in 40 years? Evolution 47:1637–1653

Ricker WE (1938) A comparison of the seasonal growth rates of young sockeye salmon and young squawfish in Cultus Lake. Fish Res Board Canada Pacif Progr Rep 36:3–5

Ricker WE (1954) Stock and recruitment. J Fish Res Board Can 11:559–623

Ricker WE (1972) Hereditary and environmental factors affecting certain salmonid populations. The stock concept of Pacific salmon. Univ Brit Columbia, Vancouver, pp 19–160

Ricker WE (1981) Changes in the average size and average age of Pacific salmon. Can J Fish Aquat Sci 38:1636–1656

Rieger R, Michaelis A (1967) Genetisches und cytogenetisches Wörterbuch. Springer, Berlin

Rogers AR (2001) Order emerging from chaos in human evolutionary genetics. Proc Natl Acad Sci USA 98:779–780

Rogers JH (1985) The origin and evolution retroposons. Int Rev Cytol 93:187–279

Roginsky YY (1947) The regularities of spatial distribution of human blood groups: anthropology of "the outlying" ethnic groups. Tr Inst Etnogr Akad Nauk SSSR 1:216–234

Rogues S, Sévigny J-M, Bernatchez L (2001) Evidence for broadscale introgressive hybridization between two redfish (genus *Sebastes*) in the North-West Atlantic: a rare marine example. Mol Ecol 10:127–148

Rohlf FJ, Schnell GD (1971) An investigation of an isolation by distance model. Am Nat 105:295–324

Romashov DD (1931) Conditions of "equilibrium" in a population. Zh Eksp Biol A 7:442–454

Rosenkranz HS, Mersch-Sundermann V, Klopman G (1999) SOS chromotest and mutagenicity in Salmonella: evidence for mechanistic differences. Mutat Res 431:31–38

Ross KG, Shoemaker DD, Krieger MJ, DeHeer CJ, Keller L (1999) Assessing genetic structure with multiple classes of molecular markers: a case study involving the introduced fire ant *Slenopsis invicta*. Mol Biol 16:525–543

Rubin GM (2001) Comparing species. Nature 409:820–821

Ruddle FH, Roderick TH, Shows TB et al (1969) Measurement of genetic heterogeneity by means of enzyme polymorphisms in wild populations of the mouse. J Hered 60:321–322

Ruzzante DE (1998) A comparison of several measures of genetic distance and population structure with microsatellite data: bias and sampling variance. Can J Fish Aquat Sci 55:1–14

Ruzzante DE, Hansen MM. Meldrup D (2001) Distribution of individual inbreeding coefficients, relatedness and influence of stocking on native anadromous brown trout (*Salmo trutta*) population structure. Mol Ecol 10:2107–2128

Ryabova GC, Goncharova AA, Titova AYu, Altukhov YP (1978) On the factors of stationary distribution of lactate dehydrogenase and phosphoglucomutase allele frequencies in sockeye salmon populations of the Lake Azabach'ye (Kamchatka). Theses Int Symp Biol Pacific Salmon (Yuzhno-Sakhalinsk), pp 88–90

Rychkov YG (1965) Peculiarities of serological differentiation in the aboriginal peoples of Siberia. Vopr Antropol 21:18–33

Rychkov YG (1968) Response of populations to isolation. Problemy evolutsii (Problems of evolution), vol 1. Nauka, Novosibirsk, pp 212–236

Rychkov YG (1969) Some population-genetic approaches to anthropology of Siberia. Vopr Antropol 33:16–33

Rychkov YG (1973) A system of ancient human isolates in Northern Asia in view of the stability and evolution of the populations. Vopr Antropol 44:3–22

Rychkov YG (1975) Definition of natural selection on AB0 group genotypes by data of genetic population structure for humans from north Asia. In: Malyutov MB (ed) Statistic methods in population genetics, vol 49. MGU, Moscow, pp 19–44

Rychkov YG (1979) A comparative study of the genetic process in urbanized and isolated populations. Vopr Antropol 63:3–21

Rychkov YG (1984) Space and time in genogeography. Vestn Akad Med Nauk SSSR 7:11–16

Rychkov YG (ed) (1999) Population gene pool and gene geography. 1. Gene pool of the population of Russia and adjacent countries. Nauka, St Petersburg, 611 pp

Rychkov YG, Balanovskaya EV (1990a) Ethnogenetics: on alignment of adaptive and selective-neutral genetic differentiation in ethnoses. Genetika 26(3):541–49

Rychkov YG, Balanovskaya EV (1990b) Genetic differentiation of the population: can the data on DNA polymorphism be inferred from immunobiochemical polymorphism? Molecular Mechanisms of Biological Processes. Nauka, Moscow, pp 67–83

Rychkov YG, Movsesyan AA (1972) A genetical anthropological analysis of the distribution of cranial anomalies in Mongoloids of Siberia in view of their origin. Byull Mosk Ova Ispyt Prir Otd Biol 43:114–132

Rychkov YG, Sheremet'yeva VA (1976) Genetics of the circumpolar populations of Eurasia in connection with the problem of human adaptation. Resursy biosfery (Resources of the biosphere), vol 3. Nauka, Leningrad, pp 10–41

Rychkov YG, Sheremet'yeva VA (1977) The genetic process in the system of ancient human isolates in North Asia. In: Harrison GA (ed) Population structure and human variation, Fasc 2, . Cambridge Univ Press, New York, pp 47–108

Rychkov YG, Sheremet'yeva VA (1979) The genetics of circumpolar populations of Eurasia related to the problem of human adaptation. In: Milan FA (ed) The human biology of circumpolar populations. International biological programme, vol 21. Cambridge Univ Press, Cambridge, pp 37–80

Rychkov YG, Yaschuk EV (1980) Genetics and ethnogenesis. Vopr Antropol 64:23–39

Rychkov YG, Yaschuk EV (1983) Genetics and ethnogenesis: trends of genetic processes in connection with the features of development of European population. Vopr Antropol 72:3–17

Rychkov YG, Yashchuk EV (1985) Genetics and ethnogenesis. Historical order of genetic differentiation of human population: model and reality. Vopr Antropol 75:97–106

Rychkov YG, Rusakova OL, Rapoport MP (1973) Factors of genetic differentiation of the population systems of the northern Asia native populations. Genetika 9(2):136–145

Rychkov YG, Sheremet'yeva VA, Tausik T (1976) Genetics and anthropology of the populations of the taiga reindeer breeders of Siberia (the Evenks of Middle Siberia). 3. Genetic markers and genetic differentiation in the population of Evenks of Middle Siberia. Vopr Antropol 53:38–56

Rychkov YG, Yaschuk EV, Veselovskaya EV (1982) Genetics and ethnogenesis. Vopr Antropol 69:3–18

Rychkov YG, Zhukova OV, Sheremet'eva VA et al (1999) Population gene pool and gene geography. 1. Gene pool of the population of Russia and adjacent countries. Nauka, St Petersburg, 611 pp

Rychkov YG, Zhukova OV, Sheremet'eva VA et al (2000). Gene pool of the population of Russia and adjacent countries. In: Population gene pool and gene geography, vol 1. Nauka, St. Petersburg, 611 pp

Ryman H, Ståhl G (1980) Genetic changes in hatchery stocks of brown trout (*Salmo trutta*). Can J Fish Aquat Sci 37(1):82–87

Ryman N, Utter F (eds) (1987) Population genetics and fishery management. Univ Washington Press, Seattle, 420 pp

Ryskov AP (1999) Multilocus DNA fingerprinting in genetic population studies of biodiversity. Mol Biol 33(6):997–1011

Saccone C, Pesole G, Sbisa E (1991) The main regulatory region of mammalian mitochondrial DNA: structure-function model and evolutionary pattern. J Mol Evol 33:83–91

Sachko GD (1973) Genetics of lactate dehydrogenase isozymes in Pacific Salmon. Biokhimicheskaya genetika ryb (Fish biochemical genetics). Nauka, Leningrad, pp 155–160

Saiki RK, Gelfand DH, Stoffel S et al (1988) Premier-directed enzymatic amplification of DNA with thermostable DNA polymerase. Science 239:487–491

Salmenkova EA (1989) General results and aims of population-genetic studies of salmonid fishes. In: Kirpichnikov VS (ed) Genetika v akvakulture (Genetics in aquaculture). Nauka, Leningrad, pp 7–29

Salmenkova EA, Omel'chenko VT (1978) Polymorphism of proteins in the populations of diploid and tetraploid fishes. Biol Morya 4:67–74

Salmenkova EA, Volokhonskaya LG (1973) Biochemical polymorphism in the populations of diploid and tetraploid fishes. Biokhimicheskaya genetika ryb (Fish biochemical genetics). Nauka, Leningrad, pp 54–61

Salmenkova EA, Omelchenko VT, Pobedonostseva EY, Milishnikov AN, Nikonorov SI, Altukhov YP (1983) Population-genetic analysis of the effectiveness of transferring Kuril chum salmon eggs to southwestern Sakhalin. Genetika 19(10):1660–1667

Salmenkova EA, Altukhov YP, Viktorovsky RM, Omel'chenko VT, Bachevskaya LT, Ermolenko LN et al (1986) The genetic structure of chum salmon (Oncorhynchus keta) populations spawning in the rivers of the Far East and north-east of the USSR. Zh Obshch Biol 47(4):529–548

Salmenkova EA, Omelchenko VT, Altukhov YP (1992) Genogeographic analysis of chum salmon, *Oncorhynchus keta* (Walbaum), populations of the Asian part of species range. Genetika 28(1):76–92

Sanchez JA, Clabby C, Ramos D et al (1996. Protein and microsatellite single locus variability in *Salmo salar* L (Atlantic salmon). Heredity 77:423–432

Sang JH (1949a) The ecological determinants of population growth in a Drosophila culture. 3. Larval and pupal survival. Physiol Zool 22:183–202

Sang JH (1949b) The ecological determinants of population growth in a Drosophila culture. 4. The significance of successive batches of larvae. Physiol Zool 22:202–210

Sansing RC, Chinnici JP (1976) Optimal and discriminating birth weights in human populations. Ann Hum Genet 40:123–131

Santos F, Tyler-Smith C (1996) Reading the human Y-chromosome: emerging DNA markers and human genetic history. Braz J Genet 18:669–672

Sarsenbayev NA (1980) The Astrakhan tracery and genetic peculiarities of Karakul lambs referred to the "average" and "extreme" types. Cand Sci Thesis, Inst Gen Genet USSR Acad Sci, Moscow

Sarsenbayev NA, Afanas'ev KI (1984) Genetic evaluation of the origin of karakul sheep. Skh Biol 7:95–97

Satoh C, Mohrenweiser HW (1979) Genetic heterogeneity within an electrophoretic phenotype of phosphoglucose isomerase in Japanese population. Skh Biol 42:283–292

Satoh C, Takahashi N, Asakawa J-I et al (1984) Electrophoretic variants of blood proteins in Japanese. 1. Phosphoglucomutase 2 (PGM2). Jpn J Hum Genet 29:89–104

Schierwater B, Ender A, Schroth W et al (1997) Arbitrarily amplified DNA in ecology and evolution. In: Caetano-Anolles G, Gressnoff PM (eds) DNA markers: protocols, applications and overview. Wiley-VCH, New York, pp 313–330

Schilthuizen M (1999) Cloning Odysseus and the seed of speciation. Trends Ecol Evol 14:90–91

Schlötterer C (2004) The evolution of molecular markers – just a matter of fashion? Nature Genet 5:63–69

Schneider EL (1999) Aging in the third millennium. Science 283:796–797

Schneider S, Roessli D, Excoffier L (2000) Arlequin: a software for population genetics data analysis. Ver 2000. Genetics and biometry Lab, Dept Anthropology, Univ Geneva, Geneva

Schröder JH (1983) The guppy as a model for evolutionary studies in genetics, behavior, and ecology. Ber Natur-Med Verein Innsbruk 70:249–279

Schull WJ, Neel JV (1965) The effects of inbreeding on Japanese children. Harper and Row, New York, 417 pp

Schultz RJ (1961) Reproductive mechanism of unisexual and bisexual strains of the viviparous fish *Poeciliopsis*. Evolution 15:302–325

Schultz RJ (1966) Hybridization experiments with an all-female fish of the genus *Poeciliopsis*. Biol Bull 130:415–429

Schultz RJ (1967) Gynogenesis and triploidy in the viviparous fish *Poeciliopsis*. Science 157:1564–1567

Schultz RJ (1969) Hybridization, unisexuality, and polyploidy in the teleost *Poeciliopsis* and other vertebrates. Am Nat 103:605–619

Schultz RJ (1973) Unisexual fish: laboratory synthesis of a "species". Science 179:180–181

Schultz RJ (1980) Role of polyploidy in the evolution of fishes. In: Lewis WH (ed) Polyploidy: biological relevance. Plenum Press, New York, pp 313–340

Schultz RJ, Kallman KD (1968) Triploid hybrids between the all-female teleost *Poecilia formosa* and *Poecilia sphenops*. Nature 219:280–282

Schwartz OA, Armitage CB (1980) Genetic variation in social mammals: the marmot model. Science 207:665–667

Scribner KT, Arntzen JW, Burke T (1994) Comparative analysis of intra- and interpopulation genetic diversity in *Bufo bufo*, using allozyme, single-locus microsatellite, minisatellite and multilocus minisatellite data. Mol Biol Evol 11:737–748

Seielstad MT, Minch E, Cavalli-Sforza LL (1998) Genetic evidence for a higher female migration rate in humans. Nat Genet 20:278–280

Selander RK (1970) Behaviour and genetic variation in natural populations. Am Zool 10:53–66

Selander RC, Yang SY (1969) Protein polymorphism and genic heterozygosity in a wild population of the house mouse (*Mus musculus*). Genetics (US) 63:653–667

Seppa P, Pamilo P (1995) Gene flow and population viscosity in *Myrmica* ants. Heredity 74:200–209

Serebrovsky AS (1927) Genetic analysis of hens breeding on the farms of Daghestan highlanders. Zh Eksp Biol A 3(1/2):62–146; 3(4):125–146

Serebrovsky AS (1930) Problems and methods of genegeography, vol 2. Trudy s'ezda po genetike i selektsii, Moscow, pp 71–86

Serebrovsky AS (1935) The genegeography of hens in Armenia. Usp Zootekh Nauk 1(3):317–348

Severtsov AS (1981) Vvedeniye v teoriyu evolutsii (An introduction to the theory of evolution). Izdat Mosk Gosudarstvennogo Univ, Moscow, 318 pp

Shaffer HB, Fellers GM, Magee A et al (2000) The genetics of amphibian declines: population substructure and molecular differentiation in the Yosemite toad, *Bufo canorus* based on single-strand conformation polymorphism analysis (SSCP) and mitochondrial DNA sequence data. Mol Ecol 9:245–257

Shaklee JB, Salini J (1993) Electrophoretic characterizations of multiple genetic stocks of barramundi perch in Queensland, Australia. Trans Am Fish Soc 122:685–701

Shaklee JB, Klaybor DC, Houng S, White BA (1991) Genetic stock structure of odd-year pink salmon, *Oncorhynchus gorbuscha* (Walbaum), from British Columbia and potential mixed-stock fisheries applications. J Fish Biol 39 [Suppl A]:21–34

Shapoval SN, Kratirov OV, Altukhov YP (1992) Recommendations on using the modal selection method for accelerating stabilization of morphological and commercial traits in cotton selection lines. Zaitsev Cotton Selection and Seed Production Research and Production Association, Tashkent, 12 pp

Shapvil' F, Enni AL (1977) Biosintez belka (Protein biosynthesis). Mir, Moscow, 315 pp

Shaw CR, Barto E (1963) Genetic evidence for the subunit structure of lactate dehydrogenase isozymes, vol 50. Mir, Moscow, pp 211–214

Shaw PW, Turan C, Wright JM, O'Connell, Carvalho GR (1999) Microsatellite DNA analysis of population structure in Atlantic herring (*Clupea harengus*), with direct comparison to allozyme and mtDNA RFLP analysis. Heredity 83:490-499

Sheen SJ (1970) Peroxidases in the genus *Nicotiana*. Theor Appl Genet 40:18–25

Shelby J, Tindall KR (1997) Mammalian germ cell mutagenicity of ENU, IPMS and MMS, chemicals selected for a transgenic mouse collaborative study. Mutat Res 388:99-109

Sheppard PM (1967) Natural selection and heredity. Hutchinson Univ Library

Sheremetyeva VA, Rychkov YG (1978) Populyatsionnaya genetika narodov Severo-Vostochnoy Azii (Population genetics of peoples of northeast Asia). Moscow Univ Press, Moscow. 151 pp

Shevchenko VA (2001) Evolution of the ideas of genetic danger of ionising radiation for humans. Radiat Biol Radioekol 41(5):615–626

Shevlyakov EA (2001. Trends of the number dynamics in rheophilic and lake sockeye salmon *Oncorhynchus nerka* from the Azabach'e Lake basin. Biol Morya 27(5):347–351

Shigaeva MK, Akhmatullina NB, Abilev SK (1994) Environmental mutagens and comutagens. Almaty, Gylym, 254 pp

Shilenko BV (1974) Mechanisms stabilizing the allele contents of esterase locus in *Drosophila melanogaster*. Genetika 10(3):76–84

Shilov IA (1977) Ekologo-fiziologicheskie osnovi populyatsionnikh otnosheniy u zhivotnikh (Ecologic and physiologic bases of population relations among animals). MGU, Moscow, 263 pp

Shirinsky MA (1962) Selection for the curl size and adequate feeding of ewes in yean prove to be important factors in Karakul sheep breeding. Ovtsevodstvo 4:23–27

Shirinsky MA (1975) Scientific research in Karakul sheep breeding in Kazakhstan. Dokl Uchastnikov III Mezhdunarodnogo Simp po Karakulevodstvu. Samarkand, pp 108–119

Shmal'hauzen II (1938) Organizm kak tseloye v indvidual'nom i istoricheskom razvitii (The body as a whole in individual and historical development). Izdat Akad Nauk SSSR, Moscow, Leningrad, 144 pp

Shmal'hauzen II (1969) Faktory evolutsii. Teoriya stabiliziruyuschego otbora (Factors of evolution. The theory of stabilizing selection). Nauka, Moscow, 451 pp

Shows TB, Ruddle FH (1968) Malate dehydrogenase: evidence for tetrameric structure in *Mus musculus*. Science 160:1356–1357

Shram R (1981) The necessary and sufficient test-system for evaluating the genetic risks of environmental pollution. Genetika i blagosostoyanie chelovechestva (Genetics and the well being of mankind). Nauka, Moscow, pp 220–236

Shram R, Kuleshov NP (1984) Monitoring of industrial mutagens by means of cytogenetic analysis of human peripheric lymphocytes. Nasledstvennost' cheloveka i okruzhayushchaya sreda (Man's heredity and the environment). Nauka, Moscow, pp 34–55

Shriver MD, Jin L, Boerwinkle LE et al (1995) A novel measure of genetic distance for highly polymorphic tandem repeat loci. Mol Biol Evol 12:914–920

Shurkhal AV (1982) Analysis of the structure of reproductive population of a big city on the basis of a set of biochemical gene markers. Cand Sci Thesis, Inst Gen Genet USSR Acad Sci, Moscow, 22 pp

Simongulyan NG (1975) The influence of selection on hereditability of traits in hybrid populations of cotton. Genetika i selektsiya rastenii (Genetics and plant selection). Fan, Tashkent, pp 73–79

Sinclair AH, Berta P, Palmer MS et al (1990) A gene from the human sex-determining region encodes a protein with a conserved DNA-binding motif. Nature 346:240–244

Singh SM, Zouros E (1978) Genetic variation associated with growth rate in American oyster (*Crassostrea virginica*). Evolution 32:342–353

Singer MF (1982) SINEs and LINEs: highly repeated short and long interspersed sequences in mammalian genomes. Cell 28:433–434

Singer M, Berg P (1991) Genes and genomes. University Science Books Mill Valley, California

Singh RS (1979) Genetic heterogeneity within electrophoretic "alleles" and the pattern of variation among loci in *Drosophila pseudoobscura*. Genetics (US) 93:997–1018

Singh RS, Hubby JL, Throckmorton LH (1974) The study of genic variation by electrophoretic and heat denaturation techniques at the octanol dehydrogenase locus in members of the *Drosophila virilis* group. Genetics (US) 80:637–650

Singh RS, Lewontin RC, Felton AA (1976) Genetic heterogeneity within electrophoretic "alleles" of xanthine dehydrogenase in *Drosophila pseudoobscura*. Genetics (US) 85:609–629

Slatkin M (1993) Isolation by distance in equilibrium and non-equilibrium populations. Evolution 47:264–279

Slatkin M (1995) A measure of population subdivision based on microsatellite allele frequencies. Genetics 139:457–462

Small MP, Beacham TD, Withler RE, Nelson RJ (1998) Discriminating Coho salmon (*Oncorhynchus kisutch*) populations within the Fraser River. Mol Ecol 7:141–155

Smith MF (1979) Geographic variation in genic and morphological characters in *Peromyscus californicus*. J Mammal 60:705–722

Smithies O (1955) Zone electrophoresis in starch gels: group variations in the serum proteins of normal human adults. Biochem J 61:629–641

Snyder RD, Green JW (2001) A review of the genotoxicity of marketed pharmaceuticals. Mutat Res 488:151–169

Sokolov VE (1979) A study of higher vertebrates in biosphere reserves. Biosphernye Zapovedniki (Biosphere reserves). Gidrometeoizdat, Leningrad, pp 77–80

Sokolov V (1981) The biosphere reserve concept in the USSR. AMBIO 10(2/3):97–101

Solbrig OT, Solbrig DJ (1979) Introduction to population biology and evolution. Addison-Wesley, Menlo Park, 510 pp

Somes G (1985) International registry of poultry genetic stocks. Connect Exp Stn Bull 469:1–95

Soulé ME (ed) (1987) Viable populations for conservation. Cambridge Univ Press, New York

Soulé ME (1987) Viable populations for conservations. Cambridge Univ Press, Cambridge, 211 pp

Southern EM (1975) Detection of specific DNA fragments separated by gel electrophoresis. J Mol Biol 98:503–517

Sozinov AA (1985) Polimorphizm belkov i ego znacheniye v genetike i selektsii (Protein polymorphism and its significance in genetics and selection). Nauka, Moscow, 272 pp

Sozinov AA, Poperelya FA (1979) Polymorphism of prolamins and selection. Vestn Skh Nauki 10:21–34

Sozinov AA, Poperelya FA (1980) Genetically determined plant protein polymorphism and plant breeding. Well-being of mankind and genetics: Proc XIV Int Congr Genet Moscow, Mir 1b(2):230–248

Sozinov AA, Poperelya FA, Stakanova AI (1973) Intravarietal polymorphism of gliadine in some varieties of wheat. Dokl VASKHNIL 6:8–11

Sperlich D, Pfriem P (1986) Chromosomal polymorphism in natural and experimental populations. Genetics and biology of *Drosophila*, vol 3. Academic Press, New York, pp 257–309

Spieth PT (1974) Gene flow and population differentiation. Genetics (US) 78:961–972

Spight TM (1974) Sizes of population of a marine snail. Ecology 55:712–729

Spritz RA, Forget BG (1983) The thalassemias: molecular mechanisms of human genetic disease. Am J Hum Genet 35:333–361

Stabell OB (1984) Homing and olfaction in salmonids: a critical review with special reference to the Atlantic salmon. Biol Rev 59:333–388

Stacey PB, Johnson VA, Taper ML (1997) Migration within metapopulations: the impact upon local population dynamics. In: Hanski LA, Gilpin ME (eds) Metapopulation biology. Academic Press, Boston, pp 267–291

Ståhl G (1981) Genetic differentiation among natural populations of Salmo salar in Northern Sweden. Fish gene pools. Ecol Bull 34:95–105

Ståhl G (1983) Differences in the amount and distribution of genetic variation between natural populations and hatchery stocks of Atlantic salmon. Aquaculture 33:23–32

Ståhl G (1987) Genetic population structure of Atlantic salmon. In: Ryman N, Utter FM (eds) Population genetics and fishery managements. Univ Washington Press, Seattle, pp 121–140

Starikovskaya YB, Sukernik RI, Schurr TG et al (1998) mtDNA diversity in Chukchi and Siberian Eskimos: implications for the genetic history of ancient Beringia and the peopling of the New World. Am J Hum Genet 63:1473–1491

Starobogatov YI (1975) A review of the book by YP Altukhov "Populyatsionnaya genetika ryb", Pischevaya Promyshlennost', Moscow, 1974. Zh Obshch Biol 36(4):626–628

Stegnii VN (1979) Restructuring of the interphase nuclei in onto- and philogenesis of malaria mosquitoes. Dokl Akad Nauk SSSR 249(5):1231–1234

Stegnii VN (1980a) The chromosome mechanisms of speciation and genetic adaptation of sibling-species of malaria mosquito. Problemy populyatsionnoi i evolutsionnoy tsitogenetiki rastenii i zhivotnykh (Problems of the population and evolution cytogenetics of plants and animals). Izdat Tomskogo Univ, Tomsk, pp 77–83

Stegnii VN (1980b) Spatial and temporal stationarity of the inversion polymorphism in malaria mosquito Anopheles messae. Tr Zool Inst Akad Nauk SSSR 95:59–64

Stegnii VN (1984) The evolutionary meaning of chromosomal inversions. Zh Obshch Biol 45(1):3–15

Stegnii VN (2001) Systemic reorganization of the genome at speciation. Evolutionary biology, vol 1. Tomsk State Univ, Tomsk, pp 128–137

Stormont C (1965) Mammalian immunogenetics. Genetics today, vol 3. Pergamon Press, New York, pp 716–722

Stow AJ, Sunnucks P (2004) Inbreeding avoidance in Cunningham's skins (*Egernia cunninghami*) in natural and fragmented habitat. Mol Ecol 13:443-447

Strobeck C, Morgan K (1978) The effect of intragenic recombination on the number of alleles in finite population. Genetics (US) 88:829–844

Strunnikov VA (1974) Appearance of the compensating complex of genes as one of the causes of heterosis. Zh Obshch Biol 35(5):666–677

Strunnikov VA (1983) Control of silkworm reproduction: development and sex. Mir, Moscow, 280 pp

Strunnikov VA (1986) Genetic bases for heterosis and the combinative ability in *Bombyx mori*. Genetika 22(2):229–243

Strunnikov VA, Strunnikova LV (2000) The nature of heterosis, methods of its enhancements and fixing in subsequent generations without hybridization. Izvestiya Ross Akad Nauk Ser Biol 6:679–687

Sunyaev SR, Lathe WC III, Ramensky VE, Bork P (2000) SNP frequencies in human genes an excess of rare alleles and differing modes of selection. Trends Genet 16:335–337

Sutton HE (1980) An introduction to human genetics, 3rd edn Saunders, Philadelphia, 460 pp

Sutton HE, Harris MI (eds) (1972) Mutagenic effects of environmental chemicals. Academic Press, New York, 511 pp

Swanson WJ, Vacquer VD (2002) The rapid evolution of reproductive proteins. Nat Genet 31:137–144

Tabachnik WJ, Powell JR (1978) Genetic structure of East African domestic population of *Aedes aegypti*. Nature 272:535–537

Taggart JB, Ferguson A (1990) Hypervariable minisatellite DNA single locus probes for the Atlantic salmon, *Salmo salar* L. J Fish Biol 37:991–993

Takhtadjan AL (1983) Macro-evolutionary processes in the history of flora. Bot Zh 68(12):1593–1603

Tatarinov LP (1987) Ocherki po teorii evolutsii (Essays on the theory of evolution). Nauka, Moscow, 251 pp

Tautz D (1989) Hypervariability of simple sequences as a general source for polymorphic DNA markers. Nucleic Acids Res 17:6463–6471

Taylor JS, Peer YV, Meyer A (2001) Genome duplication, divergent resolution and speciation. Trends Genet 17(6):299–301

Templeton AR (1986) Coadaptation and outbreeding depression. In: Soule ME (ed) Conservation biology: the sciences of scarcity and diversity. Sinauer, Sunderland, MA, pp 105–116

Terrenato L (1983) Natural selection associated with birth weight. 4. USA data from 1950 to 1976. Ann Hum Genet 47:67–71

Terrenato L, Gravina MF, Ulizzi L (1981a) Natural selection associated with birth weight. 1. Selection intensity and selection deaths from birth to one month of life. Ann Hum Genet 45:55–63

Terrenato L, Gravina MF, San Martini A et al (1981b) Natural selection associated with birth weight. 3. Changes over the last twenty years. Ann Hum Genet 45:267–278

Tessier N, Bernatchez L, Presa P, Angers B (1995) Gene diversity analysis of mitochondrial DNA, microsatellites and allozymes in landlocked Atlantic salmon. J Fish Biol 47 [Suppl A]:156–163

Tessier N, Bernatchez L, Wright JM (1997) Population structure and impact of a supportive breeding program inferred from mitochondrial and microsatellite DNA analyses in sympatric landlocked Atlantic salmon populations. Mol Ecol 6:735–750

Thorpe JE (1988) Salmon enhancement: stock discreteness and choice of material for stocking. Atlantic salmon: planning for the future. Timber, Portland, OR, pp 373–388

Thorpe JE, Koonce JF, Borgeson D (1981) Assessing and managing man's impact on fish genetic resources. Can J Fish Aquat Sci 38:1899–1907

Timofeeff-Ressovsky HA, Timofeeff-Ressovsky NW (1927) Genetische Analyse einer freilebenden *Drosophila melanogaster*-population. Roux' Arch Entwicklungsmech Org 109(1):70–109

Timofeeff-Ressovsky NV, Yablokov AV, Glotov NV (1973) Ocherk ucheniya o populatsii (Essay of teaching on population). Nauka, Moscow, 277 pp

Timofeeff-Ressovsky NV, Vorontsov NN, Yablokov AV (1977) Kratkii ocherk teorii evolutsii (A short essay on the theory of evolution). Nauka, Moscow, 301 pp

Ting C-T, Tsaur S-C, Wu M-L, Wu C-I (1998) A rapidly evolving homeobox at the site of a hybrid sterility gene. Science 282:1501–1504

Tinkle DW (1965) Population structure and effective size of a lizard populations. Evolution 19:569–573

Tishkin VV, Glotov NV (1983) General fitness and the quantitative morphological traits in *Drosophila melanogaster*. Genetika 19(4):622–627

Tobari YN, Kojima K (1972) A study of spontaneous mutation rates at ten loci detectable by starch gel electrophoresis in *Drosophila melanogaster*. Genetics (US) 70:397–403

Torroni A, Sukernik RI, Schurr TG et al (1993) mtDNA variation of aboriginal Siberians reveals distinct genetic affinities with Native Americans. Am J Hum Genet 53:591–608

Tregenza T, Wedell N (2000) Genetic compatibility, mate choice and patterns of parentage: invited review. Molecular Ecol 9:1013–1027

Trujillo JM, Walden B, O'Neill P et al (1967) Inheritance and subunit composition of haemoglobin in the horse, donkey and their hybrids. Science 213:88–90

Tsigenopoulos CS, Rab P, Naran D, Berrebi P (2002) Multiple origins of polyploidy in the phylogeny of southern African barbs (Cyprinidae) as inferred from mtDNA markers. Heredity 88:466–473

Tsuno K (1981) Studies of mutation at esterase loci in *Drosophila virilis*. Jpn J Genet 56:155–174

Tsuyuki H, Ronald AP (1970) Existence in salmonid hemoglobins of molecular species with three and four different polypeptides. J Fish Res Board Can 27:1326–1328

Ulizzi L, Gravina MF, Terrenato L (1981) Natural selection associated with birth weight. 2. Stabilizing and directional components. Ann Hum Genet 45:207–212

Underhill PA, Jin L, Zemans R et al (1996) A pre-Columbian Y chromosome-specific transition and its implications for human evolutionary history. Proc Natl Acad Sci USA 93:196–200

Underhill PA, Shen P, Lin AA et al (2000) Y chromosome sequence variation and the history of human populations. Nat Genet 26:358–361

Ushakov BP (1958) On conservatism of species specific proteins in the poikilotherms. Zool Zh 37(5):693–706

Ushakov BP (1959a) Physiology of cell and a problem of species in zoology. Tsitologiya 1(5):541–645

Ushakov BP (1959b) Thermal resistance of tissues as a species specific trait of poikilothermal animals. Zool Zh 38(9):1292–1302

Ushakov BP (1964) Thermostability of cells and proteins of poikilotherms and its significance in speciation. Physiol Rev 44:518–560

Ushakov BP (1982a) Physiological population structure arising in the course of thermal selection. Genetika 18(5):773–781

Ushakov BP (1982b) Evolutionary importance of thermal adaptations in animals. Usp Sovrem Biol 93(2):302–319

Ushakov BP (1984) Physiological analysis of the changes hi the fitness of an individual in ontogenesis and selective advantages of heterozygotes. Zh Obshch Biol 45(5):602–614

Ushakov BP (1989) Cell physiology and the problem of species in poikilotherms. Nauka, Leningrad, 232 pp

Utter F, Milner G, Ståhl G, Teel D (1989) Genetic population structure of Chinook salmon, *Oncorhynchus tshawytscha*, in the Pacific Northwest. Fish Bull US 87:239–264

Utter PM, Aebersold P, Helle J, Wymans G (1984) Genetic characterization of populations in the southeastern range of sockeye salmon. Walton JM, Houston DB (eds) Peninsula College Press, Washington, DC, pp 17–32

Van Alphen JJM, Seehausen O (2001) Sexual selection, reproductive isolation and the genic view of speciation. J Evol Biol 14:874–875

Van Hooft WF, Groen AF, Prins HHT (2002) Phylogeography of the African buffalo based on mitochondrial and Y-chromosomal loci: Pleistocene origin and population expansion of the Cape buffalo subspecies. Mol Ecol 11:267–279

Van Ommen G-JB, Boer PH, Groot GS et al (1980) Mutations affecting RNA splicing and the interaction of gene expression of the yeast mitochondrial loci cob and oxi-3. Cell 20:173–183

Varnavskaya NV (1984) Adaptive genetic structure and its relationship with intrapopulation differentiation by sex, age and growth rate in Pacific sockeye salmon, *Oncorhynchus nerka* (Walb). Cand Sci Thesis, NI Vavilov Inst Gen Genet USSR Acad Sci, Moscow, 21 pp

Varnavskaya NV, Beacham TD (1992) Biochemical genetic variation in odd-year pink salmon (*Oncorhynchus gorbuscha*) from Kamchatka. Canad J Zool 70:2115–2120

Vasilyev VP (1977) Polyploidy in fish and some aspects of salmon karyotypes evolution. Zh Obshch Biol 38(3):380–392

Vasilyev VP (1985) Evolutsionnaya kariologiya ryb (Evolutionary karyology of fish). Nauka, Moscow, 300 pp

Vasin BN (1939) The main methodical principles of Karakul lamb breeding. Sov Zootekh 2/3:121–126

Vasin BN (1971) Astrakhan of Karakul lambs. Rukovodstvo po Karakulevodstvu (Guide to Karakul breeding). Kolos, Moscow, pp 109–124

Vaughan E (1947) Time of appearance of pink salmon runs in southeastern Alaska. Copeia 1:40–50

Vavilov NI (1926) On the origin of cultivated plants. Novoe v agronomii (The latest in agronomy). Gos RSFSR, Moscow, pp 76–85

Vavilov NI (1927) World centres possessing a wealth of varieties (genes) of cultivated plants. Izv Gos Inst Opyt Agron 5:339–351

Vernon EH (1957) Morphometric comparison of three races of kokance (*Oncorhynchus nerka*) within a large British Columbia Lake. J Fish Res Board Can 14:573–598

Verspoor E (1988) Reduced genetic variability in first generation hatchery populations of Atlantic salmon (*Salmo salar*). Can J Fish Aquat Sci 45:1686–1690

Viktorovsky RM (1969) On a possibility of polyploidy in evolution of fish. Genetika, selektsiya i gibridizatsiya ryb (Fish genetics, selection and hybridization). Nauka, Moscow, pp 98–104

Viktorovsky RM (1979) Mekhanizmy vidoobrazovaniya u gol'tsov Kronotskogo ozera (Speciation mechanisms in chars of Lake Kronotsky). Nauka, Moscow, 110 pp

Viktorovsky RM, Glubokovsky MK (1977) Mechanisms and rates of speciation in loach Salvelinus (Salmonidae, Pisces). Dokl Akad Nauk SSSR 235(4):946–949

Vogel F (1970) Monitoring of human populations. Chemical mutagenesis in mammals and man. Springer, Berlin Heidelberg New York, pp 445–452

Vogel F, Motulsky AG (1997) Human genetics, 3rd edn Springer, Berlin Heidelberg New York, 851 p

Volkova TV (1988) Akseleratsiya naseleniya SSSR (Acceleration in human population of the USSR). Izdat MGU, Moscow, 71 pp

Volobuev V, Pasteur G, Ineich I, Dutrillaux B (1993) Chromosomal evidence for a hybrid origin of diploid parthenogenetic females from the unisexual-bisexual *Lepidodactylus lugubris* complex (Reptilia, Gekkonidae). Cytogenet Cell Genet 63:194–199

Vorontsov NN (1966) Evolution of karyotype. Rukovodstvo po tsitologii (Guide to cytology), vol 2. Nauka, Leningrad, pp 359–389

Vorontsov NN (1980) Synthetic theory of evolution: origin, main postulates and unsolved problems. Zh Vses Khim Ova im DI Mendeleeva 25(3):295–314

Vorontsov NN (1984) Teoriya evolutsii: istoki, postulaty i problemy (The theory of evolution: sources, postulates and problems). Znaniye, Moscow, 64 pp

Vorontsov NN (1999) The development of evolutionary ideas in Russia. Progress-Tradition, Moscow, 640 pp

Vorontsov NN, Lyapunova EA (1984) Explosive chromosomal speciation in seismic active regions. Chrom Today 8:279–294

Waddington CH (1942) Canalization of development and the inheritance of acquired character. Nature 150:563–565

Waddington CH (1957) The strategy of the genes. Alien and Unvin, London, 262 pp

Waddington CH (1970) Osnovnie biologicheskie kontseptsii. Na puti k teoreticheskoi biologii, I: Prolegomeny (Towards a theoretical biology. I. Prolegomena). Mir, Moscow, pp 11–38

Wahlund S (1928) Zusammensetzung von Populationen und Korrelationserscheinungen vom Standtpunkt der Vererbungslehre aus betrachtet. Hereditas 11:65–106

Wade MJ, Goodnight CJ (1998) The theories of Fisher and Wright in the context of metapopulations: when nature does many small experiments. Evolution 52:1537–1553

Waldren CA (1983) Mutational analysis in cultured human-hamster hybrid cells. Chem Mutagens 8:235–260

Waldren C, Correll L, Sognier MA et al (1986) Measurement of low levels of X-ray mutagenesis in relation to human disease. Proc Natl Acad Sci USA 83:4839–4843

Walker JT (1964) Modal selection in Upland cotton. Heredity 19:559–583

Wallace B (1966) On the dispersal of *Drosophila*. Am Nat 100:551–562

Wallace B (1979) The migration of a mutant gene into isolated populations of *Drosophila melanogaster*. Genetica (Netherl) 50:[illegible]–72

Wallace B, Madden C (1953) The frequencies of sub- and supervitals in experimental populations of *Drosophila melanogaster*. Genetics (US) 38:456–470

Wallis GP (1999) Do animal mitochondrial genomes recombine? Trends Ecol E14:209–210

Walton KE, Styer D, Gruenstein EA (1979) Genetic polymorphism in normal human fibroblasts as analysed by two-dimensional polyacrylamide gel electrophoresis. J Biol Chem 254:7951–7960

Wang DG, Fan J-B, Siao CJ et al (1998) Large-scale identification, mapping and genotyping of single-nucleotide polymorphism in the human genome. Science 280:1077–1082

Waples RS (1990a) Conservation genetics of Pacific salmon. II. Effective population size and the rate of loss of genetic variability. J Hered 81:267–276

Waples RS (1990b) Conservation genetics of Pacific salmon. III. Estimating effective population size. J Hered 21:277–89

Ward FJ, Larkin PA (1968) Cyclic dominance in Adams River sockeye salmon. Prog Rep Int Pac Salmon Fish Commiss 11:1–116

Ward RD, Grewe P (1994) Appraisal of molecular genetic techniques in fisheries. Rev Fish Biol Fish 4:300–325

Ward RDM, Skibinski DOF, Woodwark M (1992) Protein heterozygosity, protein structure and taxonomic differentiation. Evol Biol 26:73–159

Ward RD, Woodwark M, Skibinski DOF (1994) A comparison of genetic diversity levels in marine, freshwater and anadromous fishes. J Fish Biol 44:213–232

Waser NM, Price MV (1989) Optimal outcrossing in Ipomopsis aggregata: seed set and offspring fitness. Evolution 43:1097–1109

Waser NM, Price MV (1991) Outcrossing distance effects in *Delphinium nelsonii*: pollen load, pollen tubes, and seed set. Ecology 72:171–179

Waser NM, Price MV (1994) Crossing-distance effects in *Delphinium nelsonii*: outbreeding and inbreeding deression in progeny fitness. Evolution 48:842–852

Waser NM, Price MV, Shaw RG (2000) Outbreeding depression varies among cohorts of *Ipomopsis aggregata* planted in nature. Evolution 54:485–491

Watt WB (1972) Intragenic recombination as a source of population genetic variability. Am Nat 106:737–753

Wayne RK, Jenks SM (1991) Mitochondrial DNA analysis supported extensive hybridization of the endangered wolf (*Canis rufus*). Nature 351:565–568

Weaver RF, Hedrick PW (1997) Genetics. WCB Publishers, Dubuque, 638 pp

Weber JL, Wong C (1993) Mutation of human short tandem repeats. Hum Mol Genet (US) 2:1123–1128

Weir BS (1990) Genetic data analysis. Sinauer Associates, Sunderland, MA, USA

Weir BS (1995) Genetic data analysis. Mir, Moscow, 400 pp

Weir BS, Cockerham CC (1984) Estimating F-statistics for the analysis of population structure. Evolution 38:1358–1370

Weir BS, Allard RW, Kahler AL (1974) Further analysis of complex allozymes polymorphisms in a barley population. Genetics (US) 78:911–919

Weldon WFR (1901) A first study of natural selection in *Clausilia laminata* (Montagu). Biometrika 1:109–124

White MJD (1954) Animal cytology and evolution. Cambridge Univ Press, Cambridge, 434 pp

White MJD (1968) Models of speciation. Science 159:1065–1070

White MJD (1973) Animal cytology and evolution. Clowes, Cambridge, 961 pp

White MJD (1978) Modes of speciation. Freeman, San Francisco, 455 pp

Whitlock MC, Barton NH (1997) The effective size of subdivided population. Genetics (US) 146:427–441

Whitt GS (1981) Developmental genetics of fishes: isozymic analyses of differential gene expression. Am Zool 21:549–572

Whitt GS (1987) Species differences in isozyme tissue patterns: their utility for systematic and evolutionary analyses. Curr Top Biol Med Res 15:1–26

WHO (1989) Guidelines for short-term tests for isolation of mutagenic and carcinogenic chemical substances. Genetic Criteria of the State of the Environment, no 51. WHO, Geneva, 212 pp

Wilding CS, Beaumont AR, Latchford, JW Mitochondrial DNA variation in the scallop *Pecten maximus* (L.) assessed by a PCR-RFLP method // Heredity. 1997. Vol. 79. P. 178-189

Wilkins NP (1970) The subunit composition of the haemoglobins of the Atlantic salmon (*Salmo salar* L.). Biochim Biophys Acta 214:52–63

Wilkinson GS, Chapman AM (1991) Length and sequence variation in evening bat D-loop mitochondrial DNA. Genetics (US) 128:607–617

Williams JGK, Hanafey MK, Rafalski JA, Tingey SV (1993) Genetic analysis using random amplified polymorphic DNA markers. In: Wu R (ed) Recombinant DNA. Methods in enzymology, vol 218. Academic Press, San Diego, pp 704–740

Williams MA, Blouin AG, Noor MAF (2001) Courtship songs of *Drosophila pseudoobscura* and *D. persimilis*. II. Genetics of species differences. Heredity 86:68–77

Wilmot RL, Everett RJ, Spearman WJ, Baccus R, Varnavskaya NV, Putivkin SV (1994) Genetic stock structure of western Alaska chum salmon and comparison with Russian Far East stock. Can J Fish Aquat Sci 51 [Suppl 1]:84–94

Wilson CC, Bernatchez L (1998) The ghost of hybrid past: fixation of Arctic charr (*Salvelinus alpinus*) mitochondrial DNA in introgressed population of lake trout (*S. namaycush*). Mol Ecol 7:127–132

Wilson EO (ed) (1988) Biodiversity. National Academy Press, Washington, DC, 521 pp

Winans GA, Aebersold PB, Urawa S, Varnavskaya NV (1994) Determining continent of origin of chum salmon (*Oncorhynchus keta*) using genetic stock identification techniques: status of allozyme baseline in Asia. Can J Fish Aquat Sci 51 [Suppl 1]:95–113

Winters JB, Waser PM (2003) Gene dispersal and outbreeding in a philopatric mammal. Mol Ecol 12:2251–2259

Wirth T, Bernatchez L (2001) Genetic evidence against panmixia in the European eel. Nature 409:1037–1040

Wise CA, Sraml M, Easteal S (1998) Departure from neutrality at the mitochondrial NADP dehydrogenase subunit 2 gene in humans, but in chimpanzees. Genetics (US) 148:400–421

Withler FG (1982) Transplanting Pacific salmon. Can Techn Rep Fish Aquat Sci 1079, 27 pp

Wolf PG, Soltis PS (1992) Estimates of gene flow among populations, geographic races, and species in the *Ipomopsis aggregata* complex. Genetics (US) 130:639–647

Wolf U, Engel W, Faust J (1970) Zum Mechanismus der Diploidisirung in der Wierbeltierevolution: Koexistenz von tetrasomen and disomen Genloci der Isocitrat-Dehydrogenasen bei der Regenbogenforelle (*Salmo irideus*). Hum Genet 9:150–156

Wolfsberg TG, Landsman D (1997) A comparison of expressed sequence tags (ESTs) to human genomic sequences. Nucleic Acids Res 26:1626–1632

Woodruff DS (1989) Genetic anomalies associated with Cerion hybrid zone: the origin and maintenance of new electromorphic variants called hybrizymes. Biol J Linn Soc 36:281–294

World Conservation Strategy (1980) Living Resource Conservation for Sustainable Development. International Union for Conservation of Nature and Natural Resources, Gland

Wright J (1993) DNA fingerprinting in fishes. In: Hochachka P, Mommsen T (eds) Biochemistry and molecular biology of fishes, vol 2. Elsevier, Amsterdam, pp 57–91

Wright JE, Atherton L (1968) Genetic control of interallelic recombination at the LDH B locus in brook trout. Genetics (US) 60:240

Wright JM (1994) Mutation at VNTRs: are minisatellites the evolutionary progeny of microsatellites? Genome 37:345–347

Wright JM, Bentzen P (1994) Microsatellites: genetic markers for the future. Rev Fish Biol Fish 4:384–388

Wright S (1921) Systems of mating. Genetics (US) 6:111–178

Wright S (1931) Evolution in Mendelian populations. Genetics (US) 16:97–159

Wright S (1932) The roles of mutation, inbreeding, crossbreeding and selection in evolution. Proc VI Int Congr Genet Ithaca, vol 1, pp 356–366

Wright S (1938) Size of populations and breeding structure in relation to evolution. Science 87:430–431

Wright S (1940) Breeding structure of populations in relation to speciation. Am Nat 74:232–246

Wright S (1941) On the probability fixation of reciprocal translocations. Am Nat 65:513–522

Wright S (1943a) Isolation by distance. Genetics (US) 28:114–138

Wright S (1943b) An analysis of local variability of flower color *Linanthus parryae*. Genetics (US) 28:139–156

Wright S (1951) The genetical structure of populations. Ann Eugenics 15:323–354

Wright S (1969) Evolution and genetics of populations. 2. The theory of gene frequencies. Univ Chicago Press, Chicago, 511 pp

Wright S (1970) Random drift and shifting balance theory of evolution. In: Kojima K (ed) Mathematical topics in population genetics. Springer, Berlin

Wright S (1977) Evolution and genetics of populations. 3. Experimental results and evolutionary deductions. Univ Chicago Press, Chicago, 613 pp

Wright S (1978) Evolution and the genetics of populations. 4. Variability within and among natural populations. Univ Chicago Press, Chicago, 580 pp

Wright S (1980) Genic and organismic selection. Evolution 34:825–843

Wright S, Dobzhansky T (1946) Genetics of natural populations. 12. Experimental reproduction of some of the changes caused by natural selection in certain populations of *Drosophila pseudoobscura*. Genetics (US) 31:125–150

Wu C-I (2001) The genic view of the process of speciation. J Evol Biol 14:851–865

Yablokov AV (1986) Population biology. Mir, Moscow, 303 pp

Yarbrough K, Kojima K (1967) The mode of selection at the polymorphic esterase-6 locus in cage populations of *Drosophila melanogaster*. Genetics (US) 57:677–689

Youngson AF, Martin SAM, Jordan WC, Verspoor E (1989) Genetic protein variation in farmed Atlantic salmon in Scotland: comparison of farmed with source populations. Scott Fish Res Rep 42:1–12

Yudin VM (1955) Methods for the selection of black Karakul sheep. D Sci Thesis VASKHNIL, Moscow

Zaloi D, Richard M, Lecomte J, Massot M, Clobert J (2004) Multiple paternity in clutches of common lizard *Lacerta vivipara*: data from microsatellite markers. Mol Ecol 13:719–723

Zane L, Bargelloni L, Patarnello T (2002) Strategies for microsatellite isolation: a review. Mol Ecol 11:1–16

Zhang De-Xing, Hewitt GM (2003) Nuclear DNA analyses in genetic studies of populations: practice, problems and prospects. Mol Ecol 12:563–584

Zhirmunsky AV, Kuzmin VI (1982) Kriticheskiye urovni v protsessakh razvitiya biologicheskikh sistem (Critical levels in the development processes of biological systems). Nauka, Moscow, 179 pp

Zhivotovsky LA (1981a) Dynamics of polygenic systems under the action of selection. Matematicheskiye modeli v ekologii i genetike (Mathematical models in ecology and genetics). Nauka, Moscow, pp 120–148

Zhivotovsky LA (1981b) On Fisher's theorem. Genetika 17(2):324–331

Zhivotovsky LA (1982) Integration of polygenic systems in populations and problems of analyzing the complex of traits. D Sci Thesis, NI Vavilov Inst Gen Genet, Moscow

Zhivotovsky LA (1984) Integratsiya poligennykh sistem v populyatsiyakh (Integration of polygenic systems in populations). Nauka, Moscow, 184 pp

Zhivotovsky LA (1991) Populyatsionnaya biometriya (Population biometrics). Nauka, Moscow, 271 pp

Zhivotovsky LA (1999) A new genetic distance with application to constrained variation at microsatellite loci. Mol Biol Evol 16:467–471

Zhivotovsky LA, Altukhov YP (1980) Method of the identification of morphologically "average" and "extreme" types on the basis of a set of quantitative characters. Dokl Akad Nauk SSSR 251(2):473–476

Zhivotovsky LA, Feldman MW (1992) On the difference between mean and optimum of quantitative characters under selection. Evolution 46:1574–1578

Zhivotovsky LA, Feldman MW (1995) Microsatellite variability and genetic distance. Proc Natl Acad Sci USA 92:11549–11552

Zhivotovsky LA, Afanas'ev SI, Rubtsova GA (1987) Selection for the enzymatic loci in pink salmon, *Oncorhynchus gorbuscha* (Walb.). Genetika 23(10):1876–1883

Zhivotovsky LA, Glubokovsky MK, Viktorovsky RM, Bronevsky AM, Afanas'ev KI, Efremov VV, Ermolenko LN, Kalabushkin BA, Kovalyev VG, Makoedov AN, Malinina TV,

Pustovoyt SP, Rubtzova GA (1989) Genetic differentiation in pink salmon. Genetika 25(7):1261–1274

Zhomova VK (1965) Analysis of the circle of marriages within the Russian population. Vopr Antropol 21:111–114

Zhuze AN (1959) Paleogeography of the Sea of Okhotsk. Izv Akad Nauk SSSR, Ser Geogr 2:12–27

Zhuchenko AA, Korol' AB (1985) Rekombinatsiya v evoluisii i selektsii (Recombination in evolution and selection). Nauka, Moscow, 400 pp

Zolotareva IM (1980) Estimation of breeding effectivity of pink salmon in Sakhalin salmon hatcheries. In: Skarlato OA (ed) Lososevidniye ryby (Salmonid fishes). Nauka, St. Petersburg, pp 259–261

Zouros E, Foltz DW (1987) The use of allelic isozyme variation for the study of heterosis. Isozymes, vol 13. Liss, New York, pp 1–59

Zouros E, Krimbas CB, Tsakas S et al (1974) Genic versus chromosomal variation in natural populations of *Drosophila subobscura*. Genetics (US) 78:1223–1244

Zouros E, Singh SM, Miles HE (1980) Growth rate in oysters: an overdominant phenotype and its possible explanations. Evolution 3:856–867

Zurabyan AS, Timofeeff-Ressovsky NV (1967) On heterozygous polymorphism of ebony mutation in model populations of *Drosophila melanogaster*. Zh Obshch Biol 28(4):612–617

Printing: Krips bv, Meppel
Binding: Stürtz, Würzburg